Student's t (comparing two means)

$$t = \frac{(\bar{y}_1 - \bar{y}_2) - D_0}{s\sqrt{\dfrac{1}{n_1} + \dfrac{1}{n_2}}}$$

pooled estimate of σ^2

$$s^2 = \frac{\sum\limits_{i=1}^{n_1} (y_i - \bar{y}_1)^2 + \sum\limits_{i=1}^{n_2} (y_i - \bar{y}_2)^2}{n_1 + n_2 - 2}$$

correlation coefficient

$$r = \frac{SS_{xy}}{\sqrt{SS_x SS_y}}$$

where

$$SS_y = \sum_{i=1}^{n} (y_i - \bar{y})^2 = \sum_{i=1}^{n} y_i^2 - \frac{\left(\sum\limits_{i=1}^{n} y_i\right)^2}{n}$$

$$SS_x = \sum_{i=1}^{n} (x_i - \bar{x})^2 = \sum_{i=1}^{n} x_i^2 - \frac{\left(\sum\limits_{i=1}^{n} x_i\right)^2}{n}$$

$$SS_{xy} = \sum_{i=1}^{n} (x_i - \bar{x})(y_i - \bar{y})$$

$$= \sum_{i=1}^{n} x_i y_i - \frac{\left(\sum\limits_{i=1}^{n} x_i\right)\left(\sum\limits_{i=1}^{n} y_i\right)}{n}$$

least squares estimators of β_0 and β_1

$$\hat{\beta}_1 = \frac{SS_{xy}}{SS_x} \quad \text{and} \quad \hat{\beta}_0 = \bar{y} - \hat{\beta}_1 \bar{x}$$

least squares line (single independent variable)

$$\hat{y} = \hat{\beta}_0 + \hat{\beta}_1 x$$

Mann-Whitney U

$$U_A = n_1 n_2 + \frac{n_1(n_1 + 1)}{2} - T_A$$

and

$$U_B = n_1 n_2 + \frac{n_2(n_2 + 1)}{2} - T_B$$

chi-square statistic for contingency tables

$$X^2 = \sum_{i=1}^{k} \frac{(n_i - np_i)^2}{np_i}$$

Spearman's rank correlation coefficient

$$r_s = 1 - \frac{6 \sum\limits_{i=1}^{n} d_i^2}{n(n^2 - 1)}$$

Introduction

to Probability

and Statistics

Introduction

to Probability

and Statistics

Sixth Edition

William Mendenhall

Duxbury Press · **Boston**

PWS PUBLISHERS

Prindle, Weber & Schmidt • ♣ • Duxbury Press • ♠ • PWS Engineering • ⚊
Statler Office Building • 20 Park Plaza • Boston, Massachusetts 02116

PWS Publishers is a division of Wadsworth, Inc.

Library of Congress Cataloging in Publication Data

Mendenhall, William.
    Introduction to probability and statistics.

    Includes index.
    1. Mathematical statistics.   2. Probabilities.
I. Title.
QA276.M425   1983   519.2   82-14731
ISBN 0-87150-365-4

Copy editing by Nancy Blodgett. Cover and text design by Trisha Hanlon. Cover photo by Erich Hartmann/MAGNUM. Art by ANCO/Boston. Composition by Syntax International. Covers printed by Lehigh Press Lithographers. Text printed and bound by Halliday Lithograph.

ISBN 0-87150-365-4
ISBN 0-534-98050-3 INTERNATIONAL

Printed in the United States of America

87   86   85   —   10   9   8   7   6   5

Preface

The teaching objective of this sixth edition of *Introduction to Probability and Statistics* is to present a connected introduction to statistics that presents statistical inference, *the objective of statistics*, as its theme. To achieve this objective, the sixth edition retains the features of earlier editions, slightly modified, with some substantial changes. These changes include the following:

Changes in this Edition

1. Each chapter is now introduced by an extensive case study, which serves as motivation for the topics developed in the chapter. Questions or problems suggested by the case study are solved at the conclusion of the chapter. Instructors have the option of lecturing on the case study or assigning it for student reading.

2. We have partitioned the material in Chapter 4, "Probability," and Chapter 5, "Random Variables and Probability Distributions," into two different tracks for instructors who desire different levels of comprehension from their students. One level offers the basic concepts of probability that are necessary for further work in inference. A second level offers the coverage necessary if the student is to be able to solve problems using probability theory. Sections and exercise sets for each track are clearly indicated.

3. Statistical inference has been reorganized in this edition, and large sample estimation is developed in a separate Chapter 8. Large sample hypothesis testing is now covered in Chapter 9.

4. The concept of a *p* value has been introduced in Chapter 9 as an alternative means for interpreting the results of a statistical test of an hypothesis.

5. New to this edition is a chapter covering multiple regression, including extensive incorporation of printouts from statistical computing packages.

6. Many new, applied exercises have been added to the exercise sets throughout the text.

7. Numerous less obvious changes in details have been made throughout the text in response to suggestions made by past and current users of the fifth edition.

Approach of this Text

As stated in the preface for earlier editions, the intent of the original text, and all subsequent editions, is to provide a cohesive, connected presentation of statistics that identifies inference as its objective and stresses the relevance of statistics in learning about the world in which we live.

To accomplish this goal, we think that the student should understand how statistical inference-making procedures work and why they function better than our own intuitive inference-making machinery. Thus, we believe that an elementary but clear understanding of the role of probability in making inferences and assessing their reliabilities is essential to an intelligent use of statistical methods. Consequently, a strong effort has been made to weld the theories of probability and inference throughout the book.

The student should understand the importance of each chapter in the text and how each relates to the objective of the course, statistical inference. Too often, students see statistics as the computation of means and standard deviations, as probability, as tests of hypotheses and estimation, but they do not see these topics as connected and forming a coherent picture of statistics. To create this connectivity, we attempt to explain at the beginning of each chapter its relevance to statistical inference and its relation to preceding chapters.

Applications to Many Fields

The student should also see that the same statistical problems occur in business, biology, engineering, and the social sciences and should realize that the basic concepts of statistics and inferential methods apply in all fields. Only the frequency of occurrence of specific types of problems varies from one field to another. Thus, I think that a first course in statistics should stress the unity of concepts, rather than being field-oriented. Specialized courses can follow in the student's field of application.

Three distinct points of emphasis in the text are a result of this philosophy. The objective of statistics—inference—is stated in Chapter 1, and the introduction to every subsequent chapter identifies chapter objectives and explains the role of the chapter material in making inferences and measuring their reliabilities. Second, not only is probability presented in a very elementary (but thorough) manner, based on the concept of the sample space, but it is employed at the earliest opportunity (Chapter 6) in two very practical inferential problems, as well as in subsequent discussions when the complexity of the probabilistic problem is within the grasp of the student. Finally, a large number of examples and exercises with practical application in a variety of fields indicate the broad applicability of statistical methodology.

Suggested Course Outlines

The chapters on regression analysis, the analysis of enumerative data, important considerations in the design of experiments, the analysis of variance, and nonparametric statistics provide flexibility in the construction of an introductory course. For example, a one-quarter or one-semester course might cover the material contained in Chapters 1 through 11. This course can be followed by a specialized course in a student's field of application; or the student can take a second quarter covering the remainder of this text. For example, business majors might follow the introductory course with a second-semester course on regression analysis, index numbers, and time-series analysis. A course oriented toward the social sciences might utilize some combination of Chapters 12, 13, 14, and 15; an elementary course in the theory of probability, with some emphasis on inference, might include Chapters 1 through 9 and Chapter 15.

Exercises are graduated in difficulty so that all students can solve some of the exercises, a substantial number can solve most, and a few of the best students are challenged to solve all without error. Symbols are used to identify areas of application in business, industry, and the sciences. A thorough knowledge of definitions and concepts is essential before the students attempt the exercises. Unannounced quizzes are very effective in helping the student learn the new language of statistics.

A semiprogrammed study guide to help students who have individual difficulties with the subject matter is available. This study guide, now in its sixth edition, benefits from considerable user experience.

Acknowledgments

The preparation of a text and its subsequent new editions is always a team effort, and I am fortunate to have had the assistance of many talented people. William Miller and the late Paul Benson of Bucknell University helped develop and test the first edition and made comments concerning its revision. Numerous reviewers aided in the writing of the second, third, fourth, and fifth editions. They include John T. Webster, William Brelsford, Robert Crovelli, Douglas Chapman, Arthur Coladarci, the late Paul Meyer, Joyce Curry, Frank Deane, Roy E. Myers, P.V. Rao, Dennis D. Wackerly, and John N. Quiring. The following reviewers provided detailed comments on this manscript: Michael Burke, College of San Mateo; S. Cheng, University of Manitoba; Rudy J. Freund, Texas A & M; David Lund, University of Wisconsin, Eau Claire; Donald R. LaTorre, Clemson University; and Donald Ramirez, University of Virginia. A special note of appreciation is extended to Dennis Wackerly, University of Florida, for his detailed line-by-line critique. In addition, the fine work of Barbara Beaver who has prepared the solutions manual for this and other editions, Susan Reiland, and Faith Sincich is most appreciated. I would also like to express my appreciation to the production and editorial staffs of Duxbury Press. Thanks are also due to E.J. Pearson, A. Hald, W.H. Beyer, R.A. Wilcox, and the Biometrika Trustees for their kind permission to use tables reprinted in the Appendix. I am indebted to the typists who have given their best and endured the worst: Florence Valentine, who typed the original draft of the manuscript; Angie Anastasia and Gay Midelis, who prepared the second revision; Mary Jackson and Catherine Kennedy, who typed the third edition; Ellen Evans, who typed the fourth; Carol Rozear, who typed the fifth edition; and Brenda Dobson, who typed drafts of the current edition. Finally, as before, I acknowledge the assistance of my family and their partnership in this writing endeavor.

William Mendenhall

Contents

1. What Is Statistics?

Chapter Objectives

General Objective

The purpose of this chapter is to identify the nature of statistics, its objective, and how it plays an important role in the sciences, in industry, and, ultimately, in our daily lives.

Specific Objectives

1. To use a case study to motivate a discussion of statistics.
2. To answer the question, "What is statistics?" *Sections 1.1, 1.2, 1.3, 1.4*
3. To identify statistical inference as the objective of modern statistics. *Section 1.3*
4. To identify the contributions statistics can make to inference making based on sample data. *Section 1.4*
5. To apply the knowledge gained in this chapter to the case study. *Section 1.5*
6. To define the basic words and concepts used in statistics. *Sections 1.1–1.7*

Case Study

Reaping the Rewards of Refunds

Will court-enforced retail refunds make you rich? Not likely, according to a *Wall Street Journal* article (June 9, 1981) that reports on a court case stemming from a price-fixing indictment. According to the article, approximately 54,000 persons—customers of Saks Fifth Avenue, Bonwit Teller, and Bergdorf Goodman during the period 1968 to 1974—may be due refunds if they can resurrect their receipts and file the necessary applications. The case raises a serious question about this and other court-directed refund programs that are often the result of price-fixing cases. Who profits (other than the attorneys involved in the lawsuit)? Specifically, how many of the 54,000 customers eligible for refunds will actually find their receipts and submit the required applications? And how much of the settlement money will really be distributed to the eligible customers?

To answer these questions, you would need to either interview all 54,000 customers who are potentially eligible (an impossible task) or select a sample from the 54,000 and use statistical methods to estimate the number of people eligible for refunds and the amount of money that will eventually be refunded.

Our intention here is not to answer the question, "Who profits from refunding?" but to present a difficult problem that can be solved using a very powerful tool, statistics. In this chapter we will describe the objective of statistics. In particular we will identify the types of problems that statistical methodology can solve and explain how this valuable tool can be used to answer some very practical questions. We will return to the refunding case study in Section 1.5.

1.1 Illustrative Statistical Problems

What is statistics? How does it function? How does it help to solve certain practical problems? Rather than attempt a definition at this point, let us examine several problems that might come to the attention of the statistician. From these examples we can then select the essential elements of a statistical problem.

In predicting the outcome of a national election, pollsters interview a predetermined number of people throughout the country and record their preference. On the basis of this information a prediction is made. Similar problems are encountered in market research (What fraction of smokers prefer cigarette brand A?); in sociology (What fraction of rural homes have electricity?); in industry (What fraction of items purchased, or produced, are defective?).

The yield (production) of a chemical plant is dependent upon many factors. By observing these factors and the yield over a period of time, we can construct a prediction equation relating yield to the observed factors. How do we find a good prediction equation? If the equation is used to predict yield, the prediction will rarely equal the true yield; that is, the prediction will almost always be in error. Can we place a limit on the prediction error? Which factors are the most important in predicting yield? One encounters similar problems in the fields of education, sociology, psychology, and the physical sciences. For instance, colleges wish to predict the grade-point average of college students based upon information obtained *prior* to college entrance. Here, we might relate grade-point average to rank in high school class, the type of high school, college board examination scores, socioeconomic status, or any number of similar factors.

In addition to prediction, the statistician is concerned with decision making based upon observed data. Consider the problems of determining the effectiveness of a new cold vaccine. For simplification, let us assume that ten people have received the new cold vaccine and are observed over a winter season. Of these ten, eight survive the winter without acquiring a cold. Is the vaccine effective?

Two physically different teaching techniques are used to present a subject to two groups of students of comparable ability. At the end of the instructional period, a measure of achievement is obtained for each group. On the basis of this information, we ask: Do the data present sufficient evidence to indicate that one method produces, on the average, higher student achievement?

Consider the inspection of purchased items in a manufacturing plant. On the basis of such an inspection, each lot of incoming goods must be either accepted or rejected and returned to the supplier. The inspection might involve drawing a sample of ten items from each lot and recording the number of defectives. The decision to accept or reject the lot could then be based upon the number of defective items observed.

A company manufacturing complex electronic equipment produces some systems that function properly but also some that, for unknown reasons, do not. What makes good systems good and bad systems bad? In attempting to answer this question, we might make certain internal measurements on a system in order to find important factors which differentiate between an acceptable and an unacceptable product. From a sample of good and bad systems, data could then be collected that might shed light on the fundamental design or on production variables affecting system quality.

1.2
The Population and the Sample

The examples we have cited are varied in nature and complexity, but **each involves prediction or decision making.** In addition, **each of these examples involves sampling.** A specified number of items (objects or bits of information)— a sample—is drawn from a much larger body of data which we call the population. The pollster draws a sample of opinion (those interviewed) from the statistical population, which is the set of opinions corresponding to all the eligible voters in the country. In predicting the fraction of smokers who prefer cigarette brand *A*, we assume that those interviewed yield a representative sample of the population of all smokers. The sample for the cold vaccine experiment consists of observations made on the ten individuals receiving the vaccine. The sample is presumably representative of data pertinent to a much larger body of people—the population—who could have received the vaccine.

Which is of primary interest, the sample or the population? In all the examples given above, we are primarily interested in the population. We cannot interview all the people in the United States; hence we must predict their behavior on the basis of information obtained from a representative sample. Similarly, it is practically impossible to give all possible users a cold vaccine. The manufacturer of the drug is interested in its effectiveness in preventing colds in the purchasing public (the population) and must predict this effectiveness from information extracted from the sample. Hence, the sample may be of immediate interest, but we are primarily interested in describing the population from which the sample is drawn.

Definition

A **population** is the set representing all measurements of interest to the sample collector.

Definition

A **sample** is a subset of measurements selected from the population of interest.

As used by most people, the word "sample" has two meanings. It can refer to the set of objects on which measurements are to be taken or it can refer to the objects themselves. A similar double use could be made of the word "population."

For example, you read in the newspapers that a Gallup Poll was based on a sample of 1823 people. In this use of the word "sample," the objects selected for the sample are clearly people. Presumably each person is interviewed on a particular question, and that person's response represents a single item of data. The collection of data corresponding to the people represents a sample of data.

In a study of sample survey methods, it is important to distinguish between the objects measured and the measurements themselves. To experimenters the objects measured are called "experimental units." The sample survey statistician calls them "elements of the sample."

To avoid a proliferation of terms, we will use the word "sample" in its everyday meaning. Most of the time we will be referring to the set of measurements made on the experimental units (elements of the sample). If occasionally we use the term in referring to a collection of experimental units, the context of the discussion will clarify the meaning.

1.3
The Essential Elements of a Statistical Problem

The objective of statistics and the essential elements of a statistical problem are now vaguely apparent.

The objective of statistics is to make inferences (predictions, decisions) about a population based upon information contained in a sample.

How will we achieve this objective? We will find that every statistical problem contains five elements. The first and foremost of these is a clear specification of the question to be answered and of the population of data that is related to the question.

The second element of a statistical problem is deciding how the sample will be selected. This is called the design of the experiment or sampling procedure. This element is important because data cost money and time. In fact, it is not unusual for an experiment or a statistical survey to cost $50,000 to $500,000, and the costs of many biological or technological experiments can run into the millions. And what do these experiments and surveys produce? Numbers on a sheet of paper or, in brief, information. So planning the experiment is important. Including too many observations in the sample is often costly and wasteful; including too few is also unsatisfactory. But most importantly, you will learn that the method used to collect the sample will often affect the amount of information per observation. A good sampling design can sometimes reduce the costs of data collection to one-tenth or as little as one-hundredth of the cost of another sampling design.

The third element of a statistical problem involves the analysis of the sample data. No matter how much information the data contain about the practical question, you must use an appropriate method of data analysis to extract the desired information from the data.

The fourth element of a statistical problem is using the sample data to make an inference about the population. As you will subsequently learn, many different procedures can be employed to make an estimate or decision about some characteristic of a population or to predict the value of some member of the population. For example, ten different methods might be available to estimate human response to a new drug, but one procedure might be much more accurate than another. Therefore, you will wish to employ the best inference-making procedure when you use sample data to make an estimate or decision about a population or a prediction about some member of a population.

The final element of a statistical problem identifies what is perhaps the most important contribution of statistics to inference making. It answers the question, "How good is the inference?" To illustrate, suppose you manage a small manufacturing concern. You arrange for an agency to conduct a statistical survey for you and it estimates that your company's product will gain 34% of the market this year. How much faith can you place in this estimate? You will quickly discern that you are lacking some important information. Of what value is the estimate without a measure of its reliability? Is the estimate accurate to within 1%, 5%, or 20%? Is it reliable enough to be used in setting production goals? As you will subsequently learn, statistical estimation, decision-making, and prediction procedures enable you to calculate a measure of goodness for every inference. Consequently, in a practical inference-making situation, every inference should be accompanied by a measure which tells you how much faith you can place in the inference.

To summarize, a statistical problem involves the following:

1. A clear definition of the objective of the experiment and the pertinent population.
2. The design of the experiment or sampling procedure.
3. The collection and analysis of data.
4. The procedure for making inferences about the population based upon sample information.
5. The provision of a measure of goodness (reliability) for the inference.

It is extremely important to note that the steps in the solution of a statistical problem are sequential; that is, you must identify the population of interest and plan how you will collect the data before you can collect and analyze it. And all these operations must precede the ultimate goal, making inferences about the population based on information contained in the sample. These steps—carefully identifying the pertinent population and designing the experiment or sampling procedure—often are omitted. The experimenter may select the sample from the wrong population or may plan the data collection in a manner that intuitively seems reasonable or logical but that may be an extremely poor plan from a statistical point of view. The resulting data may be difficult or impossible to analyze, may contain little or no pertinent information, or, inadvertently, the sample may not be representative of the population of interest. This means that every experimenter should be knowledgeable in the statistical design of experiments and (or) sample surveys or should consult an applied statistician for the appropriate design *before* the data are collected.

1.4
The Role of Statistics and the Statistician

We may now ask: What is the role of the statistician in attaining the objective in what we have described as statistical problems? People have been making observations and collecting data for centuries. Furthermore, they have been using the data as a basis for prediction and decision making completely unaided by statistics. What, then, do statisticians and statistics have to offer?

Statistics is an area of science concerned with the extraction of information from numerical data and its use in making inferences about a population from which the data are obtained. In some respects the statistician quantifies information and studies various designs and sampling procedures, searching for the procedure that yields a specified amount of information in a given situation at a minimum cost. Therefore, one major contribution of statistics and statisticians is in designing experiments and surveys, thereby reducing their cost and size. The second major contribution is in the inference making itself. The statistician studies various inferential procedures, looking for the best predictor or decision-making process for a given situation. Even more importantly, the statistician provides information concerning the goodness of an inferential procedure. When we predict, we would like to know something about the error in our prediction. If we make a decision, we want to know the chance that our decision is correct. Our built-in individual prediction and decision-making systems do not provide immediate answers to these important questions. They could be evaluated only by observation over a long period of time. In contrast, statistical procedures do provide answers to these questions.

1.5
More on the Consumer Refund Case Study

Let us examine the consumer refunding problem in greater detail and identify the populations associated with the two questions posed in the case study.

How many of the 54,000 customers eligible for refunds will actually find their receipts and submit the required applications? Each of the 54,000 customers either will or will not file for a refund. If we assign a 0 to each customer who will not file for a refund and a 1 to those who will, then the population pertinent to the question consists of a large (54,000) set of 0s and 1s. The sum of the 54,000 numbers in the population (equal to the number of 1s in the population) is equal to the number of customers who will file for a refund.

Although this sum is unknown to us, we can gain insight into its value by selecting a sample from the population (that is, selecting a subset of the 54,000 customers) and, for each, recording whether or not the customer planned to apply for a refund. For example, if we were to sample 540 customers from among the 54,000, and 54 (i.e., 10 percent of the number of customers in the sample) planned to apply for a refund, we might estimate that 5400 (10 percent) of the total number of 54,000 customers would take similar actions. Thus, statistical methods could be used to estimate the unknown number of customers who would file for a refund. But, more importantly, they would enable us to evaluate the accuracy of the estimate; that is, to say something about how far the estimate might depart from the actual number of persons in the population who would apply for a refund.

How much of the refund money will actually be distributed to the eligible customers? To answer this question, we envision a large (54,000) set of dollar values, each value corresponding to the dollar refund that a specific customer will receive. Naturally, the refund associated with a customer who does not file for a refund will be 0. Those who do file will receive a refund (in dollars) that will vary from one customer to another. Then the answer to our question—the total amount of money that will be refunded—will be the sum of the 54,000 refunds in the population.

Again, this sum is unknown to us, but we can estimate its value by sampling. For example, suppose that each customer in a sample of 540 (i.e., 1 percent of the total number of potentially eligible customers) is questioned concerning the amount (if any) of refund money received from the merchandisers, and that the total amount of refund money for the sample was $4,000. Then, since the sample represents only 1 percent (1/100) of the total number of measurements in the population, we might estimate the total amount of money refunded to the 54,000 potentially eligible customers as 100($4,000), or $400,000.

How reliable is this number as an estimate of the unknown actual amount of money that will be refunded to the 54,000 customers? In performing this extrapolation from sample to population, we have made an inference that is subject to error. How large might this error be? The ability to answer this question (the topic of Chapter 8) is one of the major contributions of statistics.

1.6
Summary

Statistics is an area of science concerned with the design of experiments or sampling procedures, the analysis of data, and the making of inferences about a population of measurements from information contained in a sample. The statistician is concerned with developing and using procedures for design, analysis, and inference making which will provide the best inference at a minimum cost. In addition to making the best inference, the statistician is concerned with providing a quantitative measure of the goodness of the inference-making procedure.

A careful identification of the target population and the design of the sampling procedure are often omitted but are essential steps in drawing inferences from experimental data. Poorly designed sampling procedures will often produce data that are of little or no value (although this may not be obvious to the experimenter). And after you make an inference, cast a critical eye upon it. Be sure you acquire a measure of its reliability.

1.7
A Note to
the Reader

We have stated the objective of statistics and, hopefully, have answered the question, "What is statistics?" The remainder of this text is devoted to the development of the basic concepts involved in statistical methodology. In other words, we wish to explain how statistical techniques actually work and why.

As you proceed through this text, you may wonder why we are devoting so much space to the making of inferences about populations from sample data and so little to the design of experiments. The reason is that the principles of good design do not become apparent until you know how to make inferences. As a consequence, your first contact with a design that may greatly increase the information in an experiment is in Section 10.5. An elementary discussion of experimental design follows in Chapter 14. The purpose of this text is to introduce you to the concepts and some of the elementary methods of statistics. For assistance in designing real-life experiments or surveys, you will want to consult a professional statistician.

Statistics is a very heavy user of applied mathematics. Most of the fundamental rules (called theorems in mathematics) are developed and based upon a knowledge of the calculus or higher mathematics. Inasmuch as this is meant to be an introductory text, we omit proofs except where they can be easily derived. Where concepts or theorems can be shown to be intuitively reasonable, we shall attempt to give a logical explanation. Hence, we shall attempt to convince you with the aid of examples and intuitive arguments rather than with rigorous mathematical derivations.

You should refer occasionally to Chapter 1 and review the objective of statistics and the elements of a statistical problem. Each of the following chapters should, in some way, be directed toward answering the questions posed here. Each is essential to completing the overall picture of statistics.

References

Careers in Statistics, American Statistical Association and the Institute of Mathematical Statistics, 1974.

Tanur, J.M., et al., eds., *Statistics: A Guide to the Unknown*, 2nd ed. San Francisco: Holden-Day, 1978.

Exercises

1.1. Suppose that you wish to obtain an estimate of the gasoline consumption (miles per gallon) of the Ford Fiesta. Describe the population that would likely be of interest to you and from which you would like to select your sample.

1.2. You are a candidate for your state legislature and you wish to survey voter attitudes regarding your chances of winning. Identify the population of interest to you and from which you would like to select your sample. How will this population be dependent upon time?

1.3. A medical researcher wishes to estimate the survival time of a patient after the onset of a particular type of cancer and after a particular regimen of radiotherapy. Identify the population of interest to the medical researcher. Can you perceive some problems in sampling this population?

1.4. An educational researcher wishes to evaluate the effectiveness of a new method for teaching reading to deaf students. Achievement at the end of a period of teaching is to be measured by a student's score on a reading test. Discuss the population (or populations) that might be of interest to the researcher.

1.5. The *Orlando Sentinel Star* (April 16, 1981) describes a survey conducted by the Merit Systems Protection Board, a government agency charged with the protection of whistle blowers, federal workers who report observed fraud, waste, or mismanagement. Only 8500 of the 13,000 questioned (probably by mailed questionnaires) responded. In response to one question, 70 percent of the respondents stated that they told no one and did nothing about observed improper activity for fear of reprisal or because they thought that nothing would be done about it. As it pertains to this question,

a. Describe the population of interest to the Merit Systems Protection Board.

b. Describe the sample.

c. Explain why the sample data might give a distorted view of the nature of the population.

1.6. One deterrent to the development of new life-saving drugs is the high cost and length of time for development and the limited patent life of a drug (17 years). To illustrate, the *Philadelphia Inquirer* (March 29, 1981) reports that Timoptic, the new Merck & Co. glaucoma drug, required six years of internal testing plus a rare "fast track" eight-month test by the Food and Drug Administration (FDA) before it was available to consumers. Similarly, it required almost 14 years of testing before Tavist, an antihistamine, and

Topicort, an anti-inflammatory cream, were brought to market by Sandoz, Inc. and Hoechst-Roussell Pharmaceuticals, Inc., respectively. This allowed only approximately ten years of remaining patent life for Merck to recover the cost of its investment and make a profit, and only four years for Sandoz and Hoechst-Roussell. Suppose that you wish to research the length of time it requires to bring a new drug to market and that you plan to select a sample from among all new drugs approved by the FDA since 1970. Describe the population that is the target of your interest.

2. Useful Mathematical Notation (Optional)

Chapter Objectives

General Objective

Because functional notation and summation notation are used throughout this text, the objective of this chapter is to review (or introduce) these two important mathematical notations.

Specific Objectives

1. To develop an understanding of functional notation. This notation will be used in the subsequent discussion of summation notation. *Section 2.2*

2. To develop an understanding of the meaning of summation symbols. *Sections 2.3, 2.4*

3. To show how summation notion is used to provide instructions for the summation of numbers. Summation notation will be used in the summing of sample data in Chapter 3 and subsequent chapters. *Section 2.4*

4. To present three useful theorems that can be used to find simplified expressions for formulas that appear in Chapter 3 and subsequent chapters. *Section 2.5*

2.1
Introduction

As you will learn subsequently, the statistical analysis of data will require numerous arithmetical operations which are too cumbersome to describe in words. The solution to this difficulty is to express the operations in terms of one or more formulas. Then, to analyze a given set of sample data, you substitute the sample measurements into the appropriate set of formulas.

One of the most common operations in a statistical analysis of data is the process of summation (addition). Consequently, we need a symbol to instruct the reader to sum the sample measurements or perhaps a set of numbers computed from the sample measurements. This symbol, called **summation notation,** will be very familiar to some readers and entirely new to others.

If you are unfamiliar with the notation or need a review, it is suggested that you read this chapter and work the exercises at the ends of the sections. If you are familiar with summation notation, the chapter can be omitted.

Understanding summation notation requires that you be familiar with two topics from elementary algebra: functional notation and the notion of a sequence. Functional notation is important because we shall often wish to sum some function of the sample measurements (to give a simple example, their squares). Thus the function of the sample measurements to be summed will be given as a formula in functional notation. Sequences are important because we must order (identify by arranging in order) the sample measurements so that we can give instructions on which measurements are to be summed.

At first glance this chapter may seem more mathematical than necessary. Please bear with us. You will need an understanding of this notation if you are to understand the formulas that will appear in subsequent chapters.

2.2
Functional Notation

Let us consider two sets* of elements (objects, numbers, or anything we want to use) and their relation to one another. Let the symbol x represent an element of the first set and y an element of the second. One could specify a number of different rules defining relationships between x and y. For instance, if our elements are people, we have rules for determining whether x and y, two persons, are first cousins. Or, suppose that x and y are integers taking values 1, 2, 3, 4, ... For some reason we might wish to say that x and y are "related" if $x = y$.

Now let us direct our attention to a specific relationship useful in mathematics called a functional relation between x and y.

Definition

A function consists of two sets of elements and a defined correspondence between an element of the first set, x, and an element of the second set, y, such that for each element x there corresponds one, and only one, element y.

---

* A set is a collection of specific things.

A function may be exhibited as a collection of ordered pairs of elements, written (x, y) where x represents an element from the first set and y represents an element from the second. In fact, a function is often defined as a collection of ordered pairs of elements with the property stated in the definition.

Most often, x and y will be variables taking numerical values. The two sets of elements would represent all the possible numerical values that x and y might take, and the rule defining the correspondence between them would be an equation. For example, if we state that x and y are real numbers and that

$$y = x + 2,$$

then y is a function of x. Assigning a value to x (that is, choosing an element of the set of all real numbers), there corresponds one, and only one, value of y. When $x = 1$, $y = 3$. When $x = -4$, $y = -2$; and so on.

The area, A, of a circle is related to the radius, r, by the formula

$$A = \pi r^2.$$

Note that if we assign a value to r, the value of A can be determined from the formula. Hence A is a function of r. Likewise, the circumference of a circle, C, is a function of the radius.

As a third example of a functional relation, consider a classroom containing 20 students. Let x represent a specific body in the classroom and y represent a name. Is y a function of x? To answer the question, we examine x and y in the light of the definition. We note that each body has a name y attached to it. Hence when x is specified, y will be uniquely determined. According to our definition, y is a function of x. Note that the defining rule for functional relations does not require that x and y take numerical values. We shall encounter an important functional relation of this type in Chapter 4.

Having defined a functional relation, we may now turn to functional notation.

Mathematical writing frequently uses the same phrases or refers to a specific object many times in the course of a discussion. Unnecessary repetition wastes the reader's time, takes up valuable space, and is cumbersome to the writer. Hence the mathematician resorts to mathematical symbolism which is in some respects a type of mathematical shorthand. Rather than state "The area of a circle, A, is a function of the radius, r," the mathematician would write $A(r)$ or $A = f(r)$. The expression

$$y = f(x)$$

tells us that y is a function of the variable appearing in parentheses, namely x. Note that this expression does not tell us the specific functional relation that exists between y and x.

Consider the function of x,

$$y = 3x + 2.$$

In functional notation, this would be written as

$$f(x) = 3x + 2.$$

It is understood that $y = f(x)$.

Functional notation is especially advantageous when we wish to indicate the value of the function y when x takes a specific value, say $x = 2$. For the example above, we see that when $x = 2$,

$$
\begin{aligned}
y &= 3x + 2 \\
&= 3(2) + 2 \\
&= 8.
\end{aligned}
$$

Rather than write this, we use the simpler notation

$$f(2) = 8.$$

Similarly, $f(5)$ would be the value of the function when $x = 5$:

$$
\begin{aligned}
f(5) &= 3(5) + 2 \\
&= 15 + 2 \\
&= 17.
\end{aligned}
$$

Thus $f(5) = 17$.

To find the value of the function when x equals any value, say $x = c$, we substitute the value of c for x in the equation and obtain

$$f(c) = 3c + 2$$

Formulas for probabilities in later chapters will be expressed in functional notation.

Example 2.1

Given a function of x,

$$g(x) = \frac{1}{x} + 3x \qquad \text{where } x \neq 0.$$

(Note that x cannot equal 0 because then $1/x$ is undefined.)

a. Find $g(4)$.

$$
\begin{aligned}
g(4) &= \frac{1}{4} + 3(4) \\
&= .25 + 12 \\
&= 12.25.
\end{aligned}
$$

b. Find $g\left(\dfrac{1}{a}\right)$.

$$g\left(\frac{1}{a}\right) = \frac{1}{(1/a)} + 3\left(\frac{1}{a}\right)$$

$$= a + \frac{3}{a}$$

$$= \frac{a^2 + 3}{a}.$$

Example 2.2

Let $p(y) = (1 - a)a^y$. Find $p(2)$, $p(3)$, and $p(0)$.

$$p(2) = (1 - a)a^2;$$
$$p(3) = (1 - a)a^3;$$
$$p(0) = (1 - a)a^0 = 1 - a, \quad \text{since } a^0 = 1.$$

Example 2.3

Let $f(y) = 4$. Find $f(2)$ and $f(3)$.

$$f(2) = 4;$$
$$f(3) = 4.$$

Note that $f(y)$ always equals 4, regardless of the value of y. Hence when a value of y is assigned, the value of the function is determined and will equal 4.

Exercises

2.1. If $f(x) = x^2 - 3x + 1$, find $f(0)$, $f(1)$, and $f(2)$.

2.2. If $p(y) = (1/2)^y$, find $p(0)$, $p(1)$, and $p(4)$.

2.3. If $f(y) = y^2 + 3$, find $f(0)$ and $f(3)$.

2.4. If $f(y) = y^3$, find $f(1)$, $f(2)$, and $f(3)$.

2.5. If $f(y) = y$, find $f(2)$, $f(-3)$, and $f(-1)$.

2.6. If $g(y) = (y - 1)^2$, find $g(1)$, $g(2)$, and $g(3)$.

2.3
**Numerical
Sequences**

In statistics we shall be concerned with samples consisting of sets of measurements. There will be a first measurement, a second, and so on. Introducing the notion of a mathematical sequence at this point will provide us with a simple notation for discussion of data and will, at the same time, supply our first practical application of functional notation. This will, in turn, be used in the summation notation introduced in Section 2.4.

A set of objects, $a_1, a_2, a_3, a_4, \ldots$, ordered in the sense that we can identify the first member of the set a_1, the second a_2, etc., is called a **sequence**. Most often, $a_1, a_2, a_3, \ldots$, called **elements** of the sequence, are numbers, but this is not a requirement. For example, the numbers

$$1, 5, 4, 8, 7, 11, 10, \ldots$$

form a sequence moving from left to right. Note that the elements are ordered only in their position in the sequence and need not be ordered in magnitude. Likewise, in some card games, we are interested in a sequence of cards; for example,

10, jack, queen, king, ace.

Although nonnumerical sequences are of interest, we shall be concerned solely with numerical sequences. Specifically, data obtained in a sample from a population will be regarded as a sequence of measurements.

Since sequences will form an important part of subsequent discussions, let us turn to a shortcut method of writing sequences that utilizes functional notation. Inasmuch as the elements of a sequence are ordered in position in the sequence, it would seem natural to attempt to write a formula for a typical element of the sequence as a function of its position. For example, consider the sequence

$$3, 4, 5, 6, 7, \ldots,$$

where each element in the sequence is one greater than the preceding element. Let y be a position variable for the sequence so that y can take values 1, 2, 3, 4, for positions 1, 2, 3, 4, $\ldots$, respectively. Then we might write a formula for the element in position y as

$$f(y) = y + 2.$$

Thus the first element in the sequence would be in position $y = 1$, and $f(1)$ would equal

$$f(1) = 1 + 2 = 3.$$

Likewise, the second element would be in position $y = 2$, and the second element of the sequence would be

$$f(2) = 2 + 2 = 4.$$

A brief check convinces us that this formula works for all elements of the sequence. Note that finding a proper formula (function) is a matter of trial and error and requires a bit of practice.

Example 2.4

Given the sequence

$$1, 4, 9, 16, 25, \ldots,$$

find a formula that expresses a typical element in terms of a position variable y.

We note that each element is the square of the position variable; hence

$$f(y) = y^2.$$

Example 2.5

The formula for the typical element of the following sequence is not so obvious:

$$0, 3, 8, 15, 24, 35, \ldots.$$

The typical element would be

$$f(y) = y^2 - 1.$$

Readers of mathematical writings (and this includes the author) prefer consistency in the use of mathematical notation. Unfortunately, this is not always practical. Writers are limited by the number of symbols available and also by the desire to make their notation consistent with some other texts on the subject. Hence x and y are very often used in referring to a variable, but we could just as well use i, j, k, or z. For instance, suppose that we wish to refer to a set of measurements and denote the measurements as a variable y. We might write this sequence of measurements as

$$y_1, y_2, y_3, y_4, y_5, \ldots,$$

using a subscript to denote a particular element in the sequence. We are now forced to choose a new variable to denote the position of an element in the sequence (since y has been used). Suppose that we use the letter i. Then we could write a typical element as

$$f(i) = y_i.$$

Note that, as previously, $f(1) = y_1, f(2) = y_2$, etc.

Exercises

2.7.   If y is the position variable for a sequence and the formula for a typical element is $y^2 + 1$, give the first three elements of the sequence.

2.8.   Given the sequence

$$2, 3, 4, 5, 6, \ldots,$$

give a formula for a typical element of the sequence as a function of y, the position variable.

2.4
Summation Notation

As we shall observe in Chapter 3, in analyzing statistical data we shall often be working with sums of numbers and will need a simple notation for indicating a sum. For instance, consider the sequence of numbers

$$1, 2, 3, 4, 5, \ldots$$

and suppose that we wish to discuss the sum of the squares of the first four numbers of the sequence. Using summation notation, this would be written as

$$\sum_{y=1}^{4} y^2.$$

Interpretation of the summation notation is relatively easy. The Greek letter Σ (capital sigma), corresponding to "S" in the English alphabet (the first letter in the word "Sum"), tells us to sum elements of a sequence. A typical element of the sequence is given to the right of the summation symbol, and the position variable, called the variable of summation, is shown beneath. For our example, y^2 is a typical element, y is the variable of summation, and the implied sequence is

$$1, 4, 9, 16, 25, 36, \ldots.$$

Which elements of the sequence should appear in the sum? The position of the first element in the sum is indicated below the summation sign, the last above. The sum would include all elements proceeding *in order* from the first to last. In our example, the sum would include the sum of the elements commencing with the first and ending with the fourth:

$$\sum_{y=1}^{4} y^2 = (1)^2 + (2)^2 + (3)^2 + (4)^2$$

$$= 1 + 4 + 9 + 16$$

$$= 30.$$

Example 2.6

$$\sum_{y=2}^{4} (y-1) = (2-1) + (3-1) + (4-1)$$

$$= 6.$$

Example 2.7

$$\sum_{x=2}^{5} 3x = 3(2) + 3(3) + 3(4) + 3(5)$$

$$= 42.$$

We emphasize that the typical element is a function only of the variable of summation. All other symbols are regarded as constants.

Example 2.8

$$\sum_{i=1}^{3} (y_i - a) = (y_1 - a) + (y_2 - a) + (y_3 - a).$$

In this example, note that i is the variable of summation and that it appears as a subscript in the typical element.

Example 2.9

$$\sum_{i=1}^{2} (x - i + 1) = (x - 1 + 1) + (x - 2 + 1)$$

$$= 2x - 1.$$

Example 2.10

$$\sum_{y=2}^{4} y - 1 = (2 + 3 + 4) - 1$$

$$= 8.$$

Note the difference between Example 2.6 and Example 2.10. The quantity $(y - 1)$ is the typical element in Example 2.6, while y is the typical element in Example 2.10.

Example 2.11

Suppose 122, 129, and 124 represent the blood pressure measurements on 3 twenty-year-old females. Let y_1 represent the measurement 122 and, similarly, let $y_2 = 129$ and $y_3 = 124$. Find

$$\sum_{i=1}^{3} y_i \quad \text{and} \quad \sum_{i=1}^{3} y_i^2.$$

Solution

You can see that the expression

$$\sum_{i=1}^{3} y_i = y_1 + y_2 + y_3$$

provides a compact way of writing the sum of the sample measurements. Then,

$$\sum_{i=1}^{3} y_i = y_1 + y_2 + y_3 = 122 + 129 + 124 = 375.$$

Similarly,

$$\sum_{i=1}^{3} y_i^2$$

is a compact way of writing the sum of squares of the sample measurements. Thus,

$$\sum_{i=1}^{3} y_i^2 = y_1^2 + y_2^2 + y_3^2 = (122)^2 + (129)^2 + (124)^2$$

$$= 14{,}884 + 16{,}641 + 15{,}376 = 46{,}901.$$

Exercises

2.9. Evaluate the following summations.

a. $\displaystyle\sum_{y=0}^{5} (y - 4)$ b. $\displaystyle\sum_{y=2}^{6} (y^2 - 5)$ c. $\displaystyle\sum_{i=1}^{4} (y_i - 2)$ d. $\displaystyle\sum_{i=1}^{3} (y + 2i)$

2.10. If y is the position variable for a sequence and the formula for a typical element is $y^2 - 1$:

a. Write the first four elements of the sequence.

b. Use summation notation to write an expression for the sum of the first four terms of the sequence.

c. Find the sum, part (b).

2.11. If y is the position variable for a sequence and the formula for a typical element is $(y - 3)^2$:

a. Write the first five elements of the sequence.

b. Use summation notation to write an expression for the sum of the first five terms of the sequence.

c. Find the sum, part (b).

2.12. Consider the set of the opinions of all persons at your college or university regarding the desirability of a tax on gas-guzzling automobiles. Let a person favoring the tax be represented as a 1 and a person opposed as a 0 (for the sake of simplicity, assume that everyone has an opinion). As an ultimate objective, we wish to sample this set of opinions in order to estimate the proportion of students favoring the tax.

a. Describe the population of interest.

b. Suppose the first five measurements in the sample are $y_1 = 1$, $y_2 = 1$, $y_3 = 0$, $y_4 = 0$, and $y_5 = 1$. Use summation notation to write expressions for the sum and the sum of the squares of these five measurements.

c. Find the sum and the sum of the squares of the five measurements.

d. Suppose that the complete sample contains $n = 100$ measurements, which are represented as $y_1, y_2, \ldots, y_{99}, y_{100}$, and that 73 of these favor the tax. Find

$$\sum_{i=1}^{100} y_i, \qquad \sum_{i=1}^{100} y_i^2, \qquad \text{and} \qquad \left(\sum_{i=1}^{100} y_i\right)^2.$$

2.13. To estimate weekly loss due to theft, a clothing store recorded the total dollar loss over a period of ten weeks. The losses, recorded to the nearest ten dollars, were

$$y_1 = 360, \quad y_2 = 430, \quad y_3 = 210, \quad y_4 = 320, \quad y_5 = 550,$$
$$y_6 = 170, \quad y_7 = 240, \quad y_8 = 370, \quad y_9 = 280, \quad y_{10} = 290.$$

a. Describe the population of interest to the store manager.

b. Find $\sum_{i=1}^{10} y_i$.

c. Find $\sum_{i=2}^{4} y_i$, $\quad \sum_{i=2}^{4} y_i^2$, $\quad$ and $\quad \left(\sum_{i=1}^{3} y_i \right)^2$.

2.14. For $y_i = 2i^2 - 1$, find:

a. $\sum_{i=1}^{5} y_i$ $\qquad$ b. $\sum_{i=1}^{4} y_i^2$ $\qquad$ c. $\left(\sum_{i=1}^{4} y_i \right)^2$ $\qquad$ d. $\sum_{i=1}^{5} xy_i$.

2.5
Useful Theorems
Relating to Sums

Consider the summation

$$\sum_{y=1}^{3} 5.$$

The typical element is 5 and it does not change. The sequence is, therefore,

$$5, 5, 5, 5, \ldots,$$

and

$$\sum_{y=1}^{3} 5 = 5 + 5 + 5$$
$$= 15.$$

Theorem 2.1

Let c be a constant (an element that does not involve the variable of summation) and y be the variable of summation. Then,

$$\sum_{y=1}^{n} c = nc.$$

Proof

$$\sum_{y=1}^{n} c = c + c + c + \cdots + c,$$

where the sum involves n elements. Then,

$$\sum_{y=1}^{n} c = nc.$$

Example 2.12

$$\sum_{y=1}^{4} 3a = 4(3a) = 12a.$$

(Note that y is the variable of summation and, therefore, that a is a constant.)

Example 2.13

$$\sum_{i=1}^{3} (3x - 5) = 3(3x - 5).$$

(Note that i is the variable of summation and, therefore, that x is a constant.)

Example 2.14

$$\sum_{y=4}^{10} 3a = \sum_{y=1}^{10} 3a - \sum_{y=1}^{3} 3a = 10(3a) - 3(3a) = 21a.$$

(Note that since the sum of the elements does not commence with $y = 1$, it can be written as the difference between the sum of the elements from 1 to 10 and the sum from 1 to 3.)

A second theorem is illustrated using Example 2.7. We note that 3 is a common factor in each term. Therefore,

$$\sum_{x=2}^{5} 3x = 3(2) + 3(3) + 3(4) + 3(5)$$

$$= 3(2 + 3 + 4 + 5)$$

$$= 3 \sum_{x=2}^{5} x.$$

Thus it would appear that the summation of a constant times a variable is equal to the constant times the summation of the variable.

Theorem 2.2

Let c be a constant. Then

$$\sum_{i=1}^{n} cy_i = c \sum_{i=1}^{n} y_i.$$

Proof

$$\sum_{i=1}^{n} cy_i = cy_1 + cy_2 + cy_3 + \cdots + cy_n$$

$$= c(y_1 + y_2 + \cdots + y_n)$$

$$= c \sum_{i=1}^{n} y_i.$$

Theorem 2.3

$$\sum_{i=1}^{n} (x_i + y_i + z_i) = \sum_{i=1}^{n} x_i + \sum_{i=1}^{n} y_i + \sum_{i=1}^{n} z_i.$$

Proof

$$\sum_{i=1}^{n} (x_i + y_i + z_i) = x_1 + y_1 + z_1 + x_2 + y_2 + z_2$$

$$+ x_3 + y_3 + z_3 + \cdots + x_n + y_n + z_n.$$

Regrouping, we have

$$\sum_{i=1}^{n} (x_i + y_i + z_i) = (x_1 + x_2 + \cdots + x_n) + (y_1 + y_2 + \cdots + y_n)$$

$$+ (z_1 + z_2 + \cdots + z_n)$$

$$= \sum_{i=1}^{n} x_i + \sum_{i=1}^{n} y_i + \sum_{i=1}^{n} z_i.$$

In words we would say that the summation of a typical element which is itself a sum of a number of terms is equal to the sum of the summations of the terms.

Theorems 2.1, 2.2, and 2.3 can be used jointly to simplify summations. Consider the following examples:

Example 2.15

$$\sum_{x=1}^{3} (x^2 + ax + 5) = \sum_{x=1}^{3} x^2 + \sum_{x=1}^{3} ax + \sum_{x=1}^{3} 5$$

$$= \sum_{x=1}^{3} x^2 + a \sum_{x=1}^{3} x + 3(5)$$

$$= (1 + 4 + 9) + a(1 + 2 + 3) + 15$$

$$= 6a + 29.$$

Example 2.16

$$\sum_{i=1}^{4} (x^2 + 3i) = \sum_{i=1}^{4} x^2 + \sum_{i=1}^{4} 3i$$

$$= 4x^2 + 3 \sum_{i=1}^{4} i$$

$$= 4x^2 + 3(1 + 2 + 3 + 4)$$

$$= 4x^2 + 30.$$

Example 2.17

Suppose you have a set of sample measurements, $y_1, y_2, y_3, \ldots, y_n$, and that you subtract a constant, say c, from each measurement; then the sum of $y_1 - c, y_2 - c, \ldots, y_n - c$, is

$$\sum_{i=1}^{n} (y_i - c) = \sum_{i=1}^{n} y_i - \sum_{i=1}^{n} c = \sum_{i=1}^{n} y_i - nc.$$

Exercises

Utilize Theorems 2.1, 2.2, and 2.3 to simplify and evaluate the following summations.

2.15. $\displaystyle\sum_{x=1}^{4} (x - 2)$

2.16. $\displaystyle\sum_{y=1}^{5} (3y - 4)$

2.17. $\displaystyle\sum_{y=1}^{4} (y^2 - 4y + 1)$

2.18. $\displaystyle\sum_{y=1}^{4} 4$

2.19. $\displaystyle\sum_{y=0}^{3} p^y$

2.20. $\displaystyle\sum_{x=0}^{3} (x + y) + 1$

2.21. $\displaystyle\sum_{y=2}^{4} (y + 2)^2$

2.6
Summary

Two types of mathematical notations have been presented: functional notation and summation notation. The former is used in summation notation to express the typical element as a function of the variable of summation. Formulas will also be presented in functional notation in subsequent chapters. Summation notation will be employed in Chapter 3 and succeeding chapters.

Supplementary Exercises

[Throughout, starred (∗) exercises are optional.]

2.22. If y is the position variable for a sequence and the formula for a typical element is $y^2 + y - 1$:

a. Write the first four elements of the sequence.

b. Use summation notation to write an expression for the sum of the first four terms of the sequence.

c. Find the sum, part (b).

2.23. Given the sequence

$$0, 1, 4, 9, 16, \ldots.$$

a. Write a formula for the element in the yth position, $y = 1, 2, 3, 4, 5, \ldots.$ [*Hint:* Note that the element in the third position is $(3 - 1)^2$.]

b. Write an expression for the sum of the first five terms of the sequence.

c. Find the sum, part (b).

2.24. Given $f(y) = 4y + 3$, find

a. $f(0)$ b. $f(1)$ c. $f(2)$ d. $f(-1)$

e. $f(-2)$ f. $f(a^2)$ g. $f(-a)$ h. $f(1 - y)$

2.25. If $g(y) = (y^2 + 1)/(y + 1)$, find

a. $g(1)$ b. $g(-1)$ c. $g(4)$ d. $g(-a)$

2.26. If $p(x) = (1 - a)^x$, find $p(0)$ and $p(1)$.

2.27. If $f(x) = 3x^2 - 3x + 1$ and $g(x) = x - 3$, find

a. $f(1/2)$ b. $g(-3)$ c. $f(1/x)$

2.28. If $h(x) = 2$, find

a. $h(0)$ b. $h(1)$ c. $h(2)$

2.29. Suppose that you have a sample of five measurements, where $y_1 = 3.1, y_2 = 2.0, y_3 = 4.4, y_4 = 2.5,$ and $y_5 = 2.1$. Find

$$\sum_{i=1}^{5} y_i, \quad \sum_{i=1}^{5} y_i^2, \quad \text{and} \quad \left(\sum_{i=1}^{5} y_i \right)^2.$$

2.30. Refer to Exercise 2.29. What is the effect on the sum of the measurements of subtracting 3.1 from each measurement? Use the summation theorems to simplify the expression for

$$\sum_{i=1}^{5} (y_i - 3.1).$$

Utilize Theorems 2.1, 2.2, and 2.3 to simplify and evaluate the following summations.

2.31. $\displaystyle\sum_{y=1}^{3} y^3$ 2.32. $\displaystyle\sum_{y=1}^{3} 7$ 2.33. $\displaystyle\sum_{x=2}^{3} (1 + 2x + x^2)$

2.34. $\displaystyle\sum_{x=1}^{4} (x + xy^2)$ 2.35. $\displaystyle\sum_{i=1}^{2} (y_i - i)$ 2.36. $\displaystyle\sum_{i=1}^{n} (y_i - a)$

Using the accompanying set of measurements, calculate the sums in Exercises 2.37 through 2.42.

i	1	2	3	4	5	6	7	8	9	10	11	12	13
y_i	3	12	10	-6	0	11	2	-9	-5	8	-7	4	-5

2.37. $\displaystyle\sum_{i=1}^{13} 2y_i$ 2.38. $\displaystyle\sum_{i=1}^{13} (2y_i - 5)$ 2.39. $\displaystyle\sum_{i=1}^{10} y_i^2$

2.40. $\displaystyle\sum_{i=1}^{10} (y_i^2 + y_i)$ 2.41. $\displaystyle\sum_{i=1}^{13} (y_i - 2)^2$ 2.42. $\displaystyle\sum_{i=1}^{13} y_i^2 - \frac{1}{13}\left(\sum_{i=1}^{13} y_i\right)^2$

Verify the following identities. Each of them is a shortcut that will be used in later chapters. The symbols $\bar{x}$ and $\bar{y}$ appearing in these identities have the following definitions:

$$\bar{x} = \frac{\displaystyle\sum_{i=1}^{n} x_i}{n}; \qquad \bar{y} = \frac{\displaystyle\sum_{i=1}^{n} y_i}{n}.$$

* 2.43. $\displaystyle\sum_{i=1}^{n} (y_i - \bar{y})^2 = \sum_{i=1}^{n} y_i^2 - \frac{\left(\displaystyle\sum_{i=1}^{n} y_i\right)^2}{n}.$

* 2.44. $\displaystyle\sum_{i=1}^{n} (x_i - \bar{x})(y_i - \bar{y}) = \sum_{i=1}^{n} x_i y_i - \frac{\left(\displaystyle\sum_{i=1}^{n} x_i\right)\left(\displaystyle\sum_{i=1}^{n} y_i\right)}{n}.$

* 2.45. $\displaystyle\frac{\displaystyle\sum_{i=1}^{n} (x_i - \bar{x})(y - \bar{y})}{\displaystyle\sum_{i=1}^{n} (x_i - \bar{x})^2} = \frac{n\displaystyle\sum_{i=1}^{n} x_i y_i - \left(\displaystyle\sum_{i=1}^{n} x_i\right)\left(\displaystyle\sum_{i=1}^{n} y_i\right)}{n\displaystyle\sum_{i=1}^{n} x_i^2 - \left(\displaystyle\sum_{i=1}^{n} x_i\right)^2}.$

3. Describing Distributions of Measurements

Chapter Objectives

General Objectives

Methods for describing sets of data are needed so that (1) you will better understand data sets and thus more easily construct theories about the phenomena they represent and (2) you will be able to phrase an inference (make descriptive statements) about a population based on sample data. Consequently, the objective of this chapter is to find a single compact method for describing a set of data.

Specific Objectives

1. To develop the notion of a relative frequency distribution (or relative frequency histogram), a very good graphical method for describing a set of data. As you will subsequently see, a model for the relative frequency distribution for a population of data, a probability distribution (Chapter 5), will be the tool employed in making inferences about populations based on sample data. *Sections 3.1, 3.9*

2. To present important numerical descriptive methods, particularly measures of central tendency and variation. *Sections 3.2, 3.3, 3.4, 3.9*

3. To provide you with a method for interpreting a standard deviation so that you can construct a mental picture of the frequency distribution for a set of data. *Sections 3.5, 3.8, 3.9*

4. To present an easy way to calculate the standard deviation of a set of measurements. *Sections 3.6, 3.7*

Case Study

So You Want to Be a Millionaire?

You must have seen those newspaper advertisements shortly after passage of the Tax Act of 1981. Many full-page ads portrayed the local banker, in counseling mode, telling us how we could become millionaires. According to the ads, all one has to do is invest $2000 per year in an Individual Retirement Account. After 40 years of participation, the nest egg will have grown to over a million dollars. The opportunity for this bonanza is available to all persons earning a minimum of $3000 annually. The only handicaps are that you cannot touch your treasure until you are $59\frac{1}{2}$ years old and you must pay income taxes when money is withdrawn from the retirement account.

Of course, it is not as simple or certain as the advertisements lead you to believe. Whether you retire on a million dollars will depend on whose hencoop you put your nest egg in; that is, who manages your account. It will also depend upon the type of investment that you choose—a bank savings account, a money fund, or one of the various common stock funds. Basically, it will depend upon the annual rate of return that your account manager is able to obtain for you.

Although the rate of return on your money will vary from day to day, Table 3.1 will give you an indication of the amount of money that you might expect to accumulate in 40 years. The amounts shown are based on the assumption that you invest $2000 at the beginning of each year for a period of 40 years and that the money is compounded monthly at a fixed annual rate of interest I.

Monthly rates of return vary not only among investment vehicles, banks, stock funds, money funds, etc., but they also vary within an investment vehicle itself. For example, consider a money fund, one that invests in short-term commercial rates. Since the notes will be negotiated at different times and will vary in length of time to maturity, the rate of return for the total fund at any one point in time will depend upon the fund manager's skill in placing the loans. If the interest rate rises in the future, it pays to hold notes with a small average time to maturity. If it drops, it pays to hold notes with a large average time to maturity.

The characteristics of money funds as a vehicle for investment are imbedded in the data of Table 3.2. Table 3.2 presents the asset size (in millions of

Table 3.1 Withdrawal account for an IRA account after 40 years of participation at a fixed annual rate of return I (%)

	Interest Rate I				
	8%	10%	12%	14%	16%
Amount in account after 40 years	$559,562	$973,704	$1,718,285	$3,059,817	$5,476,957

Table 3.2 Data on 98 money market funds (assets of $100 million or more) for the period ending March 24, 1982

Fund	Assets ($ million)	Average Maturity (Days)	7-Day Average Yield (%)	30-Day Average Yield (%)
Alex Brown Cash Res.	565.0	33	13.7	13.6
Alliance Capital Reserves	1,955.2	23	13.3	13.6
Alliance Government Reserves	181.9	21	12.4	12.4
American General	355.7	20	14.2	14.2
American Liquid Trust	372.5	29	13.5	13.6
BIRR Wilson Money Fund	121.3	31	13.6	13.6
Boston Company Cash Mgmt.	284.5	26	14.0	13.8
Capital Cash Mgt. Trust	145.4	20	13.6	13.7
Capital Preservation	1,788.9	14	13.0	12.9
Capital Preservation Fund II	881.6	3	13.6	13.2
Cardinal Govt. Securities	306.1	9	13.5	13.3
Carnegie Govt. Securities	147.7	7	13.1	13.0
Cash Equivalent	3,684.7	35	14.1	13.9
Cash Equivalent Gov't Only	211.5	15	13.3	13.0
Cash Mgmt Trust	655.0	14	14.0	13.9
Cash Reserve Management	7,082.0	29	14.3	14.0
Chancellor Gov t Sec. Trust	155.1	33	13.1	13.1
Columbia Daily Income	917.6	26	13.3	13.2
Composite Cash Management	346.1	23	13.8	13.7
Current Interest	1,524.9	35	13.6	13.6
DBL Cash Fund	1,035.1	28	13.7	13.7
Daily Cash Accumulation	5,605.4	23	14.1	14.0
Daily Income	844.9	27	13.3	13.3
Dean Witter				
Active Assets Trust	391.3	38	13.8	13.8
InterCapital Liquid Asset	9,656.8	42	13.5	13.3
Delaware Cash Reserve	2,090.3	22	13.9	13.7
Dreyfus				
Govt. Series	794.3	21	12.6	12.5
Liquid Assets	9,598.0	37	13.6	13.6
ED Jones Daily Passport	769.0	32	13.7	13.5
EGT Money Market Trust	155.2	28	13.6	13.3
Eston & Howard	264.4	27	13.7	13.4
Equitable Money Mkt Account	342.3	36	13.4	13.3
Fahnestock Daily Income	150.6	25	13.3	13.3
Fidelity Group				
Cash Reserves	3,615.0	34	13.9	13.8
Daily Income	3,903.9	33	13.7	13.6
U.S. Gov't Reserves	128.9	20	13.3	12.9
Financial Daily Income	302.6	14	14.2	14.2
First Investors Cash Management	615.5	27	13.5	13.6
First Variable Rate	1,256.9	*19	13.5	13.6
Franklin Federal M.F.	121.1	10	12.1	11.8

Table 3.2 *(Continued)*

Fund	Assets ($ million)	Average Maturity (Days)	7-Day Average Yield (%)	30-Day Average Yield (%)
Franklin Money Fund	1,116.3	34	13.9	13.6
Fund/Government Investors	1,234.6	19	13.4	13.4
Government Investors Trust	601.9	16	13.5	13.3
Gradison Cash Reserves	698.2	26	13.5	13.3
Hilliard Lyons C.M. Inc.	175.1	12	13.3	13.2
I.D.S. Cash Management	1,219.6	34	13.9	13.7
INA Cash Fund	733.3	29	13.8	13.6
John Hancock Cash Mgt.	649.1	19	13.6	13.7
Kemper Money Market	3,205.9	35	14.3	14.1
Legg Mason Cash Reserve	316.1	33	13.6	13.3
Lehman Cash Mgt., Inc.	453.3	24	14.0	13.8
Lexington Money Market	291.3	22	14.5	14.6
Liquid Capital Income	2,218.3	19	13.5	13.7
Liquid Green Trust	116.5	27	13.5	13.6
Lord Abbett Cash Reserve	386.8	23	13.6	13.4
MIF/Nationwide	407.4	32	13.7	13.7
Mass. Cash Management	926.0	29	13.8	13.9
McDonald Money Market	207.0	32	13.2	13.0
Merrill Lynch				
CMA Gov't Securities	568.8	34	13.1	13.1
CMA Money Fund	13,812.8	39	13.7	13.7
Government	1,241.0	33	13.3	13.2
Institutional	1,378.3	33	13.7	13.8
Ready Assets	22,814.5	36	13.5	13.7
Midwest Income ST Govt.	260.3	19	12.6	13.0
Money Market Management	546.1	31	13.5	13.3
MoneyMart Assets	4,036.0	26	14.3	14.0
Morgan Keegan Daily Cash	103.7	26	13.6	13.2
Mutual of Omaha	468.9	24	13.5	13.6
National Liquid Reserves	2,100.9	33	13.8	13.6
N.E.L. Cash Management	725.1	31	13.8	13.7
NRTA-AARP U.S. Govt. MM Trust	3,993.3	34	12.7	12.7
Oppenheimer Money Market	1,592.0	26	14.2	14.0
Paine Webber Cashfund	6,233.3	32	13.9	13.7
Putnam Daily Dividend	365.3	36	13.7	13.7
Reserve Fund—Gov't	251.8	13	13.5	13.3
Reserve Fund—Primary	3,156.1	29	13.5	13.4
St. Paul Money Fund, Inc.	178.0	21	14.1	14.0
Scudder Cash Investment	1,352.0	28	13.9	13.8
Shearson/American Express				
Daily Dividend	5,375.1	26	13.8	13.5
FedFund	1,186.3	24	13.4	13.2
Government Agencies	649.0	26	12.6	12.4

Table 3.2 *(Continued)*

Fund	Assets ($ million)	Average Maturity (Days)	7-Day Average Yield (%)	30-Day Average Yield (%)
T-Fund	399.8	24	13.3	13.1
TempFund	4,765.3	27	13.9	13.9
Short-Term Income	372.8	23	13.3	13.3
Standby Reserve Fund	203.2	25	13.5	13.6
SteinRoe Cash Reserves	841.2	30	13.7	13.6
T. Rowe Price Prime Reserve	3,189.6	31	14.2	13.8
Transamerica Cash Reserves	306.1	22	13.8	13.7
Trust/Cash Reserves	221.3	31	13.6	13.1
Tucker Anthony Cash Mgmt.	377.0	23	13.4	13.2
USAA M.M. Fund	147.0	25	13.4	13.4
Union Cash Management	868.5	17	13.8	13.6
United Cash Management	508.0	25	14.7	14.0
Value Line Cash Fund	530.6	25	14.6	14.6
Vanguard Federal	387.1	28	13.2	13.0
Vanguard Money Market Trust	1,140.3	31	13.7	13.5
Webster Cash Reserve	1,352.4	25	13.9	13.7
Ziegler Money Market	119.3	25	13.6	13.5
Donoghue's Money Fund Average (All Funds)		31	13.5	13.4

(Average for all 161 funds reported by DMFR).

Yields represent annualized total return to shareholders for past 7-and 30-day period. Past returns not necessarily indicative of future yields.

* Average term to next rate adjustment data.

Source: *Donoghue's Money Fund Report* of Holliston, Mass. 01746.

dollars), the average maturity (in days) for notes, and the average 7-day and 30-day yields (%) for the period ending March 24, 1982, for 98 large money funds available to investors. An examination of Table 3.2 clearly indicates a statistical problem. While it is possible, by examining the data, to obtain a general feeling about the asset size, the average maturity time, and the average rate of return, it is difficult to obtain a clear picture of the characteristics of these data sets by simply scanning the data. This problem motivates the topic of Chapter 3. In this chapter we will examine methods for describing data sets. Then in Section 3.8, we will apply these techniques to describe the money fund data and will see how this descriptive information is relevant to our prospects of becoming millionaires.

3.1
A Graphical Method for Describing a Set of Data

It would seem natural to introduce appropriate graphical and numerical methods of describing sets of data through consideration of a set of real data. The data presented in Table 3.3 represent the grade-point averages of 30 Bucknell University freshmen recorded at the end of the freshman year.

A cursory examination of the data indicates that the lowest grade-point average in the sample is 1.4, the largest 2.9. How are the other 28 measurements distributed? Do most lie near 1.4, near 2.9, or are they evenly distributed over the interval from 1.4 to 2.9? To answer this question, we divide the interval into an arbitrary number of subintervals of equal length, the number depending upon the amount of data available. (As a rule of thumb, the number of subintervals chosen would range from 5 to 20, the larger the amount of data available, the more subintervals employed.) For instance we might use the subintervals 1.35 to 1.55, 1.55 to 1.75, 1.75 to 1.95, etc. Note that the points dividing the subintervals have been chosen so that it is impossible for a measurement to fall on the point of division, thus eliminating any ambiguity regarding the disposition of a particular measurement. The subintervals, called **classes** in statistical language, form cells or pockets similar to the pockets of a billiard table. We wish to determine the manner in which the measurements are distributed among the pockets, or classes. A tally of the data from Table 3.3 is presented in Table 3.4.

Table 3.3 Grade-point averages of 30 Bucknell University freshmen

1.5	2.6	1.4	2.0	1.4
1.8	2.1	2.6	2.0	1.6
2.4	2.5	2.2	2.0	1.9
2.2	2.0	1.9	2.5	2.9
2.1	2.3	2.0	2.2	2.4
2.2	2.3	1.7	2.2	1.6

Each of the 30 measurements falls in one of eight classes which, for purposes of identification, we shall number. The identification number appears in the first column of Table 3.4, and the corresponding class boundaries are given in the second column. The third column of the table is used for the tally, a mark entered opposite the appropriate class for each measurement falling in the class. For example, 3 of the 30 measurements fall in class 1, three in class 2, three in class 3, seven in class 4, etc. The number of measurements falling in a particular class, say class i, is called the **class frequency** and is designated by the symbol f_i. The class frequency is given in the fourth column of Table 3.4. The last column of this table presents the fraction of the total number of measurements falling in each class. We call this the **relative frequency.** If we let n represent the total number of measurements, for instance, in our example $n = 30$, then

Table 3.4 Tabulation of relative frequencies for a histogram

Class i	Class Boundaries	Tally	Class Frequency f_i	Relative Frequency
1	1.35–1.55	111	3	3/30
2	1.55–1.75	111	3	3/30
3	1.75–1.95	111	3	3/30
4	1.95–2.15	ℿℲℲ 11	7	7/30
5	2.15–2.35	ℿℲℲ 11	7	7/30
6	2.35–2.55	1111	4	4/30
7	2.55–2.75	11	2	2/30
8	2.75–2.95	1	1	1/30
		Total $n = 30$		1

the relative frequency for the ith class would equal f_i divided by n:

$$\text{relative frequency} = \frac{f_i}{n}.$$

The resulting tabulation can be presented graphically in the form of a **frequency histogram** (Figure 3.1). Rectangles are constructed over each class interval, their height being proportional to the number of measurements (class frequency) falling in each class interval. Viewing the frequency histogram, we

Figure 3.1 Frequency histogram

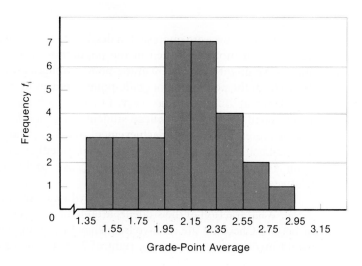

see at a glance the manner in which the grade-point averages are distributed over the interval.

It is often more convenient to modify the frequency histogram by plotting class relative frequency rather than class frequency. A relative frequency histogram is presented in Figure 3.2. Statisticians rarely make a distinction between the frequency histogram and the relative frequency histogram and refer to either as a frequency histogram or simply a histogram. If corresponding values of frequency and relative frequency are marked along the vertical axes of the graphs, then the frequency and the relative frequency histograms are identical (see Figures 3.1 and 3.2).

Figure 3.2 Relative frequency histogram

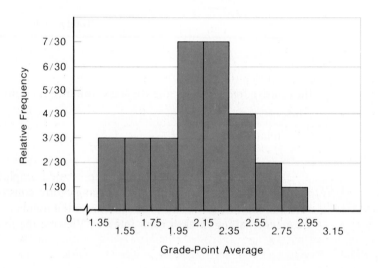

Although we were interested in describing the set of $n = 30$ measurements, we are much more interested in the population from which the sample was drawn. We might view the 30 grade-point averages as a representative sample drawn from the population of grade-point averages of the freshmen currently in attendance at Bucknell University. Or, if we are interested in the academic achievement of freshmen college students in general, we might consider our sample representative of the achievement of the population of freshmen attending Bucknell *or colleges similar to Bucknell*. In either case, if we had the grade-point averages for the entire population, we could construct a population relative frequency histogram.

Let us consider the relative frequency histogram for the sample in greater detail. What fraction of the students attained grade-point averages equal to 2.2 or better? Checking the relative frequency histogram, we see that the fraction would involve all classes to the right of 2.15. Using Table 3.4, we see that 14 students achieved grade-point averages equal to or greater than 2.2. Hence the fraction is 14/30, or approximately 47 percent. We note that this is also the percentage of the total area of the histogram, Figure 3.2, lying to the right of 2.15. Suppose that we were to write each grade-point average on a piece of paper, place the 30 slips of paper in a hat, mix them, and then draw one paper from the hat. What is the chance that this paper would contain a grade-point average equal to or greater than 2.2? Since 14 of the 30 slips are marked with numbers equal to or greater than 2.2, we would say that we have 14 chances out of 30. Or, we might say that the probability is 14/30. You have undoubtedly encountered the word "probability" in ordinary conversation and we are content to defer definition and discussion of its significance until Chapter 4.

Let us now direct our attention to the population from which the sample was drawn. What fraction of students in the population attained a grade-point average equal to or greater than 2.2? If we possessed the relative frequency histogram for the population, we could give the exact answer to this question by calculating the fraction of total area lying to the right of 2.15. Unfortunately, since we do not have such a histogram, we are forced to make an inference. We must estimate the true population fraction, basing our estimate upon information contained in the sample. Our estimate would likely be 14/30, or 47 percent. Suppose that we wish to state the chance or probability that a student drawn from the population would have a grade-point average equal to or greater than 2.2. Without knowledge of the population relative frequency histogram, we would infer that the population histogram is similar to the sample histogram and that approximately 14/30 of the measurements in the population would be equal to or greater than 2.2. Naturally, this estimate would be subject to error. We shall examine the magnitude of this estimation error in Chapter 8.

The relative frequency histogram is often called a frequency distribution because it shows the manner in which the data are distributed along the abscissa of the graph. We note that the rectangles constructed above each class are subject to two interpretations. They represent the fraction of observations falling in a given class. Also, if a measurement is drawn from the data, a particular class relative frequency is also the chance or probability that the measurement will fall in that class. The most significant feature of the sample frequency histogram is that it provides information on the population frequency histogram which describes the population. We would expect the two frequency histograms, sample and population, to be similar. Such is the case. The degree of resemblance will increase as more and more data are added to the sample. If the sample were enlarged to include the entire population, the sample and population would be synonymous and the histograms would be identical.

In the preceding discussion we showed you how to construct a frequency distribution for the grade-point average data, Table 3.3, and we explained how such a distribution could be interpreted. Before concluding this topic, we will summarize the principles that you should employ in constructing a frequency distribution for a set of data.

> ## Principles to Employ in Constructing a Frequency Distribution
>
> 1. **Determine the number of classes.** It is usually best to have from 5 to 20 classes. The larger the amount of data available, the more classes that should be employed. If the number of classes is too small, we might be concealing important characteristics of the data by grouping. If we have too many classes, too many empty classes may result and the distribution would be meaningless. The number of classes should be determined from the amount of data present and the uniformity of the data. A small sample would require fewer classes.
>
> 2. **Determine the class width.** As a general rule for finding the class width, divide the difference between the largest and smallest measurement by the number of classes desired and add enough to the quotient to arrive at a convenient figure for class width. All classes should be of equal width. This allows us to make uniform comparisons of the class frequencies.
>
> 3. **Locate the class boundaries.** To locate the class boundaries, commence with the lowest class so that you include the smallest measurement. Then add the remaining classes. Class boundaries should be chosen so that it is impossible for a measurement to fall on a boundary.

Exercises

3.1.   If you calculate the miles per gallon (mpg) of fuel consumed by your automobile by dividing the miles traveled since your last fuel stop by the number of gallons of gasoline needed for a fill-up, you will find that the mpg varies from one fill-up to another. This is because you are not filling the tank to the same level at each fill-up and you are not able to read precisely the actual number of gallons added. For example, the actual mpg calculations for 50 refills of a 1980 Oldsmobile Omega, recorded for highway driving, are shown below:

28.7	30.8	32.9	28.5	31.3
31.4	27.9	31.3	31.1	27.8
32.1	30.2	29.7	29.9	30.6
29.7	29.4	30.8	30.5	31.5
32.3	33.0	30.4	31.6	28.4
28.6	31.6	28.7	30.7	29.8
30.9	32.4	31.3	29.4	30.7
32.2	30.7	30.9	31.2	32.1
30.5	31.3	30.0	32.0	29.3
29.5	29.2	32.4	30.3	31.6

Construct a relative frequency distribution for the mpg data using the following steps:

a. Find the range for the data, the difference between the largest and smallest mpg measurements.

b. Suppose that you wish the range to approximately equal seven class intervals. Find (to the nearest tenth) the class width.

c. Use the interval 27.55 to 28.35 for the lowest class and then locate the remaining class boundaries. Construct a frequency table for the data.

d. Construct a relative frequency histogram for the data.

e. What proportion of the mpg measurements are less than 31.55? What proportion of the area under the relative frequency histogram lies to the left of 31.55?

f. The $n = 50$ mpg measurements shown above represent a sample from a population of measurements of interest to the automobile owner. Describe the population.

g. If you had chosen six or eight class intervals, instead of seven, to span the range of the measurements, it is unlikely that the resulting relative frequency histograms would be identical to the one obtained in part (b). Is this important? Explain.

3.2. In order to decide on the number of service counters needed for stores to be built in the future, a supermarket chain wished to obtain information on the length of time (in minutes) required to service customers. To obtain information on the distribution of customer service times, a sample of 1000 customers' service times were recorded. Sixty of these are shown in the accompanying tabulation.

3.6	1.9	2.1	.3	.8	.2
1.0	1.4	1.8	1.6	1.1	1.8
.3	1.1	.5	1.2	.6	1.1
.8	1.7	1.4	.2	1.3	3.1
.4	2.3	1.8	4.5	.9	.7
.6	2.8	2.5	1.1	.4	1.2
.4	1.3	.8	1.3	1.1	1.2
.8	1.0	.9	.7	3.1	1.7
1.1	2.2	1.6	1.9	5.2	.5
1.8	.3	1.1	.6	.7	.6

a. Construct a relative frequency histogram for the data.

b. What fraction of the service times are less than or equal to one minute?

3.3. The length of time (in months) between the onset of a particular illness and its recurrence was recorded for $n = 50$ patients. The times of recurrence are as follows:

2.1	4.4	2.7	32.3	9.9
9.0	2.0	6.6	3.9	1.6
14.7	9.6	16.7	7.4	8.2
19.2	6.9	4.3	3.3	1.2
4.1	18.4	.2	6.1	13.5
7.4	.2	8.3	.3	1.3
14.1	1.0	2.4	2.4	18.0
8.7	24.0	1.4	8.2	5.8
1.6	3.5	11.4	18.0	26.7
3.7	12.6	23.1	5.6	.4

a. Construct a relative frequency histogram for the data.

b. Give the fraction of recurrence times less than or equal to 10.

3.4.   Twenty-eight applicants interested in working for the Food Stamp program took an examination designed to measure their aptitude for social work. The following test scores were obtained:

79	97	86	76
93	87	98	68
84	88	81	91
86	87	70	94
77	92	66	85
63	68	98	88
46	72	59	79

a. Construct a relative frequency histogram for the test scores. (Use subintervals of width 9 beginning at 44.5.)

b. What proportion of the scores are less than 80.5?

c. Compare the proportion from part (b) with the proportion of the total area under the histogram that lies to the left of 80.5. What is the relationship between the two proportions?

3.2
Numerical Methods for Describing a Set of Data

Graphical methods are extremely useful in conveying a rapid general description of collected data and in presenting data. In many respects this supports the saying that a picture is worth a thousand words. There are, however, limitations to the use of graphical techniques for describing and analyzing data. For instance, suppose that we wish to discuss our data before a group of people and have no method of describing the data other than verbally. Unable to present the histogram visually, we would be forced to use other descriptive measures which would convey to the listeners a mental picture of the histogram. A second and not so obvious limitation of the histogram and other graphical techniques is that they are difficult to use for purposes of statistical inference. Presumably we use the sample histogram to make inferences about the shape and position of the population histogram which describes the population and is unknown to us. Our inference is based upon the correct assumption that some degree of similarity will exist between the two histograms, but we are then faced with the problem of measuring the degree of similarity. We know when two figures are identical, but this situation will not likely occur in practice. Hence, if the sample and population histograms differ, how can we measure the degree of difference or, expressing it positively, the degree of similarity? To be more specific, we might wonder about the degree of similarity between the histogram, Figure 3.2, and the frequency histogram for the population of grade-point averages from which the sample was drawn. Although these difficulties are not insurmountable, we prefer to seek other descriptive measures which readily lend themselves for use as predictors of the shape of the population frequency distribution.

The limitation of the graphical method of describing data can be overcome by the use of numerical descriptive measures. Thus we would like to use the sample data to calculate a set of numbers which will convey to the statistician a good mental picture of the frequency distribution and which will be useful in making inferences concerning the population. Numerical descriptive measures calculated from a sample are called statistics. Corresponding numerical descriptive measures for a population are called parameters.

Definition

Numerical descriptive measures computed from population measurements are called **parameters**; those computed from sample measurements are called **statistics**.

3.3

Measures of Central Tendency

In constructing a mental picture of the frequency distribution for a set of measurements, we would likely envision a histogram similar to that shown in Figure 3.2 for the data on grade-point averages. One of the first descriptive measures of interest would be a **measure of central tendency**; that is, a measure of the center of the distribution. We note that the grade-point data ranged from a low of 1.4 to a high of 2.9, the center of the histogram being located in the vicinity of 2.1. Let us now consider some definite rules for locating the center of a distribution of data.

One of the most common and useful measures of central tendency is the arithmetic average of a set of measurements. This is also often referred to as the arithmetic mean or, simply, the mean of a set of measurements.

Definition

The **arithmetic mean** of a set of n measurements $y_1, y_2, y_3, \ldots, y_n$ is equal to the sum of the measurements divided by n.*

Recall that we are always concerned with both the sample and the population, each of which possesses a mean. In order to distinguish between the two, we shall use the symbol $\bar{y}$ for the mean of the sample and μ (the Greek letter mu) for the mean of the population. Since the n sample measurements can be denoted

---

* You might prefer the use of x (rather than y) to denote the variable upon which measurements are taken to generate a sample or a population. As you will see subsequently, it is standard practice to use y for this purpose for the material covered in Chapters 11 and 12. The variable x is assigned a completely different meaning in those chapters. Consequently, for the sake of consistency we use y throughout the text to denote the variable upon which measurements are taken to generate sample data.

by the symbols $y_1, y_2, y_3, \ldots, y_n$, a formula for the sample mean would be

$$\bar{y} = \frac{\sum\limits_{i=1}^{n} y_i}{n},$$

and this can be used as a measure of central tendency for the sample.

Symbols

Sample Mean: $\bar{y} = \dfrac{\sum\limits_{i=1}^{n} y_i}{n}$

Population Mean: μ

Example 3.1

Find the mean of the set of measurements 2, 9, 11, 5, 6.

$$\bar{y} = \frac{\sum\limits_{i=1}^{n} y_i}{n} = \frac{2 + 9 + 11 + 5 + 6}{5} = 6.6.$$

Even more important than locating the center of a set of sample measurements, $\bar{y}$ will be employed as an estimator (predictor) of the value of the unknown population mean μ.

For example, the mean of the data, Table 3.3, is equal to

$$\bar{y} = \frac{\sum\limits_{i=1}^{n} y_i}{n} = \frac{62.5}{30} = 2.08.$$

Note that this falls approximately in the center of the set of measurements. The mean of the entire population of grade-point averages, μ, is unknown to us, but if we were to estimate its value, our estimate of μ would be 2.08.

A second measure of central tendency is the median.

Definition

The **median** of a set of n measurements $y_1, y_2, y_3, \ldots, y_n$ is defined to be the value of y that falls in the middle when the measurements are arranged in order of magnitude.

Example 3.2   Find the median for the set of five measurements

9, 2, 7, 11, 14.

Solution   Arranging the measurements in order of magnitude, 2, 7, 9, 11, 14, we would choose 9 as the median. If the number of measurements is even, we choose the median as the value y halfway between the two middle measurements.

Example 3.3   Find the median for the set of measurements

9, 2, 7, 11, 14, 6.

Solution   Arranged in order of magnitude, 2, 6, 7, 9, 11, 14, we would choose the median halfway between 7 and 9, which is 8.

Our rule for locating the median may seem a bit arbitrary for the case where we have an even number of measurements, but recall that we calculate the sample median either for descriptive purposes or as an estimator of the population median. If it is used for descriptive purposes, we may be as arbitrary as we please. If it is used as an estimator of the population median, "the proof of the pudding is in the eating." A rule for locating the sample median is poor or good depending upon whether it tends to give a poor or good estimate of the population median.

You will note that we have not specified a symbol for the population median. This is because most of the common methods of statistical inference suitable for an elementary course in statistics are based upon the use of the sample mean rather than the median. We say this, being wholly aware of the popularity of the median in the social sciences, but point out that it is used more often for descriptive purposes than for statistical inference. We also note that other measures of central tendency exist which have practical application in certain situations, but limitations of time and space forbid their discussion here. As we proceed in this text we will use the sample mean, exclusively, as a measure of central tendency.

3.4
Measures of Variability

Having located the center of a distribution of data, our next step is to provide a measure of the **variability** or **dispersion** of the data. Consider the two distributions shown in Figure 3.3. Both distributions are located with a center of $y = 4$, but there is a vast difference in the variability of the measurements about the mean for the two distributions. Most of the measurements in Figure 3.3(a) vary from 3 to 5, while in Figure 3.3(b) they vary from 0 to 8. Variation is a very

Figure 3.3 Variability
or dispersion of data

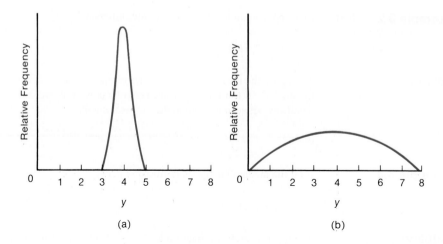

(a)   (b)

important characteristic of data. For example, if we are manufacturing bolts, excessive variation in the bolt diameter would imply a high percentage of defective product. On the other hand, if we are using an examination to discriminate between good and poor accountants, we would be most unhappy if the examination always produced test grades with little variation, as this would make discrimination very difficult indeed. In addition to the practical importance of variation in data, it is obvious that a measure of this characteristic is necessary to the construction of the mental image of the frequency distribution. Numerous measures of variability exist, and we shall discuss a few of the most important.

The simplest measure of variation is the range.

Definition

The **range** of a set of n measurements $y_1, y_2, y_3, \ldots, y_n$ is defined to be the difference between the largest and smallest measurement.

For our grade-point data, Table 3.3, we note that the measurements varied from 1.4 to 2.9. Hence the range is equal to $(2.9 - 1.4) = 1.5$. The range is easy to calculate, easy to interpret, and quite adequate as a measure of variation for small sets of data. But for large sets of data, it can be insensitive to substantial differences in data variation. This insensitivity is illustrated by the two relative frequency distributions shown in Figure 3.4. Both distributions have the same range, but the data of Figure 3.4(b) are more variable than the data of Figure 3.4(a).

We will examine a set of data, say the set of measurements 5, 7, 1, 2, 4, and use these data to calculate a more sensitive measure of data variation. These data can be displayed graphically, as in Figure 3.5, by showing the measurements as dots falling along the horizontal axis. Figure 3.5 is called a *dot diagram*.

Figure 3.4 Distribution
with equal ranges and
unequal variability

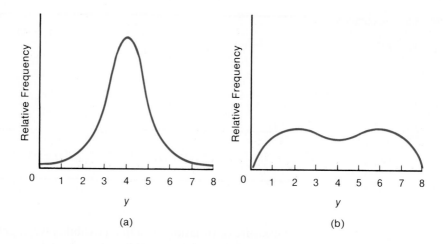

(a)   (b)

Figure 3.5 Dot diagram

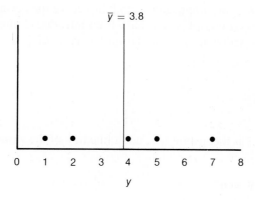

Calculating the mean as the measure of central tendency, we obtain

$$\bar{y} = \frac{\sum\limits_{i=1}^{n} y_i}{n} = \frac{19}{5} = 3.8$$

and locate it on the dot diagram. We can now view variability in terms of distance between each dot (measurement) and the mean, $\bar{y}$. If the distances are large, we can say that the data are more variable than if the distances are small. Being more explicit, we shall define the deviation of a measurement from its mean to be the quantity $(y_i - \bar{y})$. Note that measurements to the right of the mean represent positive deviations, and those to the left, negative deviations. The values of y and the deviations for our example are shown in columns 1 and 2 of Table 3.5.

  If we now agree that deviations contain information on variation, our next step is to construct a formula based upon the deviations which will provide a

Table 3.5 Computation of $\displaystyle\sum_{i=1}^{n} (y_i - \bar{y})^2$

y_i	$y_i - \bar{y}$	$(y_i - \bar{y})^2$
5	1.2	1.44
7	3.2	10.24
1	−2.8	7.84
2	−1.8	3.24
4	.2	.04
$\displaystyle\sum_{i=1}^{5} y_i = 19$	$\displaystyle\sum_{i=1}^{5} (y_i - \bar{y}) = 0$	$\displaystyle\sum_{i=1}^{5} (y_i - \bar{y})^2 = 22.80$

good measure of variation. As a first possibility we might choose the average of the deviations. Unfortunately, this will not work because some of the deviations are positive, some are negative, and the sum of the deviations is always equal to zero (unless round-off errors have been introduced into the calculations). This can be shown using the summation theorems of Chapter 2. Given n measurements $y_1, y_2, \ldots, y_n$,

$$\sum_{i=1}^{n} (y_i - \bar{y}) = \sum_{i=1}^{n} y_i - \sum_{i=1}^{n} \bar{y}.$$

But since $\bar{y}$ is a constant (i.e., not a function of the variable of summation, i),

$$\sum_{i=1}^{n} \bar{y} = n\bar{y}.$$

Then,

$$\sum_{i=1}^{n} (y_i - \bar{y}) = \sum_{i=1}^{n} y_i - \sum_{i=1}^{n} \bar{y}$$

$$= \sum_{i=1}^{n} y_i - n\bar{y}$$

$$= \sum_{i=1}^{n} y_i - n \left(\frac{\sum_{i=1}^{n} y_i}{n} \right)$$

$$= \sum_{i=1}^{n} y_i - \sum_{i=1}^{n} y_i = 0.$$

Note that the deviations, the second column of Table 3.5, sum to zero.

You will readily observe an easy solution to this problem. Why not calculate the average of the absolute values of the deviations? This method has, in fact, been employed as a measure of variability but it is difficult to interpret and it tends to be unsatisfactory for purposes of statistical inference. We prefer overcoming the difficulty caused by the sign of the deviations by working with the sum of their squares,

$$\sum_{i=1}^{n} (y_i - \bar{y})^2.$$

For a fixed number of measurements, when this quantity is large, the data will be more variable than when it is small.

Definition

The **variance of a population** of N measurements $y_1, y_2, \ldots, y_N$ is defined to be the average of the square of the deviations of the measurements about their mean μ. The population variance is denoted by σ^2 (σ is the lowercase Greek letter sigma) and is given by the formula

$$\sigma^2 = \frac{\sum_{i=1}^{N} (y_i - \mu)^2}{N}.$$

Note that we use N to denote the number of measurements in the population and n to denote the number of measurements in the sample.

In defining the variance of the sample measurements, we will modify our averaging procedure, dividing the sum of squares of deviations by $(n - 1)$ rather than n.

Definition

The **variance of a sample** of n measurements $y_1, y_2, \ldots, y_n$ is defined to be the sum of the squared deviations of the measurements about their mean $\bar{y}$ divided by $(n - 1)$. The sample variance is denoted by s^2 and is given by the forumla

$$s^2 = \frac{\sum_{i=1}^{n} (y_i - \bar{y})^2}{n - 1}.$$

This is done because, ultimately, we will want to use the sample variance s^2 to estimate the population variance σ^2. Dividing the sum of squares of deviations by n produces estimates that tend to underestimate σ^2. Division by $(n - 1)$ eliminates this difficulty.

For example, we can calculate the variance for the set of $n = 5$ measurements presented in Table 3.5. The square of the deviation of each measurement is recorded in the third column of Table 3.5. Adding, we obtain

$$\sum_{i=1}^{5} (y_i - \bar{y})^2 = 22.80.$$

The sample variance would equal

$$s^2 = \frac{\sum_{i=1}^{5} (y_i - \bar{y})^2}{n - 1} = \frac{22.80}{5 - 1} = 5.70.$$

At this point, you might be understandably disappointed with the practical significance attached to the variance as a measure of variability. Large variances imply a large amount of variation, but this only permits comparison of several sets of data. When we attempt to say something specific concerning a single set of data, we are at a loss. For example, what can be said about the variability of a set of data with a variance equal to 100? This question cannot be answered with the facts at hand. We shall remedy this situation by introducing a new definition and, in Section 3.5, a theorem and a rule.

Definition

The **standard deviation** of a set of n measurements $y_1, y_2, y_3, \ldots, y_n$ is equal to the positive square root of the variance.

The variance is measured in terms of the square of the original units of measurement. Thus, if the original measurements were in inches, the variance would be expressed in square inches. Taking the positive square root of the variance, we obtain the standard deviation, which returns our measure of variability to the original units of measurement. The sample standard deviation is denoted by the symbol s and the population standard deviation by the symbol σ.

$$s = \sqrt{s^2} = \sqrt{\frac{\sum_{i=1}^{n} (y_i - \bar{y})^2}{n - 1}}$$

Having defined the standard deviation, you might wonder why we bothered to define the variance in the first place. Actually, both the variance and the standard deviation play an important role in statistics, a fact that you must accept on faith at this stage of our discussion.

Symbols

Sample Standard Deviation: $s = \sqrt{\dfrac{\sum_{i=1}^{n}(y_i - \bar{y})^2}{n-1}}$

Population Standard Deviation: σ

3.5
On the Practical Significance of the Standard Deviation

We now introduce an interesting and useful theorem developed by the Russian mathematician, Tchebysheff. Proof of the theorem is not difficult, but we omit it from our discussion.

Theorem 3.1

Tchebysheff's Theorem: Given a number k greater than or equal to 1 and a set of n measurements $y_1, y_2, \ldots, y_n$, at least $(1 - 1/k^2)$ of the measurements will lie within k standard deviations of their mean.

Tchebysheff's Theorem applies to any set of measurements, and for purposes of illustration we could refer to either the sample or the population. We shall use the notation appropriate for populations, but the reader should realize that we could just as easily use $\bar{y}$ and s, the mean and standard deviation for the sample.*

The idea involved in Tchebysheff's Theorem is illustrated in Figure 3.6. An interval is constructed by measuring a distance of $k\sigma$ on either side of the mean μ. Note that the theorem is true for any number we wish to choose for k as long as it is greater than or equal to 1. Then, computing the fraction $1 - 1/k^2$, we see that Tchebysheff's Theorem states that *at least* that fraction of the total number, n, of measurements will lie in the constructed interval.

Let us choose a few numerical values for k and compute $1 - 1/k^2$ (see Table 3.6). When $k = 1$, the theorem states that at least $1 - 1/(1)^2 = 0$ of the

---

* The proof of Tchebysheff's Theorem for a finite number of measurements is based on a variance defined as

$$s'^2 = \frac{\sum_{i=1}^{n}(y_i - \bar{y})^2}{n},$$

that is, with a divisor of n rather than the $(n-1)$ used in s^2. Since s^2 will always be larger than s'^2, because $n > (n-1)$, Tchebysheff's Theorem will always hold when s is used to form the intervals about $\bar{y}$. In any case, there will be very little numerical difference in the values of s and s' for moderate to large values of n.

Figure 3.6 Illustrating Tchebysheff's Theorem

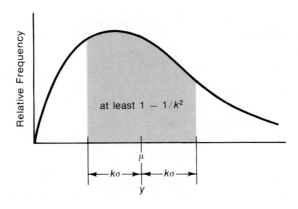

Table 3.6 Illustrative values of $1 - 1/k^2$

k	$1 - 1/k^2$
1	0
2	3/4
3	8/9

measurements lie in the interval $\mu - \sigma$ to $\mu + \sigma$, a most unhelpful and uninformative result. However, when $k = 2$, we observe that *at least* $1 - 1/(2)^2 = 3/4$ of the measurements will lie in the interval $\mu - 2\sigma$ to $\mu + 2\sigma$. At least 8/9 of the measurements will lie within three standard deviations of the mean; that is, in the interval $\mu - 3\sigma$ to $\mu + 3\sigma$. Let us consider an example where we will use the mean and standard deviation (or variance) to construct a mental image of the distribution of measurements from which the mean and standard deviation were obtained.

Example 3.4

The mean and variance of a sample of $n = 25$ measurements are 75 and 100, respectively. Use Tchebysheff's Theorem to describe the distribution of measurements.

Solution

We are given $\bar{y} = 75$ and $s^2 = 100$. The standard deviation is $s = \sqrt{100} = 10$. The distribution of measurements is centered about $\bar{y} = 75$, and Tchebysheff's Theorem states:

a. *At least* 3/4 of the 25 measurements lie in the interval $\bar{y} \pm 2s = 75 \pm 2(10)$, that is, 55 to 95.

b. *At least* 8/9 of the measurements lie in the interval $\bar{y} \pm 3s = 75 \pm 3(10)$, that is, 45 to 105.

We emphasize the "at least" in Tchebysheff's Theorem because the theorem is very conservative, applying to *any* distribution of measurements. In most situations, the fraction of measurements falling in the specified interval will exceed $1 - 1/k^2$.

We now state a rule that describes accurately the variability of a particular bell-shaped distribution and describes reasonably well the variability of other mound-shaped distributions of data. The frequent occurrence of mound-shaped and bell-shaped distributions of data in nature and hence the applicability of our rule leads us to call it the Empirical Rule.

The Empirical Rule

Given a distribution of measurements that is approximately bell-shaped (see Figure 3.7), the interval

1. $\mu \pm \sigma$ will contain approximately 68 percent of the measurements.
2. $\mu \pm 2\sigma$ will contain approximately 95 percent of the measurements.
3. $\mu \pm 3\sigma$ will contain all or almost all of the measurements.

The bell-shaped distribution, Figure 3.7, is commonly known as the normal distribution and will be discussed in detail in Chapter 7. The Empirical Rule applies specifically to data that possess a normal distribution, but it also provides an excellent description of variation for many other types of data.

Figure 3.7 The normal distribution

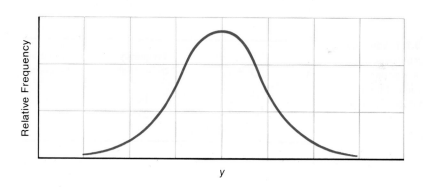

y

Example 3.5

A time study was conducted to determine the length of time necessary to perform a specified operation in a manufacturing plant. The length of time in minutes necessary to complete the operation was measured for each of $n = 40$ workmen. The mean and standard deviation were found to equal 12.8 and 1.7, respectively. Describe the data.

Solution

To describe the data, we calculate the intervals

$$\bar{y} \pm s \ \ = 12.8 \pm 1.7, \text{ or } 11.1 \text{ to } 14.5;$$

$$\bar{y} \pm 2s = 12.8 \pm 2(1.7), \text{ or } 9.4 \text{ to } 16.2;$$

$$\bar{y} \pm 3s = 12.8 \pm 3(1.7), \text{ or } 7.7 \text{ to } 17.9.$$

Although we have no prior information on the distribution of the data, there is a very good chance that it will be mound-shaped and that the Empirical Rule will provide a good description of the data. According to the Empirical Rule, we would expect approximately 68 percent of the measurements to fall in the interval 11.1 to 14.5, 95 percent in the interval 9.4 to 16.2, and all or almost all in the interval 7.7 to 17.9.

If we doubt that the distribution of measurements is mound-shaped or wish, for some reason, to be conservative, we can apply Tchebysheff's Theorem and be absolutely certain of our statements. Tchebysheff's Theorem would tell us that at least 3/4 of the measurements fell in the interval 9.4 to 16.2 and at least 8/9 in the interval 7.7 to 17.9.

How well does the Empirical Rule apply to the grade-point data of Table 3.3? We will show, in Section 3.6, that the mean and standard deviation for the $n = 30$ measurements are $\bar{y} = 2.08$ and $s = .37$. The appropriate intervals were calculated and the number of measurements falling in each interval recorded. The results are shown in Table 3.7 with k in the first column and the interval $\bar{y} \pm ks$ in the second column, using $\bar{y} = 2.08$ and $s = .37$. The frequency, or number of measurements falling in each interval, is given in the third column and the relative frequency in the fourth column. Note that the observed relative frequencies are quite close to the relative frequencies specified in the Empirical Rule.

Table 3.7 Frequency of measurements lying within k standard deviations of the mean for the data, Table 3.3

k	Interval $\bar{y} \pm ks$	Frequency in Interval	Relative Frequency
1	1.71–2.45	19	.63
2	1.34–2.82	29	.97
3	.97–3.19	30	1.0

Another way to see how well the Empirical Rule and Tchebysheff's Theorem apply to the grade-point data is to mark off the intervals $\bar{y} \pm s$, $\bar{y} \pm 2s$, and $\bar{y} \pm 3s$ on the relative frequency histogram for the data. This is shown in Figure 3.8. Now recall that the area under the histogram over an interval is proportional to the number of measurements falling in the interval and visually observe the proportion of the area above the interval, $\bar{y} \pm s$. You will observe that this proportion is near the .68 specified by the Empirical Rule. Similarly, you will note that almost all of the area lies above the interval $\bar{y} \pm 2s$. Clearly both the Empirical Rule and Tchebysheff's Theorem, using $\bar{y}$ and s, provide a good description for the grade-point data.

To conclude, **note that Tchebysheff's Theorem is a fact that can be proved mathematically, and it applies to any set of data. It gives a *lower* bound to the**

Figure 3.8 Histogram for the grade-point data (Figure 3.1) with intervals $\bar{y} \pm s$, $\bar{y} \pm 2s$, and $\bar{y} \pm 3s$ superimposed

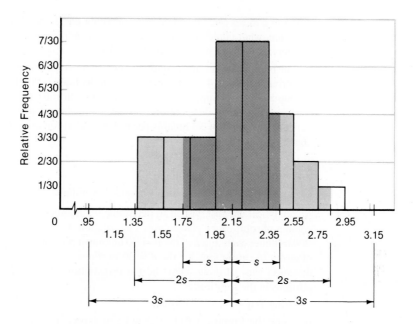

fraction of measurements to be found in an interval $\bar{y} \pm ks$, where k is some number greater than or equal to 1. In contrast, the Empirical Rule is an arbitrary statement about the behavior of data, a "rule of thumb." Although the percentages contained in the rule come from the area under the normal curve (Figure 3.7), the same percentages hold approximately for distributions with varying shapes as long as they tend to be roughly mound-shaped (the data tend to pile up near the center of the distribution). We have shown this to be true for the grade-point data. If you need further convincing, calculate $\bar{y}$ and s for a set of data of your choosing and check the fraction of measurements falling in the intervals $\bar{y} \pm s$, $\bar{y} \pm 2s$, and $\bar{y} \pm 3s$. We think you will find that the observed relative frequencies are reasonably close to the values specified in the Empirical Rule.

Exercises

3.5. Given $n = 6$ measurements, 3, 0, 2, 2, 1, 4:

a. Find $\bar{y}$.   b. Find s.

3.6. Given $n = 5$ measurements, 2, 4, 0, 3, 1:

a. Find $\bar{y}$.   b. Find s.

3.7. Suppose that you wish to create a mental picture of the relative frequency histogram for a large data set consisting of 1000 observations and that you know that the mean and standard deviation of the data set were equal to 36 and 3, respectively.

a. If you are fairly certain that the relative frequency distribution of the data was mound-shaped, how might you picture the relative frequency distribution? (*Hint:* Use the Empirical Rule.)

b. If you had no prior information concerning the shape of the relative frequency distribution, what could you say about the relative frequency histogram? (*Hint:* Construct intervals $\bar{y} \pm ks$ for several choices of k.)

3.8. Is your breath rate normal? Actually, there is no standard breath rate for humans. It can vary from as low as 4 breaths per minute to as high as 70 or 75 for a person engaged in strenuous exercise. Suppose that the resting breath rates for college-age students possess a relative frequency distribution that is mound-shaped with a mean equal to 12 and a standard deviation of 2.3 breaths per minute. What fraction of all students would possess breath rates in the interval:

a. 9.7 to 14.3 breaths per minute?

b. 7.4 to 16.6 breaths per minute?

c. more than 18.9 or less than 5.1 breaths per minute?

3.9. An analytical chemist wished to determine the number of moles of cupric ions in a given volume of solution by electrolysis. The solution was partitioned into $n = 30$ portions of .200 milliliter each. Each of the $n = 30$ unknown portions was tested. The average number of moles of cupric ions for the $n = 30$ portions was found to be .17 mole; the standard deviation was .01 mole. Describe the distribution of the measurements for the $n = 30$ portions of the solution:

a. Using Tchebysheff's Theorem.

b. Using the Empirical Rule. (Would you expect the Empirical Rule to be suitable for describing these data?)

c. Suppose that the chemist had only employed $n = 4$ portions of the solution for the experiment and obtained the readings .15, .19, .17, and .15. Would the Empirical Rule be suitable for describing the $n = 4$ measurements? Why?

3.10. According to the Environmental Protection Agency (EPA), chloroform, which in its gaseous form is suspected of being a cancer-causing agent, is present in small quantities in all of the country's 240,000 public water sources. If the mean and standard deviation of the amounts of chloroform present in the water sources are 34 and 53 micrograms per liter, respectively, describe the distribution for the population of all public water sources.

3.11. Reporting on a Columbia University survey of 2239 American authors, the *New York Times* (June 15, 1981) notes that the "average" American author's writings net less than $5000 per year. Each of the authors in the survey had published at least one book, and the books covered a wide range of topics—everything from fact to fiction. The median hourly writing income for the 2239 authors in 1979 was only $4.90 per hour, and very few of the authors had annual incomes equal to $100,000 per year or more.

a. What proportion of the authors in the sample had an hourly income from writing in 1979 that was less than or equal to $4.90?

b. Based on the survey results, the population of the incomes of all American authors (who had published at least one book) could conceivably possess a relative frequency distribution with a mean equal to $5000 and a standard deviation equal to $20,000. If this were true, approximately what proportion of American authors would have annual salaries less than or equal to $45,000?

3.12. Trichinosis, a disease derived from improperly cooked pork products (main source: pork sausage), may be on the increase. Two hundred eighty-four (284) cases of trichinosis, including one fatality, were reported to the Communicable Disease Center in 1975. This number was 2 1/2 times higher than the mean number of cases reported during the previous five years, and it represents the highest annual incidence since 1961.* From your knowledge about data variation, do you think these data confirm an increase in the per capita rate of incidence of trichinosis? Explain. (*Note:* This exercise is only intended to stimulate discussion. More clear-cut decisions will be derived from more complete sets of data in later chapters.)

3.13. The College Board's verbal and mathematics scholastic aptitude tests are scored on a scale from 200 to 800. Although originally designed to produce mean scores approximately equal to 500, the mean verbal and math scores have been as low as 463 and 493, respectively, in recent years, and have been trending downward. It seems reasonable to assume that a distribution of all GRE scores, either verbal or math, is mound-shaped. If σ is the standard deviation of one of these distributions, what is the largest value (approximately) that σ might assume? Explain.

3.6
A Short Method for Calculating the Variance

The calculation of the variance and standard deviation of a set of measurements is no small task regardless of the method employed, but it is particularly tedious if one proceeds, according to the definition, by calculating each deviation individually as shown in Table 3.5. We shall use the data of Table 3.5 to illustrate a shorter method of calculation. The tabulations are presented in Table 3.8 in two columns, the first containing the individual measurements and the second containing the squares of the measurements.

Table 3.8 Table for simplified calculations of $\sum_{i=1}^{n} (y_i - \bar{y})^2$

y_i	y_i^2
5	25
7	49
1	1
2	4
4	16
19	95

We now calculate

$$\sum_{i=1}^{n} y_i^2 - \frac{\left(\sum_{i=1}^{n} y_i \right)^2}{n} = 95 - \frac{(19)^2}{5}$$

$$= 95 - \frac{361}{5} = 95 - 72.2$$

$$= 22.8$$

---

* "VM 1362 Trichinosis," *Veterinary Medicine Newsletter*, Florida Cooperative Extension Service (August 1977), p. 6.

and notice that it is exactly equal to the sum of squares of the deviations, $\sum_{i=1}^{n} (y_i - \bar{y})^2$, given in the third column of Table 3.5.

This is no accident. We will show that the sum of squares of the deviations is always equal to

$$\sum_{i=1}^{n} (y_i - \bar{y})^2 = \sum_{i=1}^{n} y_i^2 - \frac{\left(\sum_{i=1}^{n} y_i\right)^2}{n}.$$

The proof is obtained by using the summation theorems, Chapter 2:

$$\sum_{i=1}^{n} (y_i - \bar{y})^2 = \sum_{i=1}^{n} (y_i^2 - 2\bar{y}y_i + \bar{y}^2)$$

$$= \sum_{i=1}^{n} y_i^2 - 2\bar{y} \sum_{i=1}^{n} y_i + \sum_{i=1}^{n} \bar{y}^2$$

$$= \sum_{i=1}^{n} y_i^2 - \frac{2 \sum_{i=1}^{n} y_i}{n} \sum_{i=1}^{n} y_i + n\bar{y}^2$$

$$= \sum_{i=1}^{n} y_i^2 - \frac{2}{n}\left(\sum_{i=1}^{n} y_i\right)^2 + n\left(\frac{\sum_{i=1}^{n} y_i}{n}\right)^2$$

or

$$\sum_{i=1}^{n} (y_i - \bar{y})^2 = \sum_{i=1}^{n} y_i^2 - \frac{\left(\sum_{i=1}^{n} y_i\right)^2}{n}.$$

We call this formula the shortcut method of calculating the sum of squares of deviations needed in the formula for the variance and standard deviation. Comparatively speaking, it is short because it eliminates all the subtractions required for calculating the individual deviations. A second and not so obvious advantage is that it tends to give better computational accuracy than the method utilizing the deviations. The beginning statistics student frequently finds the variance that he has calculated at odds with the answer in the text. This is usually caused by rounding off decimal numbers in the computations. We suggest that rounding off be held at a minimum since it may seriously affect the results of computation of the variance. A third advantage is that the shortcut method is especially suitable for use with many electronic calculators, some of which accumulate $\sum_{i=1}^{n} y_i$ and $\sum_{i=1}^{n} y_i^2$ simultaneously.

Shortcut Formula for Calculating $\sum_{i=1}^{n} (y_i - \bar{y})^2$

$$\sum_{i=1}^{n} (y_i - \bar{y})^2 = \sum_{i=1}^{n} y_i^2 - \frac{\left(\sum_{i=1}^{n} y_i\right)^2}{n}$$

Before leaving this topic, we will calculate the standard deviation for the $n = 30$ grade-point averages, Table 3.3. You can verify the following:

$$\sum_{i=1}^{n} y_i = 62.5,$$

$$\sum_{i=1}^{n} y_i^2 = 134.19.$$

Using the shortcut formula,

$$\sum_{i=1}^{n} (y_i - \bar{y})^2 = \sum_{i=1}^{n} y_i^2 - \frac{\left(\sum_{i=1}^{n} y_i\right)^2}{n}$$

$$= 134.19 - \frac{(62.5)^2}{30} = 134.19 - 130.21$$

$$= 3.98.$$

It follows that the standard deviation is

$$s = \sqrt{\frac{\sum_{i=1}^{n} (y_i - \bar{y})^2}{n-1}} = \sqrt{\frac{3.98}{29}} = .37.$$

Example 3.6   Calculate $\bar{y}$ and s for the measurements 85, 70, 60, 90, 81.

Solution

y_i	y_i^2
85	7,225
70	4,900
60	3,600
90	8,100
81	6,561
386	30,386

$$\bar{y} = \frac{386}{5} = 77.2.$$

$$\sum_{i=1}^{n} (y_i - \bar{y})^2 = \sum_{i=1}^{n} y_i^2 - \frac{\left(\sum_{i=1}^{n} y_i\right)^2}{n}$$

$$= 30,386 - \frac{(386)^2}{5}$$

$$= 30,386 - 29,799.2$$

$$= 586.8.$$

$$s = \sqrt{\frac{\sum_{i=1}^{n} (y_i - \bar{y})^2}{n-1}} = \sqrt{\frac{586.8}{4}} = \sqrt{146.7}$$

$$= 12.1.$$

3.7

A Check on the Calculation of s

Tchebysheff's Theorem and the Empirical Rule can be used to detect gross errors in the calculation of s. Thus, we know that at least 3/4 or, in the case of a mound-shaped distribution, nearly 95 percent of a set of measurements will lie within two standard deviations of their mean. Consequently, most of the sample measurements will lie in the interval $\bar{y} \pm 2s$, and the range will approximately equal $4s$. This is, of course, a very rough approximation, but from it we can acquire a useful check that will detect large errors in the calculation of s.

Letting R equal the range,

$$R \approx 4s,$$

then s is approximately equal to $R/4$; that is,

$$s \approx \frac{R}{4}.$$

The computed value of *s* using the shortcut formula should be of roughly the same order as the approximation.

Example 3.7

Use the approximation above to check the calculation of *s* for Example 3.6.

Solution

The range of the five measurements is

$$R = 90 - 60 = 30.$$

Then,

$$s \approx \frac{R}{4} = \frac{30}{4} = 7.5.$$

This is of the same order as the calculated value, $s = 12.1$.

You should note that the range approximation is not intended to provide an accurate value for *s*.* Rather, its purpose is to detect gross errors in calculating such as the failure to divide the sum of squares of deviations by $(n - 1)$ or the failure to take the square root of s^2. Both of these errors yield solutions that are many times larger than the range approximation to *s*.

Example 3.8

Use the range approximation to determine an approximate value for the standard deviation for the data of Table 3.3.

Solution

The range is $R = 2.9 - 1.4 = 1.5$. Then

$$s \approx \frac{R}{4} = \frac{1.5}{4} = .375.$$

We have shown $s = .37$ for the data of Table 3.3. The approximation is very close to the actual value of *s*.

---

* The range for a sample of *n* measurements will depend on the sample size *n*. The larger the value of *n*, the more likely you will observe extremely large or small values of *y*. The range for large samples, say $n = 50$ or more observations, may be as large as 6*s*, while the range for small samples, say $n = 5$ or less, may be as small or smaller than 2*s*.

Tips on Problem Solving

1. Always use the shortcut formula when calculating $\sum_{i=1}^{n} (y_i - \bar{y})^2$. This will help reduce rounding errors, that is,

$$\sum_{i=1}^{n} (y_i - \bar{y})^2 = \sum_{i=1}^{n} y_i^2 - \frac{\left(\sum_{i=1}^{n} y_i\right)^2}{n}$$

2. Be careful about rounding numbers. Carry your calculations of $\sum_{i=1}^{n} (y_i - \bar{y})^2$ to six significant figures.

3. After you have calculated the standard deviation s for a set of data, compare its value with the range of the data. The Empirical Rule tells you that most of the data should fall in an interval, $\bar{y} \pm 2s$, that is, a very approximate value for the range will be $4s$. Consequently, a very rough rule of thumb is that

$$s \approx \frac{\text{Range}}{4}.$$

This crude check will help you to detect large errors, for example, failure to divide the sum of squares of deviations by $(n - 1)$ or failure to take the square root of s^2.

Exercises

3.14. Given $n = 8$ measurements, 2, 1, 3, 1, 2, 1, 2, 2:
a. Calculate $\bar{y}$.
b. Use the shortcut formula, Section 3.6, to calculate s^2 and s.

3.15. Given $n = 5$ measurements, 4, 1, 1, 3, 5:
a. Calculate $\bar{y}$.
b. Use the shortcut formula, Section 3.6, to calculate s^2 and s.

3.16. Given $n = 8$ measurements, 0, -1, -3, 1, 1, 2, 4, 1:
a. Calculate $\bar{y}$.
b. Use the shortcut formula, Section 3.6, to calculate s^2 and s.

3.17. The length of time required for an automobile driver to respond to a particular emergency situation was recorded for $n = 10$ drivers. The times, in seconds, were .5, .8, 1.1, .7, .6, .9, .7, .8, .7, .8.

a. Scan the data and use the procedure of Section 3.7 to find an approximate value for s. Use this value to check your calculations in part (b).
b. Calculate the sample mean, $\bar{y}$, and the standard deviation, s. Compare with part (a).

3.18.   The number of television viewing hours per household and the prime viewing times are two factors that affect television advertising income. A random sample of 25 households in a particular viewing area produced the following estimates of viewing hours per household:

3.0	6.0	7.5	15.0	12.0
6.5	8.0	4.0	5.5	6.0
5.0	12.0	1.0	3.5	3.0
7.5	5.0	10.0	8.0	3.5
9.0	2.0	6.5	1.0	5.0

a. Scan the data and use the procedure of Section 3.7 to find an approximate value for s. Use this value to check your calculations in part (b).

b. Calculate the sample mean $\bar{y}$ and the sample standard deviation s. Compare s with the approximate value obtained in part (a).

c. Find the percentage of the viewing hours per household that fall in the interval $\bar{y} \pm 2s$. Compare with the corresponding percentage given by the Empirical Rule.

3.19.   To estimate the amount of lumber in a tract of timber, an owner decided to count the number of trees with diameters exceeding 12 inches in randomly selected 50 × 50-feet squares. Seventy 50 × 50 squares were randomly selected from the tract and the number of trees (with diameters in excess of 12 inches) were counted for each. The data are as follows:

7	8	7	10	4	8	6
9	6	4	9	10	9	8
3	9	5	9	9	8	7
10	2	7	4	8	5	10
9	6	8	8	8	7	8
6	11	9	11	7	7	11
10	8	8	5	9	9	8
8	9	10	7	7	7	5
8	7	9	9	6	8	9
5	8	8	7	9	13	8

a. Construct a relative frequency histogram to describe this data.

b. Calculate the sample mean, $\bar{y}$, as an estimate of μ, the mean number of timber trees for all 50 × 50-feet squares in the tract.

c. Calculate s for the data. Construct the intervals $\bar{y} \pm s$, $\bar{y} \pm 2s$, and $\bar{y} \pm 3s$. Count the percentage of squares falling in each of the three intervals and compare with the corresponding percentages given by the Empirical Rule and Tchebysheff's Theorem.

3.20.   Refer to the miles-per-gallon data for the 1980 Omega automobile, Exercise 3.1.

a. Find the range.

b. Use the procedure of Section 3.7 to find an approximate value for s.

c. Compute s for the data and compare with your answer for part (b).

3.21.   Refer to the miles-per-gallon data for the 1980 Omega automobile, Exercise 3.1. Examine the data and count the number of observations falling in the intervals $\bar{y} \pm s$, $\bar{y} \pm 2s$, and $\bar{y} \pm 3s$. (Use the value of s computed in Exercises 3.20.)

a. Do the fractions falling in these intervals agree with Tchebysheff's Theorem?

b. Do the fractions falling in these intervals agree with the Empirical Rule?

3.22.   Refer to the times of recurrence data, Exercise 3.3.

a. Find the range.

b. Use the procedure of Section 3.7 to find an approximate value for s.

c. Compute s for the data and compare with your answer for part (b).

3.23.   Refer to the times of recurrence data, Exercise 3.3. Examine the data and count the number of observations falling in the intervals $\bar{y} \pm s$, $\bar{y} \pm 2s$, and $\bar{y} \pm 3s$. (Use the value of s computed in Exercise 3.22.)

a. Do the fractions falling in these intervals agree with Tchebysheff's Theorem? The Empirical Rule?

b. Why might the Empirical Rule be unsuitable for describing these data?

3.24.   Suppose that some measurements occur more than once and that the data, $y_1, y_2, \ldots, y_k$, are arranged in a frequency table as shown here.

Observations	Frequency f_i
y_1	f_1
y_2	f_2
$\vdots$	$\vdots$
y_k	f_k
	n

Then,

$$\bar{y} = \frac{\sum\limits_{i=1}^{k} y_i f_i}{n} \qquad \text{where} \qquad n = \sum\limits_{i=1}^{k} f_i$$

and

$$\sum\limits_{i=1}^{n} (y_i - \bar{y})^2 = \sum\limits_{i=1}^{k} y_i^2 f_i - \frac{\left\{ \sum\limits_{i=1}^{k} y_i f_i \right\}^2}{n}$$

Although these formulas for grouped data are primarily of value when you have a large number of measurements, demonstrate their use for the sample 1, 0, 0, 1, 3, 1, 3, 2, 3, 0, 0, 1, 1, 3, 2.

a. Calculate $\bar{y}$ and $\sum\limits_{i=1}^{n} (y_i - \bar{y})^2$ directly using the formulas for ungrouped data.

b. The frequency table for the $n = 15$ measurements is shown below.

y	$f(y)$
0	4
1	5
2	2
3	4
	$n = 15$

Calculate $\bar{y}$ and $\sum\limits_{i=1}^{n} (y_i - \bar{y})^2$ using the formulas for grouped data. Compare with your answers to part (a).

3.25.   To illustrate the utility of the Empirical Rule, consider a distribution that is heavily skewed to the right, as shown in the accompanying figure.

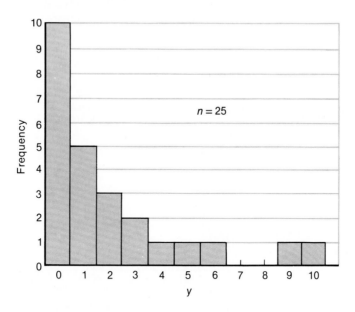

a. Calculate $\bar{y}$ and s for the data shown. (Note: There are ten 0s, five 1s, etc.)

b. Construct the intervals $\bar{y} \pm s$, $\bar{y} \pm 2s$, and $\bar{y} \pm 3s$ and locate them on the frequency distribution.

c. Calculate the proportion of the $n = 25$ measurements falling in each of the three intervals. Compare with Tchebysheff's Theorem and the Empirical Rule. Note that while the proportion falling in the interval $\bar{y} \pm s$ does not agree closely with the Empirical Rule, the proportions falling in the intervals $\bar{y} \pm 2s$ and $\bar{y} \pm 3s$ agree very well. Many times this is true, even for non-mound-shaped distributions of data.

3.8

Describing the Money Market Fund Data

A graphical description of the money fund data, Table 3.2 in the case study, is provided by the relative frequency* histograms, Figures 3.9, 3.10, 3.11, and 3.12. These histograms were plotted by computer using the SAS† statistical computer program package. The SAS package also was used to compute the means, medians, and standard deviations for the data sets.

Figure 3.9 shows the relative frequency histogram for the asset sizes of the 98 funds. You can see that this distribution is highly skewed to the right, a situation where the median is a better measure of central tendency than the mean.

From the histogram you will note that most of the funds had assets near the median, 632.25 million dollars. In contrast, the mean $\bar{y} = 1697.95$ million dollars is quite large because its value is inflated by asset values for a few large funds. More than 75 percent of the fund asset values were less than the mean. Consequently, the median gives a better measure of the asset value of a typical money fund.

Figure 3.10 gives the distribution of the average number of days to maturity for the notes held by the funds. This distribution is slightly skewed to the left although the mean, $\bar{y} = 26.0808$ days, and the median, $m = 26$ days, are, for all practical purposes identical. Surprisingly, the average number of

Figure 3.9 Assets (millions of dollars) for 98 money funds (period ending March 24, 1982)

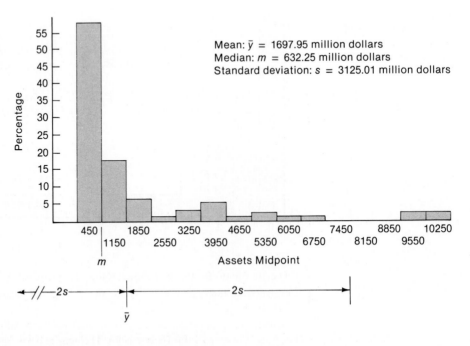

Mean: $\bar{y} = 1697.95$ million dollars
Median: $m = 632.25$ million dollars
Standard deviation: $s = 3125.01$ million dollars

* This computer constructed distribution gives the percentage (rather than the proportion) of the total number of observations in each class.

† The SAS computer program packages were developed by SAS Institute (see the references at the end of this chapter).

Figure 3.10 Average maturity (days) for notes held by 98 money funds (period ending March 24, 1982)

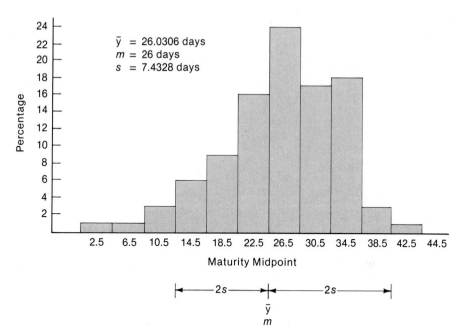

Figure 3.11 7-Day average yields for 98 money funds (period ending March 24, 1982)

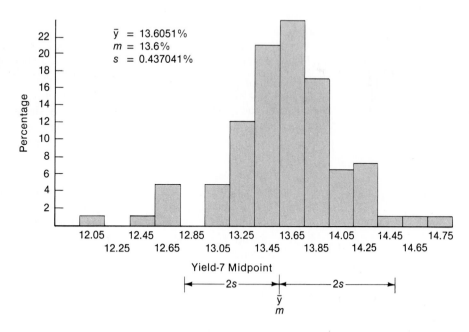

days for some of the funds was very small although most fell within $\bar{y} \pm 2s$, that is, 11.3 to 40.9 days.

The relative frequency distributions of the 7- and the 30-day average yields are shown in Figures 3.11 and 3.12, respectively. You can see that the means and the medians of both distributions of yields are near 13.6 percent.

Figure 3.12 30-Day average yields for 98 money funds (period ending March 24, 1982)

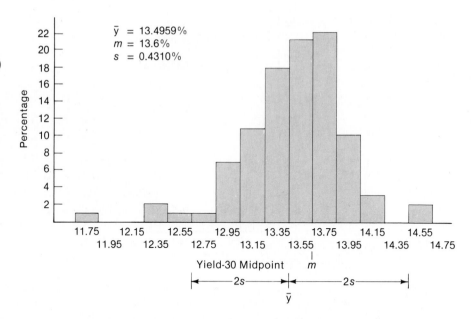

The variation in the fund yields is not very large ($2s \approx .86\%$), although a few funds performed very poorly, deviating more than 1 percentage point from the mean yield.

How well do the means and standard deviations describe the four distributions of money fund data? You can see, Table 3.9, that despite the skewness in the asset value data, the percentage of observations in the interval, $\bar{y} \pm 2s$, is very close to the 95 percent stated by the Empirical Rule. The other percentages, even the percentage corresponding to 7-day yields, are reasonably close to 95 percent. Thus, Table 3.9 provides further evidence that the Empirical Rule is a good rule of thumb for describing a set of data in terms of its mean and standard deviation.

What relevance does the information provided by the relative frequency histograms, Figures 3.9 to 3.12, have to your becoming a millionaire? Table 3.1 shows that a 2 percent change in yield on your invested money can make a substantial change in the amount of money that you will accumulate over a 40-year period. Consequently, it is important to seek investment instruments that, given comparable safety, produce the highest yields. Figures 3.9 to 3.12

Table 3.9 The percentage of observations in the interval, $\bar{y} \pm 2s$

Data Type	Percentage
Asset value	95.9
Maturity time	94.9
7-Day average yield	90.8
30-Day average yield	93.9

provide useful information for someone seeking to accumulate money for retirement via an IRA account. They show the mean rate of return that you might expect to obtain using a money fund as the vehicle for investment, and you can compare this with the current mean yields for other types of investments. They also show the substantial variation in yields within money funds and thereby emphasize the need to be selective in acquiring a good money manager.

3.9
The Effect of Coding on $\bar{y}$ and s^2 (Optional)

Data are frequently coded to simplify the calculation of $\bar{y}$ and s^2. Thus, it is easier to calculate the mean and variance of the set of measurements ($-.1$, .2, .1, 0, .2), than for (99.9, 100.2, 100.1, 100.0, 100.2). The first set was obtained by subtracting 100 from each measurement of the second set. Similarly, we might wish to simplify a set of measurements by multiplying or dividing by a constant. Thus, it is easier to work with the set (3, 1, 4, 6, 4, 2) than with (.003, .001, .004, .006, .004, .002). The first set was obtained by multiplying each element of the second set by 1000.

Data are coded by performing one or both of two operations. Thus one may subtract (or add) a constant, c, to each measurement, multiply (or divide) each measurement by a constant, k, or do both. The objective of coding is to obtain data for which we can more easily determine the mean and variance. Then we wish to use these quantities to find $\bar{y}$ and s^2, the mean and variance, of the original data.

Let $y_1, y_2, \ldots, y_n$ be the original measurements with mean and variance, $\bar{y}$ and s^2. Similarly, let $\bar{x}$ and s_x^2 be the mean and variance of the coded data. How are these quantities related for the two operations of coding? Five theorems answer this question. Proof of the theorems will be left as exercises for the interested reader.

First consider subtracting a constant from each measurement. Thus the coded variable is $x = y - c$. The following two theorems relate $\bar{x}$ and s_x^2 to $\bar{y}$ and s^2.

Theorem 3.2

Let $y_1, y_2, \ldots, y_n$ be n measurements and let $x_i = y_i - c, i = 1, 2, \ldots, n$. Then,

$$\bar{x} = \bar{y} - c.$$

The implication of Theorem 3.2 is that the difference in means between the uncoded and coded data, $\bar{y}$ and $\bar{x}$, will equal the quantity c. Thus, one can compute $\bar{x}$ for the coded data and obtain

$$\bar{y} = \bar{x} + c. \; \cdot$$

Theorem 3.3

Let $y_1, y_2, \ldots, y_n$ be n measurements and let $x_i = y_i - c, i = 1, 2, \ldots, n.$
Then,

$$s_x^2 = s^2.$$

That the variances of the coded and uncoded data should be equal is quite reasonable. A variance is a measure of the variability of a set of measurements. Adding or subtracting the same constant to each measurement would not change the distance between any pair of observations and thus would not change the variability of the data.

Example 3.9   Find $\bar{y}$ and s^2 for the sample 99.9, 100.2, 100.1, 100.0, 100.2, by coding $x_i = y_i - 100$, $i = 1, 2, \ldots, 5.$

Solution   The coded data are $-.1, .2, .1, 0, .2$. Then,

$$\sum_{i=1}^{5} x_i = 4,$$

$$\sum_{i=1}^{5} x_i^2 = .1,$$

$$\bar{x} = \frac{\sum_{i=1}^{n} x_i}{n} = \frac{.4}{5} = .08,$$

and

$$s_x^2 = \frac{\sum_{i=1}^{n} x_i^2 - \frac{\left(\sum_{i=1}^{n} x_i\right)^2}{n}}{n - 1} = \frac{.1 - \frac{(.4)^2}{5}}{4} = .017.$$

Then, by Theorem 3.2,

$$\bar{y} = \bar{x} + c = .08 + 100 = 100.08.$$

By Theorem 3.3,

$$s^2 = s_x^2 = .017.$$

The second method of coding is multiplying each measurement by a constant, k. The following theorems relate the coded and uncoded means and variances.

Theorem 3.4

Let $y_1, y_2, \ldots, y_n$ be n measurements and let $x_i = ky_i$, $i = 1, 2, \ldots, n$, where $k \neq 0$. Then,

$$\bar{x} = k\bar{y} \quad \text{and} \quad \bar{y} = \frac{\bar{x}}{k}.$$

Theorem 3.5

Let $y_1, y_2, \ldots, y_n$ be n measurements and let $x_i = ky_i$, $i = 1, 2, \ldots, n$, where $k \neq 0$. Then,

$$s_x^2 = k^2 s^2 \quad \text{and} \quad s^2 = \frac{1}{k^2} s_x^2.$$

The effect of multiplying each observation by a constant, k, on the relation between $\bar{x}$ and $\bar{y}$ was perhaps predictable. Thus, the coded mean, $\bar{x}$, is k times as large as $\bar{y}$. In agreement with this relation, Theorem 3.5 states that $s_x^2 = k^2 s^2$ or, equivalently, $s_x = ks$. Thus, like the relation between the means, the standard deviation of x is k times the standard deviation for y.

Example 3.10

Calculate $\bar{y}$ and s^2 for the sample .003, .001, .004, .006, .004, .002 by coding $x_i = 1000y_i$, $i = 1, 2, \ldots, 6$.

Solution

The coded data are 3, 1, 4, 6, 4, 2. Then

$$\sum_{i=1}^{6} x_i = 20,$$

$$\sum_{i=1}^{6} x_i^2 = 82,$$

$$\bar{x} = \frac{\sum_{i=1}^{n} x_i}{n} = \frac{20}{6} = 3.33,$$

and

$$s_x^2 = \frac{\sum_{i=1}^{n} x_i^2 - \frac{\left(\sum_{i=1}^{n} x_i\right)^2}{n}}{n - 1} = \frac{82 - \frac{(20)^2}{6}}{5} = 3.07.$$

Then, by Theorem 3.4,

$$\bar{y} = \frac{\bar{x}}{k} = \frac{3.33}{1000} = .00333.$$

By Theorem 3.5,

$$s^2 = \frac{s_x^2}{k^2} = \frac{3.07}{(1000)^2} = .00000307.$$

Both methods of coding are frequently applied to the same set of data. The effects of these operations are given in Theorem 3.6:

Theorem 3.6

Let $y_1, y_2, \ldots, y_n$ be n measurements and let

$$x_i = k(y_i - c).$$

Then,

$$\bar{x} = k(\bar{y} - c) \qquad \text{or} \qquad \bar{y} = \frac{\bar{x}}{k} + c$$

and

$$s_x^2 = k^2 s^2 \qquad \text{or} \qquad s^2 = \frac{s_x^2}{k^2}.$$

Example 3.11   Suppose that both methods of coding are applied to the data 99.9, 100.2, 100.1, 100.0, 100.2. Subtract $c = 100$ from each observation and multiply by $k = 10$. Find $\bar{y}$ and s^2.

Solution

$x_i = 10(y_i - 100)$, $i = 1, 2, \ldots, 5$, and the coded data are $-1, 2, 1, 0, 2$. Then,

$$\sum_{i=1}^{5} x_i = 4,$$

$$\sum_{i=1}^{5} x_i^2 = 10,$$

$$\bar{x} = \frac{\sum\limits_{i=1}^{n} x_i}{n} = \frac{4}{5} = .8,$$

and

$$s_x^2 = \frac{\sum\limits_{i=1}^{n} x_i^2 - \dfrac{\left(\sum\limits_{i=1}^{n} x_i\right)^2}{n}}{n-1} = \frac{10 - \dfrac{(4)^2}{5}}{4} = 1.7.$$

Applying Theorem 3.6,

$$\bar{y} = \frac{\bar{x}}{k} + c = \frac{.8}{10} + 100 = 100.08,$$

$$s^2 = \frac{s_x^2}{k^2} = \frac{1.7}{(10)^2} = .017.$$

Exercises

3.26. The mean and standard deviation for a set of measurements are equal to 20 and 3, respectively. If you were to add 4 to each measurement in the set, what would be the effect on the mean and standard deviation of the data set?

3.27. Refer to Exercise 3.26. Suppose that you were to multiply each measurement in the original data set by 2, what would be the effect on the mean and standard deviation of the data set?

3.28. Refer to Exercises 3.26 and 3.27. Suppose that you were to multiply each measurement in the original data set by 2 and then add 4, what would be the effect on the mean and standard deviation of the data set?

3.29. Given the sample measurements 3005, 2998, 3000, 3003, 2999, code by subtracting a suitable constant. Then use Theorems 3.2 and 3.3 to find the sample mean and variance. Note the reduction in computing effort achieved by the coding.

3.30. Code the sample data .020, .021, .019, .020, .021, .018, by multiplying by 1000. Use Theorems 3.4 and 3.5 to find the sample mean and variance.

3.31. Refer to Exercise 3.30. Code by subtracting .020 from each observation and then multiplying by 1000. Use Theorem 3.6 to find the sample mean and variance.

3.32.   A retail oil dealer knows that a particular customer's use during the winter months is an average of 105 gallons per week with a standard deviation of 12 gallons. Use Theorems 3.4 and 3.5 to find the mean and standard deviation of the customer's fuel oil consumption per day.

3.33.   The mean and standard deviation of the amount of gold extracted per ton of ore at a gold mine are .14 and .02 ounces, respectively. If the company mines 50 tons per day, give the mean and standard deviation of the total yield in gold per day.

3.34.   Simplify the calculation of $\bar{y}$ and s^2 and s for Exercise 3.19 by subtracting 5 from each count before performing the calculations.

3.10
Measures of Relative Standing (Optional)

Sometimes we wish to know the position of an observation relative to others in a set of data. For example, if you took a placement examination and scored 640, you might like to know the percentage of participants who scored lower than 640. Such a measure of relative standing of an observation within a data set is called a percentile.

Definition

Let $y_1, y_2, \ldots, y_n$ be a set of n measurements arranged in order of magnitude. The pth **percentile** is the value of y such that p percent of the measurements are less than that value of y and $(100 - p)$ percent are greater.

Example 3.12

Suppose that you have been notified that your score of 610 on the Verbal Graduate Record Examination placed you at the 60th percentile in the distribution of scores. Where does your score of 610 stand in relation to the scores of others who took the examination?

Solution

Scoring at the 60th percentile means that 60 percent of all the examination scores were lower than your score of 610 and 40 percent were higher.

Viewed graphically, a particular percentile—say the 60th percentile—is a point on the y axis located so that 60 percent of the area under the relative frequency histogram for the data lies to the left of the 60th percentile (see Figure 3.13) and 40 percent of the area lies to the right. Thus, by our definition, the median of a set of data is the 50th percentile because half of the measurements in a data set are smaller than the median and half are larger.

The 25th and 75th percentiles, called the *lower and upper quartiles*, along with the median (the 50th percentile), locate points that divide the data into four sets of equal number. Twenty-five percent of the measurements will be less than the lower (first) quartile; 50 percent will be less than the median (the second quartile); and 75 percent of the measurements will be less than the upper (third) quartile. Thus the median, and the lower and upper quartiles,

Figure 3.13 The 60th percentile shown on the relative frequency histogram for a data set

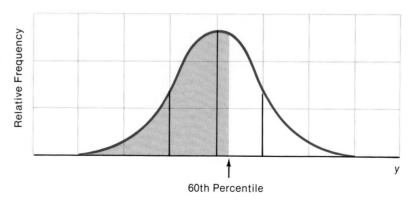

60th Percentile

are located at points on the y axis so that the area under the relative frequency histogram for the data is partitioned into four equal areas as shown in Figure 3.14. You can see (Figure 3.14) that 1/4 of the area lies to the left of the lower quartile, 3/4 to the right. The upper quartile is the value of y such that 3/4 of the area lies to the left, 1/4 to the right.

Figure 3.14 Location of quartiles

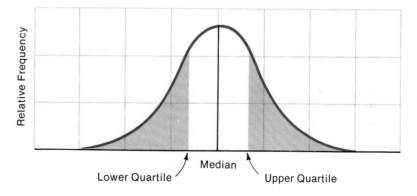

Lower Quartile   Median   Upper Quartile

Definition

Let $y_1, y_2, \ldots, y_n$ be a set of n measurements arranged in order of magnitude. The **lower quartile (first quartile)** is the value of y that exceeds 1/4 of the measurements and is less than the remaining 3/4. The **second quartile** is the median. The **upper quartile (third quartile)** is the value of y that exceeds 3/4 of the measurements and is less than 1/4.

Like the median (the 50th percentile), other percentiles may fall between two observations in a set of data. When this situation occurs, a rule, such as averaging the two observations, must be used to find the specific value of the percentile. This point is not of great importance because it is unlikely that you will ever have to calculate percentiles for a set of data. Rather, you will encounter them in the reporting of the relative standings of observations within large sociological, medical, or educational sets of data, such as the reporting of the relative standings of scores on standardized tests. Consequently, you will primarily be interested in understanding the meanings attached to percentiles, and you will need to be able to interpret them when you encounter them in the literature.

Exercises

3.35.   If you scored at the 69th percentile on a placement test, how would your score stand in relation to the others?

3.36.   According to the Internal Revenue Service, income tax returns showing incomes under $10,000 in 1980 accounted for approximately 44 percent of the 90.9 million returns, but accounted for only 4.4 percent of the total tax (*Orlando Sentinel Star*, September 13, 1980). In the distribution of all tax incomes filed, at what percentile does $10,000 stand?

3.37.   A survey sponsored by the Educational Testing Service and the Council on Learning suggests that most college students know little about foreign affairs (*Orlando Sentinel Star*, April 16, 1981). The survey of 3000 students on 185 campuses found that seniors answered only 50 percent of the questions correctly and "less than one senior in 10 scored above 67%." Identify any percentiles for the distribution of scores that can be determined from the information given above.

3.38.   According to the *Gainesville Sun* (June 20, 1980), prospective Florida public school teachers may be weak in math. The article reports on a field test of a new teacher competency test that the state may require each teacher to pass in order to acquire a state teaching certificate. Of 1186 graduating education majors at 14 colleges and universities in Florida, 44 percent scored below 80 percent on math. Twenty-five percent scored below 70 percent, a score which, in the opinion of one state official, should be significantly exceeded in order to achieve a pass. Identify any percentiles in the distribution of math scores that can be determined from the information given above.

3.11
Summary and Comments

Methods for describing sets of measurements fall into one of two categories, graphical methods and numerical methods. The relative frequency histogram is an extremely useful graphical method for characterizing a set of measurements. Numerical descriptive measures are numbers that attempt to create a mental image of the frequency histogram (or frequency distribution). We have restricted the discussion to measures of central tendency and variation, the most useful of which are the mean and standard deviation. While the mean possesses intuitive descriptive significance, the standard deviation is significant only when used in conjunction with Tchebysheff's Theorem and the Empirical Rule. The objective of sampling is the description of the population from which the sample was obtained. This is accomplished by using the sample mean, $\bar{y}$, and the quantity, s^2, as estimators of the population mean, μ, and variance, σ^2.

Many descriptive methods and numerical measures have been presented in this chapter, but these constitute only a small percentage of those which might have been discussed. In addition, many special computational techniques usually found in elementary texts have been omitted. This was necessitated by the limited time available in an elementary course and because the advent and common use of electronic calculators and computers have minimized the importance of special computational formulas. But, more importantly, the inclusion of such techniques would tend to detract from and obscure the main objective of modern statistics and this text—statistical inference.

References

Dixon, W.J., and M.B. Brown, eds. *BMDP Biomedical Computer Programs, P Series.* Berkeley, Calif.: University of California, 1979.

Freund, J.E., *Modern Elementary Statistics*, 5th ed. Englewood Cliffs, N.J.: Prentice-Hall, Inc., 1979.

Helwig, J.P., and K.A. Council, eds. *SAS User's Guide.* Cary, N.C.: SAS Institute, Inc., P.O. Box 8000, 1979.

Hoel, P.G., *Elementary Statistics*, 4th ed. New York: John Wiley & Sons, Inc., 1976.

Nie, N., C.H. Hull, J.G. Jenkins, K. Steinbrenner, and D.H. Bent, *Statistical Package for the Social Sciences*, 2nd ed. New York: McGraw-Hill Book Company, 1979.

Supplementary Exercises

[Starred (∗) exercises are optional.]

3.39. Conduct the following experiment: toss 10 coins and record y, the number of heads observed. Repeat this process $n = 50$ times, thus providing 50 values of y. Construct a relative frequency histogram for these measurements.

3.40. The following measurements represent the grade-point average of 25 college freshmen:

2.6	1.8	2.6	3.7	1.9
2.1	2.7	3.0	2.4	2.3
3.1	2.6	2.6	2.5	2.7
2.7	2.9	3.4	1.9	2.3
3.3	2.2	3.5	3.0	2.5

Construct a relative frequency histogram for these data.

3.41. Given the set of $n = 7$ measurements 3, 0, 1, 5, 3, 0, 4, calculate $\bar{y}$, s^2, and s.

3.42. Calculate $\bar{y}$, s^2, and s for the data in Exercise 3.39.

3.43. Calculate $\bar{y}$, s^2, and s for the data in Exercise 3.40.

3.44. Refer to the histogram constructed in Exercise 3.39 and find the fraction of measurements lying in the interval $\bar{y} \pm 2s$. (Use the results of Exercise 3.42) Are the results consistent with Tchebysheff's Theorem? Is the frequency histogram of Exercise 3.39 relatively mound-shaped? Does the Empirical Rule adequately describe the variability of the data in Exercise 3.39?

3.45. Repeat the instructions of Exercise 3.44 use the interval $\bar{y} \pm s$.

3.46. Refer to the grade-point data in Exercise 3.40. Find the fraction of measurements falling in the intervals $\bar{y} \pm s$ and $\bar{y} \pm 2s$. Do these results agree with Tchebysheff's Theorem and the Empirical Rule?

3.47. In contrast to aptitude tests, which are predictive measures of what one can accomplish with training, achievement tests tell what an individual can do at the time of the test. Mathematics achievement test scores for 400 students were found to have a mean and a variance equal to 600 and 4900, respectively. If the distribution of test scores was bell-shaped, approximately how many of the scores would fall in the interval 530 to 670? Approximately how many scores would be expected to fall in the interval 460 to 740?

3.48. Use the range approximation for s (Section 3.7) to check the calculated value of s for the data in Exercise 3.40.

3.49. Petroleum pollution in seas and oceans stimulates the growth of some types of bacteria. A count of petroleumlytic microorganisms (bacteria per 100 milliliters) in 10 portions of seawater gave the following readings: 49, 70, 54, 67, 59, 40, 61, 69, 71, 52.
a. Observe the data and guess the value for s by use of the range approximation.

b. Calculate $\bar{y}$ and s and compare with the range approximation of part (a).

3.50. Why do statisticians generally prefer to divide the sum of squares of deviations of the sample measurements by $(n - 1)$ rather than n when estimating a population variance σ^2?

3.51. The percentage of city telephone subscribers who use unlisted numbers is on the increase. The distribution of the percentage of unlisted numbers in cities possesses a mean and standard deviation that are near 14 and 6 percent, respectively. If you were to pick a city at random, is it likely that the percentage of unlisted numbers will exceed 20 percent? Explain. Within what limits would you expect the percentage to fall?

3.52. Find the ratio of the range to s for the data in Exercise 3.39.

3.53. The range approximation for s can be improved if it is known that the sample is drawn from a bell-shaped distribution of data. Thus, the calculated s should not differ substantially from the range divided by the appropriate ratio given in the following table:

Number of measurements	5	10	25
Expected ratio of range to s	2.5	3	4

For the data in Exercise 3.41 estimate s as suggested. Compare this estimate with the calculated s.

3.54. Given the set of $n = 6$ measurements, 0, 1, 2, 0, 3, 2, find $\bar{y}$ and s^2.

3.55. An industrial concern uses an employee screening test with average score μ and standard deviation $a = 10$. Assume that the test-score distribution is approximately bell-shaped and that a score of 65 qualifies an applicant for further consideration. What is the value of μ such that approximately 2.5 percent of the applicants qualify for further consideration?

3.56. Attendances at a high school's basketball games were recorded and found to have a sample mean and variance of 420 and 25, respectively. Calculate $\bar{y} \pm s$, $\bar{y} \pm 2s$, $\bar{y} \pm 3s$, and state the approximate fraction of measurements we would expect to fall in these intervals according to the Empirical Rule:

$\bar{y} \pm s$ _____ fraction _____

$\bar{y} \pm 2s$ _____ fraction _____

$\bar{y} \pm 3s$ _____ fraction _____

3.57. The mean duration of television commercials on a given network is 75 seconds, with a standard deviation of 20 seconds. Assuming that duration times are approximately normally distributed:

a. What is the approximate probability that a commercial will last less than 35 seconds?

b. What is the approximate probability that a commercial will last longer than 55 seconds?

3.58. A random sample of 100 foxes was examined by a team of veterinarians to determine the prevalence of a particular type of parasite. Counting the number of parasites per fox, the veterinarians found that 69 foxes possessed no parasites, 17 possessed one, etc. The following is a frequency tabulation of the data:

Number of parasites, y	0	1	2	3	4	5	6	7	8
Number of foxes, f	69	17	6	3	1	2	1	0	1

a. Construct a relative frequency histogram for y, the number of parasites per fox.

b. Calculate $\bar{y}$ and s for the sample.

c. What fraction of the parasite counts fall within two standard deviations of the mean? three? Do these results agree with Tchebysheff's Theorem? the Empirical Rule?

3.59. Consider a population consisting of the number of teachers per college at small two-year colleges. Suppose that the number of teachers per college has an average $\mu = 175$ and a standard deviation $\sigma = 15$.

a. Use Tchebysheff's Theorem to make a statement about the percentage of colleges that have between 145 and 205 teachers.

b. Assume that the population is normally distributed. What fraction of the colleges have more than 190 teachers?

3.60. From the following data, a student calculated s to be .263. On what grounds might we doubt his accuracy? What is the correct value (nearest hundredth)?

17.2	17.1	17.0	17.1	16.9
17.0	17.1	17.0	17.3	17.2
17.1	17.0	17.1	16.9	17.0
17.1	17.3	17.2	17.4	17.1

* 3.61. Code the data given in Exercise 3.40 by subtracting 2.5 from each observation. Use Theorems 3.2 and 3.3 to find $\bar{y}$ and s^2.

* 3.62. Code the data given in Exercise 3.40 by subtracting 2.5 from each observation and then multiplying by 10. Calculate $\bar{x}$ and s_x^2 for the coded data and use Theorem 3.6 to find $\bar{y}$ and s^2.

* 3.63. Code the data given in Exercise 3.60 by subtracting 17 from each observation. Use Theorems 3.2 and 3.3 to find $\bar{y}$ and s^2.

* 3.64. Code the data given in Exercise 3.60 by subtracting 17 from each observation and then multiplying by 10. Calculate $\bar{x}$ and s_x^2 for the coded data and use Theorem 3.6 to find $\bar{y}$ and s^2.

* 3.65. Use the summation theorems, 2.1, 2.2, and 2.3, to prove Theorems 3.2, 3.3, 3.4, and 3.5.

* 3.66. Use Theorems 2.1, 2.2, and 2.3 to prove Theorem 3.6.

3.67. Most purchasers of new automobiles would agree that EPA ratings for mileage per gallon (mpg) of gasoline for new cars are suspiciously high and consequently are of limited value to consumers. A record of the actual mileages per gallon for 12 fill-ups of a 1975 VW Rabbit (automatic transmission, mostly in-town driving) are as follows:

21.2	24.8	22.7
24.4	22.3	23.4
23.7	24.3	21.8
25.1	25.2	24.9

a. Describe the population from which this sample was selected. Are the sample measurements representative of the population of mpg ratings for all 1975 VW Rabbits equipped with automatic transmissions?

b. Use the mean of the data to estimate the population mean mpg for the 1975 VW Rabbit described above.

c. As you will learn in Chapter 8, the sample standard deviation will be used in assessing the accuracy of the estimate, part (b). Use the range to obtain an approximate value for s. Then calculate s and use the range approximation as a check on gross errors in your calculations.

3.68. In an article entitled, "If Fido takes a piece of the postman, Feds may take a chunk out of you," the *Wall Street Journal* (July 10, 1981) notes that the Postal Service is helping letter carriers assemble evidence needed to seek damages from dog owners for harmful anti-postman behavior of their pets. The article notes that the mean cost in medical bills and lost time per dog bite is $300; but, of course, the actual cost can run much higher. Suppose that the distribution of cost per bite possesses a mean equal to $300 and a standard deviation equal to $200.

a. Why might the relative frequency distribution of costs per dog bite be skewed to the right as shown below?

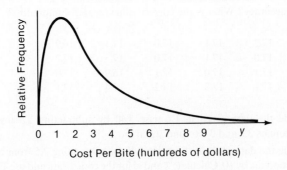

b. Approximately what percentage of the costs per bite are less than $700?

4. Probability

Chapter Objectives

General Objective

The objective of this chapter is to develop an understanding of the basic concepts of probability. These concepts will be used in subsequent chapters to make inferences and to evaluate the reliability of inferences about populations based on sample data.

Specific Objectives

1. To show the use of probability in making a decision based on sample data by giving a simple example. *Section 4.1*

2. To identify two learning objectives and explain how the chapter should be used to achieve them. *Section 4.2*

3. To present a model for the repetition of an experiment, and to explain how the model provides a straightforward mechanism for calculating the probability of an experimental outcome. *Sections 4.3, 4.4*

4. To present a method for representing an event as a composition of two or more other events and then to give two laws of probability that you can use to find the probability of the composition. *Sections 4.5, 4.6, 4.7*

5. (*Optional*) To present Bayes' Rule, an adaptation of conditional probability that is used in business decision making and is the basis for an important method of statistical inference.

6. (*Optional*) To present mathematical methods useful for counting or enumerating large numbers of items.

Case Study

Probability and Decision Making in the Congo

In his exciting novel *Congo*,* Michael Crichton describes a search by Earth Resources Technology Services (ERTS), a geological survey company, for deposits of boron-coated blue diamonds, diamonds that ERTS believed to be the key to a new generation of optical computers. In the novel, ERTS is in a race with an international consortium to find the Lost City of Zinj, a city that thrived on diamond mining and existed several thousand years ago, deep in the rain forests of eastern Zaire, according to African fable.

After the mysterious destruction of its first expedition, ERTS launched a second expedition under the leadership of Karen Ross, a twenty-four-year-old computer genius who is accompanied by Professor Peter Elliot, an anthropologist; Amy, a talking gorilla; and the famed mercenary and expedition leader, "Captain" Charles Munro. Ross's efforts to find the city are blocked by offensive actions of the consortium, by the deadly rain forest, and by hordes of "talking" killer gorillas whose perceived mission is to defend the diamond mines. Karen overcomes these obstacles by using space-age computers to evaluate the probabilities of success for all possible circumstances and all possible actions that the expedition might take. At each stage of the expedition, she is able to quickly evaluate the chances of success.

Ultimately Ross and her computer, Elliot and his knowledge of anthropods, Amy and her ability to translate gorilla language, and Munro's knowledge of the Congo overcome most of the obstacles and bring the expedition to a surprising ending. Crichton gives computers and modern technology most of the credit for the expedition members' remarkable ability to make the right decisions. While we acknowledge the computer's ability to store and process vast quantities of data at unbelievable speeds, computers would be useless without human programming. In fact, probability concepts, theory, and statistics played a fundamental role in the expedition's successful decision making. You will learn about probability and the basic theory for finding probabilities in this chapter. You will also gain an insight into the role that probability plays in making statistical inferences.

* Crichton, Michael, *Congo*, (New York: Alfred A. Knopf, 1980).

4.1
The Role of Probability in Statistics

Probability and statistics are related in a most curious way. In essence, probability is the vehicle that enables the statistician to use information in a sample to make inferences or describe the population from which the sample was obtained. We illustrate this relationship with a simple example.

Consider a balanced die with its familiar six faces. By balanced we mean that the chance of observing any one of the six sides on a single toss of the die is just as likely as any other. Tossing the die might be viewed as an experiment that could conceivably be repeated a very large number of times, thus generating a population of numbers where the measurements, y, would be either 1, 2, 3, 4, 5, or 6. Assume that the population is so large that each value of y occurs with equal frequency. Note that we do not actually generate the population; rather, it exists conceptually. Now let us toss the die once and observe the value of y. This one measurement represents a sample of $n = 1$ drawn from the population. What is the probability that the sample value of y will equal 2? Knowing the structure of the population, we realize that each value of y has an equal chance of occurring and hence the probability that $y = 2$ is 1/6. This example illustrates the type of problem considered in probability theory. The population is assumed to be known and we are concerned with calculating the probability of observing a particular sample. Exactly the opposite is true in statistical problems, where we assume the population unknown, the sample known, and we wish to make inferences about the population. **Thus probability reasons from the population to the sample, while statistics acts in reverse, moving from the sample to the population.**

To illustrate how probability is used in statistical inference, consider the following example. Suppose that a die is tossed $n = 10$ times and the number of dots, y, appearing is recorded after each toss. This represents a sample of $n = 10$ measurements drawn from a much larger body of tosses, the population, which could be generated if we wished. Suppose that all 10 measurements resulted in $y = 1$. We wish to use this information to make an inference concerning the population to tosses; specifically, we wish to infer that the die is or is not balanced. Having observed 10 tosses, each resulting in $y = 1$, we would be somewhat suspicious of the die and would likely reject the theory that the die was balanced. We reason as follows: If the die were balanced as we hypothesize, observing 10 identical measurements is most improbable. Hence, either we observed a rare event or else our hypothesis is false. We would likely be inclined to the latter conclusion. Notice that the decision was based upon the probability of observing the sample, assuming our theory to be true.

The above illustration emphasizes the importance of probability in making statistical inferences. In the following discussion of the theory of probability, we shall assume the population known and calculate the probability of drawing various samples. In doing so, we are really choosing a **model** for a physical situation because the actual composition of a population is rarely known in practice. Thus the probabilist models a physical situation (the population) with probability much as the sculptor models with clay.

Chapter 4 is divided into segments. The first, Sections 4.3 and 4.4, provides a probabilistic model for the repetition of an experiment and, at the same time, presents a clear-cut method for calculating the probability of an observed event. We call this the sample point approach. The second segment, Sections 4.5 through 4.7, presents the methodology for a second method for calculating the probability of an event, which we call the event composition approach. Acquiring an understanding of the probabilistic model and the two methods for calculating the probability of an event are the major objectives of Chapter 4.

4.2
Two Teaching Objectives

This chapter can be used to achieve varying depths of knowledge concerning the theory and applications of statistics. Only a knowledge of the basic concepts and the theory of probability are needed for an understanding of the statistical concepts that begin in Chapter 6 and are developed in succeeding chapters. Only very basic problem-solving abilities are needed for this objective.

In addition to learning the basic concepts of probability, some instructors may want their students to learn to diagnose and solve probability problems. This is a difficult objective for a beginning student to achieve because skill in problem solving is only developed by working many problems over a period of time.

To assist instructors and students in the use of this material, the exercises in Chapters 4 and 5 have been divided into two groups: those needed only as a prerequisite to the discussion of statistical topics in later chapters, and those which have been included for the instructor and students who also wish to develop skill in solving probability problems. These exercises are identified respectively as "Conceptual Level" and "Skill Level" exercises.

Students seeking only a conceptual understanding of probability should cover Sections 4.1 through 4.8 and the Conceptual Level Exercises following those sections. Instructors and students who want to develop skill in solving probability problems should devote greater time to the optional examples and the Skill Level and Supplementary Exercises. Section 4.9 (Bayes' Law) and Section 4.10 (Counting Sample Points) provide additional material to assist in developing skill in solving probability problems.

The preceding explanation should help you decide on the manner in which you will use this chapter. Let us proceed to a discussion of experiments, events, and probability.

4.3
The Sample Space

Data are obtained either by observation of uncontrolled events in nature or by controlled experimentation in the laboratory. To simplify our terminology, we seek a word that will apply to either method of data collection and hence define the term "experiment."

Definition

An **experiment** is the process by which an observation (or measurement) is obtained.

Note that the observation need not be numerical. Typical examples of experiments are:

1. Recording a test grade.
2. Making a measurement of daily rainfall.
3. Interviewing a voter to obtain his preference prior to an election.
4. Inspecting a light bulb to determine whether it is a defective or an acceptable product.
5. Tossing a coin and observing the face that appears.

A population of measurements results when the experiment is repeated many times. For instance, we might be interested in the length of life of television tubes produced in a plant during the month of June. Testing a single tube to failure and measuring length of life would represent a single experiment, while repetition of the experiment for all tubes produced during this period would generate the entire population. A sample would represent the results of some small group of experiments selected from the population.

Let us now direct our attention to a careful analysis of an experiment and the construction of a mathematical model for a population. A by-product of our development will be a systematic and direct approach to the solution of probability problems.

We commence by noting that each experiment may result in one or more outcomes which we will call events and denote by capital letters. Consider the following experiment.

Example 4.1

Experiment: Toss a die and observe the number appearing on the upper face. Some events would be:

1. Event A: observe an odd number.
2. Event B: observe a number less than 4.
3. Event E_1: observe a 1.
4. Event E_2: observe a 2.
5. Event E_3: observe a 3.
6. Event E_4: observe a 4.
7. Event E_5: observe a 5.
8. Event E_6: observe a 6.

The events detailed above do not represent a complete listing of all possible events associated with the experiment but suffice to illustrate a point. The reader will readily note a difference between events A and B and events E_1, E_2, E_3, E_4, E_5, and E_6. Event A will occur if event E_1, E_3, or E_5 occurs; that is, if we observe a 1, 3, or 5. Thus A could be decomposed into a collection of simpler events, namely, E_1, E_3, and E_5. Likewise, event B will occur if E_1, E_2, or E_3 occur and could be viewed as a collection of smaller or simpler events. In contrast, we note that it is impossible to decompose events E_1, E_2, E_3, ..., E_6. Events E_1, E_2, ..., E_6 are called simple events. A and B are compound events.

Definition

An event that cannot be decomposed is called a **simple event.** Simple events will be denoted by the symbol E with a subscript.

The events $E_1, E_2, \ldots, E_6$ represent a complete listing of all simple events associated with the experiment, Example 4.1. An interesting property of simple events is readily apparent. **An experiment will result in one and only one of the simple events.** For instance, if a die is tossed, we will observe either a 1, 2, 3, 4, 5, or 6, but we cannot possibly observe more than one of the simple events at the same time. Hence a list of simple events provides a breakdown of all possible indecomposable outcomes of the experiment. For purposes of illustration, consider the following examples:

Example 4.2

Experiment: Toss a coin. Simple events:
  E_1:   observe a head,
  E_2:   observe a tail.

Example 4.3

Experiment: Toss two coins. Simple events:

Event	Coin 1	Coin 2
E_1	Head	Head
E_2	Head	Tail
E_3	Tail	Head
E_4	Tail	Tail

It would be extremely convenient if we were able to construct a model for an experiment that could be portrayed graphically. We do this by creating a correspondence between simple events and a set of points. To each simple event we assign a point, called a **sample point.*** Thus the symbol E_i will now be associated with either simple event   or its corresponding sample point. The resulting diagram is called a Venn diagram.

---

* The preceding terminology and the use of sample points to represent simple events is consistent with W. Feller's treatment of this subject in his text on probability. Feller's introductory text on probability is considered to be a classic.

Example 4.1 may be viewed symbolically in terms of the Venn diagram shown in Figure 4.1. Six sample points are shown corresponding to the six possible simple events enumerated in Example 4.1. Likewise, a Venn diagram for the two-coin-toss experiment, Example 4.3, would represent an experiment in which there are four sample points.

Figure 4.1 Venn diagram for die tossing

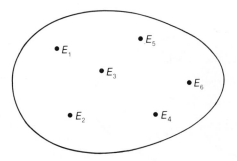

Definition

The set of all sample points for the experiment is called the **sample space** and is represented by the symbol S. We say that S is the totality of all sample points.

What is an event in terms of the sample points? We recall that event A, Example 4.1, occurred if any one of the simple events E_1, E_3, or E_5 occurred. That is, we observe A, an odd number, if we observe either a 1, 3, or 5. Event B, a number less than 4, occurs if E_1, E_2, or E_3 occurs. Thus, if we designate the sample points associated with an event (those for which the event occurs), the event is as clearly defined as if we had presented a verbal description of it. The event "observe E_1, E_3, or E_5" is obviously the same as the event "observe an odd number." Both represent event A.

Definition

An **event** is a specific collection of sample points.

To decide whether a sample point is included in a particular event, check to see if the occurrence of the sample point implies occurrence of the event. If it does, that sample point is included in the event. For example, in the die-tossing experiment, the sample point E_1 is in the event A, "observe an odd number," because if E_1 occurs, then A will occur.

Keep in mind that the foregoing discussion refers to the outcome of a single experiment and that the performance of the experiment will result in the occurrence of one and only one sample point. **A particular event will occur if any sample point in the event occurs.**

An event could be represented on the Venn diagram by encircling the sample points in that event. Events A and B for the die-tossing problem are shown in Figure 4.2. Note that points E_1 and E_3 are in both events A and B and that both A and B occur if either E_1 or E_3 occurs.

Figure 4.2 Events A and B for die tossing

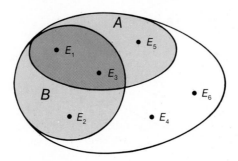

4.4
The Probability of an Event

Populations of observations are obtained by repeating an experiment a very large number of times. Some fraction of this very large number of experiments will result in E_1, another fraction in E_2, etc. From a practical point of view, we think of the fraction of the population resulting in an event A as the probability of A. Putting it another way, if an experiment is repeated a large number of times, N, and A is observed n times, the probability of A is

$$P(A) = \frac{n}{N}.$$

This practical interpretion of the meaning of probability, a view held by most laypersons, might properly be called the **relative frequency concept of probability.**

In practice, the exact composition of the population is rarely known and hence the exact values of the probabilities for various events are unknown. Mathematically speaking, we ignore this aspect of the problem and take the probabilities as given, hence providing a model for a real population. For instance, we would assume that for a large population of die tosses, Example 4.1, the numbers 1, 2, 3, 4, 5, and 6 should appear with approximately the same relative frequency and therefore that

$$P(E_1) = P(E_2) = \cdots = P(E_6) = 1/6.$$

That is, we assume that the die is perfectly balanced. Is there such a thing as a perfectly balanced die? Probably not, but we would be inclined to think that the probability of the sample points would be so near 1/6 that our assumption is quite valid for practical purposes and provides a good model for die tossing.

We complete our model for the population by adding the following:

Requirements of the Sample Point Probabilities

To each point in the sample space we assign a number called the probability of E_i, denoted by the symbol $P(E_i)$, such that:

1. $0 \leq P(E_i) \leq 1$, for all i.
2. $\sum_S P(E_i) = 1$.

The two requirements placed upon the probabilities of the sample points are necessary in order that the model conform to our relative frequency concept of probability. Thus we require that a probability be greater than or equal to 0 and less than or equal to 1 and that the sum of the probabilities over the entire sample space, S, be equal to 1. Furthermore, from a practical point of view, we would choose the $P(E_i)$ in a realistic way so that they agree with the observed relative frequency of occurrence of the sample points.

Keeping in mind that a particular event is a specific collection of sample points, we can now state a simple rule for the probability of any event, say event A.

Definition

The **probability** of an event A is equal to the sum of the probabilities of the sample points in A.

Note that the definition agrees with our intuitive concept of probability.

Example 4.4

Calculate the probability of the event A for the die-tossing experiment, Example 4.1.

Solution

Event A, "observe an odd number," includes sample points E_1, E_3, E_5. Hence

$$P(A) = P(E_1) + P(E_3) + P(E_5)$$
$$= \frac{1}{6} + \frac{1}{6} + \frac{1}{6}$$
$$= \frac{1}{2}.$$

Example 4.5

Calculate the probability of observing exactly one head in a toss of two coins.

Solution

Construct the sample space letting H represent a head, T a tail.

Event	First Coin	Second Coin	$P(E_i)$
E_1	H	H	1/4
E_2	H	T	1/4
E_3	T	H	1/4
E_4	T	T	1/4

It would seem reasonable to assign a probability of 1/4 to each of the sample points. We are interested in:

Event A: observe exactly one head.

Sample points E_2 and E_3 are in A. Hence

$$P(A) = P(E_2) + P(E_3)$$
$$= \frac{1}{4} + \frac{1}{4}$$
$$= \frac{1}{2}.$$

Example 4.6

Consider the following experiment involving two urns. Urn 1 contains two white balls and one black ball. Urn 2 contains one white ball. A ball is drawn from urn 1 and placed in urn 2. Then a ball is drawn from urn 2. What is the probability that the ball drawn from urn 2 will be white? See Figure 4.3.

Figure 4.3 Representation of the experiment in Example 4.6

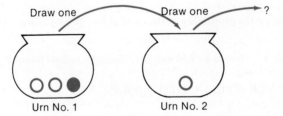

Draw one   Draw one   ?

Urn No. 1   Urn No. 2

Solution

The problem is easily solved once we have listed the sample points. For convenience, number the white balls 1, 2, and 3, with ball 3 residing in urn 2. The sample points are listed below using W_i to represent the ith white ball and B to represent the black ball.

For example, E_1 is the event that white ball 1 is drawn from urn 1, placed in urn 2, and then is drawn from urn 2.

| Event | Drawn from: | | $P(E_i)$ |
	Urn 1	Urn 2	
E_1	W_1	W_1	1/6
E_2	W_1	W_3	1/6
E_3	W_2	W_2	1/6
E_4	W_2	W_3	1/6
E_5	B	B	1/6
E_6	B	W_3	1/6

Event A, drawing a white ball from urn 2, occurs if E_1, E_2, E_3, E_4, or E_6 occurs. Once again, it would seem reasonable to assume the sample points equally likely and to assign a probability of 1/6 to each of the six sample points. Hence

$$P(A) = P(E_1) + P(E_2) + P(E_3) + P(E_4) + P(E_6)$$

$$= \frac{1}{6} + \frac{1}{6} + \frac{1}{6} + \frac{1}{6} + \frac{1}{6}$$

$$= \frac{5}{6}.$$

The sample space provides a probabilistic model for a population that possesses a great deal of utility in addition to elegance. The model provides us with a simple, logical, direct method for calculating the probability of an event or, if you like, the probability of a sample drawn from a theoretical population. A practical limitation to this (or any other) method for calculating probabilities is that the initial assignment of probabilities to the sample points is somewhat subjective. Consequently, a valid solution will require that the subjective probability assignments provide a good measure of the likelihood of occurrence of the sample points.

A second limitation, which specifically applies to the sample-point procedure, will be evident when you attempt to apply it to the solution of some probability problems. Listing the sample points can be quite tedious, and one must be certain that none has been omitted. Since the total number of sample points in S may run into the millions, it is very convenient to have counting rules to simplify the counting of sample points. Although not necessary for an understanding of the basic concepts of probability, counting rules are included in Section 4.10 (optional). If you wish to develop an ability to solve complex probability problems using the sample-point approach, you should move directly to Section 4.10.

Tips on Problem Solving: Calculating the probability of an event: the sample point approach

1. Use the following steps for calculating the probability of an event by summing the probabilities of the sample points:

   a. Define the experiment.

   b. Identify a typical simple event. List the simple events associated with the experiment and test each to make certain that they cannot be decomposed This defines the sample space, S.

   c. Assign reasonable probabilities to the sample points in S, making certain that $0 \le P(E_i) \le 1$ and $\sum_S P(E_i) = 1$.

   d. Define the event of interest, A, as a specific collection of sample points. (A sample point is in A if A occurs when the sample point occurs. Test *all* sample points in S to locate those in A.)

   e. Find $P(A)$ by summing the probabilities of sample points in A.

2. When the sample points are equiprobable, the sum of the probabilities of the sample points in A, step (e), can be acquired by counting the points in A and multiplying by the probability per sample point.

3. Calculating the probability of an event by using the five-step procedure described in part 1 is systematic and will lead to the correct solution if all the steps are correctly followed. Major sources of error are:

   a. failing to define the experiment clearly [step (a)],

   b. failing to specify simple events [step (b)],

   c. failing to list all the simple events,

   d. failing to assign valid probabilities to the sample points.

Exercises
Conceptual Level

4.1.   An experiment involves tossing a single die. Specify the sample points in the events:

   A: observe a 4.

   B: observe an even number.

   C: observe a number less than 3.

   D: observe both A and B.

   E: observe either A or B or both.

   F: observe both A and C.

Calculate the probabilities of the events D, E, and F by summing the probabilities of the appropriate sample points.

4.2.   A particular basketball player hits 70 percent of her free throws. When tossing a pair of free throws, the four possible simple events and three of their associated probabilities are:

Simple Event	Outcome of 1st Free Throw	Outcome of 2nd Free Throw	Probability
1	Hit	Hit	.49
2	Hit	Miss	?
3	Miss	Hit	.21
4	Miss	Miss	.09

a. Find the probability that the player will hit on the first free throw and miss on the second.

b. Find the probability that the player will hit on at least one of the two free throws.

4.3.   A jar contains four coins: a nickel, a dime, a quarter, and a 50-cent piece. Three coins are randomly selected from the jar.

a. List the sample points in S.

b. What is the probability that the selection will contain the 50-cent piece?

c. What is the probability that the total amount drawn will equal 60 cents or more?

4.4.   According to *Webster's New Collegiate Dictionary*, a divining rod is "a forked rod believed to indicate [divine] the presence of water or minerals by dipping downward when held over a vein." To test the claims of success by a divining rod expert, four cans are buried in the ground, two empty and two filled with water. The expert will use the divining rod to test each of the four cans and will decide which two contain water.

a. Define the experiment.

b. List the sample points in S.

c. If the rod is completely useless in locating water, what is the probability that the expert correctly identifies (by guessing) the two cans containing water?

4.5.   Patients arriving at a hospital outpatient clinic can select one of three stations for service. Suppose that physicians are randomly assigned to the stations and that the patients have no station preference. Three patients arrive at the clinic and their selection of stations is observed.

a. List the sample points for the experiment.

b. Let A be the event that each station receives a patient. List the sample points in A.

c. Make a reasonable assignment of probabilities to the sample points and find $P(A)$.

4.6.   A tea taster is required to taste and rank three varieties of tea, A, B, and C, according to the taster's preference.

a. Define the experiment.

b. List the sample points in S.

c. If the taster had no ability to distinguish a difference in taste between teas, what is the probability that the taster will rank tea type A as best? As the least desirable?

4.7.   The odds are two to one that when A and B play racquetball, A wins. Suppose A and B play three matches and that the winners of the matches are recorded. Using the letters

A and *B* to denote the winner of each match, the eight sample points are listed in the accompanying table. As you will subsequently learn, under certain conditions,

Winner of Match			Sample Point	
1	2	3	i	$P(E_i)$
A	*A*	*A*	1	8/27
A	*A*	*B*	2	4/27
A	*B*	*A*	3	4/27
A	*B*	*B*	4	2/27
B	*A*	*A*	5	4/27
B	*A*	*B*	6	?
B	*B*	*A*	7	2/27
B	*B*	*B*	8	1/27

it is reasonable to assume that the sample point probabilities are as listed in the table. Using these probabilities, find the following:

a. $P(E_6)$.

b. The probability that *A* wins at least two of the three matches.

4.8. Four union men, two from a minority group, are assigned to four distinctly different one-man jobs.

a. Define the experiment.

b. List the sample points in *S*.

c. If the assignment to the jobs is unbiased, that is, if any one ordering of assignments is as probable as any other, what is the probability that the two men from the minority group are assigned to the two least desirable jobs?

Exercises
Skill Level

4.9. Refer to Exercise 4.4. Suppose the experiment was conducted using five cans, three empty and two containing water. Answer parts (a), (b), and (c) of Exercise 4.4.

4.10. Two dice are tossed. What is the probability that the sum of the numbers shown on the dice is equal to 7? 11?

4.11. An investor has the option of investing in three of five recommended stocks. Unknown to the investor, only two will show a substantial profit within the next five years. If the investor selects the three stocks at random (giving every combination of three stocks an equal chance of selection), what is the probability that the investor selects the two profitable stocks?

4.12. A popular test to control the quality of brand-name food products is obtained by presenting three specimens identical in appearance to each member of a panel of tasters. In each case, two of the specimens are from batches of stock known to possess the desired taste, while the third specimen is from the latest batch. Each panelist is told to select the specimen that is different from the other two. Suppose that there are four tasters (designated T_1, T_2, T_3, and T_4) on the panel. Define the following events as specific collections of sample points:

a. The sample space S.

b. The event A that T_1, T_2, and T_3 are each "successful" in identifying the specimen from the latest batch.

c. The event B that exactly three of the four tasters are successful.

d. The event C that at least three of the four tasters are successful.

e. Assign reasonable probabilities to the sample points and find the probabilities of events A, B, and C. Assume that the latest batch actually does possess the desired taste so that the three specimens for each taster are identical.

4.13. Two city commissioners are to be selected from a total of five to form a subcommittee to study the city's traffic problems.

a. Define the experiment.

b. List the sample points in S.

c. If all possible pairs of commissioners have an equal probability of selection, what is the probability that commissioners Jones and Smith will be selected?

4.14. Four racquetball players participate in a match. They are randomly split into two pairs and the members of the pairs play each other. Then the winners of each pair play each other for the championship.

a. List the sample points in S.

b. If two of the players can definitely defeat either of the remaining two, what is the probability that one of the two best players will win the championship?

4.5
Compound Events

Most events of interest in practical situations are compound events that require enumeration of a large number of sample points. Actually, we find a second approach available for calculating the probability of events that obviates the listing of sample points and is therefore much less tedious and time consuming. It is based upon the classification of events, event relations, and two probability laws, which will be discussed in this section and in Sections 4.6 and 4.7, respectively. We call it the "event composition approach" for finding the probability of an event.

Compound events, as the name suggests, are formed by some composition of two or more events. Composition takes place in one of two ways, or in some combination of the two, namely, a **union** or an **intersection.**

Definition

Let A and B be two events in a sample space, S. The union of A and B is defined to be the event containing all sample points in A or B or both. In ordinary terms, a union is the event that *either* event A or event B or both A and B occur. We denote the union of A and B by the symbol $(A \cup B)$.

As an example of the union of two events, refer to Example 4.1 and define

A: $E_1, E_3, E_5,$
B: $E_1, E_2, E_3.$

The union ($A \cup B$) would be the collection of points E_1, E_2, E_3, and E_5. This is shown diagrammatically as the shaded area in Figure 4.4.

Figure 4.4 Event ($A \cup B$) in Example 4.1

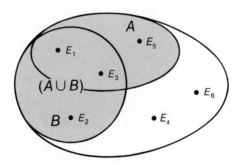

Definition

Let A and B be two events in a sample space, S. The **intersection** of A and B is the event composed of all sample points that are in both A and B. In ordinary terms, the intersection AB is the event that *both* A and B occur. The intersection of events A and B is represented by the symbol AB. (Some authors use $A \cap B$.)

The intersection of two events, A and B, would appear in a Venn diagram as the overlapping area between A and B. The intersection AB for Example 4.1 would be the event consisting of points E_1 and E_3. If either E_1 or E_3 occurs, both A and B occur. This is shown diagrammatically as the shaded area in Figure 4.5.

Figure 4.5 Intersection AB

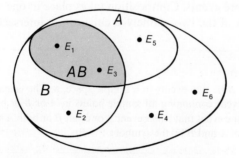

Example 4.7

Refer to the experiment, Example 4.3, where two coins are tossed, and define:

Event A: at least one head,

Event B: at least one tail.

Define events A, B, AB, and $(A \cup B)$ as collections of sample points.

Solution

Recall that the sample points for this experiment were:

E_1: HH (head on first coin, head on second),

E_2: HT,

E_3: TH,

E_4: TT.

The occurrence of sample points E_1, E_2, and E_3 implies and hence defines event A. The other events could similarly be defined:

Event B: E_2, E_3, E_4,

Event AB: E_2, E_3,

Event $(A \cup B)$: E_1, E_2, E_3, E_4.

Note that $(A \cup B) = S$, the sample space, and is thus certain to occur.

The concept of unions and intersections can be extended to more than two events. For example, the union of three events, A, B, and C, would be the event that either A or B or C occurs or any combination of these events occurs. This event, which is denoted by the symbol $A \cup B \cup C$, would be the set of sample points that are in either A or B or C or any combination of those events. Similarly, the intersection of the three events, A, B, and C, denoted as ABC, is the event that all three of the events A, B, and C occur. This event is the collection of sample points that are common to the three events, A, B, and C.

Example 4.8

Refer to the die-tossing experiment, Example 4.1, and define:

Event A: E_1, E_3, E_5,

Event B: E_1, E_2, E_3,

Event C: E_3, E_4.

Define the events $A \cup B \cup C$ and ABC as collections of sample points.

Solution

To find the union of A, B, and C, we list all sample points in events A, B, or C. A simple event cannot appear in a union more than once. Commencing with event A, we obtain the sample points E_1, E_3, and E_5. Checking the sample points in event B, we note that E_1 and E_3 are in event A but E_2 is not. Therefore, we add E_2 to our collection to obtain E_1, E_2, E_3, and E_5. Finally, examining the sample points in C, we note that we have already included E_3 in our set of sample points for $A \cup B \cup C$ but need to add E_4. Therefore, the set of sample points in $A \cup B \cup C$ is:

Event $A \cup B \cup C$: E_1, E_2, E_3, E_4, E_5.

Similarly, to find the sample points in the intersection, ABC, we need to list only those sample points that appear in all three of the events. Only one sample point, E_3, satisfies this criterion. Therefore, the intersection of events A, B, and C is:

Event ABC: E_3.

The union and intersection of events A, B, and C are shown in the Venn diagrams, Figures 4.6(a) and (b).

Figure 4.6 Events $A \cup$ $B \cup C$ and ABC, Example 4.8

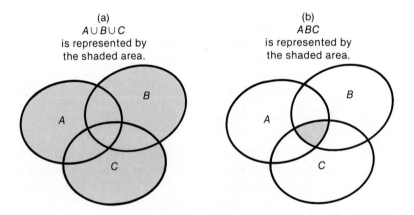

(a)
$A \cup B \cup C$
is represented by
the shaded area.

(b)
ABC
is represented by
the shaded area.

4.6
Event Relations

We shall define three relations between events: complementary, independent, and mutually exclusive events. You will have many occasions to inquire whether two or more events bear a particular relationship to one another. The test for each relationship is inherent in the definition, as we shall illustrate, and their use in the calculation of the probability of an event will become apparent in Section 4.7.

> **Definition**
>
> The **complement** of an event A is the collection of all sample points in S and not in A. The complement of A is denoted by the symbol $\bar{A}$.

Since

$$\sum_{S} P(E_i) = 1,$$

$$P(A) + P(\bar{A}) = 1,$$

and

$$P(A) = 1 - P(\bar{A}),$$

which is a useful relation for obtaining $P(A)$ when $P(\bar{A})$ is known or easily calculated.

Two events are often related in such a way that the probability of occurrence of one depends upon whether the second has or has not occurred. For instance, suppose that one experiment consists in observing the weather on a specific day. Let A be the event "observe rain" and B be the event "observe an overcast sky." Events A and B are obviously related. The probability of rain, $P(A)$, is not the same as the probability of rain given prior information that the day is cloudy. The probability of A, $P(A)$, would be the fraction of the entire population of observations that result in rain. Now let us look only at the subpopulation of observations that result in B, a cloudy day, and the fraction of these that result in A. This fraction, called the **conditional probability of A given B**, may equal $P(A)$, but we would expect the chance of rain, given that the day is cloudy, to be larger. **The conditional probability of A, given that B has occurred, is denoted as**

$P(A|B),$

where the vertical bar in the parentheses is read "given" and events appearing to the right of the bar are the events that have occurred.

We shall define the conditional probabilities of B given A and A given B as follows.

Definition

$$P(B|A) = \frac{P(AB)}{P(A)}$$

and

$$P(A|B) = \frac{P(AB)}{P(B)}.$$

By attaching some numbers to the probabilities in our weather example, you can see that this definition of conditional probability is consistent with the relative frequency concept of probability. Recall that A denotes rain on a given day; B denotes the day is cloudy. Now suppose that 10 percent of all days are rainy and cloudy [that is, $P(AB) = .10$] and 30 percent of all days are cloudy [$P(B) = .30$].

The situation that we have just described is graphically portrayed in Figure 4.7. Each sample point in event B, denoted by the large egg-shaped area, is associated with a single cloudy day. Since 30 percent of all days will be cloudy, we can regard this area as .3. Ten percent of all days, or 1/3 of all cloudy days, will also be rainy. These days are included in the color-shaded event, AB. Hence, if a single day is selected from the set of all days representing the population, what is the probability that we will select a rainy day, given that we know the day is cloudy? That is, what is $P(A|B)$?

Figure 4.7 Events A and B

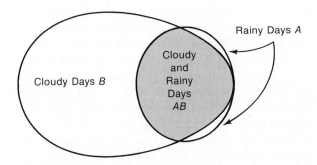

Since we already know that the day is cloudy, we know that the sample point to be selected must fall in event B (Figure 4.7). One-third of these days will result in rain. Hence the probability that we will select a rainy day is

$$P(A|B) = 1/3.$$

You can see that this result agrees with our definition for $P(A|B)$. That is,

$$P(A|B) = \frac{P(AB)}{P(B)} = \frac{.10}{.30} = 1/3.$$

Example 4.9

Calculate $P(A|B)$ for the die-tossing experiment described in Example 4.1.

Solution

Given that B has occurred, we are concerned only with sample points E_1, E_2, and E_3, which occur with equal frequency. Of these, E_1 and E_3 imply event A. Hence

$$P(A|B) = 2/3.$$

Or, we could obtain $P(A|B)$ by substituting into the equation

$$P(A|B) = \frac{P(AB)}{P(B)} = \frac{1/3}{1/2} = 2/3.$$

Note that $P(A|B) = 2/3$ while $P(A) = 1/2$, indicating that A and B are dependent upon each other.

Definition

Two events, A and B, are said to be **independent** if and only if either

$$P(A|B) = P(A)$$

or

$$P(B|A) = P(B).$$

Otherwise, the events are said to be **dependent.**

Translating this definition into words, two events are independent if the occurrence or nonoccurrence of one of the events does not change the probability of the occurrence of the other event. Note that if $P(A|B) = P(A)$, then $P(B|A)$ will also equal $P(B)$. Similarly, if $P(A|B)$ and $P(A)$ are unequal, then $P(B|A)$ and $P(B)$ will be unequal.

A third useful event relation was observed but not specifically defined in our discussion of simple events. Recall that an experiment could result in one and only one simple event. No two could occur at exactly the same time. Two events, A and B, are said to be mutually exclusive, if, when one occurs, it excludes the possibility of occurrence of the other. Another way to say this is to state that the intersection, AB, will contain no sample points. It would then follow that $P(AB) = 0$.

Definition

Two events, A and B, are said to be **mutually exclusive** if the event AB contains no sample points.

Mutually exclusive events have no overlapping area in a Venn diagram (see Figure 4.8).

Figure 4.8 Mutually exclusive events

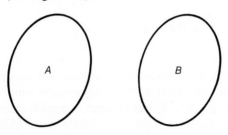

A   B

Example 4.10   Refer to the die-tossing experiment in Example 4.1. Are events A and B mutually exclusive? Are they complementary? Are they independent?

Solution

Event A: E_1, E_3, E_5,

Event B: E_1, E_2, E_3.

The event AB is the set of sample points in both A and B. You can see that AB includes points E_1 and E_3. Therefore, A and B are not mutually exclusive. They are not complementary because B is not the set of all points in S which are not in A. The test for independence lies in the definition. That is, we shall check to see if $P(A|B) = P(A)$. From Example 4.9, $P(A|B) = 2/3$. Then, since $P(A) = 1/2$, $P(A|B) \neq P(A)$ and, by definition, events A and B are dependent.

Example 4.11   Given two mutually exclusive events, A and B, with $P(A)$ and $P(B)$ not equal to zero, are A and B independent events?

Solution   Since A and B are mutually exclusive, if A occurs, B cannot occur and vice versa. Then $P(B|A) = 0$. But $P(B)$ was said to be greater than zero. Hence $P(B|A)$ is not equal to $P(B)$, and according to the definition the events are dependent.

4.7
Two Probability Laws and Their Use

As previously stated, a second approach to the solution of probability problems is based upon the classification of compound events, event relations, and two probability laws, which we now state and illustrate. The "laws" can be simply stated and taken as fact as long as they are consistent with our model and with reality. The first is called the Additive Law of Probability and applies to unions.

The Additive Law of Probability

Given two events, A and B, the probability of the union $(A \cup B)$ is equal to

$$P(A \cup B) = P(A) + P(B) - P(AB).$$

If A and B are mutually exclusive, $P(AB) = 0$ and

$$P(A \cup B) = P(A) + P(B).$$

The Additive Law conforms to reality and our model. As you will note from Figure 4.9 the sum, $P(A) + P(B)$, contains the sum of the probabilities of all sample points in $(A \cup B)$ but includes a double counting of the probabilities of all points in the intersection, AB. Subtracting $P(AB)$ gives the correct result.

Figure 4.9 The union of two events, A and B ($A \cup B$ is shaded)

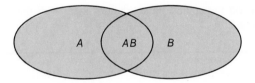

The second law of probability is called the Multiplicative Law and applies to intersections.

The Multiplicative Law of Probability

Given two events, A and B, the probability of the intersection, AB, is

$$P(AB) = P(A)P(B|A)$$
$$= P(B)P(A|B).$$

If A and B are independent, $P(AB) = P(A)P(B)$.

The Multiplicative Law follows from the definition of conditional probability.

The use of the probability laws for calculating the probability of a compound event is less direct than the listing of sample points and requires a bit of experience and ingenuity. The approach involves the expression of the event of interest as a union or intersection (or combination of both) of two or more events whose probabilities are known or easily calculated. This can often be done in many ways. The trick is to find the right combination, a task requiring no little amount of creativity in some cases. The usefulness of event relations is now apparent. If the event of interest is expressed as a union of mutually exclusive events, the probabilities of the intersections will equal zero. If they are independent, we can use the unconditional probabilities to calculate the probability of an intersection. Examples 4.12 through 4.17 illustrate the use of the probability laws and the technique described above.

Example 4.12

Calculate $P(AB)$ and $P(A \cup B)$ for Example 4.1.

Solution

Recall that $P(A) = P(B) = 1/2$ and $P(A|B) = 2/3$. Then,

$$P(AB) = P(B)P(A|B)$$
$$= (1/2)(2/3)$$
$$= 1/3,$$
$$P(A \cup B) = P(A) + P(B) - P(AB)$$
$$= 1/2 + 1/2 - 1/3$$
$$= 2/3.$$

You will notice that these solutions agree with those obtained using the sample point approach.

Example 4.13

Consider the experiment in which two coins are tossed. Let A be the event that the toss results in at least one head. Find $P(A)$.

Solution

$\bar{A}$ is the collection of sample points implying the event "two tails." Because $\bar{A}$ is the complement of A,

$$P(A) = 1 - P(\bar{A}).$$

The event $\bar{A}$ will occur if both of two independent events occur, "tail on the first coin," T_1, and "tail on the second coin," T_2. Then $\bar{A}$ is the intersection of T_1 and T_2, or

$$\bar{A} = T_1 T_2.$$

Applying the Multiplicative Law and noting that T_1 and T_2 are independent events,

$$P(\bar{A}) = P(T_1)P(T_2) = (1/2)(1/2) = 1/4.$$

Then, $P(A) = 1 - P(\bar{A}) = 1 - 1/4 = 3/4$.

Example 4.14

Refer to Example 4.13 and find the probability of exactly one head.

Solution

Event B, exactly one head, is the union of two mutually exclusive events, B_1 and B_2, where

B_1:   head on the first coin, tail on the second,
B_2:   head on the second coin, tail on the first.

Then B_1 and B_2 are both intersections of independent events:

$$B_1 = H_1 T_2$$
$$B_2 = H_2 T_1.$$

Applying the Multiplicative Law for independent events,

$$P(B_1) = P(H_1)P(T_2)$$
$$= (1/2)(1/2) = 1/4,$$
$$P(B_2) = P(H_2)P(T_1)$$
$$= (1/2)(1/2) = 1/4.$$

Then, $P(B) = P(B_1) + P(B_2)$, because B_1 and B_2 are mutually exclusive events, and

$P(B) = 1/4 + 1/4 = 1/2.$

You will note that this problem was solved in Example 4.5 using the sample point approach.

Example 4.15

A city council contains eight members, of which two are local contractors. If two councilmen are selected at random to fill vacancies on the zoning committee, what is the probability that both of the contractors will be selected?

Solution

Let D_1 be the event that the first councilman selected is a contractor and, correspondingly, let D_2 be the event that the second is a contractor. The event that both councilmen are contractors is A. Then A will occur if both D_1 and D_2 occur or, equivalently, A is the intersection of D_1 and D_2:

$A = D_1 D_2.$

Applying the Multiplicative Law,

$P(A) = P(D_1)P(D_2 | D_1).$

The probability of selecting a contractor on the first draw is $2/8$ or $1/4$. Similarly, the probability that the second councilman will be a contractor, given that the first was a contractor, is $1/7$. Then,

$P(A) = P(D_1)P(D_2 | D_1) = (1/4)(1/7) = 1/28.$

(*Note:* This solution can be easily obtained using the sample point approach.

Example 4.16

Two cards are drawn from a deck of 52 cards. Calculate the probability that the draw will include an ace and a ten.

Solution

Event A: draw an ace and a ten. Then $A = B \cup C$, where
  B: draw the ace on the first draw and the ten on the second,
  C: draw the ten on the first draw and the ace on the second.

Note that B and C were chosen so as to be mutually exclusive and also intersections of events with known probabilities. Thus,

$B = B_1 B_2$ and $C = C_1 C_2,$

where

B_1:   draw an ace on the first draw,

B_2:   draw a ten on the second draw,

C_1:   draw a ten on the first draw,

C_2:   draw an ace on the second draw.

Applying the Multiplicative Law,

$$P(B_1B_2) = P(B_1)P(B_2|B_1)$$
$$= (4/52)(4/51)$$

and

$$P(C_1C_2) = (4/52)(4/51).$$

Then, applying the Additive Law,

$$P(A) = P(B) + P(C)$$
$$= (4/52)(4/51) + (4/52)(4/51)$$
$$= \frac{8}{663}.$$

The student is cautioned to check each composition carefully to be certain that it is actually equal to the event of interest.

Example 4.17   Use the laws of probability to solve the "two-urns" problem, Example 4.6.

Solution   Define the event of interest as:

Event A: ball drawn from urn 2 is white.

Then,

$$A = (B \cup C),$$

where

B: draw a white ball from urn 1 and a white ball from urn 2,

C: draw a black ball from urn 1 and a white ball from urn 2.

Note that B and C were chosen to be mutually exclusive and that both are intersections. Thus,

$$B = B_1 A$$

and

$$C = C_1 A$$

where

B_1:   draw a white ball from urn 1,
C_1:   draw a black ball from urn 1.

Then,

$$P(A) = P(B \cup C)$$
$$= P(B) + P(C) - P(BC).$$

Since B and C are mutually exclusive, $P(BC) = 0$. Then,

$$P(A) = P(B) + P(C)$$
$$= P(B_1 A) + P(C_1 A).$$

Applying the Multiplicative Law,

$$P(B_1 A) = P(B_1)P(A|B_1)$$
$$= (2/3)(1)$$
$$= 2/3,$$
$$P(C_1 A) = P(C_1)P(A|C_1)$$
$$= (1/3)(1/2)$$
$$= 1/6.$$

Substituting,

$$P(A) = P(B_1 A) + P(C_1 A)$$
$$= 2/3 + 1/6$$
$$= 5/6.$$

Example 4.17 shows you that a problem often can be solved using either the sample point approach (Example 4.6) or the event composition approach.

Because the number of sample points was small, the sample point approach was the easier method for solving this problem. If the number of balls in the urn had been large, it would have been difficult to enumerate the large number of sample points. Then the event composition approach should have been used. The method of solution would be identical to the solution for Example 4.17. The difficulty of the solution would not depend upon the number of balls in the urn.

Tips on Problem Solving: Calculating the probability of an event: event composition approach.

1. Use the following steps for calculating the probability of an event using the event composition approach:
   a. Define the experiment.
   b. Clearly visualize the nature of the sample points. Identify a few to clarify your thinking.
   c. Write an equation expressing the event of interest, say A, as a composition of two or more events using either or both of the two forms of composition (unions and intersections). Note that this equates point sets. Make certain that the event implied by the composition and event A represent the same set of sample points.
   d. Apply the Additive and Multiplicative Laws of Probability to step (c) and find $P(A)$.
2. Be careful with step (c). You often can form many compositions that will be equivalent to event A. The trick is to form a composition in which all the probabilities appearing in step (d) will be known. Thus, you must visualize the results of step (d) for any composition and select the one for which the component probabilities are known.
3. Always write down letters to represent events described in an exercise. Then write down the probabilities that are given and assign them to events. Identify the probability that is requested in the exercise. This may help you to arrive at the appropriate event composition.

Exercises
Conceptual Level

4.15.   An experiment consists in tossing a single die and observing the number of dots shown on the upper face. The events, A, B, and C, are defined as follows:

   A: observe a number less than 4,

   B: observe a number less than or equal to 2,

   C: observe a number greater than 3.

List the sample points in the following events and find their respective probabilities:

a. *S.*   b. *A.*   c. *B.*   d. *C.*   e. *AB.*

f. *AC.*   g. *BC.*   h. *A* ∪ *B.*   i. *A* ∪ *C.*   j. *B* ∪ *C.*

4.16.   Refer to Exercise 4.15 and list the sample points in:

a. *ABC.*   b. *A* ∪ *B* ∪ *C.*

4.17.   Find the probabilities of the events listed in Exercise 4.16.

4.18.   Refer to Exercise 4.15 and find:

a. $P(\bar{A})$.   b. $P(\overline{AB})$.   c. $P(A|B)$.

d. $P(A|C)$.   e. $P(B|C)$.

4.19.   Refer to Exercise 4.15. Show that events *A* and *B* are or are not:

a. independent.   b. mutually exclusive.

4.20.   Refer to Exercise 4.15. Show that events *A* and *C* are or are not:

a. independent.   b. mutually exclusive.

4.21.   Refer to Exercise 4.15. Show that events *B* and *C* are or are not:

a. independent.   b. mutually exclusive.

4.22.   An experiment generates a sample space containing eight simple events $E_1, \ldots, E_8$ with $P(E_i) = 1/8, i = 1, \ldots, 8$. The events *A* and *B* are defined as:

$$A:\ E_1, E_4, E_6,$$
$$B:\ E_3, E_4, E_5, E_6, E_7.$$

Find the following:

a. $P(A)$.

b. $P(\bar{A})$.

c. $P(A \cup B)$.

d. $P(AB)$.

e. $P(A|B)$.

f. Are events *A* and *B* mutually exclusive? Why?

g. Are events *A* and *B* independent? Why?

4.23.   A study of the behavior of a large number of drug offenders after treatment for drug abuse suggests that the likelihood of conviction within a two-year period after treatment may depend upon the offender's education. The proportions of the total number of cases falling in four education–conviction categories are shown in the accompanying table.

Education	Status Within Two Years After Treatment		
	Convicted	Not Convicted	Totals
Ten years or more	.10	.30	.40
Nine years or less	.27	.33	.60
Totals	.37	.63	1.00

Suppose that a single offender is selected from the treatment program. Define the events:

A: the offender has ten or more years of education,

B: the offender is convicted within two years after completion of treatment.

Find the approximate probabilities for the events:

a. A.

b. B.

c. AB.

d. $A \cup B$.

e. $\overline{A}$

f. $\overline{A \cup B}$.

g. $\overline{AB}$.

h. A given that B has occurred.

i. B given that A has occurred.

4.24.   Use the probabilities of Exercise 4.23 to show that:

a. $P(AB) = P(A)P(B|A)$.

b. $P(AB) = P(B)P(A|B)$.

c. $P(A \cup B) = P(A) + P(B) - P(AB)$

4.25.   Television commercials are designed to appeal to the most likely viewing audience of the sponsored program. However, S. Ward, in "Children's Reactions to Commercials,"* notes that children often possess a very low understanding of commercials, even for those designed to appeal especially to children. Ward's studies show that the percentages of children understanding TV commercials for different age groups are as given in the accompanying table.

	Age		
	5–7	8–10	11–12
Don't understand	55%	40%	15%
Understand	45%	60%	85%

An advertising agent has shown a television commercial to a 6-year-old and another to a 9-year-old child in independent laboratory experiments to test their understanding of the commercials.

a. What is the probability that the message of the commercial is understood by the 6-year-old child?

b. What is the probability that both children demonstrate an understanding of the TV commercials?

c. What is the probability that one or the other, or both, children demonstrate an understanding of the TV commercials?

4.26.   Two surgeons in Chicago recently demonstrated that, when surgeons perform leg or arm operations, it is conceivable that they might operate on the wrong leg (the *Orlando Sentinel Star*, July 31, 1981). Suppose that two surgeons performing a leg operation are

* *Journal of Advertising Research*, April 1972.

able to choose the correct leg with probability equal to .8 and that their choice in one operation is independent of their choice in any other. If the surgeons perform two operations; and their choice of legs is recorded:

a. List the four sample points in S.

b. What is the probability that they will select the correct leg for both operations?

c. What is the probability that they will select the wrong leg for both operations?

d. What is the probability that they will select the wrong leg for one of the two operations?

e. What is the probability that they will select the wrong leg for at least one of the two operations?

4.27. Refer to Exercise 4.26 and define the events:

W_1:   surgeon selects wrong leg on first operation,

W_2:   surgeon selects wrong leg on second operation.

Suppose that the surgeons learn from experience and, if they select the wrong leg for the first operation (W_1), the probability of making the same mistake on the second operation is very small, say .05. However, if they select the correct leg for the first operation, the probability that they select the correct leg for the second operation does not improve; that is, it remains at .8.

a. Define the events $\bar{W}_1$ and $\bar{W}_2$.

b. List the sample points in S.

c. Find the probabilities of each sample point in S.

c. Find the probabilities of each sample point in S.

d. Find the probability that the surgeons select the correct leg for both operations. operations.

e. Find the probability that the surgeons choose the correct leg on the second operation.

f. Find the probability that the surgeons pick the correct leg on at least one of the two operations.

Exercises
Skill Level

4.28. Dough Hannon, a Florida professional fishing guide, estimates that probably only one out of every 5000 bass eggs lives to reach adulthood; only one in 5000 adult bass live to reach 1 1/2 pounds; and only one out of every 2000 of those live to reach 10 pounds.*

a. Find the conditional probability that a bass will live to reach 10 pounds, given that it managed to reach 1 1/2 pounds.

b. Find the probability that an adult bass will live to reach 10 pounds.

c. Find the probability that a bass egg will live to become a 10-pound fish.

4.29. Unknown to you, five of the twelve bottles in a case of wine are bad. If you were to randomly select two bottles from the case, what is the probability that:

a. both bottles are bad?

b. both bottles are good?

* Carter, W. Horace, "*Unique Guide to Trophy Bass*," *Bassmaster Annual*, 1978.

c. the first bottle selected is bad and the second is good?

d. the first bottle selected is good and the second is bad?

e. one of the two bottles in bad?

Solve using the Probability Laws of Section 4.7.

4.30.   A smoke detector system utilizes two devices, *A* and *B*. If smoke is present, the probability that it will be detected by device *A* is .95; by device *B*, .98; and by both devices, .94.

a. If smoke is present, find the probability that the device will be detected by either device *A* or *B* or both devices.

b. Find the probability that the smoke will be undetected.

4.31.   A policy requiring all hospital employees to take lie detector tests may reduce losses due to theft, but some employees regard such actions as a violation of their rights. Reporting on a particular hospital that uses this procedure, the *Orlando Sentinel Star* (August 3, 1981) notes that lie detectors have accuracy rates that vary from 92 to 99 percent. To gain some insight into the risks that employees face when taking a lie detector test, suppose that a particular lie detector erroneously concludes that a person is lying with probability equal to .05, and suppose that any pair of tests are independent.

a. What is the probability that a machine will conclude that each of three employees is lying when all are telling the truth?

b. What is the probability that the machine will conclude that at least one of the three employees is lying when all are telling the truth?

4.32.   If you are having trouble renting an apartment, it may be because you are listed as a problem tenant (*Wall Street Journal*, July 21, 1981). Checks on prospective tenants, similar to credit checks, are provided by several companies which maintain large computer listings of problem tenants for landlord subscribers. Suppose that the probability $P(A)$ that a landlord utilizes a tenant checking service is .10 and the probability $P(B)$ that you are listed as a bad tenant by the service is .2. If the landlord refuses to rent to 80 percent of the prospective renters who are listed as problem tenants, what is the probability that the next time you apply for an apartment rental, you will be refused because you are listed as a problem tenant? [*Hint:* Let *C* be the event the landlord refuses to rent to a person listed as a problem renter. Note that $P(C|AB) = .8$ and that you wish to find $P(ABC)$.]

4.33.   A certain article is visually inspected successively by two different inspectors. When a defective article comes through, the probability that it gets by the first inspector is .1. Of those that do get past the first inspector, the second inspector will "miss" 5 of 10. What fraction of the defectives will get by both inspectors?

4.34.   According to the Food and Drug Administration (FDA), each year approximately 1000 children under age 5 are rushed to hospital emergency rooms for treatment of accidental poisoning by anti-depressant drugs (*U.S. News and World Report*, March 30, 1981). The probability that one of these children must be hospitalized is approximately .5, and the probability that the dose is fatal is .01. Suppose that the probability that a poisoned child who is admitted to the emergency room will be hospitalized and die is .008. Find the approximate probability that a poisoned child who is admitted to the emergency room will die, given that the child is hospitalized.

4.35.   The *Orlando Sentinel Star* (April 10, 1981) reports that ex-officials of the Defense Department appear to receive far too many unsolicited and unnecessary consulting contracts. A survey of 256 Defense Department contracts indicates that approximately one-half of all consulting contracts are awarded to ex-Defense Department employees. Of those contracts awarded to ex-Defense employees, it is estimated that 2/3 could have been completed by current employees of the Department of Defense. Approximately what proportion of all contracts awarded could have been done by current employees but were, instead, awarded to ex-employees?

4.36. Are men subject to sexual harassment on the job? According to a study conducted by the U.S. Merit Systems Protection Board, you had better believe it (*Fortune*, June 1, 1981). The report suggests that the probability that a male may be harassed on the job is approximately .15 and, if harassed, the probability is approximately .72 that he was harassed by a woman. If you were to randomly select an employee who was surveyed by the U.S. Merit Systems Protection Board, what is the probability that the employee is being harassed on the job by a woman?

4.37. A survey of people in a given region showed that 20 percent were smokers. The probability of death due to lung cancer, given that a person smoked, was roughly ten times the probability of death due to lung cancer, given that a person did not smoke. If the probability of death due to lung cancer in the region is .006, what is the probability of death due to lung cancer given that a person is a smoker?

4.38. Suppose that the probability of exposure to the flu during an epidemic is .6. Experience has shown that a serum is 80 percent successful in preventing an inoculated person who is exposed to the flu from acquiring it. A person not inoculated faces a probability of .90 of acquiring the flu if exposed to it. Two persons, one inoculated and one not, are capable of performing a highly specialized task in a business. Assume that they are not at the same location, are not in contact with the same people, and cannot expose each other. What is the probability that at least one will get flu?

4.39. Two people enter a room and their birthdays (ignoring years) are recorded.

a. Identify the nature of the sample points in S.

b. What is the probability that the two people have a specific pair of birth dates?

c. Identify the sample points in

A: both persons have the same birthday.

d. Find $P(A)$.

e. Find $P(\bar{A})$.

4.40. If n people enter a room, find the probability that

A: none of the persons have the same birthday,

B: at least two of the persons have the same birthday.

Solve for:

a. $n = 3$.   b. $n = 4$.

[*Note:* Surprisingly, $P(B)$ increases rapidly as n increases. For example, for $n = 20$, $P(B) = .411$; for $n = 40$, $P(B) = .891$.]

4.41. A worker-operated machine produces a defective item with probability .01 if the worker follows the machine's operating instructions exactly, and probability .03 if he does not. If the worker follows the instructions 90 percent of the time, what proportion of all items produced by the machine will be defective?

4.8

More on the ERTS Case Study

As we learned in our ERTS case study, the leader of the ERTS expedition, Karen Ross, employs space-age computer techniques to calculate probabilities associated with various combinations of environmental conditions and actions that she might take. At one stage in the expedition, Ross is informed by her Houston headquarters that their computers estimate that she is 18 hours, 20 minutes behind the competing Euro–Japanese team, instead of 40 hours ahead. She changes plans and decides to have the 12 members of her team—Ross, Elliot, Munro, Amy, and eight native porters—parachute into a volcanic region near the estimated location of Zinj. As Crichton relates, "Ross had double-checked outcome probabilities from the Houston computer, and the results were unequivocal. The probability of a successful jump was .7980, meaning that there was approximately one chance in five that someone would be badly hurt. However, given a successful jump, the probability of expedition success was .9943, making it virtually certain that they would beat the consortium to the site."

Keeping in mind that this is an excerpt from a novel, let us examine the probability, .7980, of a successful jump. If you were a member of the 12-member team, what is the probability that you would successfully complete your jump? In other words, if the probability of a successful jump by all 12 team members is .7980, what is the probability that a single member could successfully complete the jump?

To answer this question, let the probability that a single person successfully completes the jump equal p. Also, assume that the event that any one person successfully completes the jump is independent of the outcome of the other 11 individual jumps. Then the probability that all 12 individuals (including Amy, the gorilla) successfully complete the jump is the intersection of 12 separate events, each one corresponding to the successful jump for a single person. We define the events as follows:

J: all members of the team successfully complete the jump,

A_1:   Ross successfully completes her jump,

A_2:   Elliot successfully completes his jump,

A_3:   Munro successfully completes his jump,

A_4:   Amy successfully completes her jump,

A_5:   Porter #1 successfully completes his jump,

$\vdots$

A_{12}:   Porter #8 successfully completes his jump.

Then J is the intersection of the events, $A_1, A_2, \ldots, A_{12}$, and

$$P(J) = P\{A_1 A_2 A_3 \cdots A_{11} A_{12}\}.$$

Since we have assumed that $A_1, A_2, \ldots, A_{12}$ are independent events,

$$P(J) = P(A_1)P(A_2) \cdots P(A_{12})$$
$$= p \cdot p \cdot p \cdots p = p^{12}.$$

Substituting $P(J) = .7980$ into this expression and solving for p, we have:

$$P(J) = .7980 = p^{12},$$

or $p = .9814$. Thus, Houston is basing its probability of a successful team jump on a probability equal to .9814 that a single individual will successfully land in the soft volcanic scree.

This probability, of course, is fictitious and bears no similarity to the actual probability of making a successful parachute jump, one that does not result in a serious injury. The American Parachute Association, Washington, D.C., does not have the data that would enable us to estimate this probability, but they note that, of the approximately 3,000,000 parachute jumps in 1981, 53 resulted in deaths. Based on these numbers, the probability of surviving a single jump is approximately $(1 - 53/3,000,000)$ or 0.9999823. The probability of jumping and not sustaining serious injury would presumably be smaller.

4.9 Bayes' Rule (Optional)

We most frequently wish to find the conditional probability of an event A, given that an event B has occurred at a prior point in time. Thus, we might wish to know the probability of rain tomorrow given that it has rained during the preceding seven days. Or, we might wish to know the probability of drawing two aces from a deck of cards given that the deck contains only two aces. Hence, we assume that some state of nature exists, and we wish to calculate the probability of some event that will occur in the future.

Equally interesting is the probability that a particular state of nature exists given that a certain sample is observed. If the first two cards in a five-card poker hand yield a pair, what is the probability that the hand is a full house? A similar but more interesting problem occurs in X-raying people for tuberculosis. A positive X-ray result may or may not imply that the person tested has tuberculosis. Given that the X-ray result for a certain patient is positive, what is the probability that he has tuberculosis? The answer to this probabilistic question suggests an approach to statistical inference. If the probability is high, we will infer that the patient has tuberculosis. This section presents an application of the Multiplicative Law and a formula derived by the probabilist, Thomas Bayes.

Consider an experiment that involves the selection of a sample from one of k populations, call them $H_1, H_2, \ldots, H_k$. The sample is observed, but it is not known from which population the sample was selected. Suppose that the sample results in an event A. Then the problem is to determine the population from which the sample was selected. This inference will be based on the conditional probabilities, $P(H_i|A)$, $i = 1, 2, \ldots, k$.

To find the probability that the sample was selected from population i given that event A was observed, $P(H_i|A)$, $i = 1, 2, \ldots, k$, note that A could have been observed if the sample were selected from population 1, population 2, or any one of the k populations, $H_1, H_2, \ldots, H_k$. The probability that population i was selected *and* that event A occurred is the intersection of the events H_i and A, or (AH_i). These events, $(AH_1), (AH_2), \ldots, (AH_k)$, are mutually exclusive and hence

$$P(A) = P(AH_1) + P(AH_2) + \cdots + P(AH_k).$$

Then the probability that the sample came from population i is

$$P(H_i|A) = \frac{P(AH_i)}{P(A)} = \frac{P(H_i)P(A|H_i)}{\sum_{J=1}^{k} P(AH_J)} = \frac{P(H_i)P(A|H_i)}{\sum_{J=1}^{k} P(H_J)P(A|H_J)}.$$

The expression for $P(H_i|A)$ is known as Bayes' Rule for the probability of causes. As you can see, it follows easily from the definition of conditional probability.

Finding $P(H_i|A)$ requires knowledge of the probabilities, $P(H_i)$ and $P(A|H_i)$, $i = 1, 2, \ldots, k$. One can often determine the probability of an event if the population is known and hence can find $P(A|H_i)$, but $P(H_i)$, $i = 1, 2, \ldots, k$, are the probabilities that certain states of nature exist and are either unknown or difficult to ascertain. This set of probabilities $P(H_i)$ is called the **prior probability distribution** because it gives the distribution of the states of nature prior to conducting the experiment.

In the absence of knowledge concerning the values of $P(H_i)$, Bayes suggested that these probabilities should be taken as equal. That is, he assumed the populations over which his experiment was defined to be equiprobable and, therefore, assigned equal probabilities to the $P(H_i)$, $i = 1, 2, \ldots, k$. We would not agree that the assignment of equal prior probabilities to the populations $H_1, H_2, \ldots, H_k$ is logical, but the procedure for selecting one of the set $H_1, H_2, \ldots, H_k$, given the observation of event A, is certainly intriguing. In many instances the experimenter does have some notion—sometimes vague, sometimes exact—of the prior probabilities for $H_1, H_2, \ldots, H_k$. Then, inferring the population—that is, which of the set $H_1, H_2, \ldots, H_k$ is the true population— can be achieved by finding $P(H_i|A)$ and selecting H as the population that gives the highest probability, $P(H_i|A)$.

Example 4.18

An individual is selected at random from a community in which 1 percent are afflicted with tuberculois, and is X-rayed to detect the presence of the disease. A positive X-ray indication of the disease can occur when a nontubercular person is tested. Let T denote the event that the person selected is tubercular and let E indicate a positive X-ray result. The probability of a positive X-ray result, given that the person selected is tubercular, is $P(E|T) = .90$. The corresponding probability for a nontubercular person is $P(E|\overline{T}) = .01$. What is the probability that a person will be tubercular given that his X-ray result is positive?

Solution

The population H_1 and H_2 corresponds to the two sets of people, those who have tuberculosis and those who do not. (Thus H_1 is equivalent to event T, H_2 to $\bar{T}$.) The prior probabilities are $P(T) = .01$ and $P(\bar{T}) = .99$. The event E, a positive X-ray result, can occur whether or not the person has tuberculosis. These two mutually exclusive events are the intersections (ET) and $(E\bar{T})$. The probability of E is the probability of the union of (ET) and $(E\bar{T})$, or

$$\begin{aligned} P(E) &= P(ET) + P(E\bar{T}) \\ &= P(T)P(E|T) + P(\bar{T})P(E|\bar{T}) \\ &= (.01)(.90) + (.99)(.01) \\ &= .0189. \end{aligned}$$

Then the probability that a person has tuberculosis, given a positive X-ray result, is

$$P(T|E) = \frac{P(ET)}{P(E)} = \frac{P(T)P(E|T)}{P(E)} = \frac{(.01)(.90)}{.0189} = .476.$$

You can see how Bayes' Rule could be used to make an inference about the population from which the X-rayed person (Example 4.18) was selected. We have shown that the probability that a person has tuberculosis, given a positive X-ray result, is .476. Similarly, we could compute $P(\bar{T}|E)$, the probability that a person does not have tuberculosis, given a positive X-ray result, to be $1 - P(T|E)$, or .524. Now suppose that you randomly select a sample of one person and find that the person's X-ray is positive. What do you infer about the population from which the person was selected? Tubercular or not? Because $P(\bar{T}|E)$ is slightly larger than $P(T|E)$, we think that you would be inclined to believe that the person is not tubercular. Thus Bayes' Rule was used to make an inference about a population based on an observed sample. Naturally, in a practical situation, additional tests would be conducted to verify whether tuberculosis was or was not present.

Another way to view Bayes' Rule is to view it as a method of incorporating the information from sample observations to adjust the probability of some event. For example, if we had no information on the result of a person's X-ray examination, we would regard the probability that he or she is tubercular as .01 (because 1 percent of all people are afflicted with tuberculosis). However, if we are given the added information that the person's X-ray was positive, the probability that he or she is tubercular is

$$P(T|E) = .476.$$

Thus, based on the X-ray information, we have adjusted the probability that the person is tubercular from a small probability, .01, to a rather high probability, .476.

Exercises
Skill Level

4.42. A study of Georgia residents suggests that those who worked in shipyards during World War II were subjected to a significantly higher risk of lung cancer (the *Wall Street Journal*, September 21, 1978). It was found that approximately 22 percent of those persons who had lung cancer worked in a shipyard at some prior time. In contrast, only 14 percent of those who had no lung cancer worked in a shipyard at some prior time. Suppose that the proportion of all Georgians living during World War II who have contracted or will contract lung cancer is .04 percent. Find the percentage of Georgians living during the same period who will contract (or have contracted) lung cancer, given that they worked in a shipyard at some prior time.

4.43. As items come to the end of a production line, an inspector chooses which items are to go through a complete inspection. Ten percent of all items produced are defective. Sixty percent of all defective items go through a complete inspection, and 20 percent of all good items go through a complete inspection. Given that an item is completely inspected, what is the probability it is defective?

4.44. Medical case histories indicate that many different illnesses may produce identical symptoms. Suppose that a particular set of symptoms, which we will denote as event H, occurs only when any one of three illnesses, A, B, or C, occurs (for the sake of simplicity, we will assume that illnesses A, B, and C are mutually exclusive). Studies show that the probabilities of getting the three illnesses are:

$$P(A) = .01,$$

$$P(B) = .005,$$

$$P(C) = .02.$$

The probabilities of developing the symptoms, H, given a specific illness are:

$$P(H|A) = .90,$$

$$P(H|B) = .95,$$

$$P(H|C) = .75.$$

Assuming that an ill person shows the symptoms, H, what is the probability that the person has illness A?

4.45. A student answers a multiple-choice examination question that possesses four possible answers. Suppose that the probability that the student knows the answer to the question is .8, and the probability that the student will guess is .2. Assume that if the student guesses, the probability of selecting the correct answer is .25. If the student correctly answers a question, what is the probability that the student really knew the correct answer?

4.46. Suppose that 5 percent of all people filing the long income tax form seek deductions which they know to be illegal, and an additional 2 percent will incorrectly list deductions because of lack of knowledge of the income tax regulations. Of the 5 percent guilty of cheating, 80 percent will deny knowledge of the error if confronted by an investigator. If a filer of the long form is confronted with an unwarranted deduction and he denies knowledge of the error, what is the probability that he is guilty?

✵ 4.47.   A particular football team is known to run 30 percent of its plays to the left, 70 percent to the right. A linebacker on an opposing team notes that the right guard shifts his stance most of the time (80 percent) when plays go the right and that he uses a balanced stance the remainder of the time. When plays go the left, the guard takes a balanced stance 90 percent of the time and the shift stance the remaining 10 percent of the time. On a particular play, the linebacker notes that the guard takes a balanced stance. What is the probability that the play will go the left?

4.10
Results Useful in Counting Sample Points (Optional)

The preceding sections cover the basic concepts of probability and provide a background that will help you understand the role that probability plays in making inferences. To introduce the concept of a sample space, we used only examples and exercises for which the total number of sample points in the sample space was small. This allowed you to list the sample points in S, to identify the sample points in the event of interest, and then to calculate its probability. Although these examples and exercises were adequate for our learning objective, most real-life problems involve many more sample points. Consequently, we include this optional section for the student who wishes to improve his or her problem-solving ability. (*Note:* Developing the ability to solve problems requires practice.)

We will present three theorems that fall in the realm of combinatorial mathematics and which can be of assistance in solving probability problems that involve a large number of sample points. For example, suppose that you are interested in the probability of an event A and you know that the sample points in S are equiprobable. Then

$$P(A) = \frac{n_A}{N},$$

where

n_A = number of sample points in A,

N = number of sample points in S.

Often we can use counting rules to find the values of n_A and N and thereby eliminate the necessity of listing the sample points in S.

The first theorem, which we call the *mn* rule, is as follows.

Theorem 4.1

With *m* elements $a_1, a_2, a_3, \ldots, a_m$ and *n* elements $b_1, b_2, \ldots, b_n$ it is possible to form *mn* pairs that contain one element from each group.

Proof

Verification of the theorem can be seen by observing the rectangular table in Figure 4.10. There will be one square in the table for each a_i, b_j combination—a total of mn squares.

Figure 4.10 Table that indicates the number of pairs (a_i, b_j)

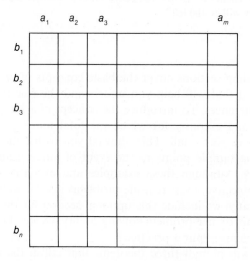

To illustrate, suppose that four companies have job openings in each of three areas, sales, manufacturing, and personnel. How many job opportunities are available to you? You can see that you have two sets of "things," companies (four) and types of jobs (three). Therefore, as shown in Figure 4.11, there are

Figure 4.11 Company–job combinations

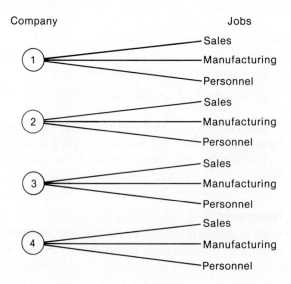

three jobs for each of the four companies, or $(4)(3) = 12$ possible pairings of companies and jobs. This example illustrates a use of the *mn* rule.

Example 4.19   Two dice are tossed. How many sample points are associated with the experiment?

Solution   The first die can fall in one of six ways; that is, $m = 6$. Likewise, the second die can fall in $n = 6$ ways. Since an outcome of this experiment involves a pairing of the numbers showing on the faces of the two dice, the total number, N, of sample points is

$$N = mn = 6(6) = 36.$$

Example 4.20   How many sample points are associated with the experiment, Example 4.6?

Solution   A ball can be chosen from urn 1 in one of $m = 3$ ways. After one of these ways has been chosen, a ball may be drawn from urn 2 in $n = 2$ ways. The total number of sample points is

$$N = mn = 3(2) = 6.$$

As noted, the *mn* rule gives the number of pairs you can form in selecting one object from each of two groups. The rule can be extended to apply to triplets formed by selecting one object from each of three groups, quadruplets formed by selecting one object from each of four groups, etc. The application to triplets is shown in Figure 4.12. If you have *m* elements in the first group, *n* in the second, and *t* in the third, the total number of triplets that you can form, taking one object from each group, is equal to *mnt*, the number of branchings shown in Figure 4.12.

A Counting Rule for Forming Pairs, Triplets, etc.

Given k groups of elements, n_1 elements in the first group, n_2 in the second, ..., and n_k in the kth group, then the number of ways of selecting one element from each of the k groups is

$$n_1 n_2 n_3 \cdots n_k.$$

Example 4.21   How many sample points are in the sample when three coins are tossed?

Solution   Each coin can land in one of two ways. Hence

$$N = 2(2)(2) = 8.$$

Figure 4.12 Forming
mnt triplets

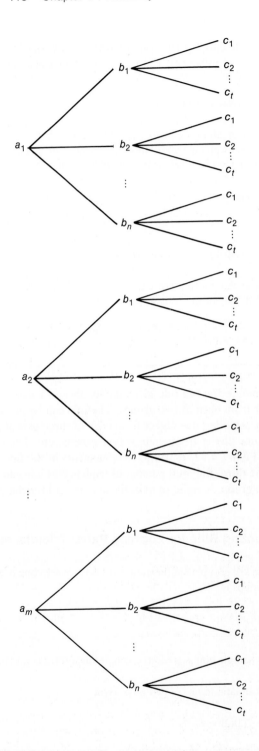

Example 4.22

A truck driver can take three routes in going from City A to City B, four from City B to City C, and three from City C to City D. If, in going from A to D, the driver must proceed from A to B to C to D, how many possible A-to-D routes are available to the driver?

Solution

Let

$$m = \text{number of routes from } A \text{ to } B = 3,$$

$$n = \text{number of routes from } B \text{ to } C = 4,$$

$$t = \text{number of routes from } C \text{ to } D = 3.$$

Then the total number of ways that you can construct a complete route, taking one sub-route from each of the three groups (A to B), (B to C), and (C to D), is

$$mnt = (3)(4)(3) = 36.$$

A second useful mathematical result is associated with orderings or permutations. For instance, suppose that we have three books, b_1, b_2, and b_3. In how many ways can the books be arranged on a shelf, taking them two at a time? We enumerate, listing all combinations of two in the first column and a reordering of each in the second column:

Combinations of Two	Reordering of Combinations
$b_1 b_2$	$b_2 b_1$
$b_1 b_3$	$b_3 b_1$
$b_2 b_3$	$b_3 b_2$

The number of permutations is six, a result easily obtained from the mn rule. The first book can be chosen in $m = 3$ ways, and, once selected, the second book can be chosen in $n = 2$ ways. The result is $mn = 6$.

In how many ways can three books be arranged on a shelf taking three at a time? Enumerating, we obtain

$$b_1 b_2 b_3 \qquad b_2 b_1 b_3 \qquad b_3 b_1 b_2$$
$$b_1 b_3 b_2 \qquad b_2 b_3 b_1 \qquad b_3 b_2 b_1,$$

a total of six. This, again, could be obtained easily by the extension of the mn rule. The first book can be chosen and placed in $m = 3$ ways. After choosing the first, the second can be chosen in $n = 2$ ways, and finally the third in $t = 1$

way. Hence the total number of ways is

$$N = mnt = 3 \cdot 2 \cdot 1 = 6.$$

Definition

An ordered arrangement of r distinct objects is called a **permutation**. The number of ways of ordering n distinct (different) objects taken r at a time will be designated by the symbol P_r^n.

Theorem 4.2

$$P_r^n = n(n-1)(n-2)\cdots(n-r+1).$$

Proof

We are concerned with the number of ways of filling r positions with n distinct objects. Applying the extension of the mn rule, the first object can be chosen in one of n ways. After choosing the first, the second can be chosen in $(n-1)$ ways, the third in $(n-2)$ ways, and the rth in $(n-r+1)$ ways. Hence the total number of ways is

$$P_r^n = n(n-1)(n-2)\cdots(n-r+1).$$

Expressed in terms of factorials,

$$P_r^n = \frac{n!}{(n-r)!}.$$

[You should recall that $n! = n(n-1)(n-2)\cdots 3 \cdot 2 \cdot 1$ and $0! = 1$. Thus $4! = 4 \cdot 3 \cdot 2 \cdot 1 = 24$.]

A Counting Rule for Permutations

The number of ways that you can arrange n distinct objects taking them r at a time is

$$P_r^n = n(n-1)(n-2)\cdots(n-r+1)$$

$$= \frac{n!}{(n-r)!}$$

where $n! = n(n-1)(n-2)\cdots(3)(2)(1)$ and $0! = 1$.

Example 4.23

Three lottery tickets are drawn from a total of 50. Assume that order is of importance. How many sample points are associated with the experiment?

Solution

The total number of sample points is

$$P_3^{50} = \frac{50!}{47!} = 50(49)(48) = 117,600.$$

Example 4.24

A piece of equipment is composed of five parts which may be assembled in any order. A test is to be conducted to determine the length of time necessary for each order of assembly. If each order is to be tested once, how many tests must be conducted?

Solution

The total number of tests would equal

$$P_5^5 = \frac{5!}{0!} = 5(4)(3)(2)(1) = 120.$$

The enumeration of the permutations of books in the previous discussion was performed in a systematic manner, first writing the combinations of n books taken r at a time and then writing the rearrangements of each combination. In many situations, ordering is unimportant and we are interested solely in the number of possible combinations. For instance, suppose than an experiment involves the selection of 5 men, a committee, from a total of 20 candidates. Then the simple events associated with this experiment correspond to the different combinations of men selected from the group of 20. How many simple events (different combinations) are associated with this experiment? Since order in a single selection is unimportant, permutations are irrelevant. Thus we are interested in the number of combinations of $n = 20$ things taken $r = 5$ at a time.

Definition

The **number of combinations** of n objects taken r at a time will be denoted by the symbol, C_r^n. [*Note:* Some authors prefer the symbol $\binom{n}{r}$.]

Theorem 4.3

$$C_r^n = \frac{P_r^n}{r!} = \frac{n!}{r!(n-r)!}.$$

Proof

The number of combinations of n objects taken r at a time is apparently related to the number of permutations since it was used in enumerating P_r^n. The relationship can be developed using the mn rule. (We shall use the symbols a and b instead of m and n since n is used in P_r^n and C_r^n.)

Let $a = C_r^n$ and let b equal the number of ways of rearranging each combination once chosen, or P_r^r. Then,

$$P_r^n = ab$$
$$= (C_r^n)(b).$$

Note that $b = P_r^r = r!$ Therefore,

$$P_r^n = C_r^n(r!)$$

or

$$C_r^n = \frac{P_r^n}{r!} = \frac{n!}{r!(n-r)!}.$$

A Counting Rule for Combinations

The number of distinct combinations of n distinct objects that can be formed, taking them r at a time, is

$$C_r^n = \frac{n!}{r!(n-r)!}.$$

Example 4.25

A radio tube may be purchased from five suppliers. In how many ways can three suppliers be chosen from the five?

Solution

$$C_3^5 = \frac{5!}{3!2!} = \frac{(5)(4)}{2} = 10.$$

The following example illustrates the use of the counting rules in the solution of a probability problem.

Example 4.26

Five manufacturers, of varying but unknown quality, produce a certain type of electronic tube. If we were to select three manufacturers at random, what is the chance that the selection would contain exactly two of the best three?

Solution   Without enumerating the sample points, we would likely agree that each point, that is, any combination of three, would be assigned equal probability. If N points are in S, then each point receives probability

$$P(E_i) = \frac{1}{N}.$$

Let n be the number of points in which two of the best three manufacturers are selected. Then the probability of including two of the best three manufacturers in a selection of three is

$$P = \frac{n}{N}.$$

Our problem is to use the counting rules to find n and N.

Since order within a selection is unimportant and is unrecorded, each selection is a combination and hence

$$N = C_3^5 = \frac{5!}{3!2!} = 10.$$

Determination of n is more difficult, but it can be obtained using the mn rule. Let a be the number of ways of selecting exactly two from the best three, or

$$C_2^3 = \frac{3!}{2!1!} = 3,$$

and let b be the number of ways of choosing the remaining manufacturer from the two poorest, or

$$C_1^2 = \frac{2!}{1!1!} = 2.$$

Then the total number of ways of choosing two of the best three in a selection of three is $n = ab = 6$.

Hence the probability, P, is equal to

$$P = \frac{6}{10}.$$

Many other counting rules are available in addition to the three presented in this section. If you are interested in this topic, you should consult one of the many texts on combinatorial mathematics.

Tips on Problem Solving

Many students have difficulty deciding which (if any) of the three counting rules to apply in a given problem. The following tips may help.

1. Look at the problem and note whether a simple event is formed by:
    a. selecting elements from each of *two* (*or more*) sets (a situation which suggests the use of the *mn* rule), or
    b. selecting *r* elements from a single set of *n* elements (a situation which suggests the use of either combinations or permutations).

2. If the situation is 1(b), you must decide whether you should use combinations or permutations. If every different ordering of the elements in the group of *r* leads to a different simple event, then use permutations. If ordering does not produce a new simple event, then use combinations. So, your diagnostic thought process should perform the checks indicated in the decision tree shown in the accompanying diagram.

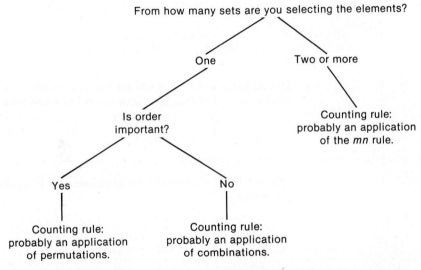

3. If you have difficulty visualizing the appropriate counting rule (or rules) to use for a problem involving a large number of sample points, construct a miniature version of the problem so that you can manually count them. This may help you to see how the more complex version can be solved.

Exercises
Skill Level

Use the Tips on Problem Solving to help you diagnose and solve the following problems.

4.48. Suppose you have six poker chips numbered 1 through 6. If you split the chips into two groups, the first containing chips 1, 2, and 3, and the second containing those numbered 4, 5, and 6, in how many ways can you select two chips, one from each group?

a. State which of the diagnostic tips identify the counting rule to be employed?

b. Use the counting rule to solve the problem.

c. Solve the problem by actually listing the different pairs of chips.

4.49. Suppose you have six poker chips numbered 1 through 6 and you select two chips from among the six. How many different pairs of chips can you select? To answer this question:

a. state which of the diagnostic tips identify the counting rule to be employed,

b. use the counting rule to solve the problem,

c. list the pairs of chips that could be selected.

4.50. Suppose you have six poker chips numbered 1 through 6. How many two-digit numbers can be formed using the numbers on the chips? To answer this question:

a. state which of the diagnostic tips identify the counting rule to be employed,

b. use the counting rule to solve the problem,

c. list the complete set of two-digit numbers that could be formed from the numbers on the chips.

4.51. A salesperson in New York is preparing an itinerary to visit six major cities. The distance traveled, and hence the cost of the trip, will depend upon which city is visited first, second, . . . , sixth. How many different itineraries (and hence trip costs) are possible?

4.52. A company wishes to fill three vice-presidential positions—sales, manufacturing, and finance—from among 24 lower-echelon company managers. How many different options do they have for filling the positions?

4.53. An experiment consists in assigning ten workmen to ten different jobs. In how many different ways can the ten men be assigned to the ten jobs?

4.54. Refer to Exercise 4.53. Suppose there are only four different jobs available for the ten men. In how many different ways can four men be selected from the ten and assigned to the four jobs?

4.55. Probability played a role in the rigging of the April 24, 1980, Pennsylvania state lottery (*Los Angeles Times*, September 8, 1980). To determine each digit of the three-digit winning number, each of the numbers 0, 1, 2, . . . , 9 is placed on a ping pong ball; the ten balls are blown into a compartment; and the number selected for the digit is the one on the ball that floats to the top of the machine. To alter the odds, the conspirators injected a liquid into all balls used in the game except those numbered 4 and 6, making it almost certain that the lighter balls would be selected and determine the digits in the winning number. They then proceeded to buy lottery tickets bearing the potential winning numbers. How many potential winning numbers were there (666 was the eventual winner)?

4.56. Refer to Exercise 4.55. Hours after the rigging of the Pennsylvania state lottery was announced, Connecticut state lottery officials were stunned to learn that their winning number for the day was 666 (*Los Angeles Times*, September 21, 1980).

a. All evidence indicates that the Connecticut selection of 666 was due to pure chance. What is the probability that a 666 would be drawn in Connecticut, given that a 666 had been selected in the April 24, 1980, Pennsylvania lottery?

b. What is the probability of drawing a 666 in the April 24, 1980, Pennsylvania lottery (remember, this drawing was rigged) *and* a 666 on the September 19, 1980, Connecticut lottery?

4.57. A study is to be conducted to determine the attitudes of nurses in a hospital to various administrative procedures that are currently employed. If a sample of ten nurses is to be selected from a total of 90, how many different samples could be selected? (Note that order within a sample is unimportant.)

4.58. Buyers of television sets are offered a choice of one of three different styles. How many different outcomes could result if one customer makes a selection? two customers? ten?

4.59. If five cards are to be selected, one after the other, in sequence, from a 52-card deck, each card being replaced in the deck before the next draw, how many different selections are possible?

4.60. Refer to Exercise 4.59. Suppose that the five cards are drawn from the 52-card deck, simultaneously and without replacement. How many different hands could be selected?

4.61. The following actual case occurred in the city of Gainesville, Florida, in 1976. The eight-member Human Relations Advisory Board considered the complaint of a woman who claimed discrimination, based upon her sex, on the part of a local surveying company. The Board, composed of five women and three men, voted 5–3 in favor of the plaintiff, the five women voting in favor of the plantiff, the three men against. The attorney representing the company appealed the Board's decision by claiming sex bias on the part of the Board members. If the vote in favor of the plaintiff was 5–3 and the Board members were not sex-biased, what is the probability that the vote would split along sex lines (five women for, three men against)?

4.62. Ten thousand tickets, each worth $5.00, are issued for a lottery.

a. If you purchase two tickets, how many different pairs of tickets are available?

b. Suppose that the lottery operators choose four tickets from the 10,000 and that each is judged a winner. What is the probability that both of your tickets will be winners?

4.63. For each of twenty questions on a multiple-choice test, a student can choose one of five possible answers.

a. How many completely different sets of answers are possible for the test?

b. If a person guessed on all of the questions, what is the probability that all of the questions would be answered correctly?

4.64. In a psychological learning experiment, a mouse is given the option of choosing one of five paths, two of which are expected to be more attractive to the mouse than the others.

a. If two mice are chosen for the experiment, how many different simple events are associated with the experiment?

b. If no learning has occurred and it is equally likely that a mouse would choose any one of the paths, what is the probability that both mice would choose one of the two "attractive" paths?

4.65. Refer to Exercise 4.64. Suppose that ten mice were employed in the experiment, that no learning had occurred, and hence that it is equally likely that a mouse would choose any one of the paths.

a. What is the probability that all ten mice would choose one of the "attractive" paths?

b. Noting the result of part (a), do you think learning occurred or do you think the observed outcome occurred purely due to chance?

4.66. A lineup of ten men is conducted to test the ability of a witness to identify three burglary suspects. Suppose that the three burglary suspects who committed the crime are in the lineup. If the witness is actually unable to identify the suspects but feels compelled to make a choice, what is the probability that the three guilty men are selected by chance? What is the probability that the witness selects three innocent men?

4.67. The sizes of wildlife populations are often estimated using the capture–recapture method.

For example, suppose that a forest contains an unknown number, N, of deer. A sample of K is caught, marked, and released. Then the capture process is repeated and the number of marked animals in the sample is observed. The reasoning is that the proportion of marked animals in the second sample must be related to N, the size of the deer population. To illustrate how N is related to the recapture probabilities (we shall use small numbers to simplify our calculations), suppose that 4 deer were originally captured, marked, and released. After a short period of time, five deer were again captured and it was found that one was marked. Calculate the probability of recapturing exactly one marked deer if the size of the deer population was equal to:

 a. $N = 8$.   b. $N = 10$.   c. $N = 15$.   d. $N = 20$.   e. $N = 25$.

Graph the recapture probabilities as a function of N.

4.68.   To understand how a single elimination tennis tournament works, we will illustrate the matchings for $n = 8$ players. The players are assigned to eight positions. Player No. 1 plays No. 2, No. 3 plays No. 4, . . . , and player No. 7 plays No. 8. Then the winner of the No. 1–No. 2 pair plays the winner of No. 3–No. 4 and the winner of the No. 5–No. 6 pair plays the winner of No. 7–No. 8 in the semifinal round. The winners of these two matches meet in the final match to decide the winner. Jones and Smith, the two best players, can definitely defeat any of the other players. If the eight players are randomly assigned to the eight starting positions, what is the probability that Jones and Smith meet in the finals? (Does it seem as though the answer should be 1/2?) (*Hint:* You can obtain insight into this problem by solving the problem for a single elimination tournament containing $n = 4$ players. This will enable you to list the simple events.)

4.69.   If you are selling a radio or televsion station, it is possible to obtain a substantial tax break if you sell to someone classified as a minority buyer. *Fortune* magazine (August 10, 1981) reports that the Storer Broadcasting Company recently qualified for this tax break following the sale of an AM radio station. One of the arguments justifying the minority status of the buyers by the Federal Communications Commission was that the buyers—a family—were Hispanic by virtue of the fact that they were descendents of Spanish Jews who were expelled from Spain in 1492. It is interesting to imagine the size of a "minority" group that might have descended from just one Spaniard (let alone 10 or 20) since 1492. Figuring 20 years to a generation, there would have been approximately 24 generations since 1492. If one original Spaniard married in 1492, and had two children, and the children and every descendent thereafter married and had two children, how many descendents would the original Spaniard have?

4.11
Summary

The theories of both probability and statistics are concerned with samples drawn from populations. Probability assumes the population known and calculates the probability of observing a particular sample. Statistics assumes the sample to be known and, with the aid of probability, attempts to describe the frequency distribution of the population that is unknown. Chapter 4 is directed toward the construction of a model, the sample space, for the frequency distribution of a population of observations. The theoretical frequencies, representing probabilities of events, can be obtained by one of two methods:

1. The summation of the probabilities of the sample points in the event of interest.
2. The joint use of event composition (compound events) and the laws of probability.

References

Cramer, H. *The Elements of Probability Theory and Some of its Applications*, 2nd ed. Melbourne, Fla.: R.E. Krieger Publishing Co., 1973.

Crichton, Micheal. *Congo*. New York: Alfred A. Knopf, 1980.

Feller, W., *An Introduction to Probability Theory and Its Applications*, Vol. 1, 3rd ed. New York: John Wiley & Sons, Inc., 1968.

Meyer, P.L., *Introductory Probability and Statistical Applications*, 2nd ed. Reading, Mass.: Addison-Wesley Publishing Company, Inc., 1970.

Parzen, E., *Modern Probability Theory and Its Applications*. New York: John Wiley & Sons, Inc., 1960.

Riordan, J., *An Introduction to Combinatorial Analysis*. New York: John Wiley & Sons, Inc., 1958.

Scheaffer, R.L., and W. Mendenhall, *Introduction to Probability: Theory and Applications*. Boston: Duxbury Press, 1975.

Supplementary Exercises

[Starred (*) exercises are optional]

4.70. Solve Exercise 4.68 for the special case when $n = 4$ by listing all sample points in S (i.e., by using the sample point approach).

4.71. Two cold tablets are accidentally placed in a box containing two aspirin tablets. The four tablets are identical in appearance. One tablet is selected at random from the box and is swallowed by patient A. A tablet is then selected at random from the three remaining tablets and is swallowed by patient B. Define the following events as specific collections of sample points:

a. The sample space S.

b. The event A that patient A obtained a cold tablet.

c. The event B that exactly one of the two patients obtained a cold tablet.

d. The event C that neither patient obtained a cold tablet.

4.72. A coin is tossed four times and the outcome is recorded for each toss.

a. List the sample points for the experiment.

b. Let A be the event that the experiment yields exactly three heads. List the sample points in A.

c. Make a reasonable assignment of probabilities to the sample points and find $P(A)$.

4.73. A retailer sells two styles of high-priced high-fidelity consoles that experience indicates are in equal demand. (Fifty percent of all potential customers prefer style 1, and 50 percent favor style 2.) If the retailer stocks four of each, what is the probability that the first four customers seeking a console all purchase the same style?

a. Define the experiment.

b. List the sample points.

c. Define the event of interest, A, as a specific collection of sample points.

d. Assign probabilities to the sample points and find $P(A)$.

(Note that this example is illustrative of a very important problem associated with product inventory.)

4.74. A boxcar contains seven complex electronic systems. Unknown to the purchaser, three are defective. Two are selected out of the seven for thorough testing and then classified as defective or nondefective.

a. List the sample points for this experiment.

b. Let A be the event that the selection includes no defectives. List the sample points in A.

c. Assign probabilities to the sample points and find $P(A)$.

4.75.   A die is tossed two times; what is the probability that the sum of the numbers observed will be greater than 9?

4.76.   Toss a die and a coin. If event A is the occurrence of a head and an even number, and event B is the occurrence of a head and a 1, find $P(A)$, $P(B)$, $P(AB)$, and $P(A \cup B)$. (Solve by listing the sample points.)

4.77.   A man takes either a bus or the subway to work with probabilities .3 and .7, respectively. When he takes the bus, he is late 30 percent of the days. When he takes the subway, he is late 20 percent of the days. If the man is late for work on a particular day, what is the probability that he took the bus?

4.78.   Refer to Exercise 4.71. Find each of the following by summing probabilities of sample points: $P(A)$, $P(B)$, $P(AB)$, $P(A \cup B)$, $P(C)$, $P(AC)$, and $P(A \cup C)$.

4.79.   Refer to Exercise 4.12. Suppose that the latest batch does actually possess the desired taste so that the three specimens for each taster are identical. Use the Multiplicative Law of Probability to find the probabilities attached to the sample points. Find each of the following by summing probabilities of sample points: $P(A)$, $P(B)$, $P(AB)$, $P(A \cup B)$, $P(C)$, $P(AC)$, and $P(A \cup C)$.

4.80.   Suppose that independent events A and B have nonzero probabilities. Show that A and B cannot be mutually exclusive.

4.81.   Refer to Exercise 4.76 and calculate $P(AB)$ and $P(A \cup B)$; use the laws of probability.

4.82.   A salesman figures that the probability of his consummating a sale during the first contact with a client is .4 but improves to .55 on the second contact if the client did not buy during the first contact. Suppose that the salesman will make one and only one callback to any client. If the salesman contacts a client, calculate:

a. The probability that the client will buy.

b. The probability that the client will not buy.

4.83.   A lie detector will show a positive reading (indicate a lie) 10 percent of the time when a person is telling the truth and 95 percent of the time when the person is lying. If two people are suspects in a one-man crime and if (for certain) one is guilty:

a. What is the probability that the detector shows a positive reading for both suspects?

b. What is the probability that the detector shows a positive reading for the guilty suspect and a negative reading for the innocent?

c. What is the probability that the detector is completely wrong, that is, that it gives a positive reading for the innocent suspect and a negative reading for the guilty?

d. What is the probability that it gives a positive reading for either or both of the two suspects?

4.84.   According to a recent study, the number of people susceptible to Chinese restaurant syndrome, a reaction linked to the food seasoner, monosodium glutamate, has been greatly exaggerated. A particular researcher reported that in a study of approximately 600 diners, only 3–5 percent reported the sensations commonly associated with the syndrome. Suppose we take 5 percent to be in fact correct. If three unrelated people dine in a particular Chinese restaurant:

a. What is the probability that all three will experience the syndrome?

b. What is the probability that only one of the three will experience it?

c. What is the probability that at least one of the three will experience it?

d. And if all three actually experienced the syndrome, what would you infer? (*Note:* We are not asking for a statistical inference. We only want your reaction to this event and the reasoning behind your statements.)

4.85.   Experience has shown that 50 percent of the time, a particular union-management contract negotiation had led to a contract settlement within a 2-week period, 60 percent of the time the union strike fund has been adequate to support a strike, and 30 percent of the time both conditions have been satisfied. What is the probability of a contract settlement given that you know that the union strike fund is adequate to support a strike? Is settlement of a contract within a 2-week period dependent on whether the union strike fund is adequate to support a strike?

4.86.   Suppose it is known that at a particular company, the probability of remaining with the company 10 years or more is 1/6. A man and a woman start work at the company on the same day.

a. What is the probability that the man will work there less than 10 years?

b. What is the probability that both the man and woman will work there less than 10 years? Assume that they are unrelated and hence their lengths of service are independent of each other.

c. What is the probability that one or the other, or both, will work longer than 10 years?

4.87.   The failure rate for a guided missile control system is 1 in 1000. Suppose that a duplicate but completely independent control system is installed in each missile so that if the first fails, the second can still take over. The reliability of a missile is the probability that it does not fail. What is the reliability of the modified missile?

4.88.   Consider the following fictitious problem. Suppose that it is known that at a particular supermarket, the probability of waiting 5 minutes or longer for checkout at the cashier's counter is .2. On a given day, a man and his wife decide to shop individually at the market, each checking out separately at different cashier counters. If they both reach cashier counters at the same time, answer the following questions:

a. What is the probability that the man will wait less than 5 minutes for checkout?

b. What is the probability that both the man and his wife will be checked out in less than 5 minutes? Assume that the checkout times for the two are independent events.

c. What is the probability that one or the other, or both, will wait 5 minutes or more?

4.89.   A quality-control plan calls for accepting a large lot of crankshaft bearings if a sample of seven is drawn and none is defective. What is the probability of accepting the lot if none in the lot is defective? If 1/10 is defective? If 1/2 is defective?

4.90.   It is said that only 40 percent of all people in a community favor the development of a mass-transit system. If four citizens are selected at random from the community, what is the probability that all four favor the mass-transit system? That none favor the mass-transit system?

4.91.   The weatherman forecasts rain with probability .6 today and .4 tomorrow. Experience has shown that in this particular locale, it rains one day in four and the probability of rain on two successive days is .15. If it is raining outside when we hear his forecast (as so often happens), what is the probability of rain tomorrow?

4.92.   A research physican compared the effectiveness of two blood-pressure drugs, *A* and *B*, by administering the two drugs to each of four pairs of identical twins. Drug *A* was given to one member of a pair, drug *B* to the other. If, in fact, there is no difference in the effect of the drugs, what is the probability that the drop in the blood-pressure reading for drug A would exceed the corresponding reading for drug B for all four pairs of twins? Suppose that drug B created a greater drop in blood pressure than drug A for each of the four pairs of twins. Do you think this provides sufficient evidence to indicate that drug B is more effective in lowering blood pressure than drug A?

4.93. To reduce the cost of detecting a disease, blood tests are conducted on a pooled sample of blood collected from a group of n people. If no indication of the disease is present in the pooled blood sample (as is usually the case), none have the disease. If analysis of the pooled blood sample indicates that the disease is present, each individual must submit to a blood test. The individual tests are conducted in sequence. If among a group of five people, one person possesses the disease, what is the probability that it requires six blood tests (including the pooled test) to detect the single diseased person? If two people possess the disease, what is the probability that it requires six tests to locate both diseased people?

4.94. A manufacturer is considering the purchase of transistors to be used in the production of an electronic system. The transistors can be purchased from any of four suppliers and it is assumed that these four products vary in quality. The manufacturer wishes to select two of the four in such a way that she is fairly certain of including at least one of the two best suppliers. Although she plans an experimental program to assist in making the choice, you are asked the following: If the choice were based on no information and the two suppliers were randomly chosen from the four, what is the probability that the selection would include at least one of the two best?

4.95. An accident victim will die unless he or she receives, in the next 10 minutes, an amount of type A Rh-positive blood which can be supplied by a single donor. It requires 2 minutes to "type" a prospective donor's blood and 2 minutes to complete the transfer of blood. A large number of untyped donors are available, and 40 percent of them have type A Rh-positive blood. What is the probability that the accident victim will be saved if there is only one blood-typing kit available?

4.96. How many times should a coin be tossed in order that the probability of observing at least one head be equal to or greater than .9?

4.97. An oil prospector will drill a succession of holes in a given area to find a productive well. The probability that he is successful on a given trial is .2.

a. What is the probability that the third hole drilled is the first that locates a productive well?

b. If his total resources allow the drilling of no more than three holes, what is the probability that he locates a productive well?

4.98. Suppose that two defective refrigerators have been included in a shipment of six refrigerators. The buyer begins to test the six refrigerators one at a time.

a. What is the probability that the last defective refrigerator is found on the fourth test?

b. What is the probability that no more than four refrigerators need be tested before both of the defective refrigerators are located?

c. Given that one of the two defective refrigerators has been located in the first two tests, what is the probability that the remaining defective refrigerator is found in the third or fourth test?

4.99. Two men each toss a coin and obtain a "match." That is, both coins are either heads or tails. If the process is repeated three times, answer the following questions:

a. What is the probability of three matches?

b. What is the probability that all six tosses (three for each man) result in tails?

c. Coin tossing provides a model for many practical experiments. Suppose that the coin tosses represented the answers given by two students for three specific true–false questions on an examination. If the two students gave three matches for answers, would the low probability found in (a) suggest collusion?

* 4.100. Two dice are tossed. Use the *mn* rule to count the total number of sample points in the sample space, S.

* 4.101. Refer to the experiment in Exercise 4.76 and use the *mn* rule to count the number of points in S.

* 4.102. How many different telephone numbers of five digits can be formed if the first digit must be a 3 or a 4?

* 4.103. Prove that $C_r^n = C_{n-r}^n$.

* 4.104. A piece of equipment can be assembled in three operations, which may be arranged in any sequence.

a. Give the total number of ways that the equipment can be assembled.

b. Comparative tests are to be conducted to determine the best assembly procedure. If each assembly procedure is to be tested and compared with every other procedure exactly once, how many tests must be conducted?

* 4.105. A chemist wishes to observe the effect of temperature, pressure, and the amount of catalyst on the yield of a particular chemical in a chemical reaction. If the experimenter chooses to use two levels of temperature, three of pressure, and two of catalyst, how many experiments must be conducted in order to run each temperature–pressure–catalyst combination exactly once?

* 4.106. How many ways can the letters in the word "charm" be arranged? in the word "church"?

* 4.107. If there are 18 players on a baseball team, how many different 9-man teams can be organized if each player can plan any position?

* 4.108. An airline has six flights from New York to California and seven flights from California to Hawaii per day. How many different flight arrangements can the airline offer from New York to Hawaii?

* 4.109. Solve Example 4.15 using the sample point approach.

* 4.110. Five cards are drawn from an ordinary 52-card bridge deck. Given two events A and B as follows:

A: all five cards are spades,

B: the five cards include an ace, king, queen, jack, and ten, all of the *same* suit.

a. Define the event AB.

b. Define the event $A \cup B$.

c. Give the probability of the event A.

d. Give the probability of the event B.

e. Give the probability of the event AB.

f. Give the probability of the event $(A \cup B)$.

[Solve parts (c) to (e) by summing probabilities of the sample points.]

* 4.111. Refer to Exercise 4.110. Find $P(B|A)$. Then calculate $P(AB)$ and $P(A \cup B)$ by the use of the laws of probability.

* 4.112. An electronic fuse is produced by five production lines in a manufacturing operation. The fuses are costly, are quite reliable, and are shipped to suppliers in 100-unit lots. Because testing is destructive, most buyers of the fuses test only a small number before deciding to accept or reject lots of incoming fuses. All five production lines produce fuses at the same rate and normally produce only 2 percent defective fuses which are randomly dispersed in the output. Unfortunately, production line 1 suffered mechanical difficulty and produced 5 percent defectives during the month of March. This situation became known to the manufacturer after the fuses had been shipped. A customer received a lot produced in March and tested three fuses. One failed. What is the probability that the lot came from one of the four other lines?

* 4.113. A student prepares for an exam by studying a list of ten problems. He can solve a certain six of these. For the exam, the instructor selects five questions at random from the list of ten. What is the probability that the student can solve all five problems on the exam?

* 4.114. Five homes and three stores have applied for telephones and only one three-party line is available. Parties are assigned to the available three-party line at random. Let the events A and B be defined as follows:

> A: the line contains two homes and one store,
> B: the line contains at least one home.

a. Find $P(A)$.   b. Find $P(B)$.

* 4.115. An employer plans to interview ten men for possible employment. Two people are to be hired. Five of the men are affiliated with fraternities and five are not.

a. In how many ways could the employer select two men, disregarding fraternal affiliations?

b. In how many ways could he select two fraternity men?

c. Assuming that the employer has no preference regarding fraternal affiliations and that the men have equal qualifications, what is the probability that two fraternity men will be hired?

* 4.116. In a sixteen-player single elimination tennis tournament, the players are assigned to sixteen positions. The player in the first position (call this player No. 1), plays player No. 2, player No. 3 plays No. 4, . . . , and player No. 15 plays No. 16. In the second round of the tournament, the winner of the first pair (No. 1 and No. 2) plays the winner of the second pair (No. 3 and No. 4), and so on. The second-round winners play the pair adjacent to them in the two semifinal matches and the winners of these matches play in the final match. If the sixteen players are randomly assigned to the sixteen positions and if the two best players will defeat any of the remaining fourteen, what is the probability that they will play each other in the final match? What is the probability a 32-player single elimination match? (This exercise is an extension of Exercise 4.68.)

* 4.117. A monkey is given 12 blocks—three shaped like squares, three like rectangles, three like triangles, and three like circles. If he draws three of each kind in order—say, three triangles, then three squares, etc.—would you suspect that the monkey associates identically shaped figures? Calculate the probability of this event.

5. Random Variables and Probability Distributions

Chapter Objectives

General Objective

In Chapter 4 we gave an example to illustrate how probability is used in making inferences about a population based on information contained in a sample, and then we presented the concepts of probability that help us find the probability of experimental outcomes. Because most samples are measurements on random variables, we need to be able to find the probabilities associated with such measurements. The objective of this chapter is to distinguish between two types of random variables and to explain how to find the probability that a random variable will assume specific values.

Specific Objectives

1. To define and give meaning to the expression "random variable." *Section 5.1*
2. To identify the two types of random variables—discrete and continuous—and to present probability models appropriate for each. *Sections 5.2, 5.3, 5.4*
3. To define the expected value of a random variable and to identify this quantity as the mean of its probability distribution. *Section 5.5*
4. To define and identify the variance of a random variable as an expectation and to explain how the expected value and variance of a random variable can be used to describe its probability distribution. *Section 5.5*
5. To note the relationship between random variables, samples, and statistical inference and to define what is meant by random sampling. We will use random sampling and introduce the topic of statistical inference as a by-product of Chapter 6. *Section 5.7*

Case Study

Thyroid Disease and Three Mile Island

Is the Three Mile Island nuclear accident responsible for an increase in hypothyroidism in the vicinity of that nuclear power plant? Hypothyroidism, a condition that results when the thyroid gland is either absent or malfunctioning, can lead to mental retardation and stunted growth unless treated quickly. The *Orlando Sentinel Star* (February 24, 1980) reports that most leading scientists believe that there is little probability that low-level radiation caused the cases, but the facts surrounding their occurrence suggest that some external and unusual causative factor is responsible.

The *Sentinel* article notes that "11 babies were born with hypothyroidism in three counties near the nuclear plant. All were in gestation March 28, 1979, the day of the nuclear accident. Ordinarily, only about three cases would be expected." If the mean number of cases expected within this three-county area is really 3, is it likely that the number of babies born with hypothyroidism would, by pure chance, reach a value as large as 11?

You will note that observation of the number of hypothyroid births within the three-county region (presumably within some specific period of time) is an experiment that produces events that are numbers. Thus, observing four cases within the region is one event; observing two cases is another. In fact, the complete set of events that you could observe corresponds to the values that the variable (number of hypothyroid births) can assume, namely 0, 1, 2, 3, ...

In this chapter we will learn more about experiments that yield numerical events, numbers that correspond to the values of some variable. We will apply the techniques of Chapter 4 to find the the probabilities associated with these events, and we will use this knowledge in Section 5.6 to analyze the evidence presented in the Thyroid Disease and Three Mile Island Case Study.

5.1
Random Variables

Observations generated by an experiment fall into one of two categories, quantitative or qualitative. For example, the daily production in a manufacturing plant would be a quantitative or numerically measurable observation, while weather descriptions, such as rainy, cloudy, sunny, would be qualitative. Statisticians are concerned with both quantitative and qualitative data, although the former are perhaps more common.

In some instances it is possible to convert qualitative data to quantitative by assigning a numerical value to each category to form a scale. Industrial production is often scaled according to first grade, second grade, etc. Cigarette tobaccos are graded and foods ranked according to preference by persons employed as taste testers.

In this text we shall be concerned primarily with quantitative observations such as the grade-point data discussed in Example 3.1. The events of interest associated with the experiment are the values that the data may take and are, therefore, numerical events. Suppose that the variable measured in the experiment is denoted by the symbol y. Recalling that an event is a collection of sample points, it would seem that certain sample points would be associated with one numerical event, say $y = 2$, another set associated with $y = 3$, etc., covering all possible values of y. Such is, in fact, the case. The variable, y, is called a random variable because the value of y observed for a particular experiment is a chance (or random) event associated with a probabilistic model, the sample space.

Note that each sample point implies one and only one value of y, although many sample points may imply the same value of y. For example, consider the experiment that consists of tossing two coins. Suppose that we are interested in y = number of heads. The random variable, y, can take three values, $y = 0$, $y = 1$, or $y = 2$. The sample points for the experiment are

$$E_1: \quad HH,$$
$$E_2: \quad HT,$$
$$E_3: \quad TH,$$
$$E_4: \quad TT.$$

The numerical event $y = 0$ includes sample point E_4; $y = 1$ includes sample points E_2 and E_3; and $y = 2$ includes sample point E_1. Note that one and only one value of y corresponds to each sample point, but the converse is not true. Two sample points, E_2 and E_3, are associated with the same value of y. Two conclusions may be drawn from this example. We note that the relationship between the random variable, y, and the sample points satisfies the definition of a functional relation presented in Chapter 2. Choose any point in the sample space, S, and there corresponds one and only one value of y. Hence, we choose the following definition for a random variable.

Definition

A random variable is a numerically valued function defined over a sample space.

You will recall (Chapter 3) that populations are described by frequency distributions. Areas under the frequency distribution are associated with the fraction of measurements in the population falling in a particular interval, or they may be interpreted as probabilities. The purpose of Chapter 4 was to construct a theoretical model for the frequency distribution, or probability distribution as it is called in probability theory, for the population. The preceding sections of Chapter 4 provide both the model and the machinery for achieving this result. To complete the picture, we need to calculate the probabilities associated with each value of *y*. These probabilities, presented in the form of a table or a formula, are called the *probability distribution* for the random variable, *y*. In reality, the probability distribution is the theoretical frequency distribution for the population. Random variables and probability distributions form the topic of Chapter 5.

5.2
A Classification of Random Variables

All the experiments described in Chapter 4 had one characteristic in common: their sample spaces contained a finite or at least countable number of sample points.* This enabled us to assign probabilities to the sample points so that the sum of their respective probabilities was equal to 1. Not all experiments possess this characteristic. Measuring the length of time for a sprinter to run the 100-yard dash is an experiment that can yield an infinitely large number of simple events (and sample points) that cannot be counted. Theoretically, with measuring equipment of perfect accuracy, we could associate each possible time measurement with a unique point contained in a line interval. Thus each of the infinitely large number of points on a line interval is a possible time for a sprinter and represents a sample point for the experiment. Since each time point represents a sample point for the experiment, the resulting sample space contains an infinitely large (and uncountable) number of sample points. Can we apportion nonzero probabilities to an infinite number of points and, at the same time, satisfy the requirement that the probabilities of the sample points sum to 1? Oddly enough, the answer is yes, but only when the infinite number of points can be counted. An example of this situation will be given in Section 5.3. In all other cases, the answer is no. This leads us to define two different types of random variables and probability models for each.

Random variables are classified as one of two types: *discrete* or *continuous*.

Definition

A **discrete random variable** is one that can assume a countable number of values.

A discrete random variable is easily identified by examining the number of values it may assume. If the number of values that the random variable may assume can be counted, it must be discrete.

Typical examples of discrete random variables are:

1. The number of defective bolts in a sample of ten drawn from industrial production.
2. The number of rural electrified homes in a township.
3. The number of malfunctions of an airplane engine over a period of time.
4. The number of people in the waiting line in a doctor's office.

---

* Countable means that you can associate the values that the random variable can assume with the integers 1, 2, 3, 4, . . . (that is, you can count them).

> **Definition**
>
> A **continuous random variable** is one that can assume the infinitely large number of values corresponding to the points on a line interval.

The word "continuous," an adjective, means proceeding without interruption. It, in itself, provides the key for identifying continuous random variables. Look for a measurement with a set of values that form points on a line with no interruptions or intervening spaces between them.

Typical examples of continuous random variables are:

1. The height of a human.
2. The length of life of a human cell.
3. The amount of sugar in an orange.
4. The length of time required to complete an assembly operation in a manufacturing process.

The distinction between discrete and continuous random variables is an important one since different probability models are required for each. Accordingly, the probability distribution for discrete and continuous random variables will be discussed in Sections 5.3 and 5.4, respectively.

Exercises
Conceptual Level

5.1. Identify the following as discrete or continuous random variables:
   a. The number of new clients acquired by a law firm in a month.
   b. The shelf life of a particular drug.
   c. The weight of a package.
   d. The velocity of a pitched baseball.
   e. The number of fatal automobile accidents at a given intersection in a 24-hour period.

5.2. Identify the following as discrete or continous random variables:
   a. The height of water in a dam.
   b. The amount of money awarded a plaintiff by a court in a damage suit.
   c. The number of people waiting for treatment at a hospital emergency room.
   d. The total points scored in a football game.
   e. The number of claims received by an insurance company during a day.

5.3. Identify the following as discrete or continous random variables.
   a. The number of incoming telephone calls at a telephone switchboard during a 5-minute interval.

    b. The amount of rainfall in Gainesville, Florida, for a 1-week period.

    c. The length of time for an automobile driver to respond when faced with an impending collision.

    d. The number of aircraft near-collisions observed by an air controller over a 24-hour period.

    e. The bacteria count per cubic centimeter in your drinking water.

5.4.   Identify the following as discrete or continuous random variables:

    a. The increase in length of life achieved by a cancer patient as a result of surgery.

    b. The tensile breaking strength, in pounds per square inch, of 1-inch-diameter steel cable.

    c. The number of deer killed per year in a state wildlife preserve.

    d. The number of overdue accounts in a department store at a particular point in time.

    e. Your blood pressure.

5.3 Probability Distributions for Discrete Random Variables

The probability distribution for a discrete random variable is a formula, table, or graph that provides the probability associated with each value of the random variable. Since each value of the variable y is a numerical event, we may apply the methods of Chapter 4 to obtain the appropriate probabilities. It is interesting to note that the events cannot overlap because one and only one value of y is assigned to each sample point, and hence that the values of y represent mutually exclusive numerical events. Summing $p(y)$ over all values of y would equal the sum of the probabilities of all sample points and, hence, equal 1. We may therefore state two requirements for a probability distribution:

Requirements of a Probability Distribution

1. $0 \leq p(y) \leq 1.$
2. $\sum_{\text{all } y} p(y) = 1.$

Example 5.1

Consider an experiment that consists of tossing two coins and let y equal the number of heads observed. Find the probability distribution for y.

Solution

The sample points for this experiment with their respective probabilities are as follows:

Sample Point	Coin 1	Coin 2	$P(E_i)$	y
E_1	H	H	1/4	2
E_2	H	T	1/4	1
E_3	T	H	1/4	1
E_4	T	T	1/4	0

Because sample point E_1 is associated with the simple event, "observe a head on coin 1 and a head on coin 2," we assign it the value $y = 2$. Similarly, we assign $y = 1$ to point E_2, etc. The probability of each value of y may be calculated by adding the probabilities of the sample points in that numerical event. The numerical event $y = 0$ contains one sample point, E_4; $y = 1$ contains two sample points, E_2 and E_3; and $y = 2$ contains one point, E_1. The values of y with respective probabilities are given in Table 5.1. Observe that $\sum_{y=0}^{2} p(y) = 1$.

Table 5.1 Probability distribution for y (y = number of heads)

y	Sample Points in y	$P(y)$
0	E_4	1/4
1	E_2, E_3	1/2
2	E_1	1/4
	$\sum_{y=0}^{2} p(y) = 1$	

The probability distribution, Table 5.1, can be presented graphically in the form of the relative frequency histogram that was discussed in Section 3.1*. The histogram for the random variable y would contain three classes, corresponding to $y = 0$, $y = 1$, and $y = 2$. Since $p(0) = 1/4$, the theoretical relative frequency for $y = 0$ is 1/4; $p(1) = 1/2$, and hence the theoretical frequency for $y = 1$ is 1/2, etc. The histogram is given in Figure 5.1.

Figure 5.1 Probability histogram showing $p(y)$ for Example 5.1

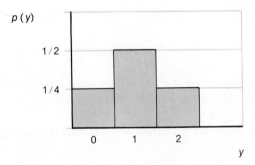

If you were to draw a sample from this population—that is, if you were to throw two balanced coins, say $n = 100$ times, and each time record the number of heads observed, y, and then construct a histogram using the 100 measurements on y—you would find that

* The probability distribution, Table 5.1, can also be presented using a formula. This formula is derived in Section 6.2.

the histogram for your sample would appear very similar to that for $p(y)$, Figure 5.1. If you were to repeat the experiment $n = 1000$ times, the similarity would be much more pronounced.

Example 5.2

Let y equal the number observed on the throw of a single balanced die. Find $p(y)$.

Solution

The sample points for this experiment are given in Table 5.2. You would assign $y = 1$ to E_1, $y = 2$ to E_2, etc. Since each value of y contains only one sample point, $p(y)$, the probability distribution for y would appear as shown in the fifth column of Table 5.2. You can also see that

$$p(y) = 1/6, \qquad y = 1, 2, 3, 4, 5, 6$$

gives the probability distribution as a formula. The corresponding histogram is given in Figure 5.2.

Table 5.2 Tossing a die: probability distribution for y

Sample Point	Number on Upper Face	$P(E_i)$	y	$P(y)$
E_1	1	1/6	1	1/6
E_2	2	1/6	2	1/6
E_3	3	1/6	3	1/6
E_4	4	1/6	4	1/6
E_5	5	1/6	5	1/6
E_6	6	1/6	6	1/6

Figure 5.2 Probability histogram for $p(y) = 1/6$ in Example 5.2

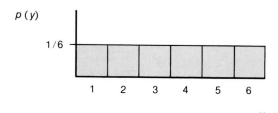

Example 5.3

As a final example, we shall consider a random variable that may assume a countable infinity of values. The experiment consists of tossing a coin until the first head appears. Let y equal the number of tosses. Find the probability distribution for y.

Solution

The sample points, infinite in number, are given below. Let the symbol $TTTH$ represent the results of the first four tosses proceeding from left to right, with H representing a head and T a tail.

$$E_1: \quad H$$
$$E_2: \quad TH$$
$$E_3: \quad TTH$$
$$E_4: \quad TTTH$$
$$E_5: \quad TTTTH$$
$$\vdots \qquad \vdots$$
$$\text{etc.} \quad \text{etc.}$$

You can see that E_{65} would be the sample point associated with the event that each of the first 64 tosses resulted in a tail and the 65th resulted in a head. Conceivably, the experiment might never end. Let us now calculate the probability of each sample point and assign values of y to each of these points. The probability of E_1, a head on the first toss, is 1/2. The probability of E_2, a tail and then a head, is an intersection of two independent events and can be obtained by use of the Multiplicative Law of Probability. Hence, $P(E_2) = (1/2)(1/2)$. Likewise, $P(E_3) = (1/2)(1/2)(1/2) = (1/2)^3$ and it would follow that $P(E_{50}) = (1/2)^{50}$. The appropriate probability distribution is given in the fourth column of Table 5.3, or by the equation $p(y) = (1/2)^y$. The frequency histogram is shown in Figure 5.3. Will $p(y)$ satisfy the second requirement of a probability distribution; that is, will

$$\sum_{y=1}^{\infty} p(y) = 1?$$

Table 5.3 Tossing a coin until the first head appears (y = number of tosses)

Sample Point	$P(E_i)$	y	$P(y)$
E_1	1/2	1	1/2
E_2	1/4	2	1/4
E_3	1/8	3	1/8
E_4	1/16	4	1/16
E_5	1/32	5	1/32
$\vdots$	$\vdots$	$\vdots$	$\vdots$
etc.			

Summing, we obtain

$$\sum_{y=1}^{\infty} p(y) = p(1) + p(2) + p(3) + \cdots$$

$$= 1/2 + 1/4 + 1/8 + \cdots.$$

Figure 5.3 $p(y)$ for Example 5.3

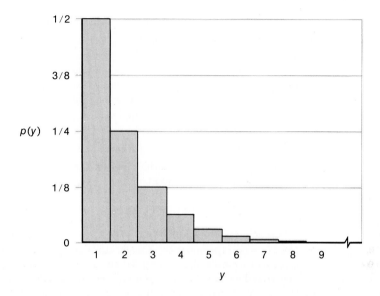

The reader familiar with high school algebra will recognize this as the sum of an infinite geometric progression with common ratio equal to 1/2. A college algebra book or mathematical handbook will verify that this sum is equal to (or, correctly, approaches as a limit) 1. Thus we observe a sample space that contains an infinite number of sample points, each of which is assigned a positive (nonzero) probability, and the sum of the probabilities over the infinite number of sample points is equal to 1. The probability distribution for y is shown in Figure 5.3.

Tips on Problem Solving

To find the probability distribution for a discrete random variable, construct a table listing each value that the random variable y can assume. Then calculate $p(y)$ for each value of y.

Exercises
Conceptual Level

5.5.   A random variable y can assume five values, 0, 1, 2, 3, and 4. A portion of the probability distribution is shown below:

y	0	1	2	3	4
$p(y)$	.1	.3	.3	?	.1

a. Find $p(3)$.

b. Construct a probability histogram for $p(y)$.

c. Simulate the experiment by marking ten poker chips (or coins), one with a 0, three with a 1, three with a 2, etc. Mix the chips thoroughly, draw one, and record the observed value of y. Repeat the process 100 times. Construct a relative frequency histogram for the 100 values of y and compare with the probability histogram, part (b).

5.6.   A jar contains two black and two white balls. If the balls are thoroughly mixed and then two are randomly selected from the jar:

a. List all sample points for this experiment and assign appropriate probabilities to each.

b. Let y equal the number of white balls in the selection. Then assign the appropriate value of y to each sample point.

c. Calculate the values of $p(y)$ and display them in tabular form. Show that $\sum_{y=0}^{2} p(y) = 1$.

d. Construct a probability histogram for $p(y)$.

e. Simulate the experiment by actually drawing two balls (coins, etc.) from a jar that contains two black and two white balls. Repeat the drawing process 100 times, each time recording the value of y that was observed. Construct a relative frequency histogram for the 100 observed values of y and compare with the probability histogram, part (d).

5.7.   A salesman figures that each contact results in a sale with probability equal to .2. During a given day, he contacts two prospective clients and records, for each, whether a sale has been made.

a. List all sample points for this experiment and find their respective probabilities (assume that the contacts represent independent events).

b. Let y equal the number of clients who sign a sales contract. Then assign the appropriate value of y to each sample point.

c. Calculate the values of $p(y)$ and display them in tabular form.

d. Construct a probability histogram for $p(y)$.

5.8.   A key ring contains four office keys that are identical in appearance. Only one will open your office door. Suppose you randomly select one key and try it. If it does not fit, you randomly select one of the three remaining keys. If that does not fit, you randomly select one of the last two. Each different sequence that could occur in selecting the keys represents one of a set of equiprobable simple events.

a. List the sample points in S and assign probabilities to the sample points.

b. Let y equal the number of keys that you try before you find the one that opens the door ($y = 1, 2, 3, 4$). Then assign the appropriate value of y to each sample point.

c. Calculate the values of $p(y)$ and display them in tabular form.

d. Construct a probability histogram for $p(y)$.

Exercises
Skill Level

5.9.   If you toss a pair of dice, the sum T of the number of dots appearing on the upper faces of the dice can assume the value of an integer in the interval $2 \leq T \leq 12$.

a. Find the probability distribution for T and display it in tabular form.

b. Construct a probability histogram for $p(T)$.

5.10. A piece of electronic equipment contains six transistors, two of which are defective. Three transistors are selected at random, removed from the piece of equipment, and inspected. Let y equal the number of defectives observed, where $y = 0, 1$, or 2. Find the probability distribution for y. Express your results graphically as a probability histogram.

5.11. Simulate the experiment described in Exercise 5.10 by marking six marbles (or coins, or cards) so that two represent defectives and four represent nondefectives. Place the marbles in a hat, mix, draw three, and record y, the number of "defectives observed." Replace the marbles and repeat the process until a total of $n = 100$ observations on y have been recorded. Construct a relative frequency histogram for this sample and compare it with with the population probability distribution, Exercise 5.10.

5.12. In order to verify the accuracy of their financial accounts, companies utillize auditors on a regular basis to verify accounting entries. Suppose that the company's employees make erroneous entries 5 percent of the time. If an auditor randomly checks three entries:

a. Find the probability distribution for y, the number of errors detected by the auditor.

b. Construct a probability histogram for $p(y)$.

c. Find the probability that the auditor will detect more than one error.

5.13. Past experience has shown that on the average only one in ten wells drilled hits oil. Let y be the number of drillings until the first success (oil is struck). Assume that the drillings represent independent events.

a. Find $p(1)$, $p(2)$, and $p(3)$.

b. Give a formula for $p(y)$.

c. Graph $p(y)$.

5.14. Two tennis professionals, A and B, are scheduled to play a match in which the winner is determined by the first player to win three sets in a total that cannot exceed five sets. The event that A wins any one set is independent of the event that A wins any other, and the probability that A wins any one set is equal to .6. Let y equal the total number of sets in the match, that is, $y = 3, 4$, or 5. Find $p(y)$.

5.15. A rental agency, which leases heavy equipment by the day, found that an expensive piece of equipment is leased, on the average, only one day in ten. If rental on one day is independent of rental on any other day, find the probability distribution of y, the number of days between a pair of rentals.

5.4
Probability Distributions for Continuous Random Variables

Recall that we cannot assign probabilities to the sample points associated with a continuous random variable and that a completely different population model is required. We invite you to refocus your attention on the discussion of the relative frequency histogram, Section 3.1., and specifically to the histogram for the 30 student grade-point averages, Figure 3.2. If more and more measurements were obtained, we might reduce the width of the class interval. The outline of the histogram would change slightly, for the most part becoming less and less irregular. When the number of measurements becomes very large and the intervals very small, the relative frequency would appear, for all practical purposes, as a smooth curve. (See Figure 5.4.)

The relative frequency associated with a particular class in the population is the fraction of measurements in the population falling in that interval and,

Figure 5.4 A relative frequency histogram for a population

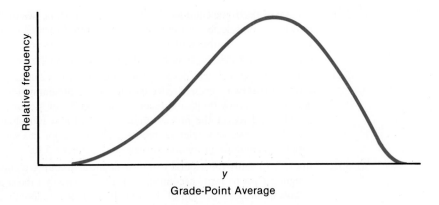

also, is the probability of drawing a measurement in that class. If the total area under the relative frequency histogram were adjusted to equal 1, then areas under the frequency curve would correspond to probabilities. Indeed, this was the basis for the application of the Empirical Rule in Chapter 3.

Let us now construct a model for the probability distribution for a continuous random variable. Assume that the random variable, y, may take values on a real line as in Figure 5.4. We will then distribute one unit of probability along the line much as a person might distribute a handful of sand, each measurement in the population corresponding to a single grain. The probability, grains of sand or measurements, will pile up in certain places, and the result will be a probability distribution, which might appear like the one shown in Figure 5.5. The depth or density of probability, which varies with y, may be represented by a mathematical equation $f(y)$, called the **probability distribution** (or the **probability density function**) for the random variable y. The function, $f(y)$, represented graphically in Figure 5.5, provides a mathematical model for the population relative frequency histogram which exists in reality. The density function, $f(y)$, is defined so that the total area under the curve is equal to 1 and therefore that the area lying above a given interval will equal the probability

Figure 5.5 Probability distribution, $f(y)$

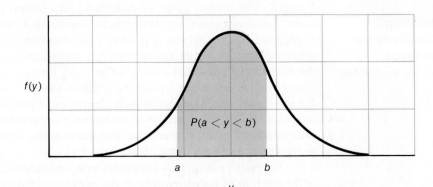

that y will fall in that interval. Thus, the probability that $a < y < b$ (a is less than y and y is less than b) is equal to the shaded area under the density function between the two points a and b.

Two puzzling equations remain. How do we choose the model, that is, the probability distribution, $f(y)$, appropriate for a given physical situation? And then, how can we calculate the area under the curve corresponding to a given interval? The first question is the more difficult because the answer will depend upon the type of measurement involved. To grasp the reasoning better, let us consider the model for the discrete random variable. Using *good judgment*, we assigned realistic probabilities to the sample points associated with the experiment and then, applying the techniques of Chapter 4, derived the probability distribution for the random variable of interest. A similar approach is available for some continuous random variables but it is, unfortunately, beyond the scope of this course. The important thing to note, however, is that good judgment was required in assigning the probabilities to the sample points and that, although inexact, the probabilities were sufficiently accurate for practical purposes. Thus they provided a model for reality. Similarly, for continuous random variables, we utilize all available information and then use our best judgment in choosing the model, $f(y)$. In some cases the approximate distribution may be derived using the theory of mathematical statistics. In others, we may rely on the frequency histograms for samples drawn from the population or histograms for data drawn from similar populations. The form of the sample histogram will often suggest a reasonable choice for $f(y)$.

Once the model, $f(y)$, has been chosen, we may calculate areas under the curve by using the integral calculus or by applying rectangular approximations, or, as a last resort, we could make an approximation of the area by visual inspection. We will leave this problem to the mathematician. The areas associated with most useful probability distributions have been calculated and appear in tabular form. (See the Appendix.)

We conclude with a comment to the "doubting Thomas" concerned because of the lack of agreement between $f(y)$ and the true relative frequency curve for the population. Can useful results be obtained from an inaccurate population model? We need only look to the physicist, the chemist, and the engineer. The equations, formulas, and various numerical expressions used in all the sciences are simply mathematical models which, fortunately, provide good approximations to reality. The engineer uses his equations to determine the size and location of members in a bridge or an airplane wing and is concerned only that the resulting structure perform the task for which it was designed. Similarly, in statistics the attainment of the end result is the criterion by which we measure the worth of a statistical technique. Does the technique provide good inferences, that is, predictions or decisions, concerning the population from which the sample was drawn? Many statistical techniques possess this desirable property, a fact that can be shown experimentally and confirmed by the many extremely useful applications of statistics in industry and in the physical and social sciences.

5.5

**Mathematical
Expectation**

The probability distribution, described in Sections 5.3 and 5.4, provides a model for the theoretical frequency distribution of a random variable and hence must possess a mean, variance, standard deviation, and other descriptive measures associated with the theoretical population that it represents. Recalling that both the mean and variance are averages (Sections 3.3 and 3.4), we shall confine our attention to the problem of calculating the mean value of a random variable defined over a theoretical population. This mean is called the **expected value** of the random variable.

The method for calculating the population mean or expected value of a random variable can be more easily understood by considering an example. Let y equal the number of heads observed in the toss of two coins. For convenience, we give $p(y)$:

y	$p(y)$
0	1/4
1	1/2
2	1/4

Let us suppose that the experiment is repeated a large number of times, say $n = 4,000,000$ times. Intuitively, we would expect to observe approximately 1 million zeros, 2 million ones, and 1 million twos. Then the average value of y would equal

$$\frac{\text{sum of measurements}}{n} = \frac{1,000,000(0) + 2,000,000(0) + 1,000,000(2)}{4,000,000}$$

$$= \frac{1,000,000(0)}{4,000,000} + \frac{2,000,000(1)}{4,000,000} + \frac{1,000,000(2)}{4,000,000}$$

$$= (1/4)(0) + (1/2)(1) + (1/4)(2).$$

Note that the first term in this sum is equal to $(0)p(0)$, the second is equal to $(1)p(1)$, and the third is equal to $(2)p(2)$. The average value of y is then equal to

$$\sum_{y=0}^{2} yp(y) = 1.$$

You will observe that this result was not an accident and that it would be intuitively reasonable to define the expected value of y for a discrete random variable as follows.

Definition

Let y be a discrete random variable with probability distribution $p(y)$ and let $E(y)$ represent the **expected value of** y. Then

$$E(y) = \sum_y yp(y),$$

where the elements are summed over all values of the random variable y.

Note that if $p(y)$ is an accurate description of the relative frequencies for a real population of data, then $E(y) = \mu$, the mean of the population. We shall assume this to be true and let $E(y)$ be synonymous with μ. That is, we shall assume that

$$\mu = E(y).$$

The method for calculating the expected value of y for a continuous random variable is rather similar from an intuitive point of view, but, in practice, it involves the use of the calculus and is therefore beyond the scope of this text.

Example 5.4

Consider the random variable y representing the number observed on the toss of a single die. The probability distribution for y is given in Example 5.2. Then the expected value of y would be

$$E(y) = \sum_{y=1}^{6} yp(y) = (1)p(1) + (2)p(2) + \cdots + (6)p(6)$$

$$= (1)(1/6) + (2)(1/6) + \cdots + (6)(1/6)$$

$$= \frac{1}{6} \sum_{y=1}^{6} y = \frac{21}{6} = 3.5.$$

Note that this value, $\mu = E(y) = 3.5$, exactly locates the center of the probability distribution, Figure 5.2.

Example 5.5

Eight thousand tickets are to be sold at $5.00 each in a lottery conducted to benefit the local fire company. The prize is a $12,000.00 automobile. If you purchase two tickets, what is your expected gain?

Solution

Your gain, y, may take one of two values. Either you will lose $10.00 (that is, your gain will be $-$10.00) or win $11,990 with probabilities 7998/8000 and 2/8000, respectively.

The probability distribution for the gain, y, is as follows:

y	$p(y)$
$-\$10.00$	$\dfrac{7998}{8000}$
$\$11,990.00$	$\dfrac{2}{8000}$

The expected gain will be

$$E(y) = \sum_y yp(y)$$

$$= (-\$10.00)\left(\frac{7998}{8000}\right) + (\$11,990.00)\left(\frac{2}{8000}\right)$$

$$= -\$7.00.$$

Recall that the expected value of y is the average of the theoretical population that would result if the lottery were repeated an infinitely large number of times. If this were done, your average or expected gain per lottery would be a loss of $7.00.

Example 5.6

Determine the yearly premium for a $1000.00 insurance policy covering an event which, over a long period of time, has occurred at the rate of two times in 100. Let y equal the yearly financial gain to the insurance company resulting from the sale of the policy and let C equal the unknown yearly premium. We will calculate the value of C such that the expected gain, $E(y)$, will equal zero. Then C is the premium required to break even. To this the company would add administrative costs and profit.

Solution

The first step in the solution is to determine the values which the gain, y, may take and then to determine $p(y)$. If the event does not occur during the year, the insurance company will gain the premium of $y = C$ dollars. If the event does occur, the gain will be negative. That is, the company will lose $1000 less the premium of C dollars already collected. Then $y = -(1000 - C)$ dollars. The probabilities associated with these two values of y are 98/100 and 2/100, respectively. The probability distribution for the gain would be:

y = gain	$p(y)$
C	$\dfrac{98}{100}$
$-(1000 - C)$	$\dfrac{2}{100}$

Since we want the insurance premium C such that, in the long run (for many similar policies), the mean gain will equal 0, we will set the expected value of y equal to zero and solve for C. Then

$$E(y) \quad \sum_y yp(y)$$

$$= C\left(\frac{98}{100}\right) + [-(1000 - C)]\left(\frac{2}{100}\right) = 0$$

or

$$\frac{98}{100}C + \left(\frac{2}{100}\right)C - 20 = 0.$$

Solving this equation for C, we obtain

$$C = \$20.$$

Concluding, if the insurance company were to charge a yearly premium of $20, the average gain calculated for a large number of similar policies would equal zero. The actual premium would equal $20 plus administrative costs and profit.

Just as we used numerical descriptive measures to describe a relative frequency distribution (Chapter 3), we wish to use the mean and variance (ultimately, the standard deviation) of a random variable to describe its probability distribution. Knowing μ and σ, we could use Tchebysheff's Theorem or the Empirical Rule to describe $p(y)$. The variance σ^2 of a random variable is defined to be the mean value of the square of the deviation of y from its mean. That is,

$$\sigma^2 = E[(y - \mu)^2].$$

This leads us to the problem of finding the mean value of a function of a random variable, namely the expected value of $(y - \mu)^2$.

There are many other situations where we might only know the probability distribution of a random variable y but want to know the mean value of some function of y. For example, if a life insurance company has issued a fixed number of policies in a given year, then the profit per year P is a function of y, the number of policyholders who die. Now suppose we only know the equation relating P to y and the probability distribution for y. We want to know the mean value of P. It is this problem that we now consider—finding the expected value of a function of a random variable.

The rule for finding the expected value of a function of y, say $g(y)$, can be obtained by considering a simple example, the coin-tossing problem, Example 5.1, where y equals the number of heads observed when tossing two coins. Suppose that we wish to find the expected value of $g(y) = y^2$.

In accordance with the definition of a function, Section 2.2, only one value of the function corresponds to each value of y. Hence the quantity y^2 would represent a numerical event that varies over the sample space with only one value of y^2 being assigned to each sample point. Without proceeding further, it is reasonably clear that any function of a random variable, y, will itself be a random variable. The probability distribution for y, with associated values of y^2, is as follows:

y	y^2	$p(y)$
0	0	1/4
1	1	1/2
2	4	1/4

Repeating the experiment a large number of times, say $n = 4{,}000{,}000$, we would expect approximately 1,000,000 values of $y^2 = 0$, 2,000,000 values of $y^2 = 1$, and 1,000,000 values of $y^2 = 4$. The average value for y^2 would be

$$\frac{\text{sum of measurements}}{n} = \frac{1{,}000{,}000(0) + 2{,}000{,}000(1) + 1{,}000{,}000(4)}{4{,}000{,}000}$$

$$= (1/4)(0) + (1/2)(1) + (1/4)(4)$$

$$= \sum_{y=0}^{2} y^2 p(y).$$

You may obtain the expected value for other functions of y and other random variables using the technique employed above. The results will agree with the following definition.

Definition

Let y be a discrete random variable with probability distribution $p(y)$, and let $g(y)$ be a numerical valued function of y. Then the **expected value of $g(y)$** is

$$E[g(y)] = \sum g(y)p(y).$$

As noted earlier, the variance of a random variable, y, the expected value of $(y - \mu)^2$, is of particular interest to us because it, along with μ, can be used to describe the probability distribution for y.

Definition

Let y be a discrete random variable with probability distribution $p(y)$; then the variance* of y is

$$\sigma^2 = E[(y - \mu)^2] = \sum_y (y - \mu)^2 p(y).$$

We conclude with two examples.

Example 5.7

Find the variance, σ^2, for the population associated with Example 5.1, the tossing of two coins. The expected value of y, μ, was shown to equal 1.

Solution

The variance is equal to the expected value of $(y - \mu)^2$, or

$$\sigma^2 = E[(y - \mu)^2] = \sum_y (y - \mu)^2 p(y)$$

$$= (0 - 1)^2 p(0) + (1 - 1)^2 p(1) + (2 - 1)^2 p(2)$$

$$= 1(1/4) + 0(1/2) + 1(1/4)$$

$$= 1/2.$$

Calculating σ^2 can be facilitated by using the following table. Note that σ^2 is the total of column 4.

y	$(y - \mu)^2$	$p(y)$	$(y - \mu)^2 p(y)$
0	1	1/4	1/4
1	0	1/2	0
2	1	1/4	1/4
			1/2

---

* Three theorems, directly analogous to the three summation theorems of Section 2.5, exist for expectations. These theorems can be used to show that

$$\sigma^2 = \sum_y (y - \mu)^2 p(y) = \sum_y y^2 p(y) - \mu^2 = E(y^2) - \mu^2,$$

a result that is analogous to the shortcut formula for the sum of squares of deviations, Chapter 3. See Exercises 5.49 and 5.50.

Example 5.8   Let y be a random variable with probability distribution given by the following table:

y	$p(y)$
-1	.05
0	.10
1	.40
2	.20
3	.10
4	.10
5	.05

Find μ, σ^2, and σ. Graph $p(y)$ and locate the interval, $\mu \pm 2\sigma$, on the graph. What is the probability that y will fall in the interval $\mu \pm 2\sigma$?

Solution

$$\mu = E(y) = \sum_{y=-1}^{5} yp(y)$$

$$= (-1)(.05) + (0)(.10) + (1)(.40) + \cdots + (4)(.10) + (5)(.05)$$

$$= 1.70$$

$$\sigma^2 = E[(y - \mu)^2] = \sum_{y=-1}^{5} (y - \mu)^2 p(y)$$

$$= (-1 - 1.7)^2(.05) + (0 - 1.7)^2(.10) + \cdots + (5 - 1.7)^2(.05)$$

$$= 2.11$$

and

$$\sigma = \sqrt{\sigma^2} = \sqrt{2.11} = 1.45.$$

The interval $\mu \pm 2\sigma$ is $1.70 \pm (2)(1.45)$ or -1.20 to 4.60.

Figure 5.6 The probability histogram for $p(y)$, Example 5.8.

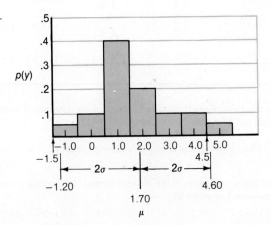

The graph of $p(y)$ and the interval, $\mu \pm 2\sigma$, are shown in Figure 5.6. You can see that $y = -1, 0, 1, 2, 3, 4$ fall in the interval. Therefore,

$$P[\mu - 2\sigma < y < \mu + 2\sigma] = p(-1) + p(1) + p(2) + \cdots + p(4)$$
$$= (.05) + (.10) + (.40) + (.20) + (.10) + (.10)$$
$$= .95.$$

Tips on Problem Solving

To find the expected value of a discrete random variable y, construct a table containing three columns, the first for y and the second for $p(y)$. Then multiply each y value by its corresponding probability and enter in the third column. The sum of this third column, the sum of $yp(y)$, will give you the expected value of y.

To find the expected value of a function $g(y)$ of a discrete random variable y, start with a table containing four columns, the first for y, the second for $g(y)$, the third for $p(y)$, and the fourth for the cross products, $g(y)p(y)$. First calculate the value of $g(y)$ for each value of y and enter in column 2. Then obtain the cross product $g(y)p(y)$ for each value of y and enter in column 4. The sum of column 4 will give the expected value of $g(y)$.

Exercises

Conceptual Level

5.16.   Let y be a random variable with probability distribution given by the following table:

y	$p(y)$
0	.1
1	.3
2	.4
3	.2

Find the expected value and variance of y. Construct a graph of the probability distribution. Do μ and σ permit a rough description of $p(y)$?

5.17.   Let y represent the number of times a customer visits a grocery store in a 1-week period. Assume that the following is the probability distribution of y.

y	$p(y)$
0	.1
1	.4
2	.4
3	.1

Find the expected value of y. This is the average number of times a customer visits the store.

5.18.   The probability distribution for a random variable y is as shown in the table.

y	0	1	2	3	4	5
$p(y)$	.05	.3	.3	.2	.1	.05

a. Find $E(y)$.

b. Find σ^2.

c. Sketch $p(y)$ and locate the interval $\mu \pm 2\sigma$ on the graph.

d. Find the probability that y falls in the interval $\mu \pm 2\sigma$.

5.19.   A $50,000 diamond is insured to its total value by paying a premium of D dollars. If the probability of theft in a given year is estimated to be .01, what premium should the insurance company charge if it wishes the expected gain to equal $1000?

5.20.   The maximum patent life for a new drug is 17 years. Subtracting the length of time required by the Food and Drug Administration for testing and approval of the drug, you obtain the actual patent life of the drug; that is, the length of time that a company has to recover research and development costs and make a profit. Suppose that the distribution of the length of patent life for new drugs is as shown below:

Years y	3	4	5	6	7	8	9	10	11	12	13
$p(y)$	.03	.05	.07	.10	.14	.20	.18	.12	.07	.03	.01

a. Find the expected number of years of patent life for a new drug.

b. Find the standard deviation of y.

c. Find the probability that y falls in the interval $\mu \pm 2\sigma$.

Exercises
Skill Level

5.21.   A manufacturing representative is considering the option of taking out an insurance policy to cover possible losses incurred by marketing a new product. If the product is a complete failure, the representative feels that a loss of $80,000 would be incurred; if it is only moderately successful, a loss of $25,000 would be incurred. Insurance actuaries have determined from market surveys and other available information that the probabilities that the product will be a failure or only moderately successful are .01 and .05, respectively. Assuming that the manufacturing representative would be willing to ignore all other possible losses, what premium should the insurance company charge for the policy in order to break even?

5.22.   A manufacturing company ships its product in two different sizes of truck trailers, an $8 \times 10 \times 30$ and an $8 \times 10 \times 40$. If 30 percent of its shipments are made using the 30-foot trailer and 70 percent using the 40-foot trailer, find the mean volume shipped per trailer load (assume that the trailers are always full).

5.23.   The number N of residential homes that a fire company can serve depends on the distance r (on city blocks) that a fire engine can cover in a specified (fixed) period of time. If we assume that N is proportional to the area of a circle r blocks from the firehouse, then

$$N = C\pi r^2,$$

where C is a constant, $\pi = 3.1416\ldots$, and r, a random variable, is the number of blocks that a fire engine can move in the specified time interval. For a particular fire company, $C = 8$, the probability distribution for r is as shown in the accompanying table, and $p(r) = 0$, $r \le 20$, and $r \ge 27$.

r	21	22	23	24	25	26
$p(r)$	.05	.20	.30	.25	.15	.05

Find the expected value of N, the number of homes that the fire department can serve.

5.24. The probability that a tennis player A can win a set from tennis player B is one measure of the comparative abilities of the two players. In Exercise 5.14, you found the probability distribution for y, the number of sets required to play a best-of-five-sets match, given that the probability that A wins any one set—call this $P(A)$—is .6.

a. Find the expected number of sets required to complete the match for $P(A) = .6$.

b. Find the expected number of sets required to complete the match when the players are of equal ability; that is, $P(A) = .5$.

c. Find the expected number of sets required to complete the match when the players differ greatly in ability; that is, say $P(A) = .9$.

5.25. The probability that a gambler will win on a single spin of a wheel is .49. To overcome these less-than-even odds of winning, the gambler decides to bet $10 on the first spin to either lose or double her money. If she wins, she quits, but if she loses, she doubles her bet for the second spin, gambling $20 to win $40. She plans to continue this procedure, doubling her bet on each spin of the wheel, until she wins or until she runs out of her $310 stake. Each time she gambles, she will always employ a $310 stake. What is the gambler's expected gain when she gambles?

5.6
Analysis of the Number of Hypothyroid Births near Three Mile Island

You will recall in our case study that 11 hypothyroid births were observed within a three-county area in the vicinity of Three Mile Island and that all of the babies were in gestation at the time of the nuclear accident there. If the mean number of hypothyroid births for the three-county area and for a corresponding period of time is 3, can we conclude that 11 such births is a highly improbable number? If the answer is yes, we would conclude that either some external factor(s), such as the nuclear accident, caused this unusually large number of hypothyroid cases or that we have observed a very rare event (rare events do happen!).

A reasonable model for the relative frequency distribution of the number of rare events that occur randomly in a specified unit of space, distance, time etc. is given by the Poisson probability distribution. This distribution, named after the French mathematician and probabilist, S.D. Poisson (1781–1840), is given by the equation:

$$p(y) = \frac{\mu^y e^{-\mu}}{y!}$$

where,

μ = mean value of y

e = 2.718 . . .

y = number of rare events observed per unit of space, distance, time, etc.

= 0, 1, 2, 3, . . .

and $y!$ is the product $y(y - 1)(y - 2) \ldots 3 \cdot 2 \cdot 1$
(i.e., $6! = 6 \cdot 5 \cdot 4 \cdot 3 \cdot 2 \cdot 1 = 720$ and, by definition, $0! = 1$).

Poisson probability distributions provide good approximations to the distributions of many variables observed in nature, such as the number of deaths per year in a given region due to a rare disease, the number of accidents in a manufacturing plant per month, the number of air collisions per month, the number of worms observed on an ear of corn, and the number of emissions per second from a radioactive source. Since the number y of hypothyroid cases was measured for a specified period of time and for a specified area, the three-county region, it is likely that a Poisson probability distribution with $\mu = 3$ would provide a reasonable model for the relative frequency distribution of y. This probability distribution is shown in Figure 5.7.

Figure 5.7 Poisson probability distribution, mean = μ = 3

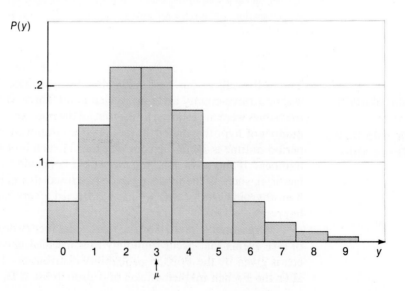

You can see at a glance (Figure 5.7) that the probability that $y = 11$ cases or more is nearly 0 if, in fact, the mean number of cases in the three-county area for a comparable period of time is equal to 3. Calculating the exact value of this probability would be tedious because we would have to calculate the individual values of $p(y)$ by substituting values of y into the formula for a Poisson probability distribution with $\mu = 3$. An easier way to confirm our visual conclusion is to calculate the standard deviation for the distribution and determine how many standard deviations lie between $y = 11$ and $\mu = 3$.

It can be shown (proof omitted) that the standard deviation for a Poisson random variable is equal to,

$$\sigma = \sqrt{\mu}.$$

Thus, for our distribution (Figure 5.7),

$$\sigma = \sqrt{\mu} = \sqrt{3} = 1.73.$$

You can see that the distance between $y = 11$ and $\mu = 3$ is $8/1.73 = 4.62$ standard deviations.

From our knowledge of the Empirical Rule and Tchebysheff's Theorem, we know that the probability that an observation will fall more than 3 standard deviations from its mean μ is very small. Therefore, we conclude that $y = 11$ hypothyroid births in the three-county area is a very improbable event. This suggests that something *caused* the mean rate of hypothyroid births to rise during the period of time following the Three Mile Island nuclear accident. The cause may have been low-level radiation emitted during the nuclear accident or it may have been some other factor that existed unnoticed during that particular period of time.

5.7
Random Sampling

Since probability distributions are theoretical models for population relative frequency distributions, samples selected from populations can be viewed as observations on random variables. As noted in earlier chapters, the way a sample is selected from a population affects the quantity of information in the sample (the topic of Chapter 12) as well as the probabilities of observing particular samples. Since the probability of observed sample outcomes is the basis for inference making, a topic that is introduced in Chapters 6, 7, 8, and 9, it is important that we define one of the most commonly employed sampling procedures at this point. It is called **simple random sampling.**

Simple random sampling gives every different sample in the population an equal chance of being selected. To illustrate, suppose that we wish to select a sample of $n = 2$ from a population containing $N = 4$ elements (we are choosing a small value of N to simplify our discussion). If the four elements are identified by the symbols y_1, y_2, y_3, and y_4, then there are six different samples that could be selected from the population, namely:

Sample	Observations in Sample
1	y_1, y_2
2	y_1, y_3
3	y_1, y_4
4	y_2, y_3
5	y_2, y_4
6	y_3, y_4

If the sample was selected in such a way that each of these six samples had an equal chance of selection (probability equal to 1/6), the sample would be called a simple random (or, simply, "random") sample.

It can be shown* that the number of ways of selecting $n = 2$ elements from a set of $N = 4$, denoted by the symbol C_2^4, is

$$C_2^4 = \frac{4!}{2!2!} = \frac{4 \cdot 3 \cdot 2 \cdot 1}{(2 \cdot 1)(2 \cdot 1)} = 6.$$

As explained in Section 5.6, the symbol $n!$ (read "n factorial") is used to denote the product, $n(n - 1)(n - 2) \cdots 3 \cdot 2 \cdot 1$. Thus $5! = 5 \cdot 4 \cdot 3 \cdot 2 \cdot 1 = 120$. [The quantity 0! is defined to be equal to 1.] In general, the number of ways of selecting n elements from a set of N is

$$C_n^N = \frac{N!}{n!(N - n)!}.$$

For example, if you wish to conduct an opinion poll of 5000 people based on a sample of $n = 100$, there are

$$C_{100}^{5000}$$

different combinations of people who could be selected in the sample. If the sampling is conducted in such a way that each of these combinations has an equal probability of being selected, then the sample is called a simple random sample.

---

* A derivation of this result along with examples and applications is given in optional Section 4.10. An understanding of the derivation is not essential to our discussion.

Definition

Let N and n represent the numbers of elements in the population and sample, respectively. If the sampling is conducted in such a way that each of the C_n^N samples has an equal probability of being selected, the sampling is said to be random and the result is said to be a **simple random sample.**

Perfect random sampling is difficult to achieve in practice. If the population is not too large, we might write each of the N numbers on a poker chip, mix the total, and select a sample of n chips. The numbers on the poker chips would specify the measurements to appear in the sample. Other techniques are available when the population is large. One of these, the use of a random-number table, is discussed in Section 14.3.

In many situations, the population is conceptual, as in an observation made during a laboratory experiment. Here the population is envisioned to be the infinitely large number of measurements obtained when the experiment is repeated over and over again. If we wish a sample of $n = 10$ measurements from this population, we repeat the experiment 10 times and hope that the results represent, to a reasonable degree of approximation, a random sample.

Although the primary purpose of this discussion was to clarify the meaning of a random sample, we would like to mention that some sampling techniques are partly systematic and partly random. For instance, if we wish to determine the voting preference of the nation in a presidential election, we would not likely choose a random sample from the population of voters. Just due to a pure chance, all the voters appearing in the sample might be drawn from a single city, say San Francisco, which might not be at all representative of the population. We would prefer a random selection of voters from smaller political districts, perhaps states, allotting a specified number to each state. The information from the randomly selected subsamples drawn from the respective states would be combined to form a prediction concerning the entire population of voters in the country. The purpose of systematic sampling, as in the design of experiments in general, is to obtain a maximum of information for a fixed sample size. This, we recall, was one of the five elements of a statistical problem discussed in Chapter 1.

Exercises
Conceptual Level

5.26. Evaluate:

    a. 4!   b. C_2^5 c. C_0^6 d. C_6^6

5.27. How many different samples of $n = 2$ elements can be selected from $N = 6$? List them.

5.28. How many different samples of $n = 10$ could be selected from a population containing $N = 100$ elements?

5.8
Summary

Random variables, representing numerical events defined over a sample space, may be classified as discrete or continuous random variables depending upon whether the number of sample points in the sample space is or is not countable. The theoretical population frequency distribution for the discrete random variable is called a probability distribution and often may be derived using the techniques of Chapter 4. The model for the frequency distribution for a continuous random variable is a mathematical function, $f(y)$, called a probability distribution or probability density function. This function, usually a smooth curve, is defined over a line interval and is chosen such that the total area under the curve is equal to 1. The probabilities associated with a continuous random variable are given as areas under the probability distribution, $f(y)$.

A mathematical expectation is the average of a random variable calculated for the theoretical population defined by its probability distribution.

As noted in earlier discussions, inferences about a population will be based on the probability of an observed sample and this probability will depend upon how the sample was selected from the population. The probability distribution for a random variable plays a role in inference making because it enables us to calculate the probability that a single observation, randomly selected from the population, will assume one or more specific values. The most common method for sampling more than one observation from a population is called simple random (or simply "random") sampling. A simple random sample is selected from a population in such a way that every possible different sample has an equal probability of selection. Simple random sampling will be used in Chapter 6 and in subsequent chapters in this text.

References

Freund, J.E., *Mathematical Statistics*, 2nd ed. Englewood Cliffs, N.J.: Prentice-Hall, Inc., 1971.

Mendenhall, W., R.L. Scheaffer, and D. Wackerly *Mathematical Statistics with Applications*. 2nd ed. Boston: Duxbury Press, 1981.

Mood, A.M., et al., *Introduction to the Theory of Statistics*, 3rd ed. New York: McGraw-Hill Book Company, 1973.

Mosteller, F., R.E.K. Rourke, and G.B. Thomas, Jr., *Probability with Statistical Applications*, 2nd ed. Reading, Mass.: Addison-Wesley Publishing Company, Inc., 1970.

Supplementary Exercises

[Starred (*) exercises are optional.]

5.29. Identify the following as continuous or discrete random variables:

a. The number of homicides in Detroit during a 1-month period.

b. The length of time between arrivals at an outpatient clinic.

c. The number of typing errors on a page of typing.

d. The number of defective light bulbs in a packet containing four bulbs.

e. The time required to finish an examination.

5.30. Given the discrete random variable y and its probability distribution, $p(y)$:

y	$p(y)$
0	1/8
2	1/4
3	1/2
4	1/8

a. Find $\mu = E(y)$.   b. Find $E(y^2)$.   c. Find σ^2 and σ.

5.31. Let y equal the number of dots observed when a single die is tossed, and let $p(y)$ equal 1/6, $y = 1, 2, 3, \ldots, 6$. We found in Example 5.4 that $\mu = E(y) = 3.5$. Find σ^2 and show that $\sigma = 1.71$. Then find the probability that y will fall in the interval, $\mu \pm 2\sigma$.

5.32. Refer to Exercise 5.10. Find μ, the expected value of y, for the theoretical population by using the probability distribution obtained in Exercise 5.10. Find the sample mean, $\bar{y}$, for the $n = 100$ measurements generated in Exercise 5.11. Does $\bar{y}$ provide a good estimate of μ?

5.33. Find the population variance, σ^2, for Exercise 5.10 and the sample variance, s^2, for Exercise 5.11. Compare them.

5.34. Draw a sample of $n = 50$ measurements from the die-throw population of Example 5.2 by tossing a die 50 times and recording y after each toss. Calculate $\bar{y}$ and s^2 for the sample. Compare $\bar{y}$ with the expected value of y, Example 5.4, and s^2 with the variance of y obtained in Exercise 5.31. Do $\bar{y}$ and s^2 provide good estimates of μ and σ^2?

5.35. Use the probability distribution obtained in Exercise 5.10 to find the fraction of the total population of measurements that lies within two standard deviations of the mean. Compare with Tchebysheff's Theorem. Repeat for the sample in Exercise 5.11.

5.36. A heavy-equipment salesman can contact either one or two customers per day with probability 1/3 and 2/3, respectively. Each contact will result in either no sale or a $50,000 sale with probability 9/10 and 1/10, respectively. What is the expected value of his daily sales?

5.37. A county containing a large number of rural homes is thought to have 60 percent insured against a fire. Four rural home owners are chosen at random from the entire population and y are found to be insured against a fire. Find the probability distribution for y. What is the probability that at least three of the four will be insured?

5.38. A fire-detection device utilizes three temperature-sensitive cells acting independently of each other in such a manner that any one or more may actuate the alarm. Each cell possesses a probability of $p = .8$ of actuating the alarm when the temperature reaches 100 degrees or more. Let y equal the number of cells actuating the alarm when the temperature reaches 100 degrees. Find the probability distribution for y. Find the probability that the alarm will function when the temperature reaches 100 degrees.

5.39. Find the expected value and variance for the random variable y defined in Exercise 5.38.

5.40. If you toss a pair of dice, the sum T of the numbers of dots appearing on the upper faces of the dice can assume the value of an integer in the interval, $2 \leq T \leq 12$.

a. Find $E(T)$.

b. Find σ^2.

c. Use your graph of $P(T)$, found in Exercise 5.9, and locate the interval, $\mu \pm 2\sigma$.

d. What is the probability that T will assume a value in the interval, $\mu \pm 2\sigma$?

5.41.   Accident records collected by an automobile insurance company give the following information. The probability that an insured driver has an automobile accident is .15. If an accident has occurred, the damage to the vehicle amounts to 20 percent of its market value with probability .80, 60 percent of its market value with probability .12, and a total loss with probability .08. What premium should the company charge on a $4000 car so that the expected gain by the company is zero?

5.42.   If you own common stock in a corporation, you can sometimes increase the return on your investment by selling an option. For a stated price, the purchaser of the option gains the right to buy your stock at any time up to a specified expiration date. Suppose you purchased 200 shares of a stock at $25 per share and you sell an option for the option purchaser to buy the stock at any time within the next six months for $30 per share. If the stock reaches $30 per share within the next six months; your stock will be sold and you will gain $5 per share plus $2 per share (from the dividends and the sale of the option) less $109 commission to the broker for selling your stock. If you do not sell, you will gain $2 per share. If the probability of the stock reaching $30 per share within the next six months is .7, what is the expected return (in dollars) from your 200 shares of stock? What is the expected annual rate of return, in percentage, on your investment? (*Note:* We have ignored a $15 commission on the sale of the option.)

* 5.43.   Express the probability distribution. Exercise 5.10 as a formula. (The distribution is known to statisticians as a **hypergeometric probability distribution.**)

* 5.44.   The **Poisson probability distribution** provides a good model for count data when y, the count, represents the number of "rare events" observed in a given unit of time, distance, area, or volume. The number of bacteria per small volume of fluid and the number of traffic accidents at a given intersection during a given period of time possess, approximately, a Poisson probability distribution which is given by the formula

$$p(y) = \frac{\mu^y e^{-\mu}}{y!},$$

where $y = 0, 1, 2, 3, \ldots,$; $e = 2.718 \ldots$; and μ is the population mean. If the average number of accidents over a specified section of highway is two per week:

a. Construct a probability histogram for y. (Use Table 2, in the Appendix to evaluate $e^{-\mu}$.)

b. Calculate the probability that no accidents will occur during a given week.

c. It can be shown that the standard deviation of a Poisson random variable is

$$\sigma = \sqrt{\mu}.$$

Construct the interval $\mu \pm 2\sigma$ and locate it on the graph, part (a). What is the probability that y will fall in the interval $\mu \pm 2\sigma$?

* 5.45.   A case similar to the hyperthyroid births (Case Study) occurred at the Lawrence Livermore Laboratory in California (*Gainesville Sun*, May 10, 1980). Eighteen employees of the nuclear weapons research facility contracted the deadly skin cancer, melanoma, during the period 1972 to 1977, in comparison with a mean of 2.3 cases per five years during the

previous 20-year period. If the mean rate of melanoma cases per five-year period is really equal to 2.3, is it likely that you could observe as many as 18 cases during one five-year period? Solve using the procedure and logic employed in Section 5.6.

* 5.46. The number y of people entering the intensive care unit at a particular hospital on any one day possesses a Poisson probability distribution with mean equal to five persons per day.

    a. What is the probability that the number of people entering the intensive care unit on a particular day is equal to 2? Less than or equal to 2?

    b. Is it likely that y will exceed 10? Explain. (*Note:* Use the information about the Poisson probability distribution given in Exercise 5.44.)

* 5.47. If a drop of water is placed on a slide and examined under a microscope, the number y of a particular type of bacteria present has been found to possess a Poisson probability distribution. Suppose that the maximum permissible count per water specimen for this type of bacteria is 5. If the mean count for your water supply is 2 and you test a single specimen, is it likely that the count will exceed the maximum permissible count? Explain. (Use the information about the Poisson probability distribution given in Exercise 5.44. *Note:* This exercise requires very little computation.)

* 5.48. The probability distribution for the number of well drillings y until the first success (Exercise 5.13) possesses a geometric probability distribution that is given by the formula

$$p(y) = p(1 - p)^{y-1},$$

where p is the probability of success for a single drilling. Further, it can be shown (proof omitted) that the mean and variance of p are equal to

$$E(y) = \frac{1}{p} \quad \text{and} \quad \sigma^2 = \frac{1 - p}{p^2}.$$

Suppose the company, Exercise 5.13, commenced drilling for oil. Would it be unlikely that it would take as many as 20 drillings before the company encountered its first success? Explain.

* 5.49. Use the summation theorems, Theorems 2.1, 2.2, and 2.3, to prove the following theorems:

    a. Theorem 5.1: Let c be a constant and let y be a discrete random variable with probability distribution $p(y)$. Then the expected value of c is c. [That is, $E(c) = c$.]

    b. Theorem 5.2: Let c be a constant and let y be a discrete random variable with probability distribution $p(y)$. Then,

$$E(cy) = cE(y).$$

    c. Theorem 5.3: Let y be a discrete random variable with probability distribution $p(y)$ and let $g_1(y)$ and $g_2(y)$ be functions of the random variable y. Then,

$$E[g_1(y) + g_2(y)] = E[g_1(y)] + E[g_2(y)].$$

Note that this theorem implies that the expected value of a sum of functions of a random variable is equal to the sum of their respective expected values.

* 5.50.   Use Theorems 5.1, 5.2, and 5.3 to show that

$$\sigma^2 = E[(y - \mu)^2] = E(y^2) - \mu^2.$$

(Hint: $E[(y - \mu)^2] = E(y^2 - 2\mu y + \mu^2)$.)

* 5.51.   Use the formula given in Exercise 5.50 to find σ^2 for Example 5.7.
* 5.52.   Use the formula given in Exercise 5.50 to find σ^2 for Exercise 5.30.
* 5.53.   Refer to Exercise 5.30 and find:
    a. $P[y = 2 | y > 0]$.   b. $P[y \geq 3 | y > 0]$.
    (Hint: Refer to the definition of a conditional probability, Section 4.6.)

6. The Binomial

Probability

Distribution

Chapter Objectives

General Objectives

A method for finding the probabilities associated with specific values of random variables was presented in Chapter 5. Specifically, we learned that discrete and continuous random variables required different probability distributions and that these distributions were subject to different interpretations. Now we turn to some specific applications of these notions and present a useful discrete random variable and its probability distribution. In concluding the chapter, we will show you how this probability distribution can be used in making inferences.

Specific Objectives

1. To describe the characteristics of a binomial experiment, to indicate types of data which represent measurements on binomial random variables, to give the formula for the binomial probability distribution and to show how it is used to calculate probabilities for a binomial random variable.   *Sections 6.1, 6.2*

2. To present the formulas for the expected value and variance of a binomial random variable. These quantities will be used to describe a binomial probability distribution and will be used in Chapter 7 for a simple procedure for approximating binomial probabilities.   *Section 6.3*

3. To show you how the binomial probability distribution is used to make inferences about a binomial population based on the information contained in a sample. *Sections 6.4, 6.5, 6.6*

4. To introduce the concepts involved in a statistical test of an hypothesis.   *Sections 6.6, 6.7, 6.8*

Case Study

The Fear of Flying

Are you afraid to fly? A study by the Boeing Aircraft Company, reported in the *Orlando Sentinel Star* (June 28, 1981), reveals some surprising statistics on consumer attitudes towards flying. It is estimated that nearly 25,000,000 Americans are afraid to fly and that one in ten who fly do so in a state of fear. Fear of flying is more prevalent among blacks than whites and among lower socioeconomic groups. Fifteen percent of those who have flown consider flying to be dangerous, as compared to 29 percent of those who have not flown.

Thousands of businessmen, businesswomen, and vacationers fly only when it is absolutely necessary. These people, along with those who refuse to fly, represent an enormous loss of income to the airlines, particularly considering that in the first six months of 1981 the airlines were operating at 56.5 percent of capacity. Consequently, the need to learn more about the fear of flying, its causes, and the effects of measures to cope with it, motivates sample surveys of the type conducted by the Boeing Aircraft Company.

Sample surveys conducted by mail, telephone, or personal interview, are usually employed to sample consumer attitudes or opinions. While sometimes quantitative, the answers solicited by the survey questions are often of the dichotomous, yes-or-no variety, such as the answer to the question, "Are you afraid of flying?". Typical surveys by the Gallup, Harris, or various consumer polling organizations usually involve sampling and questioning 1000 to 2000 consumers. A summary of these dichotomous responses to a question yields the number of yeses (or the number of noes) in the sample.

Ultimately, the polling organization will use the number of yeses in the sample to estimate the proportion of yeses in the entire population, and will state how far the sample estimate is likely to depart from the population proportion. Note that y, the number of yeses in, say, a sample of 1000 people, will vary from one sample to another and will cause the sample proportions to vary in a similar manner. Consequently, the key to evaluating the reliability of an estimate of a population proportion is the probability distribution of the discrete random variable y.

In this chapter you will learn that sample surveys generate data that are similar to data generated by many experiments in the biological, physical, and social sciences. We will identify the characteristics common to these experiments, find the probability distribution for y, and determine its properties. Then in Section 6.4 we will examine the probability distribution for the number of yes responses to a question in a sample survey.

6.1
The Binomial Experiment

One of the most elementary, useful, and interesting discrete random variables is associated with the coin-tossing experiment described in Examples 4.3, 4.5, and 5.1. In an abstract sense, numerous coin-tossing experiments of practical importance are conducted daily in the social sciences, physical sciences, and industry.

To illustrate, we might consider a sample survey conducted to predict voter preferences in a political election. Interviewing a single voter bears a similarity, in many respects, to tossing a single coin because the voter's response may be in favor of our candidate—a "head"—or it may be opposed (or indicate indecision)—a "tail." In most cases, the fraction of voters favoring a particular candidate will not equal one-half, but even this similarity to the coin-tossing experiment is satisfied in national presidential elections. History demonstrates that the fraction of the total vote favoring the winning presidential candidate in most national elections is very near one-half.

Similar polls are conducted in the social sciences, in industry, and in education. The sociologist is interested in the fraction of rural homes that have electricity; the cigarette manufacturer desires knowledge concerning the fraction of smokers who prefer his brand; the teacher is interested in the fraction of students who pass his course. Each person sampled is analogous to the toss of an unbalanced (since the probability of a "head" is usually not 1/2) coin.

Firing a projectile at a target is similar to a coin-tossing experiment if a "hit the target" and a "miss the target" are regarded as a head and a tail, respectively. A single missile will result in either a successful or unsuccessful launching. A new drug will prove either effective or noneffective when administered to a single patient, and a manufactured item selected from production will be either defective or nondefective. Although dissimilar in some respects, the experiments described above will often exhibit, to a reasonable degree of approximation, the characteristics of a **binomial experiment.**

Definition

A **binomial experiment** is one that possesses the following properties:

1. The experiment consists of n identical trials.
2. Each trial results in one of two outcomes. For lack of a better nomenclature, we will call the one outcome a success, S, and the other a failure, F.
3. The probability of success on a single trial is equal to p and remains the same from trial to trial. The probability of a failure is equal to $(1 - p) = q$.
4. The trials are independent.
5. We are interested in y, the number of successes observed during the n trials.

Example 6.1

Suppose that there are approximately 1,000,000 adults in a county and that an unknown proportion, p, favor, and $(1 - p)$ oppose the Equal Rights Amendment (ERA). A random sample of 1000 adults will be selected from among the 1,000,000 in the county and each will be asked whether he or she favors the Equal Rights Amendment. (The ultimate objective of this survey is to estimate the unknown proportion, p, a problem that we will learn how to solve in Chapter 8.) Is this a binomial experiment?

Solution

To decide whether this is a binomial experiment, we must see if the sampling satisfies the five characteristics described in the definition.

1. The sampling consists of $n = 1000$ identical trials. One trial represents the selection of a single adult from the 1,000,000 adults in the county.

2. Each trial will result in one of two outcomes. A person will either favor the amendment or will not. These two outcomes could be associated with the "success" and "failure" of a binomial experiment.*

3. The probability of a success will equal the proportion of adults favoring the ERA. For example, if 500,000 of the 1,000,000 adults in the county favor the ERA, then the probability of selecting an adult favoring the ERA out of the 1,000,000 in the county is $p = .5$. For all practical purposes, this probability will remain the same from trial to trial even though adults selected in the earlier trials are not replaced as the sampling continues.

4. For all practical purposes, the probability of a success on any one trial will be unaffected by the outcome on any of the others (it will remain very close to p).

5. We are interested in the number y of adults in the sample of 1000 who favor the Equal Rights Amendment.

Because the survey satisfies the five characteristics reasonably well, for all practical purposes it can be viewed as a binomial experiment.

Example 6.2

A purchaser, who has received a boxcar containing 20 large electronic computers, wishes to sample three of the computers to see if they are in working order before he unloads the shipment. The three nearest the door of the boxcar are removed for testing and, afterward, are declared either defective or nondefective. Unknown to the purchaser, two of the computers are defective. Is this a binomial experiment?

Solution

As for Example 6.1, we check the sampling procedure against the characteristics of a binomial experiment.

1. The experiment consists of $n = 3$ identical trials. Each trial represents the selection and testing of one computer from the total of 20.

---

* Although it is traditional to call the two possible outcomes of a trial "success" and "failure," they could have been called "head" and "tail," "red" and "white," or any other pair of words. Consequently, the outcome called a success need not be viewed as a success in the ordinary usage of the word.

2. Each trial results in one of two outcomes. Either a computer is defective (call this a "success") or it is not (a "failure").

3. Suppose that the computers were randomly loaded into the boxcar so that any one of the 20 computers could have been placed near the boxcar door. Then, the unconditional probability of drawing a defective computer on a given trial will be 3/20.

4. The condition of independence between trials *is not* satisfied because the probability of drawing a defective computer on the second and third trials will be dependent on the outcome of the first trial. For example, if the first trial results in a defective computer, then there are only two defectives left of the remaining 19 in the boxcar. Therefore, the conditional probability of success on trial 2, given a success on trial 1, is 2/19. This differs from the unconditional probability of a success on the second trial (which is 3/20). Therefore, the trials are dependent and the sampling does not represent a binomial experiment.

Example 6.1 illustrates an important point. Very few real-life situations will perfectly satisfy the requirements stated above, but this is of little consequence as long as the lack of agreement is moderate and does not affect the end result. For instance, the probability of drawing a voter favoring a particular candidate in a political poll remains approximately constant from trial to trial as long as the population of voters is relatively large in comparison with the sample. If 50 percent of a population of 1000 voters prefer candidate A, then the probability of drawing an A on the first interview will be 1/2. The probability of an A on the second draw will equal 499/999 or 500/999, depending upon whether the first draw was favorable or unfavorable to A. Both numbers are near 1/2, for all practical purposes, and would continue to be for the third, fourth, and nth trial as long as n is not too large. Hence $P(A)$ remains approximately 1/2 from trial to trial and, for all practical purposes, we could regard the trials as independent. On the other hand, if the number in the population is 10, and 5 favor candidate A, then the probability of A on the first trial is 1/2; the probability of A on the second trial is 4/9 or 5/9, depending upon whether A was or was not drawn on the first trial. Thus, for small populations, the trial outcomes will represent dependent events and the resulting experiment will not be a binomial experiment.

6.2
The Binomial Probability Distribution

Having defined the binomial experiment and suggested several practical applications, we now turn to a derivation of the probability distribution for the random variable, y, the number of successes observed in n trials. Rather than attempt a direct derivation, we will obtain $p(y)$ for experiments containing $n = 1, 2,$ and 3 trials and leave the general formula to your intuition.

For $n = 1$ trial, we have two sample points, E_1 representing a success, S, and E_2 representing a failure, F, with probabilities p and $q = (1 - p)$, respectively. Since y is the number of successes for the one ($n = 1$) trial, and since E_1

Table 6.1 $p(y)$ for a binomial experiment, $n = 1$

Sample Points		$P(E_i)$	y
E_1	S	p	1
E_2	F	q	0

y	$p(y)$
0	q
1	p

$$\sum_{y=0}^{1} p(y) = q + p = 1$$

implies a success, we assign $y = 1$ to E_1. Similarly, since E_2 represents a failure for the single trial, we assign $y = 0$ to this sample point. The resulting probability distribution for y is given in Table 6.1. (*Note:* S represents a success on a single trial, F denotes a failure.)

The probability distribution for an experiment consisting of $n = 2$ trials is derived in a similar manner and is presented in Table 6.2. The four sample points associated with the experiments are presented in the first column with the notation SF implying a success on the first trial and a failure on the second.

The probabilities of the sample points are easily calculated because each point is an intersection of two independent events, namely, the outcomes of the first and second trials. Thus $P(E_i)$ can be obtained by applying the Multiplicative Law of Probability:

$$P(E_1) = P(SS) = P(S)P(S) = p^2,$$

$$P(E_2) = P(SF) = P(S)P(F) = pq,$$

$$P(E_3) = P(FS) = P(F)P(S) = qp,$$

$$P(E_4) = P(FF) = P(F)P(F) = q^2.$$

The value of y assigned to each sample point is given in the third column. You will note that the numerical event $y = 0$ contains sample point E_4, the event $y = 1$ contains sample points E_2 and E_3, and $y = 2$ contains sample point E_1. The probability distribution $p(y)$, presented to the right of Table 6.2, reveals a most interesting consequence: the probabilities $p(y)$ are terms of the expansion of $(q + p)^2$.

Summing, we obtain

$$\sum_{y=0}^{2} p(y) = q^2 + 2pq + p^2$$

$$= (q + p)^2 = 1.$$

The point that we wish to make is now quite clear; the probability distribution for the binomial experiment consisting of n trials is obtained by expanding $(q + p)^n$. The proof of this statement is omitted but may be obtained by using

Table 6.2 $p(y)$ for a binomial experiment, $n = 2$

Sample Points		$P(E_i)$	y
E_1	SS	p^2	2
E_2	SF	pq	1
E_3	FS	qp	1
E_4	FF	q^2	0

y	$p(y)$
0	q^2
1	$2pq$
2	p^2

$$\sum_{y=0}^{2} p(y) = q^2 + 2pq + p^2$$
$$= (q + p)^2$$
$$= (1)^2 = 1$$

Table 6.3 $p(y)$ for a binomial experiment, $n = 3$

Sample Points		$P(E_i)$	y
E_1	SSS	p^3	3
E_2	SSF	p^2q	2
E_3	SFS	p^2q	2
E_4	SFF	pq^2	1
E_5	FSS	p^2q	2
E_6	FSF	pq^2	1
E_7	FFS	pq^2	1
E_8	FFF	q^3	0

y	$p(y)$
0	q^3
1	$3pq^2$
2	$3p^2q$
3	p^3

$$\sum_{y=0}^{3} p(y) = q^3 + 3pq^2 + 3p^2q + p^3$$
$$= (q + p)^3 = (1)^3$$
$$= 1$$

combinatorial mathematics (optional Section 4.10) to count the appropriate sample points or can be proved by mathematical induction, a task which we leave to the student of mathematics. Those more easily convinced may acquire further evidence of the truth of our statement by observing the derivation of the probability distribution for a binomial experiment consisting of $n = 3$ trials presented in Table 6.3.

Since the probability associated with a particular value of y is simply the term involving p to the power y in the expansion of $(q + p)^n$, we may write the probability distribution for the binomial experiment as follows:

Binomial Probability Distribution

$$p(y) = C_y^n p^y q^{n-y},$$

where y may take values $0, 1, 2, 3, 4, \ldots, n$ and C_y^n is a symbol used to represent

$$\frac{n!}{y!(n-y)!}.$$

Figure 6.1 Binomial
probability distributions

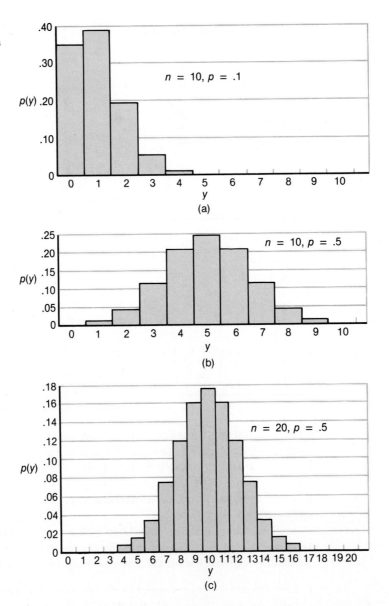

You may recall that the factorial notation $n!$ (defined in Section 5.6) is used to represent the product $n(n-1)(n-2)\cdots(3)(2)(1)$. For example, $5! = (5)(4)(3)(2)(1) = 120$, and $0!$ is defined to be equal to 1. The notation C_y^n is shorthand for $n!/y!(n-y)!$ an expression that appears in the formula for the binomial probability distribution.

Graphs of three binomial probability distributions are shown in Figure 6.1, the first for $n = 10$, $p = .1$, the second for $n = 10$, $p = .5$, and the third for $n = 20$, $p = .5$.

Let us now consider examples that illustrate applications of the binomial distribution.

Example 6.3

Over a long period of time it has been observed that a given rifleman can hit a target on a single trial with probability equal to .8. If he fires four shots at the target:

a. What is the probability that he will hit the target exactly two times?

Solution

Assuming that the trials are independent and that p remains constant from trial to trial, $n = 4$, $p = .8$, and

$$p(y) = C_y^4(.8)^y(.2)^{4-y},$$

$$p(2) = C_2^4(.8)^2(.2)^{4-2}$$

$$= \frac{4!}{2!2!}(.64)(.04) = \frac{4(3)(2)(1)}{2(1)(2)(1)}(.64)(.04)$$

$$= .1536.$$

b. What is the probability that he will hit the target at least two times?

Solution

$$\begin{aligned}
P(\text{at least two}) &= p(2) + p(3) + p(4) \\
&= 1 - p(0) - p(1) \\
&= 1 - C_0^4(.8)^0(.2)^4 - C_1^4(.8)(.2)^3 \\
&= 1 - .0016 - .0256 \\
&= .9728.
\end{aligned}$$

c. What is the probability that he will hit the target exactly four times?

Solution

$$p(4) = C_4^4(.8)^4(.2)^0$$

$$= \frac{4!}{4!0!}(.8)^4(1)$$

$$= .4096.$$

Note that these probabilities would be incorrect if the rifleman could observe the location of each hit on the target; in that case, the trials would be dependent and p would likely increase from trial to trial.

Example 6.4

Large lots of incoming product at a manufacturing plant are inspected for defectives by means of a sampling scheme. Ten items are to be examined and the lot rejected if two or more defectives are observed. If a lot contains exactly 5 percent defectives, what is the probability that the lot will be accepted? rejected?

Solution   If y equals the number of defectives observed, then $n = 10$ and the probability of observing a defective on a single trial will be $p = .05$. Then,

$$p(y) = C_y^{10}(.05)^y(.95)^{10-y}$$

and

$$P(\text{accept}) = p(0) + p(1) = C_0^{10}(.05)^0(.95)^{10} + C_1^{10}(.05)^1(.95)^9$$
$$= .914,$$

$$P(\text{reject}) = 1 - P(\text{accept})$$
$$= 1 - .914$$
$$= .086.$$

Although in a practical situation we would not know the exact value of p, we would want to know the probability of accepting bad lots (lots for which p is large) and good lots for which p is small). This example shows how we can calculate this probability of acceptance for various values of p.

***

Example 6.5   A new serum was tested to determine its effectiveness in preventing the common cold. Ten people were injected with the serum and observed for a period of one year. Eight survived the winter without a cold. Suppose that it is known that when serum is not used, the probability of surviving a winter without a cold is .5. What is the probability of observing eight or more survivors, given that the serum is ineffective in increasing the bodily resistance to colds?

Solution   Assuming that the vaccine is ineffective, the probability of surviving the winter without a cold is $p = .5$. The probability distribution for y, the number of survivors, is

$$p(y) = C_y^{10}(.5)^y(.5)^{10-y} = C_y^{10}(.5)^{10},$$

$$P(8 \text{ or more}) = p(8) + p(9) + p(10)$$
$$= C_8^{10}(.5)^{10} + C_9^{10}(.5)^{10} + C_{10}^{10}(.5)^{10}$$
$$= .055.$$

***

Examples 6.3, 6.4, and 6.5 illustrate the use of the binomial probability distribution in calculating the probability associated with values of y, the number of successes in n trials defined for the binomial experiment. Thus we note that the probability distribution, $p(y) = C_y^n p^y q^{n-y}$, provides a simple formula for calculating the probabilities of numerical events, y, applicable to a broad class of experiments that occur in everyday life. This statement must be accompanied by a word of caution. The important point, of course, is that each physical application must be carefully checked against the defining characteristics of the binomial experiment, Section 6.1, to determine whether the binomial experiment is a valid model for the application of interest.

You will note that Examples 6.3, 6.4, and 6.5 were problems in probability rather than statistics. The composition of the binomial population, characterized by p, the probability of a success on a single trial, was assumed known and we were interested in calculating the probability of certain numerical events. As we proceed in this chapter, we will reverse the procedure; that is, we will assume that we possess a sample from the population and wish to make inferences concerning p. The physical settings for Examples 6.4 and 6.5 supply excellent practical situations in which the ultimate objective was statistical inference. We shall consider these two problems in greater detail in the succeeding sections.

Exercises

6.1. A jar contains five balls—three red and two white. Two balls are randomly selected without replacement from the jar and the number y of red balls is recorded. Explain why y is or is not a binomial random variable. (*Hint:* Compare the characteristics of this experiment with the characteristics of a binomial experiment given in Section 6.1.) If the experiment is binomial, give the values of n and p.

6.2. Answer Exercise 6.1 assuming that the sampling was conducted with replacement; that is, the first ball was selected from the jar and observed. It was then replaced and the balls were mixed before the second ball was selected.

6.3. Ralph Nader's Health Research Group has petitioned the Occupational Safety and Health Administration to reduce the maximum allowable exposure to ethylene oxide from 50 parts per million parts of air to 1 part per million per 8 hours of exposure. (*Washington Post*, August 14, 1981). Ethylene oxide is a gas used as a pesticide and fumigant, and it is used in the manufacture of some cosmetics and chemicals. Let y represent the number of parts of ethylene oxide in one million parts of air. Explain why y is or is not a binomial random variable.

6.4. Calculate $p(y)$ for $y = 0, 1, 2, \ldots, 5, 6$ for $n = 6$ and $p = .1$. Graph $p(y)$. Repeat for binomial probability distributions for $p = .5$ and $p = .9$. Compare the graphs. How does the value of p affect the shape of $p(y)$? (*Note:* This exercise does not require extensive calculations. For example, you need calculate the coefficients C_y^n only once because they will be the same for all three probability distributions.)

6.5. Calculate the value of $p(y)$ and construct a probability histogram for:
   a. $n = 5, p = .2$.   b. $n = 5, p = .5$.   c. $n = 5, p = .8$.

6.6. A study of significant high school football injuries showed that approximately 1/3 of the injuries involved the knee (*U.S. News and World Report*, March 23, 1981). Suppose that a 40-man football squad sustains 12 significant injuries during a season and seven are injuries of the knee. If y is the number of significant injuries that are injuries of the knee, explain why y is or is not a binomial random variable. If y is a binomial random variable, give the values of n and p.

6.7.   Many employers are finding that some of the people they hire are not who and what they claim to be. Detecting falsification of job application information has spawned some new businesses, credential checking services. *U.S. News and World Report* (July 13, 1981), reporting on this problem, notes that, in a two-month period, one service found that that 35 percent of all credentials examined were falsified. Suppose that you hired five new employees last week and that the probability that a single employee would falsify information on his or her application form is .35. What is the probability that at least one of the five application forms has been falsified? Two or more?

6.8.   Records show that 30 percent of all patients admitted to a medical clinic fail to pay their bills and eventually the bills are forgiven. Suppose that $n = 4$ new patients represent a random selection from the large set of prospective patients served by the clinic. Find the probability that:

   a. All the patients' bills will eventually have to be forgiven.

   b. One will have to be forgiven.

   c. None will have to be forgiven.

6.9.   Refer to Exercise 6.8 and let y equal the number of patients in the sample of $n = 4$ whose bills will have to be forgiven. Construct a probability histogram for $p(y)$.

6.10.   Approximately 10 percent of all thunderstorms tracked over the past five years by the National Oceanic and Atmospheric Administrative Severe Storms Laboratory were "mesocyclones," storms that breed or turn into tornadoes (*Washington Post*, December 7, 1976). Suppose a particular community was subjected to seven thunderstorms and two are judged to be "mesocyclonic." What is the probability of observing two mesocyclones out of a total of seven thunderstorms? two or more? Would you regard two or more as a "rare event," an event that would occur with small probability if in fact the probability of observing a mesocyclonic thunderstorm is only 10 percent?

6.11.   Consider a metabolic defect that occurs in approximately 1 of every 100 births. If four infants are born in a particular hospital on a given day, what is the probability that:

   a. None has the defect?

   b. No more than one has the defect?

6.12.   Early in the United States missile development program, government and industry defense officials were proclaiming that our missiles were highly reliable, and probabilities of successful firings in the neighborhood of .999 . . . were quoted. Such statements were made even though many missile firings resulted in failure (such as the Navy *Vanguard* missile of the 1950s). If the reliability (probability of a successful launch) is even as high as .9, what is the probability of observing as many as three failures in a total of four firings? As many as one failure? If you observed two or more failures out of four, what would you think about the high claims of reliability of the missiles produced in the 1950s?

6.13.   A new surgical procedure is said to be successful 80 percent of the time. If the operation is performed five times and if we can assume that the results are independent of one another:

   a. What is the probability that all five operations are successful?

   b. Exactly four?

   c. Less than two?

6.14.   Refer to Exercise 6.13. If less than two operations were successful, how would you feel about the performance of the surgical team? (We shall have more to say about this type of reasoning in Section 6.6.)

6.15. Many utility companies have begun to promote energy conservation by offering discounted rates to consumers who keep their energy usage below certain established subsidy standards. A recent EPA report notes that 70 percent of the island residents of Puerto Rico have reduced their electricity usage sufficiently to qualify for discounted rates. If five residential subscribers are randomly selected from San Juan, Puerto Rico, find the probability that:

a. All five qualify for the favorable rates.

b. At least four qualify for the favorable rates.

6.16. A survey in a certain state indicated that nine out of ten automobiles carry automobile liability insurance. If four autos in that state are involved in accidents, what is the probability that:

a. No more than two of the four drivers have liability insurance?

b. Exactly two have liability insurance?

6.3
The Mean and Variance for the Binomial Random Variable

Because calculation of $p(y)$ becomes very tedious for large values of n, it is convenient to describe the binomial probability distribution using its mean and standard deviation. This will enable us to identify values of y that are highly improbable simply by using our knowledge of Tchebysheff's Theorem and the Empirical Rule. A more precise method for approximating binomial probabilities will be presented in Chapter 7, and this method will rely on knowledge of the mean and standard deviation of y, namely μ and σ. Consequently, we need to find the expected value and variance of the binomial random variable, y.

Our approach to the acquisition of formulas for μ and σ^2 will be similar to that employed in Section 6.2 for the determination of the probability distribution, $p(y)$. Since $\mu = E(y)$ and $\sigma^2 = E[(y - \mu)^2]$, we shall use the methods of Chapter 5 to obtain these expectations; that is, $E(y)$ and $E(y - \mu)^2$, for the simple cases $n = 1, 2,$ and 3, and then give the formulas for the general case without proof. The interested reader can derive the formulas for the general case involving n trials by using the summation theorems of Chapter 2, some algebraic manipulation, and a bit of ingenuity (see Mendenhall, Scheaffer, and Wackerly, Chapter 3).

The expected value of y for $n = 1$ and 2 can be derived using the probabilities given in Tables 6.1 and 6.2 along with the definition of an expectation presented in Section 5.6. Thus for $n = 1$ we obtain

$$\mu = E(y) = \sum_{y=0}^{1} yp(y) = (0)(q) + (1)(p) = p.$$

For $n = 2$,

$$\mu = E(y) = \sum_{y=0}^{2} yp(y) = (0)(q^2) + (1)(2pq) + (2)(p^2)$$

$$= 2p(q + p) = 2p.$$

Using Table 6.3, you can quickly show that $E(y)$ for $n = 3$ trials is equal to $3p$ and surmise that this pattern holds in general. Indeed, it can be shown that the expected value of y for a binomial experiment consisting of n trials is

---

Mean for a Binomial Random Variable

$$\mu = E(y) = np.$$

Similarly, we may obtain the variance of y for $n = 1$ and 2 trials as follows. For $n = 1$,

$$\sigma^2 = E(y - \mu)^2 = \sum_{y=0}^{1} (y - \mu)^2 p(y) = (0 - p)^2(q) + (1 - p)^2(p)$$

$$= p^2 q + q^2 p = pq(q + p).$$

Or, since $q + p = 1$,

$$\sigma^2 = pq.$$

For $n = 2$,

$$\sigma^2 = E(y - \mu)^2 = \sum_{y=0}^{2} (y - \mu)^2 p(y)$$

$$= (0 - 2p)^2(q^2) + (1 - 2p)^2(2pq) + (2 - 2p)^2(p^2)$$

$$= 4p^2 q^2 + (1 - 4p + 4p^2)(2pq) + 4(1 - p)^2 p^2.$$

Using the substitution $q = 1 - p$ and a bit of algebraic manipulation, we find that the above reduces to

$$\sigma^2 = 2pq.$$

You may verify that the variance of y for $n = 3$ trials is $3pq$. In general, for n trials it can be shown that the variance of y is

---

Variance and Standard Deviation for a Binomial Random Variable

$$\sigma^2 = npq.$$

$$\sigma = \sqrt{npq}.$$

As we have previously mentioned, the formulas for μ and σ^2 can now be employed to compute the mean and variance for the binomial random variable, y, for a practical binomial experiment. Thus we can say something concerning the center of the distribution as well as the variability of y.

Example 6.6

How do you evaluate scores on a multiple-choice test? A score of 0 on an objective test (questions requiring complete recall of the material) indicates that the person was unable to recall the test material at the time the test was given. In contrast, a person with little or no recall knowledge of the test material can score higher on a multiple-choice test because the person only needs to recognize (in contrast to recall) the correct answer and because some questions will be answered correctly, just by chance, even if the person does not know the correct answers. Consequently, the no-knowledge score for a multiple-choice test may be well above 0. If a multiple-choice test contains 100 questions, each with six possible answers, what is the ex ected score for a person who possesses no knowledge of the test material? Within what limits would you expect a no-knowledge score to be?

Solution

Let p equal the probability of a correct choice on a single question and let y equal the number of correct responses out of the $n = 100$ questions. We will assume that no-knowledge means that a student will randomly select one of the six possible answers for each question and hence that $p = 1/6$. Then for $n = 100$ questions, a no-knowledge student's expected score would be $E(y)$, where,

$$E(y) = np = 100(1/6) = 16.7 \text{ correct questions.}$$

To evaluate the variation of no-knowledge scores, we need to know σ, where

$$\sigma = \sqrt{npq} = \sqrt{(100)(1/6)(5/6)} = 3.7.$$

Then from our knowledge of Tchebysheff's Theorem and the Empirical Rule,* we would expect most no-knowledge scores to lie in the interval $\mu \pm 2\sigma$, or $16.7 \pm (2)(3.7)$, or from 9.3 to 24.1. This compares with a score of 0 for a no-knowledge student taking an objective recall test.

Exercises

6.17. Find the mean and standard deviation for binomial distributions when:

    a. $n = 100, p = .1$.   b. $n = 100, p = .5$.

    c. $n = 100, p = .9$.   d. $n = 100, p = .01$.

6.18. Use the values of μ and σ calculated in Exercise 6.17 to make rough sketches of the four distributions. In each case, locate the distribution over a y axis marked from 0 to 100. How does the distribution, part (d), differ from those in parts (a), (b), and (c)?

---

* A histogram of $p(y)$ for $n = 100$, $p = 1/6$ will be mound-shaped. Hence we would expect the Empirical Rule to work very well. The reason for this will be explained in Chapter 7.

6.19. In an article on hang gliding the *Wall Street Journal* (July 29, 1981) notes that hang gliding can be dangerous. Of the estimated 36,000 hang gliding enthusiasts in the United States, 29 were killed last year in hang gliding accidents, and it is estimated that thousands more were injured. Suppose that a hang glider pilot, randomly selected from among all hang glider pilots, has a probability equal to .0006 of being killed in any one year. If there are 40,000 pilots next year,

   a. What is the expected number of fatalities?

   b. What is the standard deviation of the number y of fatalities?

   c. Is it likely that the number of fatalities would exceed 40? (*Hint:* Use Tchebysheff's Theorem and the Empirical Rule to help answer this question.)

6.20. In Exercise 4.31 we mentioned that some hospitals are requiring all employees to take lie detector tests in an effort to combat rising losses due to theft. As a case in point, the *Orlando Sentinel Star* (August 3, 1981) reports that the Fish Memorial Hospital, New Smyrna Beach, Florida, a hospital with approximately 320 employees, has instituted this policy. Suppose that the tests on Fish Memorial's lie detector are independent experiments and that the probability that the machine erroneously judges an employee to be lying is equal to .01. If all 320 of the hospital's employees responded truthfully during the lie detector test, and if y is the number that will be judged to be lying,

   a. Find the expected value of y.

   b. Find the standard deviation of y.

   c. Suppose that the number of employees judged to be lying is equal to 11. What would you conclude?

6.21. Let us consider the medical payment problem (Exercise 6.8) in a more realistic setting. We were given the fact that 30 percent of all patients admitted to a medical clinic fail to pay their bills and that the bills are eventually forgiven. If the clinic treats 2000 different patients over a period of one year, what is the mean (expected) number of bills that would have to be forgiven? If y is the number of forgiven bills in the group of 2000 patients, find the variance and standard deviation of y. What can you say about the probability that y will exceed 700? (*Hint:* Use the values of μ and σ, along with Tchebysheff's Theorem, to answer this question.)

6.22. A national poll of 1502 adults by pollster Louis Harris (*Environmental News*, EPA, September 1977) indicated that 71 percent "would rather live in an environment that is clean rather than in an area with a lot of jobs." If people really had no preference for environment over jobs, or vice versa, comment on the probability of observing this survey result (that is, observing 71 percent or more of a sample of 1502 in favor of environment over jobs). What assumption must be made about the sampling procedure in order to calculate this probability? (*Hint:* Recall Tchebysheff's Theorem and the Empirical Rule.)

6.23. The administrator of the North Florida Regional Hospital (a private hospital) applied to the health planning council in 1975 for permission to make a 100-bed addition to the hospital, an application that was subsequently denied because of the uncertain impact that the addition would have on the existing public county hospital (*Gainesville Sun*, February 12, 1977). To support their application, and, presumably, those applications to be resubmitted in the future, the hospital administrators conducted a telephone survey of 180 people in the Gainesville area. In response to the question, "Should North Florida be allowed to expand ... even if (there are) empty beds at AGH (the county hospital)," 62 percent answered yes. If the public is really split 50–50 on this question, is it likely that the telephone poll would result in as many as 62 percent in favor of the North Florida expansion? In order for the preceding probability calculation to be correct, what must you know about the sampling procedure? Do you think the poll supports North Florida's request? Explain. (*Hint:* Recall Tchebysheff's Theorem and the Empirical Rule.)

6.24. The Energy Policy Center of the EPA reports that 75 percent of the homes in New England are heated by oil-burning furnaces. If a certain New England community is known to have 2500 homes, find the expected number of homes in the community that are heated by oil furnaces. If y is the number of homes in the community that are heated by oil, find the variance and standard deviation of y. Use Tchebysheff's Theorem to describe limits within which one could expect y to fall.

6.25. Suppose that 80 percent of a breed of hogs is infected with trichinosis. If a random sample of 1000 hogs is examined by a state inspector:

a. What is the expected value of y, the number of infected hogs in the sample of 1000?

b. What is the standard deviation of y?

c. If the inspector observed $y = 900$ infected pigs in the sample, does it appear that the percentage of infected pigs in the population is really 80 percent? Explain.

6.26. If a political candidate possesses exactly 50 percent of the popular vote and the early election returns show a total of 10,000 votes:

a. What is the expected value of y, the number of votes in favor of the candidate, if the 10,000 votes can be regarded as a random sample from the total population of voters?

b. What is the standard deviation of y?

c. Suppose that $y = 4700$. Is this a value of y that might be expected with reasonable probability? How might you explain this observed result?

6.27. It is known that 10 percent of a brand of television tubes will burn out before their guarantee has expired. If 1000 tubes are sold, find the expected value and variance of y, the number of original tubes that must be replaced. Within what limits would y be expected to fall? (*Hint:* Use Tchebysheff's Theorem.)

6.4
More on the Fear of Flying Survey

As we noted in our case study, estimation of the proportion of potential airline customers who possess some common characteristic is dependent upon the probability distribution of y, the number of yes (or no) responses to a survey question. Since the number of persons contacted in a survey will constitute a random sample from among a very large number of persons, y will possess, for all practical purposes, a binomial probability distribution.

To illustrate, suppose that 2000 potential airline travelers are randomly selected from among a specific group—say, all potential travelers who are 18 to 22 years old—and that, in actual fact, 5 percent of the people in this group are afraid to fly. Then the probability distribution for y, the number of persons who would indicate a fear of flying (yes responses), would be binomial with mean and standard deviation equal to

$$\mu = np = (2000)(.05) = 100$$
$$\sigma = \sqrt{npq} = \sqrt{(2000)(.05)(.95)} = 9.75.$$

Thus, if our suggestion is true, that 5 percent of all people in the 18-to-22-year-old age group have a fear of flying, we would expect the number y of yes responses to assume a value in the interval:

$$\mu \pm 2\sigma = 100 \pm 2(9.75)$$

or, from 80.5 to 119.5.

Suppose, in fact, that the response to a survey was $y = 69$ persons who indicated a fear of flying. What could you conclude? Either you have observed a highly improbable event (a very small number of yes responses, $y = 69$, considering that $p = .05$) or you could conclude that p is less than .05.

If you reach the latter conclusion, be careful how you interpret the result. The fraction p of persons in the 18-to-22-year-old bracket who fear flying may be less than .05. Or perhaps you are sampling a population of yes and no responses that are of no interest to you. For example, suppose that some people would be embarrassed to admit that they were afraid to fly and hence would respond "no" to the survey question. Then the population would contain a higher fraction of noes than if the respondents had given truthful answers and would yield a population that would not truly characterize the phenomenon of interest to you. Methods for coping with questions that might prove embarrassing and yield false responses, called randomized response techniques, have been devised to cope with this problem (see Scheaffer, Mendenhall, and Ott, 1980).

You can inadvertently sample from the wrong population of yeses and noes when you sample by mail or telephone. For example, business people, many of whom fly and have no fear of flying, may not be willing to take the time to respond to the survey question. The exclusion of many of these respondents from the sample changes the nature of the population and is likely to reduce the proportion of noes. In general, sampling procedures that elicit responses from only a portion of the persons originally selected for the sample may alter the nature of the sampled population and lead to invalid inferences.

The point to note is that the causes of improbable sample results must be examined carefully. They may arise because you have made incorrect assumptions about one or more population parameters (in this case p), but it is also possible that they may have been caused by an inappropriate sampling procedure.

6.5
Making Decisions: Lot Acceptance Sampling for Defectives

A manufacturing plant may be regarded as an operation that transforms raw materials into a finished product, the raw materials entering the rear door of the plant and the product moving out the front. In order to operate efficiently, a manufacturer desires to minimize the amount of defective raw material received and, in the interest of quality, minimize the number of defective products shipped to customers. To accomplish this objective, the manufacturer

Figure 6.2 Screening for defectives

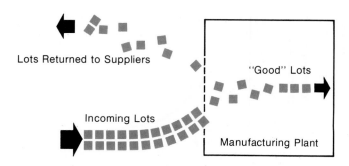

Lots Returned to Suppliers

"Good" Lots

Incoming Lots

Manufacturing Plant

will erect a "screen" at both doors in an attempt to prevent defectives from passing through.

To simplify our discussion, consider only the screening of incoming raw materials consisting of large lots (boxes) of items such as bolts, nails, or bearings. Diagrammatically, the screen would function in the manner indicated in Figure 6.2. The lots would proceed to the rear door of the plant, would be accepted if the fraction defective were small, and would be returned to the supplier if the fraction defective were large.

A screen could be constructed in a number of ways, the most obvious and seemingly perfect solution being a complete and careful inspection of each single item received. Unfortunately, the cost of total inspection is often enormous and hence economically unfeasible. A second disadvantage, not readily apparent, is that, even with complete inspection, defective items seem to slip through the screen. Humans become bored and lose perception when subjected to long hours of inspecting, particularly when the operation is conducted at high speed. Thus nondefective items are often rejected and defective items accepted.

A final disadvantage of total inspection is that some tests are by their very nature destructive. Testing a photoflash bulb to determine the quantity of light produced destroys the bulb. If all bulbs were tested in this manner, the manufacturer would have none left to sell.

A second type of screen, relatively inexpensive and lacking the tedium of total inspection, involves the use of a **statistical sampling plan** similar to the plan described in Example 6.4. A sample of n items is chosen at random from the lot of items and inspected for defectives. If the number of defectives, y, is less than or equal to a predetermined number, a, called the **acceptance number,** the lot is accepted. Otherwise the lot is rejected and returned to the supplier. The acceptance number for the plan described in Example 6.4 was $a = 1$.

You will note that this sampling plan operates in a completely objective manner and results in an inference concerning the population of items contained in the lot. If the lot is rejected, we infer that the fraction defective, p, is too large; if the lot is accepted, we infer that p is small and acceptable for use in the manufacturing process. Lot acceptance sampling plans provide an example of a statistical decision process which is, indeed, a method of statistical inference.

Figure 6.3 Typical operating characteristic curve for a sampling plan

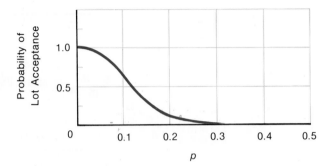

Our discussion of lot acceptance sampling plans would be incomplete if we were to neglect some comment concerning the "goodness" of our inference-making procedure. Although the sampling plan described above is a decision-making process, it is not unique. We could change the number in the sample, n, change a, the acceptance number, or, for that matter, we could use some non-statistical decision-making process, a procedure that is not uncommon in practice. In actuality, each individual is a decision maker relying on individual whims, preferences, and tastes. How can we compare these decision-making processes? The answer is immediately at hand; we choose the decision-making process which makes the correct decision most frequently, or, alternatively, makes the incorrect decision the smallest fraction of the time.

Quality control engineers characterize the goodness of a sampling plan by calculating the probability of lot acceptance for various lot fractions defective. This result is presented in a graphic form and is called the **operating characteristic curve** for the sampling plan. A typical operating characteristic curve is shown in Figure 6.3. In order for the screen to operate satisfactorily, we would like the probability of accepting lots with a low fraction defective to be high and the probability of accepting lots with a high fraction defective to be low. You will note that the probability of acceptance always drops as the fraction defective increases, a result that is in agreement with our intuition.

For instance, suppose that the supplier guarantees that lots will contain less than 1 percent defective and that the manufacturer can operate satisfactorily with lots containing less than 5 percent defective. Then the probability of accepting lots with less than 1 percent defective should be high. Otherwise the supplier will raise his price to cover the cost of "good lots" (less than 1 percent defective) which have been returned or will charge the manufacturer a fee for reinspection. On the other hand, the probability of accepting lots with 5 percent or more defective should be very low.

Example 6.7

Calculate the probability of lot acceptance for a sampling plan with sample size $n = 5$ and acceptance number $a = 0$ for lot fraction defective $p = .1, .3,$ and $.5$. Sketch the operating characteristic curve for the plan.

Figure 6.4 Operating characteristic curve, $n = 5$, $a = 0$

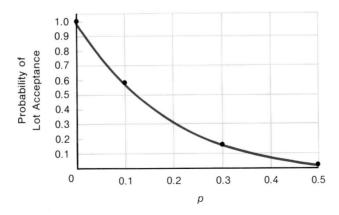

Solution

$$P(\text{accept}) = p(0) = C_0^5 p^0 q^5 = q^5,$$

$$P(\text{accept} \,|\, p = .1) = (.9)^5 = .590,$$

$$P(\text{accept} \,|\, p = .3) = (.7)^5 = .168,$$

$$P(\text{accept} \,|\, p = .5) = (.5)^5 = .031.$$

A sketch of the operating characteristic curve can be obtained by plotting the three points obtained from the above calculation. In addition, we know that the probability of acceptance must equal 1 when $p = 0$ and must equal zero when $p = 1$. The operating characteristic curve is given in Figure 6.4.

Calculating the binomial probabilities is a tedious task when n is large. To simplify our calculations, the sum of the binomial probabilities from $y = 0$ to $y = a$ is presented in the Appendix, Table 1, for sample sizes $n = 5, 10, 15, 20,$ and 25.

Why do we give the sum $\sum_{y=0}^{a} p(y)$ instead of the individual values of $p(y)$? The answer is that we will most often need sums of values of $p(y)$ to solve practical problems. If we gave individual values of $p(y)$ rather than the sums, you would be required to do more calculating when solving exercises.

To see how you use Table 1, suppose that you wish to find the sum of the binomial probabilities from $y = 0$ to $y = 3$ for $n = 5$ trials and $p = .6$. That is, you wish to find

$$\sum_{y=0}^{3} p(y) = p(0) + p(1) + p(2) + p(3)$$

where

$$p(y) = C_y^5(.6)^y(.4)^{5-y}$$

Turn to Table 1(a), the Appendix, for $n = 5$. Since the tabulated values in the table give

$$\sum_{y=0}^{a} p(y),$$

you seek the tabulated value in the row corresponding to $a = 3$ and the column for $p = .6$. The tabulated value, .663, is shown below as it appears in Table 1(a):

$n = 5$														
							p							
a	0.01	0.05	0.10	0.20	0.30	0.40	0.50	0.60	0.70	0.80	0.90	0.95	0.99	a
0	—	—	—	—	—	—	—	—	—	—	—	—	—	0
1	—	—	—	—	—	—	—	—	—	—	—	—	—	1
2	—	—	—	—	—	—	—	—	—	—	—	—	—	2
3	—	—	—	—	—	—	—	.663	—	—	—	—	—	3
4	—	—	—	—	—	—	—	—	—	—	—	—	—	4

Thus, $\sum_{y=0}^{3} p(y)$, the sum of the binomial probabilities from $y = 0$ to $a = 3$ (for $n = 5$, $p = .6$) is .663.

Table 1, the Appendix, can also be used to find an individual binomial probability, say $p(3)$ for $n = 5$, $p = .6$. We would calculate

$$p(3) = [p(0) + p(1) + p(2) + p(3)] - [p(0) + p(1) + p(2)]$$

$$= \sum_{y=0}^{3} p(y) - \sum_{y=0}^{2} p(y)$$

$$= .663 - .317$$

$$= .346.$$

Thus an individual value of $p(y)$ is equal to the difference between two sums which are adjacent entries given in a column of the table.

We will use Table 1 in the following example.

Figure 6.5 Operating
characteristic curve,
$n = 15, a = 1$

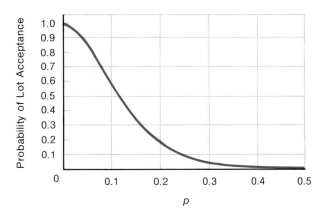

Example 6.8

Construct the operating characteristic curve for a sampling plan with $n = 15, a = 1$.

Solution

The probability of lot acceptance will be calculated for $p = .1, .2, .3, .5$.

$$P(\text{accept}) = p(0) + p(1) = \sum_{y=0}^{a=1} p(y),$$

$P(\text{accept} | p = .1) = .549,$

$P(\text{accept} | p = .2) = .167,$

$P(\text{accept} | p = .3) = .035,$

$P(\text{accept} | p = .5) = .000.$

The operating characteristic curve for the sampling plan is given in Figure 6.5.

    Sampling inspection plans are widely used in industry. Each sampling plan possesses its own unique operating characteristic curve which characterizes the plan and, in a sense, describes the size of the holes in the screen. The quality control engineer will choose the plan that satisfies the requirements of his or her situation. Increasing the acceptance number increases the probability of acceptance and hence increases the size of the holes in the screen. Increasing the sample size provides more information upon which to base the decision and hence improves the discriminatory power of the decision procedure. Thus, when n is large, the operating characteristic curve will drop rapidly as p increases. You may verify these remarks by working the exercises at the end of the chapter.

    Note that acceptance sampling is an example of statistical inference because the procedure implies a decision concerning the lot fraction defective, p. If you accept a lot, you infer that the true lot fraction defective, p, is some relatively small, acceptable value. If you reject a lot, it is clear that you think p

is too large. Consequently, lot acceptance sampling is a procedure that implies inference making concerning the lot fraction defective. The operating characteristic curve for the sampling plan provides a measure of the goodness of this inferential procedure.

Tips on Problem Solving

Remember that lots are accepted when the number y of defectives is a small number; that is,

$$y = 0, 1, 2, \ldots, a,$$

where a is the acceptance number. Therefore, the probability of lot acceptance is always equal to the summation of the values of $p(y)$ from $y = 0$ to $y = a$. For values of $n = 5, 10, 15, 20,$ and 25, this partial sum can be read directly from Table 1, the Appendix. The partial sums for other values of n can be found in tables listed in the references.

Exercises

6.28.   Use Table 1, the Appendix, to find the sum of the binomial probabilities from $y = 0$ to $y = a$ for:
   a. $n = 10, p = .1, a = 3$.
   b. $n = 15, p = .6, a = 7$.
   c. $n = 25, p = .5, a = 14$.

6.29.   Use the formula for the binomial probability distribution to calculate the values of $p(y)$ for $n = 5, p = .5$ (this was done in Exercise 6.5). Next find $\sum_{y=0}^{a} p(y)$ for $a = 0, 1, 2, 3, 4$ using the values of $p(y)$ that you computed. Then compare these sums with the values given in Table 1, the Appendix.

6.30.   Use the information given in Table 1, the Appendix, to find $p(3)$ for $n = 5, p = .5$. Then compare this answer with the value of $p(3)$ calculated in Exercise 6.29.

6.31.   Use the information given in Table 1, the Appendix, to find $p(3) + p(4)$ for $n = 5, p = .5$. Verify this answer using the values of $p(y)$ that you calculated in Exercise 6.29.

6.32.   A buyer and seller agree to use a sampling plan with sample size $n = 5$ and acceptance number $a = 0$. What is the probability that the buyer will accept a lot having the following fractions defective?
   a. $p = .1$ b. $p = .3$ c. $p = .5$ d. $p = 0$ e. $p = 1$
   Construct the operating characteristic curve for this plan.

6.33.   Repeat Exercise 6.32 for $n = 5, a = 1$.

6.34.   Repeat Exercise 6.32 for $n = 10, a = 0$.

6.35.   Repeat Exercise 6.32 for $n = 10, a = 1$.

6.36.   Graph the operating characteristic curves for the four plans given in Exercises 6.32, 6.33, 6.34, and 6.35 on the same sheet of graph paper. What is the effect of increasing the acceptance number, a, when n is held constant? What is the effect of increasing the sample size, n, when a is held constant?

6.37. A radio and television manufacturer who buys large lots of transistors from an electronics supplier selects $n = 25$ transistors from each lot shipped by the supplier and notes the number of defectives.

   a. On the same sheet of graph paper, construct the operating characteristic curves for the sampling plans $n = 25$ with $a = 1, 2,$ and 3.

   b. Which sampling plan best protects the supplier from having acceptable lots rejected and returned by the manufacturer?

   c. Which sampling plan best protects the manufacturer from accepting lots for which the fraction of defectives is exceedingly large?

   d. How might the sampling inspector arrive at an acceptance level that compromises between the risk to the producer and the risk to the consumer?

6.38. Refer to Exercise 6.37 and assume that the manufacturer wishes the probability to be at least .90 of his accepting lots containing 1 percent defective, and the probability to be about .90 of rejecting any lot with 10 percent or more defective. If the manufacturer's sampling inspector samples $n = 25$ items from the supplier's incoming shipments, what is the acceptance number (a) that more nearly meets these requirements?

6.39. A buyer and a seller agree to use sampling plan ($n = 15, a = 0$) or sampling plan ($n = 25, a = 1$). Sketch the operating characteristic curves for the two sampling plans. If you were a buyer, which of the two sampling plans would you prefer? Why?

6.6
Making Decisions: A Test of an Hypothesis

The cold vaccine problem, Example 6.5, is illustrative of a **statistical test of an hypothesis.** The practical question to be answered concerns the effectiveness of the vaccine. Do the data contained in the sample present sufficient evidence to indicate that the vaccine is effective?

    The reasoning employed in testing an hypothesis bears a striking resemblance to the procedure used in a court trial. In trying a man for theft, the court assumes the accused innocent until proved guilty. The prosecution collects and presents all available evidence in an attempt to contradict the "not guilty" hypothesis and hence to obtain a conviction. The statistical problem portrays the vaccine as the accused. The hypothesis to be tested, called the **null hypothesis,** is that the vaccine is ineffective. The evidence in the case is contained in the sample drawn from the population of potential vaccine customers. The experimenter, playing the role of the prosecutor, believes that his vaccine is really effective and hence attempts to use the evidence contained in a sample to reject the null hypothesis and thereby to support his contention that the vaccine is, in fact, a very successful cold vaccine. You will recognize this procedure as an essential feature of the scientific method, where all theories proposed must be compared with reality.

    Intuitively, we would select the number of survivors, y, as a measure of the quantity of evidence in the sample. If y were very large, we would be inclined to reject the null hypothesis and conclude that the vaccine is effective. On the other hand, a small value of y would provide little evidence to support the rejection of the null hypothesis. As a matter of fact, if the null hypothesis were true and the vaccine were ineffective, the probability of surviving a winter without a cold would be $p = 1/2$ and the average value of y would be

$$E(y) = np = 10(1/2) = 5.$$

Most individuals utilizing their own built-in decision makers would have little difficulty arriving at a decision for the case $y = 10$ or $y = 5, 4, 3, 2,$ or 1, which, on the surface, appear to provide substantial evidence to support rejection or nonrejection, respectively, of the null hypothesis. But what can be said concerning less obvious results, say $y = 7, 8,$ or 9? Clearly, whether we employ a subjective or an objective decision-making procedure, we would choose the procedure that gave the smallest probability of making an incorrect decision.

The statistician would test the null hypothesis in an objective manner similar to our intuitive procedure. A **decision maker,** commonly called a **test statistic,** is calculated from information contained in the sample. In our example, the number of survivors, y, would suffice for a test statistic. We would then consider all possible values that the test statistic may assume, for example, $y = 0, 1, 2, \ldots, 9, 10$. These values would be divided into two groups, as shown in Figure 6.6, one called the **rejection region** and the other the **acceptance region.** An experiment is then conducted and the decision maker, y, observed. If y takes a value in the rejection region, the hypothesis is rejected. Otherwise, it is accepted. (*Caution*: **As you will subsequently learn, you will reject or accept the null hypothesis only if the risks of a wrong decision are known and are small for these two actions.**) For example, we might choose $y = 8, 9,$ or 10 as the rejection region and assign the remaining values of y to the acceptance region. Since we observed $y = 8$ survivors, we would reject the null hypothesis that the vaccine is ineffective and conclude that the probability of surviving the winter without a cold is greater than $p = 1/2$ when the vaccine is used. What is the probability that we will reject the null hypothesis when, in fact, it is true? The probability of falsely rejecting the null hypothesis is the probability that y will equal 8, 9, or 10, given that $p = 1/2$, and this is indeed the probability computed in Example 6.5 and found to equal 0.55. Since we have decided to reject the null hypothesis and note that this probability is small, we are reasonably confident that we have made the correct decision.

Upon reflection, you will observe that the cold vaccine manufacturer is faced with two possible types of error. On the one hand, he might reject the null hypothesis and falsely conclude that the vaccine was effective. Proceeding with

Figure 6.6 Possible values for the test statistic, y

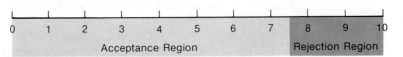

Acceptance Region   Rejection Region

y

a more thorough and expensive testing program or a pilot plant production of the vaccine would result in a financial loss. On the other hand, he might decide not to reject the null hypothesis and falsely conclude that the vaccine was ineffective. This error would result in the loss of potential profits which could be derived through the sale of a successful vaccine.

Definition

Rejecting the null hypothesis when it is true is called a **type I error** for a statistical test. The probability of making a type I error is denoted by the symbol α.

The probability, α, will increase or decrease as we increase or decrease the size of the rejection region. Then why not decrease the size of the rejection region and make α as small as possible? For example, why not choose $y = 10$ as the rejection region? Unfortunately, decreasing α increases the probability of not rejecting when the null hypothesis is false and some alternative hypothesis is true. This second type of error is called the type II error for the statistical test and its probability is denoted by the symbol β.

Definition

Accepting the null hypothesis when it is false is called a **type II error** for a statistical test. The probability of making a type II error when some specific alternative is true is denoted by the symbol β.

For a fixed sample size, n, α and β are inversely related; as one increases the other decreases. Increasing the sample size provides more information upon which to base the decision and hence reduces both α and β. In an experimental situation, the probabilities of the type I and type II errors for a test measure the risk of making an incorrect decision. The experimenter selects values for these probabilities, and the rejection region and sample size are chosen accordingly.

Example 6.9

Refer to the cold vaccine study and the statistical test based on the rejection region shown in Figure 6.6 (i.e., $y = 8, 9, 10$).

a. State the null hypothesis and the alternative hypothesis for the test.

b. Find α for the test.

c. Find β, the probability of accepting the null hypothesis when the probability of survival for a vaccinated person is $p = .9$.

Solution

a. The null hypothesis is that $p = .5$ or, equivalently, that the vaccine is ineffective. The alternative hypothesis is that $p > .5$; that is, that the vaccine is effective.

Figure 6.7 Binomial probability distribution, $n = 10$, $p = .5$

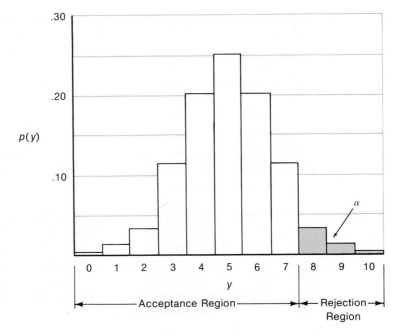

Figure 6.8 Binomial probability distribution, when the alternative hypothesis is true: $n = 10$, $p = .9$

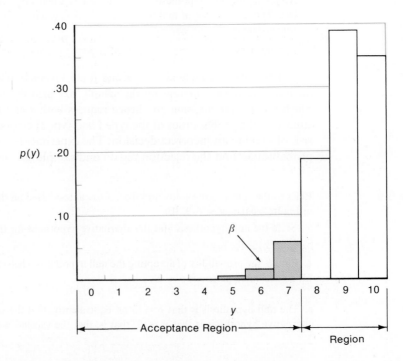

b. The probability of rejecting the null hypothesis when it is true ($p = .5$) is

$$\alpha = P(y = 8, 9, 10 \text{ given that } p = .5) = \sum_{y=8}^{10} p(y)$$

where $p(y)$ is a binomial probability distribution with $n = 10$ and $p = .5$. Then,

$$\alpha = \sum_{y=8}^{10} C_y^{10}(.5)^y(.5)^{10-y}$$

can be found using Table 1, the Appendix. Since Table 1 gives the sum of the binomial probabilities from $y = 0$ to $y = a$; that is, $\sum_{y=0}^{a} p(y)$,

$$\alpha = \sum_{y=8}^{10} p(y)$$
$$= \sum_{y=0}^{10} p(y) - \sum_{y=0}^{7} p(y)$$
$$= 1 - .945$$
$$= .055.$$

The binomial distribution for $n = 10$, $p = .5$ is shown in Figure 6.7; α is represented by the shaded portion of the probability distribution.

c. We wish to find the value of β when the null hypothesis is false (p is not equal to .5) and, in fact, $p = .9$. When $p = .9$, the probability distribution for y, the number of survivors, is as shown in Figure 6.8. Since

$$\beta = P(\text{accepting the null hypothesis when } p = .9)$$
$$= P(y = 0, 1, 2, \ldots, 6, 7, \text{ given that } p = .9),$$

β will equal the shaded portion of $p(y)$, Figure 6.8, the portion falling in the acceptance region. Thus,

$$\beta = \sum_{y=0}^{7} p(y), \qquad \text{given } p = .9,$$

where $p(y)$ is a binomial probability distribution with $p = .9$. This quantity can be obtained directly from Table 1, the Appendix, as

$$\beta = \sum_{y=0}^{7} C_y^{10}(.9)^y(.1)^{10-y} = .07.$$

To summarize the implications of (a), (b), and (c), Example 6.9, $\alpha = .055$ and $\beta = .07$ give measures of the risks of making the two (and only two) types of errors for this statistical test. The probability that the test statistic will, by chance, fall in the rejection region when the null hypothesis is true is only .055. That is, the probability of concluding that the vaccine is somewhat effective, when in fact it is worthless, is only .055. But suppose that the vaccine is really effective and the probability of surviving a winter without a cold, when the vaccine has been used, is .9. What is the probability of accepting the null hypothesis that "the vaccine is ineffective"? We have shown that the risk of making this type II error is only $\beta = .07$.

Tips on Problem Solving

When calculating α and β for a statistical test of an hypothesis:

1. Clearly identify the values of y that fall in the rejection region and the remainder that fall in the acceptance region.
2. Calculate α as the sum of the values of $p(y)$ for the values of y in the *rejection region*. The binomial probability p used to calculate the values of $p(y)$ should be the value specified in the null hypothesis.
3. Calculate β as the sum of the values of $p(y)$ for the values of y in the *acceptance region*. The binomial probability p used to calculate the values of $p(y)$ should be some alternative value of p (other than the one specified in the null hypothesis) that you wish to detect.

6.7
Choosing the Null Hypothesis

The reasoning employed in a statistical test of an hypothesis runs counter to our everyday method of thinking. That is, it is similar to the mathematical method of proof by contradiction. The hypothesis that the scientist wishes to "prove" (that is, support) is the alternative hypothesis (often called the "research hypothesis" by social scientists). To do this, he tests the converse (opposite) of the research hypothesis, the null hypothesis. He hopes that the data will support its rejection because this implies support for the alternative or "research hypothesis," which was his research objective. You can see that this is exactly what we did with the cold vaccine experiment. We showed support for the effectiveness of the cold vaccine by rejecting the null hypothesis that the vaccine was not effective.

Why employ this reverse type of thinking, gaining support for a theory by showing that there is little evidence to support its converse? Why not test the alternative or research hypothesis? The answer lies in the problem of evaluating the probabilities of incorrect decisions.

If the research hypothesis is true, the sample data will tend to support rejection of the null hypothesis (the converse of the research hypothesis). Then the probability of making an incorrect decision is readily at hand. It is α, a probability that was specified in setting up the rejection region. Thus if we reject the null hypothesis (which is what the researcher hopes will occur), we immediately know the probability of making an incorrect decision. This gives us a measure of confidence in our conclusion.

Suppose that we had taken the opposite tack, testing the alternative (research) hypothesis that the vaccine is effective. If the research hypothesis is true, the test statistic will most probably fall in the acceptance region, $y = 8, 9, 10$ (instead of the rejection, $y = 0, 1, \ldots, 7$). Now, to find the probability of an incorrect decision, we must evaluate β, the probability of accepting the null hypothesis when it is false. Although this is not an insurmountable task for the cold vaccine problem, it is extra work. And for many statistical tests, it is very difficult to calculate β.

So, to summarize, it is a lot easier to follow the route of "proof by contradiction." Thus, the statistician will select the converse of the research hypothesis as the null hypothesis and hope the test leads to its rejection. If it does, the statistician knows α and immediately has a measure of the confidence that he or she can place in this conclusion.

Tips on Problem Solving

When defining the null and alternative hypotheses, note that:

1. The null hypothesis will always specify p as a single value, say $p = .1$.
2. The alternative hypothesis will specify p as a set of values in an interval. If H_0 is $p = .1$, then H_a can be one of three possibilities:
   a. $H_a : p > .1$ for the case where you wish to detect a value of p larger than .1 (a one-tailed test, upper tail).
   b. $H_a : p < .1$ for the case when you wish to detect a value of p less than .1 (a one-tailed test, lower tail).
   c. $H_a : p \neq .1$ for the case where you wish to detect values of p that are either smaller than or larger than .1 (a two-tailed test).
3. Always examine a problem to see what it is you wish to detect: a value of p larger than some value (for example, larger than .1), a value of p smaller than some value (smaller than say .1), or a value of p that differs from some value (say differs from .1). This will help you locate the value of p that appears in H_0 and will help you decide on the nature of the alternative hypothesis; that is, whether it is 2(a), 2(b), or 2(c).

6.8

A General Comment

A discussion of the theory of tests of hypotheses may seem a bit premature at this point, but it provides an introduction to a line of reasoning that is sometimes difficult to grasp and which is best presented when it is allowed to incubate in the mind of the student over a period of time. Thus, some of the exercises at the end of Chapter 6 involve the use of the binomial probability distribution and, at the same time, lead the student to utilize the reasoning involved in statistical tests of hypotheses. We shall take occasion to expand upon these ideas through examples and exercises in Chapter 7 and will discuss in detail the topic of statistical tests of hypotheses in Chapter 9 and succeeding chapters.

In closing, we direct your attention to the similarity of the lot acceptance sampling problem and the statistical test of an hypothesis. Theoretically, they are equivalent because each involves an inference, formulated as a decision, concerning the value of p, the unknown parameter of a binomial population.

Exercises

6.40.   An experiment is conducted to test the hypothesis that a coin is balanced against the alternative hypothesis that the probability of observing a head, p, is not equal to .5. To test the hypothesis, a coin will be tossed four times and the number of heads observed. The hypothesis will be rejected if zero or four heads appear.

   a. Give the values of p involved in the null and alternative hypotheses.

   b. Given the objective of this experiment (that is, what it is attempting to detect) should this be a one or a two-tailed test?

   c. What is the probability of the type I error for this test?

   d. If the coin really is biased and the probability of observing a head on a single trial is .7, what is the probability of the type II error for the test?

6.41.   Refer to Exercise 6.40. Suppose that you know that if the coin is unbalanced it can only be biased in favor of a head.

   a. State the values of p involved in the null and alternative hypotheses.

   b. Does the alternative hypothesis imply a one or a two-tailed test? Explain.

   c. Suppose that the rejection region is chosen as $y = 4$ heads. Find α.

   d. If the rejection region is $y = 4$, find the probability of making a type II error if, in fact, $p = .9$.

6.42.   A pair of beetles is expected to produce black-eyed offspring 30 percent of the time. To test this theory, a researcher randomly selects ten offspring and observes y, the number of black-eyed offspring.

   a. State the null and alternative hypotheses that should be employed in the test.

   b. Should the researcher employ a one- or a two-tailed test? Explain.

   c. Choose an appropriate rejection region for the test so that α will equal .05 or less.

   d. If only one of the ten offspring has black eyes, is there sufficient evidence to indicate that the theory is wrong?

6.43.   A manufacturer of floor wax has developed two new brands, A and B, which he wishes to subject to consumer evaluation to determine which of the two is superior. Both waxes, A and B, are applied to floor surfaces in each of 15 homes.

   a. If there is actually no difference in the quality of the brands, what is the probability that ten or more consumers would state a preference for brand A?

   b. For either brand A or brand B?

6.44.   Continuing Exercise 6.43, let p equal the probability that a consumer will choose brand B in preference to A and suppose that we wish to test the hypothesis that there is no observable difference between the brands—in other words that $p = 1/2$. Let y, the number of times that B is preferred to A, be the test statistic.

a. Calculate the value of α for the test if the rejection region is chosen to include $y = 0, 1, 2, 3, 12, 13, 14,$ and 15.

b. If p is really equal to .8, what is the value of β for the test defined in (a)? (Note that this is the probability that $y = 4, 5, \ldots, 10, 11$ given that $p = .8$.)

6.45.   Continuing Exercise 6.44, suppose that the rejection region is enlarged to include $y = 0, 1, 2, 3, 4, 11, 12, 13, 14,$ and 15.

a. What is the value of α for the test? Should this probability be larger or smaller than the answer given in Exercise 6.44?

b. If p is really equal to .8, what is the value of β for the test? Compare with your answer to part (b), Exercise 6.44.

6.46.   The number of defective fuses proceeding from each of two production lines, A and B, was recorded daily for a period of ten days with the following results:

Day	Line A	Line B
1	172	201
2	165	179
3	206	159
4	184	192
5	174	177
6	142	170
7	190	182
8	169	179
9	161	169
10	200	210

Assume that both production lines produced the same daily output. Compare the number of defectives produced by A and B each day and let y equal the number of days when B exceeded A. Do the data present sufficient evidence to indicate that production line B produces more defectives, on the average, than A? State the null and alternative hypotheses. Use y as a test statistic.

6.47.   Experimental evidence suggests that the anti-cancer drug, cyclophosphamide, may be effective in treating an often fatal disease of the blood vessels (*U.S. News and World Report*, December 10, 1979). Sixteen patients suffering from necrotizing vasculitus were treated with the drug and 13 of the 16 patients recovered from the disease. This compares with a 48 percent survival rate using traditional methods of treatment. Do these data provide sufficient evidence to indicate that the probability of recovering using cyclophosphamide exceeds the probability of recovering when using the traditional method of treatment? To answer this question:

a. State the null and alternative hypotheses.

b. Conduct the test using a value of α near .05.

6.48. In 1975, the General Accounting Office (GAO) examined 15 school districts in 14 states and concluded that it was debatable whether the multibillion dollar Title I program, aimed primarily at improving the reading ability of poor children, was effective. In particular, the GAO noted that the gap between the reading abilities of educationally deprived and of average children increased while the students were in the program (*Orlando Sentinel Star*, December 28, 1975). If the Title I program was completely ineffective, the change in the level of reading ability scores (before and after Title I) would depend on the random variation of individual student scores so that it would be reasonable to assume that $P(\text{increase in level}) = P(\text{decrease in level}) = p = .5$. Test the hypothesis that $p = .5$ using y, the number of school districts showing an increase in the reading ability gap, as a test statistic.

  a. State the null and alternative hypotheses.

  b. Locate a rejection region for a value of α near .05 (assume that $n = 15$).

  c. Based on the GAO report, it appears that $y = 15$. What do you conclude concerning the effectiveness of the Title I program?

6.49. A number of psychological experiments are conducted as follows: A rat is attracted to the end of a ramp that divides, leading to one of two doors. The objective of the experiment, essentially, is to determine whether the rat possesses or acquires a preference for one of the two paths. For a given experiment consisting of six runs, the following results were observed:

Run	Door Chosen
1	2
2	1
3	2
4	2
5	2
6	2

  a. State the null hypothesis to be tested. State the alternative hypothesis.

  b. Let y equal the number of times the rat chose the second door. What is the value of α for the test if the rejection region includes $y = 0$ and $y = 6$?

  c. What is the value of β for the alternative, $p = .8$?

6.50. According to the National Cancer Institute, the 10-year survival rate for breast cancer patients is 52 percent. One cancer treatment clinic, which employs a combination of non-radical surgery and radiation therapy, has discovered that 11 of 15 of their breast cancer patients have survived at least 10 years. Do these data provide sufficient evidence to suggest that the clinic has a higher 10-year survival rate than the rate published by the National Cancer Institute? (For ease of computation, assume that the value of p given by the National Cancer Institute is .50.) Test using α approximately equal to .05.

6.51. Establishing the value of an art object is subjective in nature and difficult at best. To determine whether two art appraisers tend to give different levels of appraisals, two appraisers, A and B, were asked to appraise each of 7 art objects. If appraiser A tends to give smaller appraisals (or give larger appraisals) than appraiser B, the probability p that the appraisal of A will exceed the appraisal for B will be less than 1/2 (or greater than 1/2).

Under the assumption that there is no difference in the appraisal techniques of the two appraisers, $p = 1/2$. Let y, the number of art objects for which A's appraisal exceeds B's appraisal, be a test statistic.

a. Find an appropriate rejection region to test the null hypothesis, $p = 1/2$, for $\alpha \approx .10$.

b. If A tends to give conservative evaluations, so that p actually is equal to .9, calculate β for the test.

c. If A appraises 5 of the 7 art objects to be worth more than B appraises them for, what do you conclude?

6.52. Most weather forecasters seem to protect themselves very well by attaching probabilities to their forecasts ("the probability of rain today is 40 percent"). Then if a particular forecast is incorrect, you are expected to attribute the error to the random behavior of the weather rather than the inaccuracy of the forecaster. To check the accuracy of a particular forecaster, records were checked only for days in which the forecaster predicted rain "with 30 percent probability." A check of 25 of these days indicated that it rained on 10 of the 25. Do these data disagree with the forecast of a "30 percent probability of rain"? To answer this question, you will want to detect a value of p that differs from .3, i.e., either $p > .3$ or $p < .3$. Consequently, you will reject the null hypothesis, $p = .3$, if you observe either very large or very small values of y. Suppose that you select $y \leq 3$ or $y \geq 12$ as the rejection region for the test.

a. Find α.

b. Complete the test and state your conclusions.

6.53. Refer to Exercise 6.52. If the probability of rain on a given day is really .6 when the weather forecaster forecasts .3, what is the probability that you will reject the null hypothesis, $p = .3$, based on a random sample of 25 such days? What is the value of β for the test?

6.54. A packaging experiment was conducted by placing two different package designs for a breakfast food side by side on a supermarket shelf. The objective of the experiment was to see if buyers indicated a preference for one of the two package designs. In a given day, five customers purchased a package from the supermarket, with one choosing package design 1 and four choosing design 2.

a. State the hypothesis to be tested. (*Hint*: The null hypothesis should imply equal preference for the two designs.)

b. State the alternative hypothesis and justify your selection.

c. Let y equal the number of buyers who choose the second package design. What is the value of α for the test if the rejection region includes $y = 0$ and $y = 5$?

d. What is the value of β for the alternative $p = .9$ (that is, 90 percent of the buyers actually favor the second package design)?

e. In the context of our problem, give a practical interpretation of the type I error and the type II error.

6.55. Refer to Exercise 6.42. Suppose the researcher employed only $n = 3$ offspring in the sample. Show that it would be impossible to acquire sufficient evidence to refute the theory using a sample size that is as small as $n = 3$. (*Hint*: Show that the value of α would be too large.)

6.9
Summary

A binomial experiment is typical of a large class of useful experiments encountered in real life which satisfy, to a reasonable degree of approximation, the five defining characteristics stated in Section 6.1. The number of successes, y, observed in n trials is a discrete random variable with probability distribution

$$p(y) = C_y^n p^y q^{n-y},$$

where $q = 1 - p$ and $y = 0, 1, 2, 3, \ldots, n$. Statistically speaking, we are interested in making inferences concerning p, the parameter of a binomial population, as exemplified by the lot acceptance sampling, Section 6.5, and the test of the effectiveness of the cold vaccine, Section 6.6.

References

Beyer, W.C. *Handbook of Tables for Probability and Statistics*, 2nd ed. Cleveland, Ohio: The Chemical Rubber Co., 1968.

Chapman, D.G., and R.A. Schaufele, *Elementary Probability Models and Statistical Inference*. New York: John Wiley & Sons, Inc., 1970.

Feller, W., *An Introduction to Probability Theory and Its Applications*, Vol. 1, 3rd ed. New York: John Wiley & Sons, Inc., 1968.

Mendenhall, W., R.L. Scheaffer, and D. Wackerly, *Mathematical Statistics with Applications*, 2nd ed. Boston: Duxbury Press, 1981.

Mosteller, F., R.E.K. Rourke, and G.B. Thomas, Jr., *Probability with Statistical Applications*, 2nd ed. Reading, Mass.: Addison-Wesley Publishing Company, Inc., 1970.

National Bureau of Standards, *Tables of the Binomial Probability Distribution*. Washington, D.C.: Government Printing Office, 1949.

Supplementary Exercises

[Starred (*) exercises are optional.]

6.56.   List the five identifying characteristics of the binomial experiment.

6.57.   A balanced coin is tossed three times. Let y equal the number of heads observed.

   a. Use the formula for the binomial probability distribution to calculate the probabilities associated with $y = 0, 1, 2$, and 3.

   b. Construct a probability distribution similar to Figure 6.1.

   c. Find the expected value and standard deviation of y, using the formulas

$$E(y) = np,$$

$$\sigma = \sqrt{npq}.$$

   d. Using the probability distribution, (b), find the fraction of the population measurements lying within one standard deviation of the mean. Repeat for two standard deviations. How do your results agree with Tchebysheff's Theorem and the Empirical Rule?

6.58.   Refer to Exercise 6.57. Suppose that the coin was definitely unbalanced and that the probability of a head was equal to $p = .1$. Follow instructions (a), (b), (c), and (d). Note that the probability distribution loses its symmetry and becomes skewed when p is not equal to $1/2$.

6.59.   Suppose that the four engines of a commercial aircraft were arranged to operate independently and that the probability of in-flight failure of a single engine is .01. What is the probability that, on a given flight:

a. No failures are observed?

b. No more than one failure is observed?

6.60.   Use Table 1, the Appendix, to find the partial sum

$$\sum_{y=0}^{a} p(y)$$

for:

a. $n = 10, p = .7, a = 8.$ b. $n = 15, p = .05, a = 1.$ c. $n = 20, p = .9, a = 14.$

6.61.   Use Table 1, the Appendix, to find $p(y)$ for:

a. $n = 10, p = .6, y = 6.$ b. $n = 15, p = .5, y = 5.$ c. $n = 20, p = .2, y = 3.$

6.62.   Use Table 1, the Appendix, to find

$$\sum_{y=a}^{b} p(y)$$

for:

a. $n = 10, p = .1, a = 1, b = 10.$

b. $n = 10, p = .8, a = 7, b = 9.$

c. $n = 15, p = .4, a = 4, b = 15.$

6.63.   The ten-year survival rate for bladder cancer is approximately 50 percent. If 20 bladder cancer patients are properly treated for the disease, what is the probability that:

a. At least one will survive ten years?

b. At least ten will survive ten years?

c. At least 15 will survive ten years?

6.64.   A city commissioner claims that 80 percent of all people in the city favor garbage collection by contract to a private concern (in contrast to collection by city employees). To check the theory that the proportion of the people in the city favoring private collection is .8, you randomly sample 25 people and find that y, the number of people who support the commissioner's claim, is 22.

a. What is the probability of observing at least 22 who support the commissioner's claim if, in fact, $p = .80$?

b. What is the probability that y is exactly equal to 22?

6.65.   The proportion of residential households in Burlington, Vermont, that are heated by natural gas is approximately .2. A randomly selected city block within the Burlington city limits has 20 residential households. Assume that the properties of a binomial experiment are satisfied and find the probability that:

a. None of the households are heated by natural gas.

b. No more than four of the 20 are heated by natural gas.

c. Why might the binomial experiment not provide a good model for this sampling situation?

6.66.   If a person is given the choice of a number from 0 to 9, is it more likely that the person will choose a number near the middle of the sequence? To answer this question, 20 persons are asked to select a number from 0 to 9 and eight choose a 4, 5, or 6. If the choice of any one number is as likely as any other, what is the probability of this event? What is the probability of observing eight or more choices of the interior numbers, 4, 5, or 6?

6.67. Refer to Exercise 6.48 and the General Accounting Office's report on the effectiveness of the Title I program designed to improve the reading ability of poor children. The report mentions that an analysis of student records showed the gap in reading ability between poor and average children increased for 60 percent of the poor children in the program, the gap was maintained for 6 percent, and the gap decreased for 34 percent. If the GAO examined 6000 records, what is the expected value of y, the number showing an increase in the gap, if the program was really ineffective (i.e., $p = .5$)? If p is really equal to .5, is it likely that 60 percent of 6000 poor students would show an increase in the gap? Since the purpose of the Title I program is to reduce the gap, you would expect to observe a relatively small value of y. Can you find an explanation for the peculiar results ($y = 3600$) observed by the GAO?

6.68. A recent survey suggests that Americans anticipate a reduction in living standards and that a steadily increasing consumption no longer may be as important as it was in the past. Suppose that a poll of 2000 people indicated 1373 in favor of forcing a reduction in the size of American automobiles by legislative means. Would you expect to observe as many as 1373 in favor of this proposition if, in fact, the general public was split 50–50 on the issue? Why?

6.69. Suppose, as noted in Exercise 6.7, that approximately 35 percent of all applicants for jobs falsify the information on their application forms. If a company possesses 2300 employees:

a. What is the expected value of the number y of application forms that have been falsified?

b. Find the standard deviation of y.

c. Calculate the interval, $\mu \pm 2\sigma$.

d. Suppose that the company had a credential-checking firm verify the information on the 2300 application forms and that 249 application forms contained falsified information. Do you think that the company's application falsification rate is consistent with the contention that 35 percent of all job applicants falsify information on their application blanks? Explain.

6.70. A quality-control engineer wishes to study the alternative sampling plans $n = 5, a = 1$ and $n = 25, a = 5$. On the same sheet of graph paper, construct the operating characteristic curves for both plans, making use of acceptance probabilities at $p = .05, p = .10, p = .20,$ $p = .30,$ and $p = .40$ in each case.

a. If you were a seller producing lots with fraction defective ranging from $p = 0$ to $p = .10,$ which of the two sampling plans would you prefer?

b. If you were a buyer wishing to be protected against accepting lots with fraction defective exceeding $p = .30$, which of the two sampling plans would you prefer?

6.71. Consider a lot acceptance plan with $n = 20, a = 1$. Calculate the probability of accepting lots having fraction defective

a. $p = .01,$ b. $p = .05,$ c. $p = .10,$ d. $p = .20.$

Sketch the operating characteristic curve for the plan.

6.72. Suppose that early state-wide election returns indicate totals of 33,000 votes in favor of candidate A versus 27,000 for candidate B, and that these early returns can be regarded as a random sample selected from the population of all 10,000,000 eligible voters in the state.

a. If the state-wide vote will be split 50–50 (i.e., the probability that A will win is .5), find the expected number y of votes for A in the sample of 60,000 early returns.

b. Find the standard deviation of y.

c. Is the observed value, $y = 33,000$, consistent with the theory, part (a), that the vote will split (i.e., $p = .5$), or is y a highly unlikely value?

d. Do you think that the value, $y = 33,000$, is sufficient evidence to indicate that A will win?

6.73. A psychiatrist believes that 80 percent of all people who visit doctors have problems of a psychosomatic nature. He decides to select 25 patients at random to test his theory.

a. Assuming that the psychiatrist's theory is true, what is the expected value of y, the number of the 25 patients who have psychosomatic problems?

b. What is the variance of y, assuming that the theory is true?

c. Find $p(y \leq 14)$. (Use tables and assume that the theory is true.)

d. Based on the probability in part (c), if only 14 of the 25 sampled had psychosomatic problems, what conclusions would you make about the psychiatrist's theory? Explain.

6.74. A particular type of radar set has a probability of .2 of detecting and tracking an aircraft within a radius of 200 miles. If n sets are available for operation, how many sets would have to be operating simultaneously in order that the probability of detecting an aircraft would equal .9? (Assume that the sets operate independently and that each has a probability of detection equal to .2.)

6.75. A student government states that 80 percent of all students favor an increase in student fees to subsidize a new recreational area. A random sample of $n = 25$ students produced 15 in favor of increased fees. What is the probability that 15 or fewer in the sample would favor the issue if student government is correct? Do the data support the student government's assertion, or does it appear that the percentage favoring an increase in fees is less than 80 percent?

6.76. According to estimates, 1/4 of all workers in the United States are dissatisfied with their jobs (*U.S. News and World Report*, April 6, 1981). One company, believing that a much smaller fraction of its employees were dissatisfied, surveyed the attitudes of a random sample of 25 of its workers and found that only three were dissatisfied. If p is the fraction of the company's workers who are dissatisfied:

a. What is the company attempting to detect?

b. State the null and alternative hypotheses that the company would use in a statistical test for p.

c. Do the data provide sufficient evidence to support the company's belief? Test using a value of α that is near $\alpha = .05$.

* 6.77. Derive the formula for the binomial probability distribution by using mathematical induction.

* 6.78. Derive the formula for the binomial probability distribution using the sample point approach and combinatorial mathematics.

* 6.79. Use the formula for the binomial probability distribution along with the expectation theorems given in Exercises 5.49 and 5.50 to prove

$$E(y) = np,$$

variance of $y = npq$.

* 6.80.   Prove that the binomial probabilities always sum to 1; that is,

$$\sum_{y=0}^{n} p(y) = 1.$$

* 6.81.   The manager of a large motor pool wished to compare the wearing qualities of two different types (type A and type B) of automobile tires. On each of 400 cars he replaced one rear tire with a new tire of type A and the other rear tire with a new tire of type B. When a given car had been driven 10,000 miles, he determined which of the two rear tires experienced the greater wear. Let y be the number of cars out of the 400 cars on which tire A showed the greater wear. Let p be the probability that on a given car the tire of type A will experience the greater wear. Using Tchebysheff's Theorem, find an upper bound to α for a test of the hypothesis that $p = 1/2$ if the rejection region includes the outcomes $y = 0, 1, 2, \ldots, 150$, 250, 251, \ldots, 400.

* 6.82.   The "power" of a test is defined to be $1 - \beta$. Find a lower bound to the power of the test given in Exercise 6.81 if $p = .8$.

7. The Normal Probability Distribution

Chapter Objectives

General Objectives

An important discrete random variable, the binomial, and its probability distribution were presented in Chapter 6. This chapter presents the normal random variable, one of the most important and most common continuous random variables. We explain why the normal random variable occurs so frequently in practice, give its probability distribution, and show how the probability distribution can be used. We will take advantage of the normal distribution to reinforce your understanding of the basic concepts involved in a statistical test of an hypothesis.

Specific Objectives

1. To explain why normally distributed random variables occur so frequently in nature. The Central Limit Theorem is one of the reasons presented for this situation. *Sections 7.1, 7.5*

2. To present the normal probability distribution and to explain how to find the probability that a random variable will fall in a particular interval. *Sections 7.2, 7.3*

3. To introduce the concept of the sampling distribution of a statistic. *Section 7.4*

4. To present the Central Limit Theorem as a reason for studying the normal probability distribution and to use it in developing the properties of the sampling distribution of a sample mean. *Section 7.5*

5. To demonstrate the applicability of the Central Limit Theorem by showing how the normal probability distribution can be used to approximate binomial probabilities when the number n of trials is large. *Section 7.6*

6. To use examples to reinforce the inferential ideas introduced in Chapter 6. *Section 7.6*

7. To introduce the large sample statistics that we will encounter in Chapters 8 and 9 and to comment on their sampling distributions. *Section 7.7*

Case Study

The Long and the Short of It

If you were the boss, would height play a role in your selection of a successor for your job? In his FORTUNE Magazine column, *Keeping Up*, Daniel Seligman discussed his ideas concerning height as a factor in Deng Xiaoping's choice of Hu Yaobang for his replacement as Chairman of the Chinese Communist Party.* As Seligman notes, the fact surrounding the case are enough to arouse suspicions when examined in the light of statistics.

Deng, it seems, is only five feet tall, a height that is short even in China. Therefore, the choice of Hu Yaobang, who is also five feet tall, raised (or lowered) some eyebrows because, as Seligman notes, "the odds against a 'height blind' decision producing a chairman as short as Deng are about 40 to 1." In other words, if we possessed the relative frequency distribution of the heights of all Chinese males, only 1 in 41 (i.e., 2.4 percent) of them would posses heights less than or equal to five feet. To calculate these odds, the author of the article made some interesting assumptions concerning the relative frequency distribution of the heights of Chinese males, most notably that the distribution follows the "normal bell-shaped Gaussian curve (as it does in the U.S.)."

The normal curve, used as the model for the relative frequency distributions for many continuous random variables, is the topic of this chapter. We will examine its properties, learn how it can be used to calculate probabilities, and, in Section 7.3, see how the author of the FORTUNE article used it to arrive at the 40 to 1 odds.

7.1 Introduction

Continuous random variables, as noted in Section 5.4, are associated with sample spaces representing the infinitely large number of sample points contained on a line interval. The heights and weights of humans, laboratory experimental measurement errors, and the length of life of light bulbs are typical examples of continuous random variables. Reviewing Section 5.4, we note that the probabilistic model for the frequency distribution of a continuous random variable involves the selection of a curve, usually smooth, called the probability distribution or probability density function. Although these distributions may assume a variety of shapes, a very large number of random variables observed in nature possess a frequency distribution that is approximately bell-shaped or, as the statistician would say, is approximately a normal probability distribution.

Mathematically speaking, the **normal probability density function,**

$$f(y) = \frac{e^{-\frac{(y-\mu)^2}{2\sigma^2}}}{\sigma\sqrt{2\pi}} \qquad (-\infty < y < \infty),$$

is the equation of the bell-shaped curve shown in Figure 7.1.

Figure 7.1 Normal probability density function

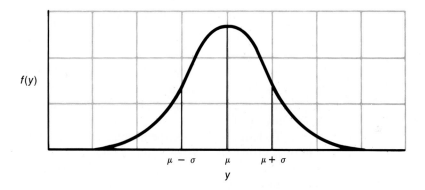

The symbols e and π represent irrational numbers whose values are approximately 2.7183 and 3.1416, respectively; μ and σ are the population mean and standard deviation. The equation for the density function is constructed such that the area under the curve will represent probability. Hence the total area under the curve is equal to 1.

In practice, we seldom encounter variables that range in value from "minus infinity" to "plus infinity," whatever meaning we may wish to attach to these phrases. Certainly the height of humans, the weight of a species of beetle, or the length of life of a light bulb do not satisfy this requirement. Nevertheless, a relative frequency histogram plotted for many types of measurements will generate a bell-shaped figure that may be approximated by the function shown in Figure 7.1. Why this particular phenomenon exists is a matter for conjecture. However, one explanation is provided by the Central Limit Theorem, a theorem that may be regarded as the most important in statistics. This theorem is discussed in Section 7.5.

7.2
**Tabulated Areas
of the Normal
Probability
Distribution**

You will recall (Section 5.4) that the probability that a continuous random variable assumes a value in the interval, a to b, is the area under the probability density function between the points a and b (see Figure 7.2).

The probability model for a continuous random variable differs greatly from the model for a discrete random variable when we talk about the probability that y equals some particular value, say a. Since the area lying over any particular point, say $y = a$, is 0, it follows from our probability model that the

Figure 7.2 The probability $P(a < y < b)$ for a continuous random variable

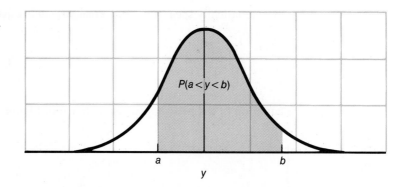

probability that $y = a$ is 0. This means that the expression $P(y \leq a)$ is the same as $P(y < a)$ because $P(y = a) = 0$. Similarly, $P(y \geq a) = P(y > a)$. This is, of course, not true for a discrete random variable because $P(y = a)$ may not equal 0.

To find areas under the normal curve, first note that the equation for the normal probability distribution, Section 7.1, is dependent upon the numerical values of μ and σ, and that by supplying various values for these parameters, we could generate an infinitely large number of bell-shaped normal distributions. A separate table of areas for each of these curves is obviously impractical; rather we would like one table of areas applicable to all. The easiest way to do this is to work with areas lying within a specified number of standard deviations of the mean as was done in the case of the Empirical Rule. For instance, we know that approximately .68 of the area will lie within one standard deviation of the mean, .95 within two, and almost all within three. What fraction of the total area will lie within .7 standard deviations, for instance? This question, as well as others, will be answered by Table 3, the Appendix.

Inasmuch as the normal curve is symmetrical about the mean, half of the area under the curve will lie to the left of the mean and half to the right (see Figure 7.3). Also, because of the symmetry, we can simplify our table of areas by listing the areas between the mean and a specified number z of standard

Figure 7.3 Standarized normal distribution

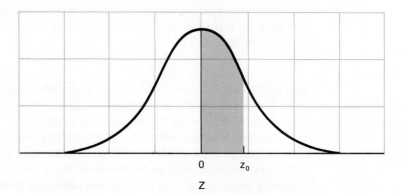

deviations to the right of μ. Areas to the left of the mean can be calculated by using the corresponding and equal area to the right of the mean. The distance from the mean to a given value of y is $(y - \mu)$. Expressing this distance in units of standard deviation σ, we obtain

$$z = \frac{y - \mu}{\sigma}.$$

Note that there is a one-to-one correspondence between z and y and particularly that $z = 0$ when $y = \mu$. z will be positive when y lies above the mean and negative when y lies below the mean. The probability distribution for z is often called the **standardized normal distribution,** because its mean is equal to zero and its standard deviation is equal to 1. It is shown in Figure 7.3. The area under the normal curve between the mean, $z = 0$, and a specified value of $z > 0$, say z_0, is the probability $P(0 \le z \le z_0)$. This area is recorded in Table 3, the Appendix, and is shown as the shaded area in Figure 7.3.

An abbreviated version of Table 3, the Appendix, is shown in Table 7.1. Note that z, correct to the nearest tenth, is recorded in the left-hand column. The second decimal place for z, corresponding to hundredths, is given across the top row. Thus the area between the mean and a point $z = .7$ standard deviation to the right, located in the second column of the table opposite $z = .7$, is found to equal .2580. Similarly, the area between the mean and $z = 1.0$ is .3413. The area lying within one standard deviation on either side of the mean would be two times the quantity .3413, or .6826. The area lying within two standard deviations of the mean, correct to four decimal places, is $2(.4772) = .9544$. These numbers provide the approximate values, 68 percent and 95 percent, used in the Empirical Rule, Chapter 3. To find the area $z = .57$ standard deviation to the right of the mean, proceed down the left-hand column to the 0.5 row. Then move across the top row of the table to the .07 column.

Table 7.1 Abbreviated version of Table 3, the Appendix

z	.00	.01	.02	.03	.04	.05	.06	.07	.08	.09
0.0	.0000	.0040	.0080	.0120	.0160	.0199	.0239	.0279	.0319	.0359
0.1	.0398	.0438	.0478	.0517	.0557	.0596	.0636	.0675	.0714	.0753
0.2	.0793	.0832	.0871	.0910	.0948	.0987	.1026	.1064	.1103	.1141
0.3	.1179	.1217	.1255	.1293	.1331	.1368	.1406	.1443	.1480	.1517
0.4	.1554	.1591	.1628	.1664	.1700	.1736	.1772	.1808	.1844	.1879
0.5	.1915	.1950	.1985	.2019	.2054	.2088	.2123	.2157	.2190	.2224
0.6	.2257	:	:	:	:	:	:	:	:	:
0.7	.2580	:								
:	:									
1.0	.3413									
:	:									
2.0	.4772									

The intersection of this row-column combination gives the approximate area, .2157. We conclude this section with some examples.

Example 7.1

Find $P(0 \leq z \leq 1.63)$. This probability corresponds to the area between the mean ($z = 0$) and a point $z = 1.63$ standard deviations to the right of the mean (see Figure 7.4).

Figure 7.4 Probability required for Example 7.1

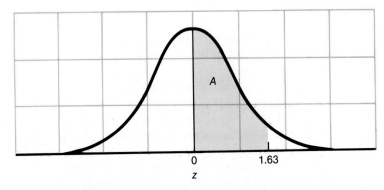

Solution

The area is shaded and indicated by the symbol A in Figure 7.4. Since Table 3, the Appendix, gives areas under the normal curve to the right of the mean, we need only find the tabulated value corresponding to $z = 1.63$. Go down the left-hand column of the table to the row corresponding to $z = 1.6$ and across the top of the table to the column marked .03. The intersection of this row and column combination gives the area, $A = .4484$. Therefore, $P(0 < z < 1.63) = .4484$.

Example 7.2

Find $P(-.5 \leq z \leq 1.0)$. This probability corresponds to the area between $z = -.5$ and $z = 1.0$, as shown in Figure 7.5.

Solution

The area required is equal to the sum of A_1 and A_2 shown in Figure 7.5. From Table 3, the Appendix, we read $A_2 = .3413$. The area A_1 would equal the corresponding area

Figure 7.5 Area under the normal curve in Example 7.2

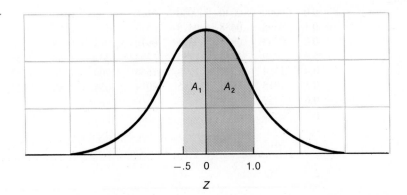

between $z = 0$ and $z = .5$, or $A_1 = .1915$. Thus the total area is

$$A = A_1 + A_2$$
$$= .1915 + .3413$$
$$= .5328.$$

Example 7.3

Find the value of z, say z_0, such that exactly (to four decimal places) .95 of the area is within $\pm z_0$ standard deviations of the mean.

Solution

Half of the area, .95, will lie to the left of the mean and half to the right because the normal distribution is symmetrical. Thus we seek the value z_0 corresponding to an area equal to .475. The area .475 falls in the row corresponding to $z = 1.9$ and the .06 column. Hence $z_0 = 1.96$. Note that this is very close to the approximate value, $z = 2$, used in the Empirical Rule.

Example 7.4

Let y be a normally distributed random variable with mean equal to 10 and standard deviation equal to 2. Find the probability that y will lie between 11 and 13.6.

Solution

As a first step, we must calculate the values of z corresponding to $y_1 = 11$ and $y_2 = 13.6$. Thus,

$$z_1 = \frac{y_1 - \mu}{\sigma} = \frac{11 - 10}{2} = .5,$$

$$z_2 = \frac{y_2 - \mu}{\sigma} = \frac{13.6 - 10}{2} = 1.80.$$

The probability desired, P, is therefore the area lying between z_1 and z_2, as shown in Figure 7.6. The areas between $z = 0$ and z_1, $A_1 = .1915$, and $z = 0$ and z_2, $A_2 = .4641$,

Figure 7.6 Area under the normal curve in Example 7.4

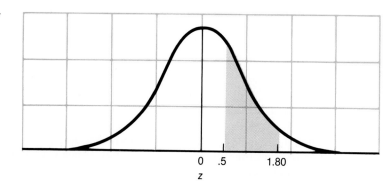

0   .5   1.80
z

are easily obtained from Table 3, the Appendix. The probability, P, is equal to the difference between A_1 and A_2; that is,

$$P = A_2 - A_1$$
$$= .4641 - .1915 = .2726.$$

Example 7.5

Studies show that gasoline usage for compact cars sold in the United States is normally distributed with a mean usage of 30.5 miles per gallon (mpg) and a standard deviation of 4.5 mpg. What percentage of compacts obtain 35 or more miles per gallon?

Solution

The proportion P of compacts obtaining 35 or more miles per gallon is given by the shaded area in Figure 7.7.

Figure 7.7 Area under the normal curve for Example 7.5

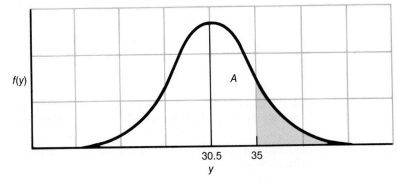

First we must find the z value corresponding to $y = 35$ mpg. Substituting into the formula for z, we obtain

$$z = \frac{y - \mu}{\sigma} = \frac{35 - 30.5}{4.5} = 1.0.$$

The area A to the right of the mean corresponding to $z = 1.0$ is .3413 (Table 3, the Appendix). Then the proportion of compacts having an mpg rating equal to or greater than 35 is equal to the entire area to the right of the mean, .5, less the area A. Thus,

$$P = .5 - A = .5 - .3413 = .1587,$$

and the percentage exceeding 35 mpg is

$$100(.1587) = 15.87 \text{ percent.}$$

Example 7.6

Refer to Example 7.5. If a manufacturer wishes to develop a compact car which outperforms 95 percent of the current compacts in fuel economy, what must be the gasoline usage rate for the new car?

Solution

Let y be a normally distributed random variable with mean equal to 30.5 and standard deviation equal to 4.5. We want to find the value y_0 such that

$$P(y < y_0) = .95.$$

As a first step, we find

$$z_0 = \frac{y_0 - \mu}{\sigma} = \frac{y_0 - 30.5}{4.5},$$

and note that our required probability is the same as the area to the left of z_0 for the standardized normal distribution. Therefore,

$$P(z \leq z_0) = .95,$$

and from Table 3, the Appendix, we find

$$z_0 = \frac{y_0 - 30.5}{4.5} = 1.645,$$

and $y_0 = 37.9$. The manufacturer's new compact car must, therefore, obtain fuel economy of 37.9 mpg to outperform 95 percent of the compact cars currently available on the U.S. market.

Exercises

7.1.   Using Table 3, the Appendix, calculate the area under the normal curve between
    a. $z = 0$ and $z = 1.4$.   b. $z = 0$ and $z = 1.75$.

7.2.   Repeat Exercise 7.1 for
    a. $z = 0$ and $z = .75$.   b. $z = 0$ and $z = -.75$.

7.3.   Repeat Exercise 7.1 for
    a. $z = -1.2$ and $z = 1.6$.   b. $z = .7$ and $z = 1.3$.

7.4.   Repeat Exercise 7.1 for
    a. $z = -1.0$ and $z = 1.0$.   b. $z = -2.0$ and $z = 2.0$.
    c. $z = -3.0$ and $z = 3.0$.

7.5.   Repeat Exercise 7.1 for
    a. $z = -1.43$ and $z = .68$.   b. $z = .44$ and $z = 1.55$.
    c. $z = -1.55$ and $z = -.44$.

7.6.   Repeat Exercise 7.1 for
    a. $z = -1.60$ and $z = .47$.   b. $z = -.38$ and $z = 1.40$.
    c. $z = .40$ and $z = 1.40$.

7.7.   Find a z_0 such that $P(z > z_0) = .025$.

7.8.   Find a z_0 such that $P(z < z_0) = .8051$.

7.9.   Find a z_0 such that $P(z < z_0) = .1314$.

7.10.   Find a z_0 such that $P(z > z_0) = .7088$.

7.11.   Find a z_0 such that $P(z > z_0) = .9750$.

7.12.   Find a z_0 such that $P(z > z_0) = .3300$.

7.13.   Find a z_0 such that $P(-z_0 < z < z_0) = .6046$.

7.14.   Find a z_0 such that $P(-z_0 < z < z_0) = .5408$.

7.15.   Find a z_0 such that $P(-z_0 < z < z_0) = .8230$.

7.16.   Find a z_0 such that $P(z < z_0) = .1660$.

7.17.   Find a z_0 such that $P(z < z_0) = .9525$.

7.18.   Find a z_0 such that $P(z < z_0) = .05$.

7.19.   Find a z_0 such that $P(-z_0 < z < z_0) = .90$.

7.20.   Find a z_0 such that $P(-z_0 < z < z_0) = .99$.

7.21.   One method of arriving at economic forecasts is to use a consensus approach. A forecast is obtained from each of a large number of analysts and the average of these individual forecasts is the consensus forecast. Suppose that the individual 1983 January prime interest rate forecasts of all economic analysts are approximately normally distributed with mean equal to 14 percent, and with a standard deviation of 2.6 percent. If a single analyst is randomly selected from among this group, what is the probability that the analyst's forecast of the prime interest rate will:

   a. Exceed 18 percent?   b. Be less than 16 percent?

7.22.   The scores on a national achievement test were approximately normally distributed with a mean of 540 and a standard deviation of 110.

   a. If you achieved a score of 680, how far, in standard deviations, did your score depart from the mean?

   b. What percentage of those who took the examination scored higher than you?

7.23.   Suppose that you must establish regulations concerning the maximum number of people who can occupy an elevator. A study of elevator occupancies indicates that if eight people occupy the elevator, the probability distribution of the total weight of the eight people possesses a mean equal to 1200 pounds and a variance equal to 9800 (pounds)^2. What is the probability that the total weight of eight people exceeds 1300 pounds? 1500 pounds? (Assume that the probability distribution is approximately normal.)

7.24.   The discharge of suspended solids from a phosphate mine is normally distributed, with a mean daily discharge of 27 milligrams per liter (mg/l) and a standard deviation of 14 mg/l. What proportion of days will the daily discharge exceed 50 mg/l?

7.25.   Philatelists (stamp collectors) often buy stamps at or near retail prices, but when they sell, the price is considerably lower. For example, it may be reasonable to assume that (depending on the mix of a collection, condition, demand, economic conditions, etc.) a collection might be expected to sell at y percent of retail price where y is normally distributed with a mean equal to 45 percent and a standard deviation of 4.5 percent. If a philatelist has a collection to sell that has a retail value of $30,000, what is the probability that the philatelist receives:

   a. More than $15,000 for the collection?

   b. Less than $15,000 for the collection?

   c. Less than $12,000 for the collection?

7.26.   The number of times y an adult human breathes per minute when at rest depends on the age of the human and varies greatly from person to person. Suppose that the probability distribution for y is approximately normal with mean equal to 16 and standard deviation equal to 4. If a person is selected at random and the number y of breaths per minute while at rest is recorded, what is the probability that y will exceed 22?

7.27. The daily sales (excepting Saturday) at a small restaurant has a probability distribution that is approximately normal with mean μ equal to $530 per day and standard deviation σ equal to $120.

a. What is the probability that the sales will exceed $700 on a given day?

b. The restaurant must have at least $300 sales per day in order to break even. What is the probability that on a given day the restaurant will not break even?

7.28. The length of life of a type of automatic washer is approximately normally distributed with mean and standard deviation equal to 3.1 and 1.2 years, respectively. If this type of washer is guaranteed for one year, what fraction of original sales will require replacement?

7.29. Suppose that the counts on the number of a particular type of bacteria in 1 milliliter of drinking water tend to be approximately normally distributed with a mean of 85 and a standard deviation of 9. What is the probability that a given 1-ml sample will contain more than 100 bacteria?

7.30. A grain loader can be set to discharge grain in amounts that are normally distributed with mean μ bushels and standard deviation equal to 25.7 bushels. If a company wishes to use the loader to fill containers that hold 2000 bushels of grain and wants to overfill only one container in 100, at what value of μ should the company set the loader?

7.31. A publisher has discovered that the number of words contained in a new manuscript is normally distributed with a mean equal to 20,000 words in excess of that specified in the author's contract and a standard deviation of 10,000 words. If the publisher wants to be almost certain (say with a probability of .95) that the manuscript will be less than 100,000 words, what number of words should the publisher specify in the contract?

7.3
Fortune's Forty-to-one Odds

In supporting the contention expressed in the case study, that the "height-blind" selection of a 5-foot-tall Chinese male is a highly improbable event, the author of the FORTUNE article notes that the Chinese equivalent of the U.S. Health Service does not exist and hence that health statistics on the current population of China are difficult to acquire. However, the author notes that "it is generally held that a boy's length at birth represents 28.6 percent of his final height" and that, in pre-revolutionary China, the average length of a Chinese boy at birth was 18.9 inches. From this, the author deduces that the mean height of mature male Chinese is

$$\frac{18.9}{.286} = 66.08 \text{ inches or 5 feet, 6.08 inches.}$$

The author then assumes that the distribution of the heights of males in China follows a normal distribution "(as it does in the U.S.)" with a mean of 66 inches and a standard deviation equal to 2.7 inches, "a figure that looks about right for that mean."

If we are willing to accept the assumption, that the heights of adult Chinese males are normally distributed with

$$\mu = 66 \text{ inches}$$

$$\sigma = 2.7 \text{ inches,}$$

we are ready to calculate the probability that a single adult Chinese male, chosen at random, will have a height that is less than or equal to 5 feet or, equivalently, 60 inches. This probability will be the area under a normal curve lying to the left of 60 inches (see Figure 7.8).

Figure 7.8 The assumed distribution of heights for adult male Chinese

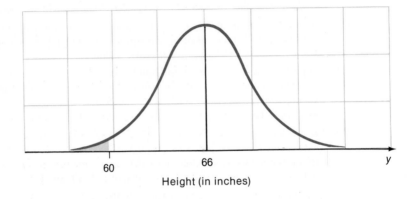

Height (in inches)

The z value corresponding to 60 inches is

$$z = \frac{y - \mu}{\sigma} = \frac{60 - 66}{2.7} = -2.22$$

and the area between $z = 0$ and $z = 2.22$ (see Table 3, the Appendix) is .4868. Hence the area between $z = -2.22$ and $z = 0$ is also .4868 and the tail area to the left of $y = 60$ is

$$P = .5 - .4868$$
$$= .0132.$$

This probability, the probability of randomly selecting a Chinese male less than 5 feet tall, is approximately 1 in 76, which corresponds to odds of 75 to 1.

Odds of 75 to 1 certainly agree with our intuition because it is difficult to believe that the proportion of 5-foot-tall adult male Chinese is very large. Nevertheless, the validity of our calculated odds depends upon the validity of our assumptions. The author of the article makes a good case for assuming that the distribution of the heights of Chinese males is approximately normal and we would be willing to accept the claim that the mean height is equal to 66 inches or larger. (If, in fact, the mean height of adult Chinese males is larger than 66 inches, the probability of randomly selecting a male with a height less than or equal to 5 feet is even less than .0132.) Choosing σ equal to 2.7 inches, solely because it is "a figure that looks about right for that mean" is more difficult to accept because the choice of σ greatly affects the calculated odds. For example, if σ equals 3 inches, then 60 inches lies $z = 2$ standard deviations below $\mu = 66$ inches. If $\sigma = 6$ inches, then 60 inches lies only $z = 1.0$ standard deviation below $\mu = 66$ inches. In spite of this point, we agree with the author. It is difficult to believe that σ could be very large, say larger than 4 or 5 inches, but the reason is probably because we find it difficult to believe that the proportion of 5-foot adult male Chinese is very large! Thus we have returned to our original point. Choosing a "reasonable" value for σ is equivalent to choosing "reasonable" odds.

Finally, there is another possible basic flaw in our assumptions. It is not clear that the distribution of the heights of all adult male Chinese is a good model for the distribution of the heights of potential candidates for Deng Xiaoping's replacement. Presumably, the candidates would be a very select group of senior, and elderly, members of the Chinese party. It is a well-known fact that the heights of humans decrease as they get older, particularly as they reach 60 years of age or older. Therefore, we would expect the distribution of the heights of the candidates for Deng's post to possess a mean that is less than the mean for the distribution of all adult male Chinese. We would also expect the odds of randomly selecting a person 5-feet tall or less from among the candidates to be larger than the corresponding odds of selection from among the population of heights of all Chinese adult males.

Did Deng Xiaoping take height into account in selecting his successor? The answer to this question depends upon the assumptions that you are willing to make. Consequently, we leave the answer to you. Perhaps you will have additional reasons for accepting or not accepting the Seligman's assumptions.

7.4
Sampling Distributions of Statistics

Statistics (Section 3.5), computed from the sample measurements, will be used in Chapter 8 to estimate and to make decisions about population parameters. These quantities, like the sample data from which they were computed, vary from sample to sample in a random manner. For example, if you compute the sample means for two different samples selected from the same population, they will almost certainly assume different values. Consequently, sample statistics, such as the sample mean $\bar{y}$ and the sample standard deviation s, are random variables.

If we compute a sample statistic, say the sample mean $\bar{y}$, based on a specific set of sample data, how near to the population mean μ is it likely to fall? What is the probability that the sample mean will deviate from μ by more than some specified amount? To answer these questions, we need to know the probability, or **sampling distribution,** of the sample mean.

Definition

The probability distribution for a statistic is called the **sampling distribution** of the statistic.

An approximation to the sampling distribution of the sample mean can be obtained by holding the sample size constant, drawing repeated samples, and calculating $\bar{y}$ for each. The resulting relative frequency distribution for these values of $\bar{y}$ would approximate the sampling distribution for $\bar{y}$ and would characterize the behavior of this statistic. This distribution locates the approximate mean value of $\bar{y}$ and gives a good indication of how large and how small a value of $\bar{y}$ might be if calculated from a single sample. We will learn more about this important sampling distribution in Section 7.5.

7.5
The Central Limit Theorem and the Sampling Distribution of the Sample Mean

In addition to being an important probability distribution in its own right, the normal probability distribution can often be used to approximate the probability distributions of other random variables because of the Central Limit Theorem. The Central Limit Theorem states that under rather general conditions, sums and means of samples of random measurements drawn from a population tend to possess, approximately, a bell-shaped distribution in repeated sampling. The significance of this statement is perhaps best illustrated by an example.

Consider a population of die throws generated by tossing a die an infinitely large number of times with resulting probability distribution given by Figure 7.9. Draw a sample of $n = 5$ measurements from the population by tossing a die five times and record each of the five observations as indicated in Table 7.2. Note that the numbers observed in the first sample were $y = 3, 5, 1, 3, 2$. Calculate the sum of the five measurements as well as the sample mean, $\bar{y}$. For experimental purposes, repeat the sampling procedure 100 times or preferably an even larger number of times. The results for 100 samples are given in Table 7.2 along with the corresponding values of $\sum_{i=1}^{5} y_i$ and $\bar{y}$. Construct a frequency histogram for $\bar{y}$ $\left(\text{or } \sum_{i=1}^{5} y_i \right)$ for the 100 samples and observe the resulting distribution in Figure 7.10. You will observe an interesting result, namely that although the values of y in the population ($y = 1, 2, 3, 4, 5, 6$) are equiprobable (Figure 7.9) and hence possess a probability distribution that is perfectly flat in shape, the distribution of the sample means (or sums) chosen from the population possesses a mound-shaped distribution (Figure 7.10). We shall add

Figure 7.9 Probability distribution for y, the number appearing on a single toss of a die

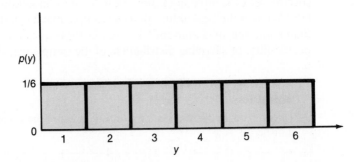

Table 7.2 Sampling from the population of die throws

Sample Number	Sample Measurements	$\sum y_i$	$\bar{y}$	Sample Number	Sample Measurements	$\sum y_i$	$\bar{y}$
1	3, 5, 1, 3, 2	14	2.8	51	2, 3, 5, 3, 2	15	3.0
2	3, 1, 1, 4, 6	15	3.0	52	1, 1, 1, 2, 4	9	1.8
3	1, 3, 1, 6, 1	12	2.4	53	2, 6, 3, 4, 5	20	4.0
4	4, 5, 3, 3, 2	17	3.4	54	1, 2, 2, 1, 1	7	1.4
5	3, 1, 3, 5, 2	14	2.8	55	2, 4, 4, 6, 2	18	3.6
6	2, 4, 4, 2, 4	16	3.2	56	3, 2, 5, 4, 5	19	3.8
7	4, 2, 5, 5, 3	19	3.8	57	2, 4, 2, 4, 5	17	3.4
8	3, 5, 5, 5, 5	23	4.6	58	5, 5, 4, 3, 2	19	3.8
9	6, 5, 5, 1, 6	23	4.6	59	5, 4, 4, 6, 3	22	4.4
10	5, 1, 6, 1, 6	19	3.8	60	3, 2, 5, 3, 1	14	2.8
11	1, 1, 1, 5, 3	11	2.2	61	2, 1, 4, 1, 3	11	2.2
12	3, 4, 2, 4, 4	17	3.4	62	4, 1, 1, 5, 2	13	2.6
13	2, 6, 1, 5, 4	18	3.6	63	2, 3, 1, 2, 3	11	2.2
14	6, 3, 4, 2, 5	20	4.0	64	2, 3, 3, 2, 6	16	3.2
15	2, 6, 2, 1, 5	16	3.2	65	4, 3, 5, 2, 6	20	4.0
16	1, 5, 1, 2, 5	14	2.8	66	3, 1, 3, 3, 4	14	2.8
17	3, 5, 1, 1, 2	12	2.4	67	4, 6, 1, 3, 6	20	4.0
18	3, 2, 4, 3, 5	17	3.4	68	2, 4, 6, 6, 3	21	4.2
19	5, 1, 6, 3, 1	16	3.2	69	4, 1, 6, 5, 5	21	4.2
20	1, 6, 4, 4, 1	16	3.2	70	6, 6, 6, 4, 5	27	5.4
21	6, 4, 2, 3, 5	20	4.0	71	2, 2, 5, 6, 3	18	3.6
22	1, 3, 5, 4, 1	14	2.8	72	6, 6, 6, 1, 6	25	5.0
23	2, 6, 5, 2, 6	21	4.2	73	4, 4, 4, 3, 1	16	3.2
24	3, 5, 1, 3, 5	17	3.4	74	4, 4, 5, 4, 2	19	3.8
25	5, 2, 4, 4, 3	18	3.6	75	4, 5, 4, 1, 4	18	3.6
26	6, 1, 1, 1, 6	15	3.0	76	5, 3, 2, 3, 4	17	3.4
27	1, 4, 1, 2, 6	14	2.8	77	1, 3, 3, 1, 5	13	2.6
28	3, 1, 2, 1, 5	12	2.4	78	4, 1, 5, 5, 3	18	3.6
29	1, 5, 5, 4, 5	20	4.0	79	4, 5, 6, 5, 4	24	4.8
30	4, 5, 3, 5, 2	19	3.8	80	1, 5, 3, 4, 2	15	3.0
31	4, 1, 6, 1, 1	13	2.6	81	4, 3, 4, 6, 3	20	4.0
32	3, 6, 4, 1, 2	16	3.2	82	5, 4, 2, 1, 6	18	3.6
33	3, 5, 5, 2, 2	17	3.4	83	1, 3, 2, 2, 5	13	2.6
34	1, 1, 5, 6, 3	16	3.2	84	5, 4, 1, 4, 6	20	4.0
35	2, 6, 1, 6, 2	17	3.4	85	2, 4, 2, 5, 5	18	3.6
36	2, 4, 3, 1, 3	13	2.6	86	1, 6, 3, 1, 6	17	3.4
37	1, 5, 1, 5, 2	14	2.8	87	2, 2, 4, 3, 2	13	2.6
38	6, 6, 5, 3, 3	23	4.6	88	4, 4, 5, 4, 4	21	4.2
39	3, 3, 5, 2, 1	14	2.8	89	2, 5, 4, 3, 4	18	3.6
40	2, 6, 6, 6, 5	25	5.0	90	5, 1, 6, 4, 3	19	3.8
41	5, 5, 2, 3, 4	19	3.8	91	5, 2, 5, 6, 3	21	4.2
42	6, 4, 1, 6, 2	19	3.8	92	6, 4, 1, 2, 1	14	2.8
43	2, 5, 3, 1, 4	15	3.0	93	6, 3, 1, 5, 2	17	3.4
44	4, 2, 3, 2, 1	12	2.4	94	1, 3, 6, 4, 2	16	3.2
45	4, 4, 5, 4, 4	21	4.2	95	6, 1, 4, 2, 2	15	3.0
46	5, 4, 5, 5, 4	23	4.6	96	1, 1, 2, 3, 1	8	1.6
47	6, 6, 6, 2, 1	21	4.2	97	6, 2, 5, 1, 6	20	4.0
48	2, 1, 5, 5, 4	17	3.4	98	3, 1, 1, 4, 1	10	2.0
49	6, 4, 3, 1, 5	19	3.8	99	5, 2, 1, 6, 1	15	3.0
50	4, 4, 4, 4, 4	20	4.0	100	2, 4, 3, 4, 6	19	3.8

Figure 7.10 Histogram of sample means for the die-tossing experiments in Section 7.5

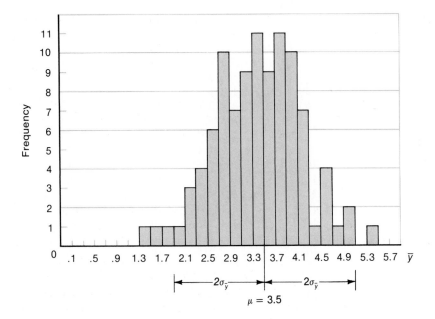

one further comment without proof. If we should repeat the experiment outlined above for a larger sample size, say $n = 10$, we would find that the distribution of the sample means tends to become more nearly bell-shaped.

Note that a proper evaluation of the form of the sampling distribution of the sample means would require an infinitely large number of samples, or, at the very least, far more than the 100 samples contained in our experiment. Nevertheless, the 100 samples illustrate the basic idea involved in the Central Limit Theorem, which may be stated as follows.

The Central Limit Theorem

If random samples of n observations are drawn from a population with finite mean, μ, and standard deviation, σ, then, when n is large, the sampling distribution of the sample mean, $\bar{y}$, will be approximately a normal distribution with mean equal to μ and standard deviation $\sigma/\sqrt{n}$. The approximation will become more and more accurate as n becomes larger and larger.

Note that the Central Limit Theorem could be restated to apply to the sum of the sample measurements, $\sum_{i=1}^{n} y_i$, which would also tend to possess a normal sampling distribution with mean equal to $n\mu$ and standard deviation $\sigma\sqrt{n}$, as n becomes large.

The Central Limit Theorem tells us that the mean and standard deviation of the distribution of sample means are definitely related to the mean and standard deviation of the sampled population as well as to the sample size n. The two distributions have the same mean μ and the standard deviation of the distribution of sample means is equal to the population standard deviation σ divided by $\sqrt{n}$ (it can be shown that this relationship is true regardless of the sample size n). Consequently, the spread of the distribution of sample means is considerably less ($1/\sqrt{n}$ as large) than the spread of the population distribution. But most importantly, the Central Limit Theorem tells us that the sampling distribution of the sample mean is approximately normal when the sample size n is large.

The significance of the Central Limit Theorem is twofold. First, it explains why some measurements tend to possess, approximately, a normal distribution. We might imagine the height of a human as being composed of a number of elements, each random, associated with such things as the height of the mother, the height of the father, the activity of a particular gland, the environment, and diet. If each of these effects tends to add to the others to yield the measurement of height, then height is the sum of a number of random variables and the Central Limit Theorem may become effective and yield a distribution of heights which is approximately normal. All of this is conjecture, of course, because we really do not know the true situation which exists. Nevertheless, the Central Limit Theorem, along with other theorems dealing with normally distributed random variables, provides an explanation of the rather common occurrence of normally distributed random variables in nature.

The second and most important contribution of the Central Limit Theorem is in statistical inference. Many estimators and decision makers that are used in making inferences about population parameters are sums or averages of the sample measurements. When this is true and when the sample size, n, is sufficiently large, we would expect the estimator or decision maker to possess a sampling distribution that, according to the Central Limit Theorem, is approximately normal. We can then use the Empirical Rule discussed in Chapter 3 to describe the behavior of the inference maker. This aspect of the Central Limit Theorem will be utilized in Section 7.6 and in later chapters dealing with statistical inference.

One disturbing feature of the Central Limit Theorem, and of most approximation procedures, is that we must have some idea as to how large the sample size, n, must be in order for the approximation to give useful results. Unfortunately, there is no clear-cut answer to this question, as the appropriate value for n will depend upon the population probability distribution as well as the use we will make of the approximation. To illustrate, we have programmed a computer to select random samples of size n, $n = 2, 5, 10,$ and 25 from each of three populations, the first possessing a normal population probability distribution, the second a uniform probability distribution, and the third a negative exponential probability distribution. These population probability distributions are shown in the top row of Figure 7.11. The computer printouts of the **approximations to sampling distributions of the sample mean $\bar{y}$ for sample sizes $n = 2$, $n = 5, n = 10,$ and $n = 25$ are shown in rows 2, 3, 4, and 5 of Figure 7.11.**

Figure 7.11 Probability distributions and approximations to the sampling distributions for three populations

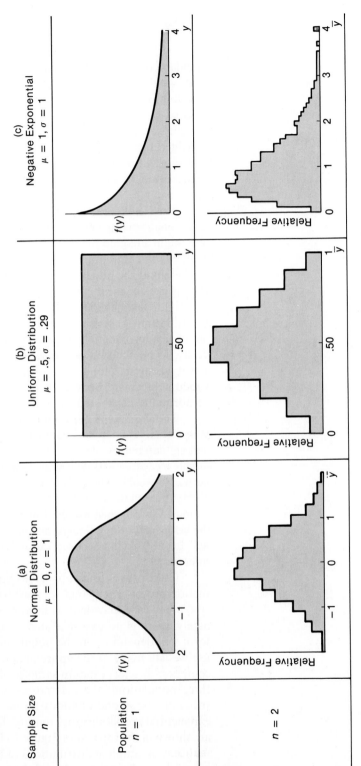

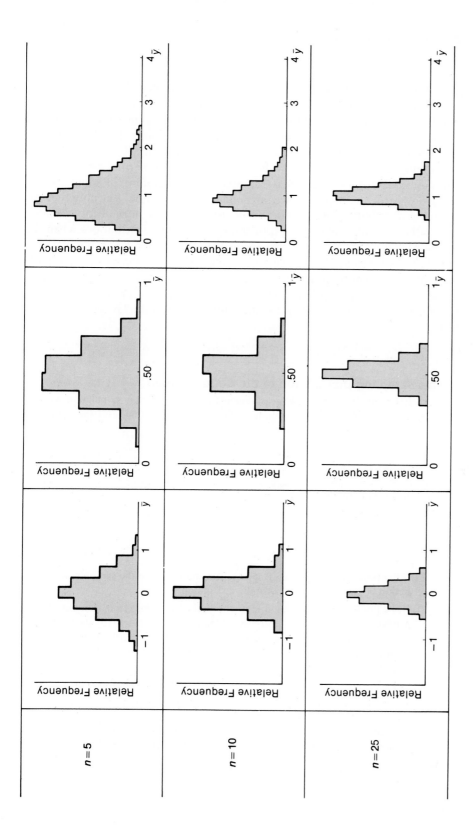

Figure 7.11 illustrates an important theorem of theoretical statistics. The sampling distribution of the sample mean is exactly normally distributed (proof omitted), regardless of the sample size, when sampling from a population that possesses a normal distribution. In contrast, the sampling distributions of $\bar{y}$, for samples selected from populations with uniform and negative exponential probability distributions, tend to become more nearly normal as the sample size n increases from $n = 2$ to $n = 25$, rapidly for the uniform distribution, and more slowly for the highly skewed exponential distribution. But note that the sampling distribution of $\bar{y}$ is normal or approximately normal for sampling from either the uniform or the exponential probability distributions when the sample size is as large as $n = 25$. This suggests that, for many populations, sampling distribution of $\bar{y}$ will be approximately normal for moderate sample sizes, but an exception to this rule will be found in Section 7.6. Consequently, we will give the appropriate sample size, n, for specific applications of the Central Limit Theorem as they are encountered in Section 7.6 and later in the text.

The Sampling Distribution of the Sample Mean $\bar{y}$

1. If a random sample of n measurements is selected from a population with mean μ and standard deviation σ, the sampling distribution of the sample mean $\bar{y}$ will possess a mean

$$\mu_{\bar{y}} = \mu$$

and a standard deviation

$$\sigma_{\bar{y}} = \frac{\sigma}{\sqrt{n}}.$$

2. If the population possesses a normal distribution, then the sampling distribution of $\bar{y}$ will be exactly normally distributed.

3. If the population distribution is nonnormal, the sampling distribution of $\bar{y}$ will be, for large samples, approximately normally distributed (by the Central Limit Theorem). Figure 7.11 suggests that the sampling distributions of $\bar{y}$ will be approximately normal for sample sizes as small as $n = 25$ for most populations of measurements.

Example 7.7

Suppose that you select a random sample of $n = 25$ observations from a population with mean $\mu = 8$ and $\sigma = .6$. Find the approximate probability that the sample mean $\bar{y}$

a. Will be less than 7.9.

b. Will exceed 7.9.

c. Will lie within .01 of the population mean $\mu = 8$.

Solution

a. Regardless of the shape of the population relative frequency distribution, the sampling distribution of $\bar{y}$ will possess a mean $\mu_{\bar{y}} = \mu = 8$ and a standard deviation

$$\sigma_{\bar{y}} = \frac{\sigma}{\sqrt{n}} = \frac{.6}{\sqrt{25}} = .12.$$

And, for a sample as large as $n = 25$, it is likely (because of the Central Limit Theorem) that the sampling distribution of $\bar{y}$ is approximately normally distributed (we will assume that it is). Therefore, the probability P that $\bar{y}$ will be less than 7.9 is approximated by the shaded area under the normal sampling distribution, Figure 7.12. To find this area, we need to calculate the value of z corresponding to $\bar{y} = 7.9$. This value of z is the distance between $\bar{y} = 7.9$ and $\mu_{\bar{y}} = \mu = 8.0$ expressed in standard deviations of the sampling distribution; that is, in units of

$$\sigma_{\bar{y}} = \frac{\sigma}{\sqrt{n}} = .12.$$

Thus,

$$z = \frac{\bar{y} - \mu}{\sigma_{\bar{y}}} = \frac{7.9 - 8.0}{.12} = -.83.$$

From Table 3, the Appendix, we find the area corresponding to $z = .83$ is .2967. Therefore,

$$P = .5 - A = .5 - .2967 = .2033.$$

[Note that we must use $\sigma_{\bar{y}}$ (not σ) in the formula for z because we are finding an area under the sampling distribution for $\bar{y}$, not under the sampling distribution for y.]

Figure 7.12 The probability that $\bar{y}$ is less than 7.9

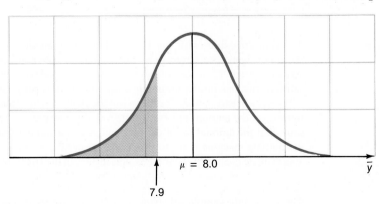

b. The event that $\bar{y}$ exceeds 7.9 is the complement of the event that $\bar{y}$ is less than 7.9. Thus, the probability that $\bar{y}$ exceeds 7.9 is,

$$P(\bar{y} > 7.9) = 1 - P\{\bar{y} < 7.9\}$$
$$= 1 - .2033$$
$$= .7967.$$

c. The probability that $\bar{y}$ lies within .01 of $\mu = 8$ is the shaded area in Figure 7.13. We found in part (a) that the area A between $\bar{y} = 7.9$ and $\mu = 8.0$ is .2967. Since the area under the normal curve between $\bar{y} = 8.1$ and $\mu = 8.0$ is identical to the area between $\bar{y} = 7.9$ and $\mu = 8.0$, it follows that

$$P(7.9 < \bar{y} < 8.1) = 2A$$
$$= 2(.2967)$$
$$= .5934.$$

Figure 7.13 The probability that $\bar{y}$ lies within .01 of $\mu = 8$

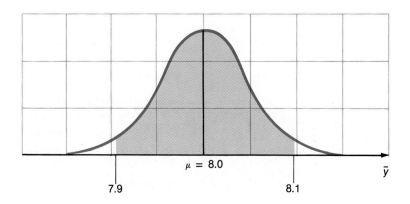

$\mu = 8.0$

7.9   8.1

$\bar{y}$

Example 7.8

To avoid difficulties with the Federal Trade Commission or state and local consumer protection agencies, a beverage bottler must make reasonably certain that 12-ounce bottles actually certain 12 ounces of beverage. To infer whether a bottling machine is working satisfactorily, one bottler randomly samples 10 bottles per hour and measures the amount of beverage in each bottle. The mean $\bar{y}$ of the 10 fill measurements is used to decide whether to readjust the amount of beverage delivered per bottle by the filling machine. If records show that the amount of fill per bottle possesses a standard deviation of .2 ounce, and if the bottling machine is set to produce a mean fill per bottle of 12.1 ounces, what is the approximate probability that the sample mean $\bar{y}$ of the 10 test bottles is less than 12 ounces?

Solution

The mean of the sampling distribution of the sample mean $\bar{y}$ is identical to the mean of the population of bottle fills, namely, $\mu = 12.1$ ounces, and the standard deviation of the sampling distribution, denoted by the symbol $\sigma_{\bar{y}}$, is

$$\sigma_{\bar{y}} = \frac{\sigma}{\sqrt{n}} = \frac{.2}{\sqrt{10}} = .063.$$

(*Note:* σ is the standard deviation of the population of bottle fills and n is the number of bottles in the sample.) Even though n is as small as 10, it is likely, for this type of data, that the sampling distribution of $\bar{y}$ will be approximately normal because of the Central Limit Theorem. Then the sampling distribution of $\bar{y}$ will appear as shown in Figure 7.14.

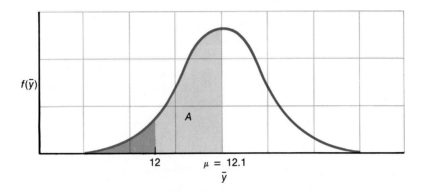

Figure 7.14 Sampling distribution of $\bar{y}$, the mean of the $n = 10$ bottle fills

The probability that $\bar{y}$ will be less than 12 ounces is approximately equal to the shaded area under the normal curve, Figure 7.14. This area will equal $(.5 - A)$, where A is the area between 12 and the mean, $\mu = 12.1$. Expressing this distance in terms of z,

$$z = \frac{\bar{y} - \mu}{\sigma_{\bar{y}}} = \frac{12 - 12.1}{.063} = \frac{-.1}{.063} = -1.59.$$

[Note that we must use $\sigma_{\bar{y}}$ (not σ) in the formula for z because we are finding an area under the sampling distribution for $\bar{y}$, not under the sampling distribution for y.] Then, the area A over the interval, $0 \leq z \leq 1.59$, is shown in Table 3, the Appendix, as .4441 and the probability that $\bar{y}$ will be less than 12 ounces is

$$\begin{aligned} P(\bar{y} < 12) &= .5 - A = .5 - .4441 \\ &= .0559 \\ &\approx .056. \end{aligned}$$

Or, if the fill machine is set to produce a mean fill of 12.1 ounces, the mean fill $\bar{y}$ of a sample of 10 bottles will be less than 12 ounces with probability approximately equal to .056. When this danger signal occurs ($\bar{y}$ is less than 12), the bottler takes a larger sample to recheck the setting of the filling machine. Note that the Central Limit Theorem plays a role in the solution of this problem because it justifies the approximate normality of the sampling distribution of the sample mean.

Tips on Problem Solving

Before attempting to calculate the probability that the statistic $\bar{y}$ falls in some interval,

1. Calculate the mean and standard deviation of the sampling distribution of $\bar{y}$.
2. Sketch the sampling distribution. Show the location of the mean μ and use the value of $\sigma_{\bar{y}}$ to locate the approximate location of the tails of the distribution.
3. Locate the interval on the sketch, part 2, and shade the area corresponding to the probability that you wish to calculate.
4. Find the z score(s) associated with the value(s) of interest in your problem. Use the table in the Appendix to find the probability.
5. When you have obtained your answer, look at your sketch of the sampling distribution and see if your calculated answer agrees with the shaded area. This provides a very rough check on your calculations.

Exercises

[Starred (*) exercises are optional.]

7.32. Looking at the histogram of Figure 7.10, guess the value of its mean and standard deviation. (*Hint:* The Empirical Rule states that approximately 95 percent of the measurements associated with a mound-shaped distribution will lie within two standard deviations of the mean.)

7.33. Let y equal the number of dots observed when a single die is tossed. The mean value of y, Example 5.4, and standard deviation (Exercise 5.31) were found to equal $\mu = 3.5$ and $\sigma = 1.71$, respectively. Suppose that the sampling experiment, Section 7.5, were repeated over and over again an infinitely large number of times, each sample consisting of $n = 5$ measurements. Find the mean and standard deviation for this distribution of sample means. (*Hint:* See the Central Limit Theorem.) Compare this solution with the solution to Exercise 7.32.

7.34. Suppose that you were to experiment by drawing thousands of samples where each sample involved tossing a die $n = 10$ times. If a histogram were constructed for the sample means, what would be the value for the mean of the distribution? The standard deviation?

7.35. Suppose that a random sample of $n = 5$ observations is selected from a population which is normally distributed with mean equal to 1 and standard deviation equal to .25.

a. Give the mean and standard deviation of the sampling distribution of $\bar{y}$.

b. Find the probability that $\bar{y}$ exceeds 1.2.

c. Find the probability that the sample mean $\bar{y}$ will be less than -1.2.

d. Find the probability that the sample mean will deviate from the population mean, $\mu = 1$, by more than .2.

7.36. Suppose that a random sample of $n = 25$ observations is selected from a population which is normally distributed with mean equal to 106 and standard deviation equal to 12.

a. Give the mean and the standard deviation of the sampling distribution of the sample mean $\bar{y}$.

b. Find the probability that $\bar{y}$ exceeds 110.

c. Find the probability that the sample mean will deviate from the population mean, $\mu = 106$, by no more than 4.

7.37. The mean and standard deviation of the population normal frequency distribution, Figure 7.11(a), are equal to 0 and 1, respectively. Examine Figure 7.11(a) and visually verify that this is true. Then, according to statistical theory, the sampling distributions of the sample mean $\bar{y}$, column 2, Figure 7.11 (a), should have a mean equal to $\mu = 0$ and standard deviation equal to

$$\sigma_{\bar{y}} = \frac{\sigma}{\sqrt{n}},$$

or, since $\sigma = 1$, $\sigma/\sqrt{n} = 1/\sqrt{n}$.

a. For $n = 2$, $\sigma_{\bar{y}} = 1/\sqrt{2} = .707$. Examine the sampling distribution of $\bar{y}$ and verify that all or almost all of the sample means fall in the interval, $\mu \pm 3\sigma_{\bar{y}}$.

b. Repeat the instructions of part (a) for $n = 5$.

c. Repeat the instructions of part (a) for $n = 10$.

d. Repeat the instructions of part (a) for $n = 25$.

7.38. The population uniform distribution, Figure 7.11(b) has a mean $\mu = .5$ and standard deviation $\sigma = .29$. Examine Figure 7.11(b) and visually verify that this is true. Then, according to statistical theory, the sampling distributions of the sample mean $\bar{y}$, column 3, Figure 7.11 (b), should have a mean equal to $\mu = .5$ and a standard deviation equal to

$$\sigma_{\bar{y}} = \frac{\sigma}{\sqrt{n}} = \frac{.29}{\sqrt{n}}.$$

a. For $n = 2$, $\sigma_{\bar{y}} = .29/\sqrt{2} = .205$. Examine the sampling distribution of $\bar{y}$ and verify that all or almost all of the sample means fall in the interval, $\mu \pm 3\sigma_{\bar{y}}$.

b. Repeat the instructions of part (a) for $n = 5$.

c. Repeat the instructions of part (a) for $n = 10$.

d. Repeat the instructions of part (a) for $n = 25$.

7.39. The population negative exponential distribution, Figure 7.11(c) has a mean $\mu = 1$ and standard deviation $\sigma = 1$. Examine Figure 7.11(c) and visually verify that this is true. Then, according to statistical theory, the sampling distributions of the sample mean $\bar{y}$, column 4, Figure 7.11 (c), should have a mean equal to $\mu = 1$ and

$$\sigma_{\bar{y}} = \frac{\sigma}{\sqrt{n}} = \frac{1}{\sqrt{n}}.$$

a. For $n = 2$, $\sigma_{\bar{y}} = 1/\sqrt{2} = .707$. Examine the sampling distribution of $\bar{y}$ and verify that all or almost all of the sample means fall in the interval, $\mu \pm 3\sigma_{\bar{y}}$.

b. Repeat the instructions of part (a) for $n = 5$.

c. Repeat the instructions of part (a) for $n = 10$.

d. Repeat the instructions of part (a) for $n = 25$.

7.40.   An important aspect of the 1981 federal economic plan is that consumers will save a substantial portion of the money that they receive from an income tax reduction. Suppose that early estimates of the portion of total tax saved, based on a random sampling of 35 economists, possessed a mean of 26 percent and a standard deviation of 12 percent.

   a. What is the approximate probability that a sample mean, based on a random sample of $n = 35$ economists, will lie within 1 percent of the mean of the population of the estimates of all economists?

   b. Is it necessarily true that the mean of the population of estimates of all economists is equal to the percent tax saving that will actually be achieved?

7.41.   A lobster fisherman's daily catch, y, is the total, in pounds, of lobster landed from a fixed number of lobster traps. What kind of probability distribution would you expect the daily catch to possess and why? If the mean catch per trap per day is 30 pounds with $\sigma = 5$ pounds, and the fisherman has 50 traps, give the mean and standard deviation of the probability distribution of the total daily catch y.

7.42.   To obtain information on the volume of freight shipped by truck over a particular interstate highway, a state highway department monitored the highway for 25 one-hour periods randomly selected throughout a one-month period. The number of truck trailers was counted for each one-hour period and $\bar{y}$ was calculated for the sample of 25 individual one-hour periods. If the number of heavy-duty trailers per hour is approximately normally distributed, with $\mu = 50$ and $\sigma = 7$:

   a. What is the approximate probability that the sample mean $\bar{y}$ for $n = 25$ one-hour periods is larger than 55?*

   b. Suppose you were to count the truck trailers for each of $n = 4$ randomly selected one-hour periods. What is the approximate probability that $\bar{y}$ would be larger than 55?*

   c. What is the approximate probability that the total number of trucks for a four-hour period would exceed 180?

7.43.   A manufacturer of paper used for packaging requires a minimum strength of 20 pounds per square inch. To check on the quality of the paper, a random sample of 10 pieces of paper is selected each hour from the previous hour's production and a strength measurement is recorded for each. The standard deviation σ of the strength measurements, computed by pooling the sum of squares of deviations of many samples, is known to equal 2 pounds per square inch.

   a. What is the approximate probability distribution of the sample mean of $n = 10$ test pieces of paper?

   b. If the mean of the population of strength samples is 21 pounds per square inch, what is the approximate probability that, for a random sample of $n = 10$ test pieces of paper, $\bar{y} < 20$?

   c. What value would you desire for the mean paper strength, μ, in order that $P(\bar{y} < 20)$ be equal to .001?

7.44.   The normal daily human potassium requirement is in the range of 2000 to 6000 milligrams with the larger amounts required during hot summer weather. The amount of potassium in food varies, depending upon the food. For example, there are approximately 7 mg in a cola drink, 46 mg in a beer, 630 mg in a banana, 300 mg in a carrot, and 440 mg in a

---

* The distribution of the sample means will be normally distributed, regardless of the sample size, for the special case when the population possesses a normal distribution.

glass of orange juice. Suppose that the distribution of potassium in a banana is normally distributed with mean equal to 630 mg and standard deviation equal to 40 mg per banana. If you eat $n = 3$ bananas per day and T is the total number of milligrams of potassium that you will receive from them,

a. Find the mean and standard deviation of T.

b. Find the probability that your total daily intake of potassium from the three bananas will exceed 2000 mg.

(*Hint:* Note that T is the sum of three random variables, y_1, y_2, and y_3, where y_1 is the amount of potassium in banana #1, etc.)

7.6
The Normal Approximation to the Binomial Distribution

In Chapter 6 we considered several applications of the binomial probability distribution, all of which required that we calculate the probability that y, the number of successes in n trials, would fall in a given region. For the most part we restricted our attention to examples where n was small because of the tedious calculations necessary in the computations of $p(y)$. Let us now consider the problem of calculating $p(y)$, or the probability that y will fall in a given region, when n is large, say $n = 1{,}000$. A direct calculation of $p(y)$ for large values of n is not an impossibility, but it does provide a formidable task that we would prefer to avoid. Fortunately, the Central Limit Theorem provides a solution to this dilemma since we may view y, the number of successes in n trials, as a sum that satisfies the conditions of the Central Limit Theorem. Each trial results in either 0 or 1 success with probability q and p, respectively. Therefore, each of the n trials may be regarded as an independent observation drawn from a simpler binomial experiment consisting of one trial, and y, the total number of successes in n trials, is the sum of these n independent observations. Then, if n is sufficiently large, the binomial variable, y, will be approximately normally distributed with mean and variance (obtained in Chapter 6) np and npq, respectively. We may then use areas under a fitted normal curve to approximate the binomial probabilities.

For instance, consider a binomial probability distribution for y when $n = 10$ and $p = 1/2$. Then

$$\mu = np = 10(1/2) = 5 \quad \text{and} \quad \sigma = \sqrt{npq} = \sqrt{2.5} = 1.58.$$

Figure 7.15 shows the corresponding binomial probability distribution and the approximating normal curve on the same graph. A visual comparison of the figures would suggest that the approximation is reasonably good, even though a small sample, $n = 10$, was necessary for this graphic illustration.

The probability that $y = 2$, 3, or 4 is exactly equal to the area of the three rectangles lying over $y = 2$, 3, and 4. We may approximate this probability with the area under the normal curve from $y = 1.5$ to $y = 4.5$, which is shaded in Figure 7.15. Note that the area under the normal curve between $y = 2$ and $y = 4$ *would not* be a good approximation to the probability that $y = 2$, 3, or 4 because it would exclude one-half of the probability rectangles corresponding

Figure 7.15 Comparison of a binomial probability distribution and the approximating normal distribution, $n = 10$, $p = 1/2$ ($\mu = np = 5$; $\sigma = \sqrt{npq} = 1.58$)

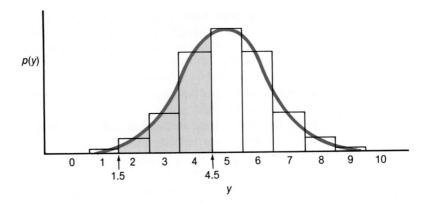

to $y = 2$ and $y = 4$. To get a good approximation you must remember to approximate the entire areas of the probability rectangles corresponding to $y = 2$ and $y = 4$ by including the area under the normal curve from $y = 1.5$ to $y = 4.5$.

Although the normal probability distribution provides a reasonably good approximation to the binomial probability distribution, Figure 7.15, this will not always be the case. When n is small and p is near 0 or 1, the binomial probability distribution will be nonsymmetrical; that is, its mean will be located near 0 or n. For example, when p is near zero, most values of y will be small, producing a distribution that is concentrated near $y = 0$ and that tails gradually toward n (see Figure 7.16). Certainly, when this is true, the normal distribution, symmetrical and bell-shaped, will provide a poor approximation to the binomial probability distribution. How, then, can we tell whether n and p are such that the binomial distribution will be symmetrical?

Figure 7.16 Comparison of a binomial probability distribution (shaded) and the approximating normal distribution, $n = 10$, $p = .1$ ($\mu = np = 1$; $\sigma = \sqrt{npq} = .95$)

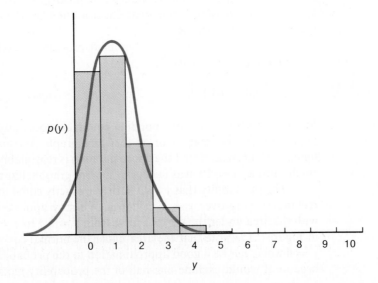

Recalling the Empirical Rule, Chapter 3, approximately 95 percent of the measurements associated with a normal distribution will lie within two standard deviations of the mean and almost all will lie within three. We would suspect that the binomial probability distribution would be nearly symmetrical if the distribution were able to spread out a distance equal to two standard deviations on either side of the mean and this is, in fact, the case. **Hence, to determine when the normal approximation will be adequate, calculate $\mu = np$ and $\sigma = \sqrt{npq}$. If the interval $\mu \pm 2\sigma$ lies within the binomial bounds, 0 and n, the approximation will be reasonably good.** Note that this criterion is satisfied for the example, Figure 7.15, but is not satisfied for Figure 7.16.

Example 7.9

Refer to the binomial experiment illustrated in Figure 7.15 where $n = 10, p = .5$. Calculate the probability that $y = 2, 3,$ or 4 correct to three decimal places using Table 1, the Appendix. Then calculate the corresponding normal approximation to this probability.

Solution

The exact probability, P_1, can be calculated using Table 1(b), the Appendix. Thus

$$P_1 = \sum_{y=2}^{4} p(y) = \sum_{y=0}^{4} p(y) - \sum_{y=0}^{1} p(y)$$

$$= .377 - .011$$

$$= .366.$$

The normal approximation would require finding the area lying between $y_1 = 1.5$ and $y_2 = 4.5$ (see Figure 7.15), where $\mu = 5$ and $\sigma = 1.58$. The corresponding values of z are

$$z_1 = \frac{y_1 - \mu}{\sigma} = \frac{1.5 - 5}{1.58} = -2.22,$$

$$z_2 = \frac{y_2 - \mu}{\sigma} = \frac{4.5 - 5}{1.58} = -.32.$$

The probability, P_2, is shown in Figure 7.17. The area between $z = 0$ and $z = 2.22$ is $A_1 = .4868$. Likewise, the area between $z = 0$ and $z = .32$ is $A_2 = .1255$. It can be seen

Figure 7.17 Area under the normal curve for Example 7.9

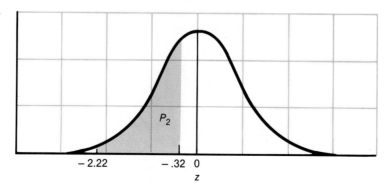

from Figure 7.17 that

$$P_2 = A_1 - A_2$$
$$= .4868 - .1255 = .3613.$$

Note that the normal approximation is quite close to the exact binomial probability obtained from Table 1.

You must be careful not to exclude half of the two extreme probability rectangles when using the normal approximation to the binomial probability distribution. This means that the y values used to calculate z values will always have a 5 in the tenths decimal place. To be certain that you include all the probability rectangles in your approximation, always draw a sketch similar to Figure 7.15.

Example 7.10

The reliability of an electrical fuse is the probability that a fuse, chosen at random from production, will function under the conditions for which it has been designed. A random sample of 1000 fuses was tested and $y = 27$ defectives were observed. Calculate the approximate probability of observing 27 or more defectives, assuming that the fuse reliability is .98.

Solution

The probability of observing a defective when a single fuse is tested is $p = .02$, given that the fuse reliability is .98. Then,

$$\mu = np = 1000(.02) = 20,$$
$$\sigma = \sqrt{npq} = \sqrt{1000(.02)(.98)} = 4.43.$$

The probability of 27 or more defective fuses, given $n = 1000$, is

$$P = P(y \geq 27),$$
$$P = P(27) + P(28) + P(29) + \cdots + P(999) + P(1000).$$

The normal approximation to P would be the area under the normal curve to the right of $y = 26.5$. (Note that we must use $y = 26.5$ rather than $y = 27$ so as to include the entire probability rectangle associated with $y = 27$.) The z value corresponding to $y = 26.5$ is

$$z = \frac{y - \mu}{\sigma} = \frac{26.5 - 20}{4.43} = \frac{6.5}{4.43} = 1.47$$

and the area between $z = 0$ and $z = 1.47$ is equal to .4292, as shown in Figure 7.18. Since the total area to the right of the mean is equal to .5,

$$P = .5 - .4292$$
$$= .0708.$$

Figure 7.18 Normal approximation to the binomial in Example 7.10

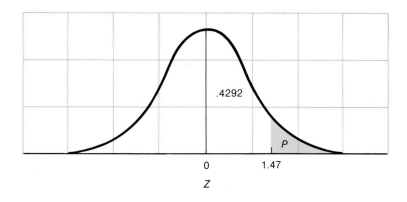

Example 7.11

A new serum was tested to determine its effectiveness in preventing the common cold. One hundred people were injected with the serum and observed for a period of one year. Sixty-eight survived the winter without a cold. Suppose that according to prior information it is known that the probability of surviving the winter without a cold is equal to .5 when the serum is not used. On the basis of the results of this experiment, what conclusions would you make regarding the effectiveness of the serum?

Solution

Translating the question into an hypothesis concerning the parameter of the binomial population, we wish to test the null hypothesis that p, the probability of survival on a single trial, is equal to .5. Assume that the content of the serum is such that it could not increase the susceptibility to colds. That is, p could not be less than 1/2. Then the alternative to the null hypothesis would be $p > 1/2$ and we would reject the null hypothesis when y, the number of survivors, is large.

Since the normal approximation to the binomial will be adequate for this example, we would interpret a large and improbable value of y to be one that lies several standard deviations away from the hypothesized mean, $\mu = np = 100(.5) = 50$. Noting that

$$\sigma = \sqrt{npq} = \sqrt{100(.5)(.5)} = 5,$$

we may arrive at a conclusion without bothering to locate a specific rejection region. The observed value of y, 68, lies more than 3σ away from the hypothesized mean, $\mu = 50$. Specifically, y lies

$$z = \frac{y - \mu}{\sigma} = \frac{68 - 50}{5} = 3.6$$

standard deviations away from the hypothesized mean. This result is so improbable, assuming the serum ineffective, that we would reject the null hypothesis and conclude that the probability of surviving a winter without a cold is greater than $p = .5$ when the serum is used. (Observe that the area to the right of $z = 3.6$ is so small that it is not included in Table 3, the Appendix.)

Rejecting the null hypothesis raises additional questions. How effective is the serum and is it sufficiently effective, from an economic point of view, to warrant commercial production? The former question leads to an estimation problem, a topic discussed in Chapter 8, while the latter, involving a business decision, would utilize the results of our experiment as well as a study of consumer demand, sales and production costs, etc., to achieve an answer useful to the drug company.

Example 7.12

The probability of a type I error, α, and location of the rejection region for a statistical test of an hypothesis are usually specified before the data are collected. Suppose that we wish to test the null hypothesis, $p = .5$, in a situation identical to the cold-serum problem in Example 7.11. Find the appropriate rejection region for the test if we wish α to be approximately equal to .05 (see Figure 7.19).

Figure 7.19 Location of the rejection region in Example 7.12

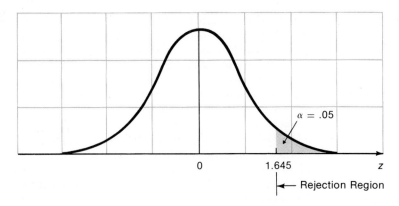

We have previously stated in Example 7.11 that y, the number of survivors, would be used as a test statistic and that the rejection region would be located in the upper tail of the probability distribution for y. Desiring α approximately equal to .05, we seek a value of y, say y_α, such that

Solution

$$P(y \geq y_\alpha) \approx .05.$$

(*Note:* The symbol $\approx$ means "approximately equal to.") This can be determined by first finding the corresponding z_α, which gives the number of standard deviations between the mean, $\mu = 50$, and y_α. Since the total area to the right of $z = 0$ is .5, the area between $z = 0$ and z_α will equal .45. Checking Table 3, we find that $z = 1.64$ corresponds to an area equal to .4495 and $z = 1.65$ to an area of .4505. Since the probability that we seek is halfway between the two probabilities found in the table, the z value will be halfway between the corresponding z values. That is,

$$z_\alpha = 1.645.$$

Recalling the relation between z and y,

$$z_\alpha = \frac{y_\alpha - \mu}{\sigma}$$

or

$$1.645 = \frac{y_\alpha - 50}{5}.$$

Solving for y_α, we obtain

$$y_\alpha = 58.225.$$

Obviously, we cannot observe $y = 58.225$ survivors and hence must choose 58 or 59 as the point where the rejection region commences. (Remember, the binomial is a discrete probability distribution; the normal is continuous.)

Suppose that we decide to reject the null hypothesis when y is greater than or equal to 59. Then the actual probability of the type I error, α, for the test is

$$P(y \geq 59) = \alpha,$$

which can be approximated by using the area under the normal curve above $y = 58.5$, a problem similar to that encountered in Example 7.10. The z value corresponding to $y = 58.5$ is

$$z = \frac{y - \mu}{\sigma} = \frac{58.5 - 50}{5} = 1.7,$$

and the tabulated area between $z = 0$ and $z = 1.7$ is .4554. Therefore,

$$\alpha = .5 - .4554$$
$$= .0446.$$

Although this method provides a more accurate value for α, there is very little practical difference between an α of .0446 and one equal to .05. When n is large, time and effort may be saved by using z as a test statistic rather than y. This method was employed in Example 7.11. We would then reject the null hypothesis when z is greater than or equal to 1.645.

Example 7.13

A cigarette manufacturer believed that approximately 10 percent of all smokers favored his product, brand A. To test this belief, 2500 smokers were selected at random from the population of cigarette smokers and questioned concerning their cigarette brand preference. A total of $y = 218$ expressed a preference for brand A. Do these data provide sufficient evidence to contradict the hypothesis that 10 percent of all smokers favor brand A? Conduct a statistical test using an α equal to .05.

Solution

We wish to test the null hypothesis that p, the probability that a single smoker prefers brand A, is equal to .1 against the alternative that p is greater than or less than .1. The rejection region corresponding to an $\alpha = .05$ would be located as shown in Figure 7.20. We would reject the null hypothesis when $z > 1.96$ or $z < -1.96$. In other words, we would reject the null hypothesis when y lies more than approximately two standard deviations away from its hypothesized mean. Note that half of α is placed in one tail of the distribution and half in the other because we wish to reject the null hypothesis when p is either larger or smaller than $p = .1$. This is called a **two-tailed statistical test**, in contrast to the **one-tailed test** discussed in Examples 7.11 and 7.12 when the alternative to the null hypothesis was only that p was larger than the hypothesized value.

Figure 7.20 Location of the rejection region in Example 7.13

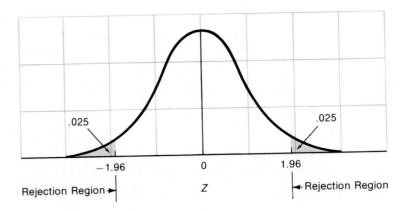

Assuming the null hypothesis to be true, the mean and standard deviation for y are

$$\mu = np = 2500(.1) = 250,$$
$$\sigma = \sqrt{npq} = \sqrt{2500(.1)(.9)} = 15.$$

The z value corresponding to the observed $y = 218$ is

$$z = \frac{y - \mu}{\sigma} = \frac{218 - 250}{15} = \frac{-32}{15} = -2.13.$$

Noting that z falls in the rejection region, we would reject the null hypothesis that $p = .1$. In fact, it appears that less than 10 percent of all smokers prefer brand A.

What is the probability that we have made an incorrect decision? The answer, of course, is either 1 or 0, depending upon whether our decision was correct or incorrect in this specific case. However, we know that if this statistical test were employed over and over again, the probability of rejecting the null hypothesis when it is true is only $\alpha = .05$. Hence we are reasonably certain that we have made the correct decision.

Exercises

7.45.   To show how the Central Limit Theorem works, compare the shape of the binomial probability distributions for $n = 10$, $p = .2$ and $n = 25$, $p = .2$. Use Table 1, the Appendix, to construct the probability histograms. Note the lack of symmetry for $n = 10$, $p = .2$. In contrast, the probability distribution for $n = 25$, $p = .2$ is more symmetric and tending to a bell shape. The larger the value of n, the more closely the binomial probability distribution will approach the normal distribution.

7.46.   Let y be a binomial random variable for $n = 25$, $p = .2$ (see Exercise 7.45).

   a. Use Table 1, the Appendix, to calculate $P(4 \leq y \leq 6)$.

   b. Find μ and σ for the binomial probability distribution and use the normal distribution to approximate the probability, $P(4 \leq y \leq 6)$. Note that this value is a good approximation to the exact value of $P(4 \leq y \leq 6)$.

7.47.   Consider a binomial experiment with $n = 20$, $p = .4$. Calculate $P(y \geq 10)$ by use of:

   a. Table 1, the Appendix.

   b. The normal approximation to the binomial probability distribution.

7.48.   According to a survey reported at the 1981 convention of the American Bar Association, there was one lawyer for every 700 Americans in 1960, one for every 600 in 1970, and one for every 410 Americans in 1981 (*Philadelphia Inquirer*, August 18, 1981). In Washington, D.C. there is one lawyer for every 64 residents.

   a. If you were to select a random sample of 1500 Americans, what is the approximate probability that the sample would contain at least one lawyer?

   b. If the sample is selected from among the residents of Washington, D.C., what is the approximate probability that the sample will contain more than 30 lawyers?

   c. If you stood on a Washington, D.C. street corner and interviewed the first 1500 persons who walked by, and 30 were lawyers, would this suggest that the density of lawyers passing the corner exceeds the density within the city? Explain.

7.49.   Briggs and King developed the technique of nuclear transplantation in which a nucleus of a cell from one of the later stages of the development of an embryo is transplanted into a zygote (a single cell, fertilized egg) to see if the nucleus can support normal development. If the probability that a single transplant from the early gastrula stage will be successful is .65, what is the probability that more than 70 transplants out of 100 will be successful?

7.50.   In a study of primary liver tumors Vana and colleagues* comment on the possible link between liver tumors and the use of oral contraceptives and report on a survey of 477 hospitals that revealed 378 cases of primary liver tumors in females. Of these, 187 were known to have used oral contraceptives. Suppose that the proportion of females in the United States who have used oral contraceptives is p and, for the sake of argument, suppose $p = .4$. If the 378 female patients with liver tumors can be viewed as a random sample from among all adult females, what is the probability that the number of patients who have used oral contraceptives is as large as 187? If p really is equal to .4, is the observation of 187 in the "random sample" of 378 a rare event? Do the 378 tumor patients represent a random sample selected from among all adult females in the United States?

7.51.   Data collected over a long period of time show that a particular genetic defect occurs in 1 out of every 1000 children. The records of a medical clinic show $y = 60$ children with the defect in a total of 50,000 examined. If the 50,000 children were a random sample

* *J. Amer. Med. Assoc.*, *238* (1977), pp. 2154–2158.

from the population of children represented by past records, what is the probability of observing a value of y equal to 60 or more? Would you say that the observation of $y = 60$ children with genetic defects represents a rare event?

7.52. A survey by Merrill Lynch Relocation Management, Inc., reported in the *Wall Street Journal* (November 20, 1979), indicates that many companies—approximately 30 percent—actively assist working spouses in continuing their careers when company employees are transferred to new locations. To determine whether this percentage is growing over time, suppose that we conducted a survey in 1982 and found that 38 of 110 companies aided working spouses in finding new jobs when the companies' employees were transferred. Does this number provide sufficient evidence to indicate that the percentage of companies giving relocation aid to working spouses increased over time?

   a. If we only wish to detect an increase in the percentage, state the null and alternative hypotheses that should be employed.

   b. Locate the rejection region corresponding to $\alpha = .05$.

   c. Conduct the test and state your conclusions.

7.53. According to an article in the *Gainesville Sun* (November 26, 1977), television may be dangerous to your diet. Psychologists believe that excessive eating may be associated with emotional states (being upset, bored, etc.) and environmental cues (TV, reading, etc.). To test this theory, suppose that we were to randomly select 60 overweight persons and match them by weight and sex in pairs. For a period of two weeks, one of each pair is required to spend evenings reading novels of interest to them. The other member of each pair spends each evening watching television. The calorie count for all snack and drink intake for the evenings is recorded for each person and we record $y = 19$, the number of pairs for which the TV watchers' calorie intake exceeded the intake of the readers. If there is no difference in the effect of TV and reading on calorie intake, the probability, p, that the calorie intake of one member of a pair exceeds that of the other member is .5. (If there is a difference, $p \neq .5$.) Do these data provide sufficient evidence to indicate a difference between the effects of TV watching and reading on calorie intake?

7.54. Airlines and hotels often grant reservations in excess of capacity to minimize losses due to no-shows. Suppose that the records of a motel show that, on the average, 10 percent of their prospective guests will not claim their reservation. If the motel accepts 215 reservations and there are only 200 rooms in the motel, what is the probability that all guests who arrive to claim a room will receive one?

7.55. In Exercise 6.23, Chapter 6, we commented on a telephone survey conducted by the administration of the North Florida Regional Hospital, Gainesville, Florida. The purpose of the survey was to determine public attitude toward the hospital's desire to expand even though the expansion might have a negative effect on the occupancy of beds in the local county hospital. Of the 180 persons polled, 112 stated that they favored the expansion. What is the probability that y, the number of people favoring the expansion, is as large or larger than 112, given that the proportion of adults in the community who favor the expansion is only .5?

7.56. Compilation of large masses of data on lung cancer show that approximately 1 of every 40 adults acquire the disease. Workers in a certain occupation are known to work in an air-polluted environment that may cause an increased rate of lung cancer. A random sample of $n = 400$ workers shows 19 with identifiable cases of lung cancer. Do the data provide sufficient evidence to indicate a higher rate of lung cancer for these workers than for the national average?

7.7
Statistics that Possess Normal Sampling Distributions

We noted in Sections 7.5 and 7.6 that the sum y and the sample mean $\hat{p} = y/n$ of the number of successes in a binomial experiment are statistics that possess, for large samples, sampling distributions that are approximately normal. In the following chapters, we will use a number of different statistics to make inferences about population parameters. Some will not possess sampling distributions that are approximately normal, but others, like those involving means or sample proportions, possess sampling distributions that are approximately normally distributed for large samples. Four of these statistics are discussed in Chapters 8 and 9.

A sample proportion,

$$\hat{p} = \frac{y}{n},$$

is equal to the number y of successes divided by the number n of trials in a binomial experiment. For example, if we were to select a random sample of 1000 consumers, and found that 600 of them prefer soap brand A, then the sample proportion,

$$\hat{p} = \frac{600}{1000} = .6,$$

is a statistic that estimates the proportion p of all consumers who prefer soap brand A.

Since y is approximately normally distributed because of the Central Limit Theorem (Section 7.6), we will not be surprised to learn that the sampling distribution of the sample proportion $\hat{p} = y/n$ is also approximately normally distributed when the sample size n is large. We will use this fact in Chapters 8 and 9 to make probability statements about $\hat{p}$ and, thereby, to evaluate the reliability of $\hat{p}$ when it is used to make inferences about the population proportion p.

We will also wish to compare the means μ_1 and μ_2 of two different populations based on independent random samples, one selected from each population. Or, we might wish to compare two parameters, p_1 and p_2, for two different binomial populations. It would seem natural to use the difference in sample means, $(\bar{y}_1 - \bar{y}_2)$, to make inferences about $(\mu_1 - \mu_2)$, and the difference between two sample proportions, $(\hat{p}_1 - \hat{p}_2)$, to estimate or to test hypotheses about $(p_1 - p_2)$. We will learn in Chapter 8 that both of these statistics, the difference in a pair of sample means, $(\bar{y}_1 - \bar{y}_2)$, and the difference in two sample proportions, $(\hat{p}_1 - \hat{p}_2)$, possess sampling distributions that are approximately normal. We will learn more about these sampling distributions in Chapter 8 and, particularly, we will learn how to use these distributions to evaluate the properties of $(\bar{y}_1 - \bar{y}_2)$ and $(\hat{p}_1 - \hat{p}_2)$.

Chapter 7 sets the stage for a study of large sample estimation in Chapter 8. Because of the Central Limit Theorem, all of the statistics discussed in Chapter 8 will possess, for large samples, sampling distributions that are approximately normal. This will enable us to use the normal curve to calculate the probability that a statistic will deviate from an estimated parameter by some specified amount and, ultimately, to evaluate the confidence that we can place in a parameter estimate.

7.8
Summary

Many continuous random variables observed in nature possess a probability distribution that is bell-shaped and that may be approximated by the normal probability distribution discussed in Section 7.1. The common occurrence of normally distributed random variables may be partly explained by the Central Limit Theorem, which states that, under rather general conditions, the sum or the mean of a random sample of n measurements drawn from a population will be approximately normally distributed in repeated sampling when n is large.

As a case in point, the number of successes, y, associated with a binomial experiment may be regarded as a sum of n sample measurements that will possess, approximately, a normal sampling distribution when n, the total number of trials, is large. This application of the Central Limit Theorem provides a method for calculating, with reasonable accuracy, the probabilities of the binomial probability distribution by using corresponding areas under the normal probability distribution. The Central Limit Theorem justifies the approximate normality of the sampling distributions of all of the statistics encountered in Chapters 8 and 9, and it provides some justification for the use of the Empirical Rule, Chapter 3.

Tips on Problem Solving

1. Always sketch a normal curve and locate the probability areas pertinent to the exercise. If the normal curve represents the sampling distribution of the sample mean $\bar{y}$, be careful to correctly calculate its mean and standard deviation. If you are approximating a binomial probability distribution, sketch in the probability rectangles as well as the normal curve.

2. Read each exercise carefully to see whether the data come from a binomial experiment or whether they possess a distribution that, by its very nature, is approximately normal. If you are approximating a binomial probability distribution, do not forget to make a half-unit correction so that you will include the half rectangles at the ends of the interval. If the distribution is not binomial, *do not* make the half-unit corrections. If you make a sketch (as suggested in step 1), you will see why the half-unit correction is or is not needed.

References

Chapman, D.G., and R.A. Schaufele, *Elementary Probability Models and Statistical Inference*. New York: John Wiley & Sons, Inc., 1970.

Hoel, P.G., *Elementary Statistics*, 4th ed. New York: John Wiley & Sons, Inc., 1976.

Huntsberger, D.V., and P. Billingsley, *Elements of Statistical Inference*, 4th ed. Boston: Allyn and Bacon, Inc., 1977.

McClave, J.T., and F.H. Dietrich. *Statistics*, 2nd ed. San Francisco: Dellen Publishing Company, 1982.

Supplementary Exercises

[Starred (*) exercises are optional.]

7.57. Using Table 3, the Appendix, calculate the area under the normal curve between:

a. $z = 0$ and $z = 1.2$.

b. $z = 0$ and $z = -.9$.

7.58. Repeat Exercise 7.57 using:

a. $z = 0$ and $z = 1.46$.

b. $z = 0$ and $z = -.42$.

7.59. Repeat Exercise 7.57 using:

a. $z = .3$ and $z = 1.56$.

b. $z = .2$ and $z = -.2$.

7.60. Find the probability that z is greater than $-.75$.

7.61. Find the probability that z is less than 1.35.

7.62. Find a z_0 such that $P(z > z_0) = .5$.

7.63. Find the probability that z lies between $z = -1.48$ and $z = 1.48$.

7.64. Find a z_0 such that $P(-z_0 < z < z_0) = .5$.

7.65. The lifespan of oil drilling bits depends upon the types of rock and soil that the drill encounters, but it is estimated that the mean length of life is 75 hours. If an oil exploration company purchases drill bits that have a lifespan that is approximately normally distributed with mean equal to 75 hours and standard deviation equal to 12 hours,

a. What proportion of the company's drill bits will fail before 60 hours of use?

b. What proportion will last at least 60 hours?

c. What proportion will have to be replaced after more than 90 hours of use?

7.66. The influx of new ideas into a college or university, introduced primarily by hiring new young faculty, is becoming a matter of concern because of the increasing ages of faculty members. That is, the distribution of faculty ages is shifting upward, due most likely to a shortage of vacant positions and an oversupply of PhD's. Thus faculty members are more reluctant to move and give up a secure position. If the retirement age at most universities is 65, would you expect the distribution of faculty ages to be normal?

7.67. A machine operation produces bearings whose diameters are normally distributed with mean and standard deviation equal to .498 and .002, respectively. If specifications require that the bearing diameter equal .500 inch plus or minus .004 inch, what fraction of the production will be unacceptable?

7.68. A used car dealership has found that the length of time before a major repair is required for the cars it sells is normally distributed with a mean equal to ten months and standard deviation of three months. If the dealer wants only 5 percent of the cars to fail before guarantee time, for how long (in months) should the cars be guaranteed?

7.69. Most users of automatic garage door openers activate their openers at distances that are normally distributed with a mean of 30 feet and a standard deviation of 11 feet. In order to minimize interference with other radio-controlled devices, the manufacturer is required to limit the operating distance to 50 feet. What percentage of the time will users attempt to operate the opener outside of its operating limit?

7.70. Consider a binomial experiment with $n = 25$, $p = .4$. Calculate $P(8 \leq y \leq 11)$ by use of:

a. The binomial probabilities, Table 1, the Appendix.

b. The normal approximation to the binomial.

7.71. The average length of time required for a college achievement test was found to equal 70 minutes, with a standard deviation of 12 minutes. When should the test be terminated if we wish to allow sufficient time for 90 percent of the students to complete the test? (Assume that the time required to complete the test is normally distributed.)

7.72. Review the die-tossing experiment, Section 7.5, where we simulated the selection of samples of $n = 5$ observations and obtained an approximation to the sampling distribution for the sample mean. Repeat this experiment, selecting 200 samples of size $n = 3$.

a. Construct the sampling distribution for $\bar{y}$. Note that the sampling distribution of $\bar{y}$ for $n = 3$ does not achieve the bell shape that you observed for $n = 5$ (Figure 7.11).

b. The mean and standard deviation of the probability distribution for y, the number of dots that appear when a single die is tossed, are $\mu = 3.5$ and $\sigma = 1.71$. What are the exact values of the mean and standard deviation of the sampling distribution of $\bar{y}$ based on samples of $n = 3$?

c. Calculate the mean and standard deviation of the simulated sampling distribution, part (a). Are these values close to the corresponding values obtained for part (b)?

7.73. Refer to the sampling experiment, Exercise 7.72. Calculate the median for each of the 200 samples of size $n = 3$.

a. Use the 200 medians to construct a relative frequency histogram that approximates the sampling distribution of the sample median.

b. Calculate the mean and standard deviation of the sampling distribution, part (a).

c. Compare the mean and standard deviation of this sampling distribution with the mean and standard deviation calculated for the sampling distribution of $\bar{y}$, Exercise 7.72(b). Which statistic, the sample mean or the sample median, appears to fall closer to μ?

7.74. The length of time required for the periodic maintenance of an automobile will usually have a probability distribution that is mound-shaped and, because some long service times will occur occasionally, is skewed to the right. Suppose that the length of time required to run a 5000-mile check and to service an automobile has a mean equal to 1.4 hours and a standard deviation of .7 hours. Suppose that the service department plans to service 50 automobiles per 8-hour day and that, in order to do so, it must spend no more service time than an average of 1.6 hours per automobile. What proportion of all days will the service department have to work overtime?

7.75. An advertising agency has stated that 20 percent of all television viewers watch a particular program. In a random sample of 1000 viewers, $y = 184$ viewers were watching the program. Do these data present sufficient evidence to contradict the advertiser's claim?

7.76. David Dreiman, writing in *Forbes* magazine (August 3, 1981), notes that senior corporation executives are not very accurate forecasters of their own annual earnings. He states that his studies of a large number of company executive forecasts "showed that the average estimate missed the mark by 15%."

a. Suppose that the distribution of these forecast errors has a mean of 15% and a standard deviation of 10%. Is it likely that the distribution of forecast errors is approximately normal?

b. Suppose that the probability is .5 that a corporate executive's forecast error exceeds 15%. If you were to sample the forecasts of 100 corporate executives, what is the probability that more than 60 would be in error by more than 15%?

7.77. A soft drink machine can be regulated so that it discharges an average of μ ounces per cup. If the ounces of fill are normally distributed with standard deviation equal to .3 ounce, give the setting for μ so that 8-ounce cups will overflow only 1 percent of the time.

* 7.78. A statistical test is to be conducted to test the hypothesis that p, the parameter of a binomial population, is equal to .1. If the sample size is $n = 400$ and we wish α to be approximately equal to .05 (two-tailed test), locate the rejection region if the test statistic is (a) z, (b) y.

* 7.79. Calculate β for the test in Exercise 7.78 if p really is equal to .15.

7.80. A manufacturing plant utilizes 3000 electric light bulbs that have a length of life that is normally distributed with mean and standard deviation equal to 500 and 50 hours, respectively. In order to minimize the number of bulbs that burn out during operating hours, all the bulbs are replaced after a given period of operation. How often should the bulbs be replaced if we wish no more than 1 percent of the bulbs to burn out between replacement periods?

7.81. The admissions office of a small college is asked to accept deposits from a number of qualified prospective freshmen so that with probability about .95 the size of the freshman class will be less than or equal to 120. Consider that the applicants comprise a random sample from a population of applicants, 80 percent of whom would actually enter the freshman class if accepted.

a. How many deposits should the admissions counselor accept?

b. If applicants in the number determined in part (a) are accepted, what is the probability that the freshman class size will be less than 105?

7.82. An airline finds that 5 percent of the persons making reservations on a certain flight will not show up for the flight. If the airline sells 160 tickets for a flight with only 155 seats, what is the probability that a seat will be available for every person holding a reservation and planning to fly?

7.83. It is known that 30 percent of all calls coming into a telephone exchange are long-distance calls. If 200 calls come into the exchange, what is the probability that at least 50 will be long-distance calls?

7.84. Suppose that the random variable y has a binomial distribution corresponding to $n = 20$ and $p = .30$. Use Table 1, the Appendix, to find

a. $P(y = 5)$.   b. $P(y \geq 7)$.

7.85. Refer to Exercise 7.84. Use the normal approximation to calculate $P(y = 5)$ and $P(y \geq 7)$. Compare with the exact values obtained from Table 1.

7.86. The safety requirements for hard hats worn by construction workers and others, established by the American National Standards Institute (ANSI), specifies that each of three hats pass the following test.* A hat is mounted on an aluminum head form. An 8-pound steel ball is dropped on the hat from a height of 5 feet and the resulting force is measured at the bottom of the head form. The force exerted on the head form by each of the three hats must be less than 1000 pounds and the average of the three must be less than 850 pounds. (The relationship between this test and actual human head damage is unknown.) Suppose that the exerted force is normally distributed and hence that a sample mean of three force measurements is normally distributed. If a random sample of three hats is

* "Job-Safety Equipment Comes Under Fire," *Wall Street Journal*, November 18, 1977.

selected from a shipment with a mean equal to 900 and $\sigma = 100$, what is the probability that the sample mean will satisfy the ANSI standard?

7.87.   A purchaser of electric relays is supplied by two suppliers, A and B. It is known that 2 of every 3 relays used by the company come from supplier A. If 75 relays are selected at random from those in use by the company, find the probability that at most 48 of these relays come from supplier A. Assume that the company uses a large number of relays.

7.88.   The maximum load (with a generous safety factor) for the elevator in an office building is 2000 pounds. The relative frequency distribution of the weights of all of the men and women using the elevator is mound-shaped (slightly skewed to the heavy weights) with mean μ equal to 150 pounds and standard deviation σ equal to 35 pounds. What is the largest number of people you can allow on the elevator if you want their total weight to exceed the maximum weight with a small probability (say near .01)? (*Hint:* If $y_1, y_2, \ldots, y_n$ are independent observations made on a random variable y, and if y possesses mean μ and variance σ^2, then the mean and variance of $\sum\limits_{i=1}^{n} y_i$ are $n\mu$ and $n\sigma^2$, respectively. This result was given in Section 7.5.)

7.89.   In Exercise 6.7, we noted the common occurrence of the falsification of information contained on job application forms. Suppose that, in a particular industry, the probability that a job applicant will falsify information on his or her application form is .35 and that a company hires 100 people over a two-month period.

   a. What is the expected number of applicants who will falsify information on their application forms?

   b. Find the standard deviation of y, the number of applicants who will falsify their application forms.

   c. Suppose that the company has all of the applications verified and finds that the number of falsified application forms is $y = 49$. Does the company have reason to suspect that it is attracting a higher percentage of imposters than other companies in the industry?

* 7.90.   If you have access to a computer and a computer program that generates random numbers, you can simulate sampling from a population that possesses a uniform probability distribution. The numbers produced by a random number generator are independent of one another, and the probability of observing any one number is the same as the probability of observing any other. Program the computer to generate a large number (say 1000) of samples of $n = 2$ observations and calculate the sample mean for each. Use a computer program to arrange these 1000 sample means in a relative frequency histogram. The resulting histogram will provide a good approximation to the sampling distribution of $\bar{y}$ for samples selected from a population which possesses a uniform relative frequency distribution. It should be similar to the sampling distribution shown in the third column of Figure 7.11 for $n = 2$.

* 7.91.   Repeat Exercise 7.90 for $n = 5$, 10, and 25. Compare with the corresponding sampling distributions shown in Figure 7.11.

* 7.92.   Repeat Exercise 7.90 for $n = 100$. Compare with the sampling distributions for $n = 2$, 5, 10, and 25.

7.93.   For a car traveling 30-miles per hour, the distance required to brake to a stop is normally distributed with mean of 50 feet and a standard deviation of 8 feet. Suppose that you are traveling 30 miles per hour in a residential area and a car moves abruptly into your path at a distance of 60 feet. If the only way to avoid a collision is to brake to a stop, what is the probability that you will avoid the collision?

8. Large-Sample Estimation

Chapter Objectives

General Objective

To present one method for making statistical inferences—estimation—and to illustrate the concept with practical examples; to use the Central Limit Theorem and the sampling distributions presented in Chapter 7 to evaluate the reliability of some estimates.

Specific Objectives

1. To relate the preceding chapters to a discussion of statistical inference, the topic of Chapter 8.   *Sections 8.1, 8.2*

2. To describe two types of estimators and to explain how we can measure their reliability.   *Sections 8.3, 8.4, 8.5*

3. To present a general formula for calculating a large-sample interval estimate of a population mean, a population proportion, or the difference between two population means or proportions.   *Sections 8.6, 8.7, 8.8, 8.9*

4. To explain how to select the sample size to obtain an estimator with a predetermined measure of reliability.   *Section 8.10*

5. To show how the concepts of estimation can be applied to evaluate the published results of an opinion poll.   *Section 8.11*

Case Study

Rating the Popularity of Sports

According to the *Gainesville Sun* (May 27, 1981), a Louis Harris poll "has established" the current favorite sports of Americans. Based on a sample of 1091 adults, the poll indicates that 36 percent of all Americans choose football as their favorite sport. The percentages of Americans choosing other sports as their favorite are shown in Table 8.1.

Table 8.1 A sports popularity poll

Sport	Percentage of Americans Choosing Sport as Favorite
Football	36
Baseball	21
Basketball	12
Tennis	5
Auto Racing	5
Golf	4
Boxing	4
Bowling	3
Horse Racing	3
Hockey	3
Soccer	2
Track	2

The article also notes that only three sports are followed by more than 52 percent of all Americans. They rank as follows: football (72 percent), baseball (61 percent), and basketball (52 percent).

A number of thoughts may have crossed your mind as you read the preceding paragraph. Knowledge of the interests and preferences of your friends and associates may lead you to question some of the percentages. And, you may wonder whether it is possible to establish the favorites of millions of Americans based on a sample of the opinions of only 1091 adults. Is it possible that some of the respondents had no interest in sports and, when questioned, chose a "favorite"? To what extent do the sample percentages establish the corresponding percentages for *all* Americans?

The proportion of all adult Americans who would rate one particular sport—say football—as a favorite is the parameter p of a binominal population. Consequently, the answer to these questions depends upon how the 1091 polled adults were selected, the nature of the poll questionnaire, and the statistic that was employed to estimate the population parameter p. In this chapter, we will examine the logic employed in selecting this and other sample statistics for use as estimators of population parameters, and we will use their sampling distributions to determine how close to a population parameter an estimate is likely to be. We will reexamine the results of this Harris poll in Section 8.11.

8.1
A Brief Summary

The preceding seven chapters set the stage for the objective of this text, developing an understanding of statistical inference and how it can be applied to the solution of practical problems. In Chapter 1 we stated that statisticians are concerned with making inferences about populations of measurements based on information contained in samples. We showed you how to phrase an inference—that is, how you describe a set of measurements—in Chapter 3. We discussed probability, the mechanism for making inferences, in Chapter 4, and we followed that with three chapters about probability distributions—a general presentation in Chapter 5, the binomial probability distribution in Chapter 6, and the normal distribution in Chapter 7.

To get you thinking about statistical inference, we introduced to you, in Chapter 6, the useful application of the binomial probability distribution to lot acceptance sampling and the test of an hypothesis concerning the effectiveness of a cold vaccine. We touched lightly on these topics again in the examples and exercises of Chapter 7.

Perhaps the most important contribution to our preparation for a study of statistical inference is the Central Limit Theorem of Chapter 7. This theorem, which justifies the approximate normality of the sampling distribution of sample means and sample sums for large samples, was used to justify the normal approximation to the binomial probability distribution in Chapter 7. But more importantly, it will be used to justify the approximate normality of the sampling distributions of numerous sample statistics in Chapter 8. As you will subsequently see, we will use these statistics to make inferences about population parameters and will use the statistics' sampling distributions to evaluate their reliabilities.

8.2
Inference: The Objective of Statistics

Inference, specifically decision making and prediction, is centuries old and plays a very important role in our individual lives. Each of us is faced daily with personal decisions and situations that require predictions concerning the future. The government is concerned with predicting short- and long-term interest rates. The broker wishes knowledge concerning the behavior of the stock market. The metallurgist seeks to use the results of an experiment to infer whether or not a new type of steel is more resistant to temperature changes than another. The housewife wishes to know whether detergent A is more effective than detergent B in her washing machine. Hopefully, these inferences are based upon relevant bits of available factual information that we would call observations or data.

In many practical situations the relevant information is abundant, seemingly inconsistent, and, in many respects, overwhelming. As a result, our carefully considered decision or prediction is often little better than an outright guess. The reader need only refer to the "Market Views" section of the *Wall Street Journal* to observe the diversity of expert opinion concerning future stock market behavior. Similarly, a visual analysis of data by scientists and engineers will often yield conflicting opinions regarding conclusions to be drawn from an experiment. Although many individuals tend to feel that their own built-in inference-making equipment is quite good, experience would suggest that most people are incapable of utilizing large amounts of data, mentally weighing each bit of relevant information, and arriving at a good inference. (You may test your individual inference-making equipment using the exercises in Chapters 8, 9, and 10. Scan the data and make an inference before using the appropriate statistical procedure. Compare the results.) Certainly, a study of inference-making systems is desirable, and this is the objective of the mathematical statistician. Although we have purposely touched upon some of the notions involved in statistical inference in preceding chapters, it will be beneficial to collect our ideas at this point as we attempt an elementary presentation of some of the basic ideas involved in statistical inference.

The objective of statistics is to make inferences about a population based upon information contained in a sample. Inasmuch as populations are characterized by numerical descriptive measures called parameters, statistical inference is concerned with making inferences about population parameters. Typical population parameters are the mean, the standard deviation, the area under the probability distribution above or below some value of the random variable, or the area between two values of the variable. Indeed, the practical problems mentioned in the first paragraph of this section can be restated in the framework of a population with a specified parameter of interest.

Methods for making inferences about parameters fall into one of two categories. We may make decisions concerning the value of the parameter, as exemplified by the lot acceptance sampling and test of an hypothesis described in Chapter 6. Or, we may estimate or predict the value of the parameter. While some statisticians view estimation as a decision-making problem, it will be convenient for us to retain the two categories and, particularly, to concentrate on estimation and tests of hypotheses.

A statement of the objective and types of statistical inference would be incomplete without reference to a measure of goodness of inferential procedures. We may define numerous objective methods for making inferences in addition to our own individual procedures based upon intuition. Certainly a measure of goodness must be defined so that one procedure may be compared with another. More than that, we would like to state the goodness of a particular inference in a given physical situation. Thus, to say we predict that the price of a stock will be $80 next Monday would be insufficient and would stimulate few of us to take action to buy or sell. Indeed, we ask whether the estimate is correct to within $\pm$ $1, $2, or $10. Statistical inference in a practical situation contains two elements: (1) the inference and (2) a measure of its goodness.

Before concluding this introductory discussion of inference, we will attempt to answer a question that frequently disturbs the beginner. Which method of inference should be used; that is, should the parameter be estimated or should we test an hypothesis concerning its value? The answer to this question is dictated by the practical question that has been posed and very often is determined by personal preference. Some people like to test theories concerning parameters; others prefer to express their inference as an estimate. We will find that there are actually two methods of estimation, the choice of which, once again, is a matter of personal preference. Inasmuch as both estimation and tests of hypotheses are frequently used in scientific literature, we want to include both topics in our discussion. The basic concepts of parameter estimation, along with some large-sample estimation procedures, will be discussed in this chapter. Chapter 9 will present a review of the concepts involved in testing an hypothesis about a population parameter. Some useful large-sample tests will demonstrate how these concepts can be applied.

8.3
Types of Estimators

Estimation procedures may be divided into two types, point estimation and interval estimation. Suppose that we wish to estimate the grade-point average of a particular student at Bucknell University. The estimate might be given as a single number, for instance 2.9, or we might estimate that the grade-point average would fall in an interval, for instance 2.7 to 3.2. The first type of estimate is called a **point estimate** because the single number, representing the estimate, may be associated with a point on a line. The second type, involving two points and defining an interval on a line, is called an **interval estimate.** We shall consider each of these methods of estimation.

A point estimation procedure utilizes information in a sample to arrive at a single number of point that estimates the parameter of interest. The actual estimation is accomplished by an estimator.

Definition

An **estimator** is a rule that tells us how to calculate the estimate based upon information in the sample and is generally expressed as a formula.

For example, the sample mean,

$$\bar{y} = \frac{\sum\limits_{i=1}^{n} y_i}{n},$$

is an estimator of the population mean, μ, and explains exactly how the actual numerical value of the estimate may be obtained once the sample values, $y_1, y_2, \ldots, y_n$, are known. On the other hand, an interval estimator uses the data in the sample to calculate *two* points that are intended to enclose the true value of the parameter estimated.

An investigation of the reasoning used in calculating the goodness of a point estimator is facilitated by considering an analogy. Point estimation is similar, in many respects, to firing a revolver at a target. The estimator, generating estimates, is analogous to the revolver, a particular estimate to the bullet, and the parameter of interest to the bull's-eye. Drawing a sample from the population and estimating the value of the parameter is equivalent to firing a single shot at the target.

Suppose that a man fires a single shot at a target and that the shot pierces the bull's-eye. Do we conclude that he is an excellent shot? Obviously, the answer is no because not one of us would consent to hold the target while a second shot was fired. On the other hand, if 1,000,000 shots in succession hit the bull's-eye, we might acquire sufficient confidence in the marksman to hold the target for the next shot, if the compensation were adequate. The point we wish to make is certainly clear. We cannot evaluate the goodness of an estimation procedure on the basis of a single estimate; rather, we must observe the results when the estimation procedure is used over and over again, many, many times—we then observe how closely the shots are distributed about the bull's-eye. In fact, since the estimates are numbers, we would evaluate the goodness of the estimator by constructing a frequency distribution of the estimates obtained in repeated sampling and note how closely the distribution centers about the parameter of interest. If the sampling process were repeated an infinitely large number of times, this relative frequency distribution would be the sampling distribution of the estimator.

This point is aptly illustrated by considering the results of the die-tossing experiment, Section 7.5, where 100 samples of $n = 5$ measurements were drawn from the die-tossing population, which possessed a mean and standard deviation equal to $\mu = 3.5$ and $\sigma = 1.71$, respectively. The distribution of the 100 sample means, each representing an estimate, is given in Figure 7.10. A glance at the distribution tells us that the estimates tend to pile up about the mean, $\mu = 3.5$, and also gives an indication as to the error of estimation that might be expected. Thus, if a sample statistic is used to estimate a population parameter, all of its properties are described by its sampling distribution.

For example, if we let the symbol θ (theta) represent a population parameter (μ, σ, or any parameter) and $\hat{\theta}$ (theta hat) represent a statistic used to estimate θ (i.e., $\hat{\theta}$ is an estimator of θ), its sampling distribution might appear as shown in Figure 8.1. If the estimator $\hat{\theta}$ is to be a "good" estimator of θ, its sampling distribution will possess two properties. First, we would like the sampling distribution to be centered (as shown in Figure 8.1) over the parameter that we are estimating. Thus we prefer that the *mean* of the sampling distribution will equal θ. Such an estimator is said to be *unbiased*. One estimate

Figure 8.1 Distribution of estimates

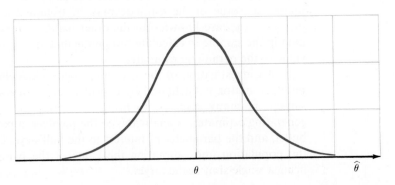

might fall above θ, another below, but if we were to use the estimator many, many times, the mean value of all the estimates would equal θ.

Definition

If $\hat{\theta}$ is an estimator of a parameter θ and if the mean of the distribution of $\hat{\theta}$ is θ, that is,

$$E(\hat{\theta}) = \theta,$$

then $\hat{\theta}$ is said to be **unbiased.** Otherwise, $\hat{\theta}$ is said to be **biased.**

The sampling distributions for an unbiased estimator and a biased estimator are shown in Figure 8.2(a) and (b). Note that the sampling distribution for the biased estimator, Figure 8.2(b), is shifted to the right of θ. This biased estimator is more likely to overestimate θ.

Figure 8.2 Distributions for unbiased and biased estimators

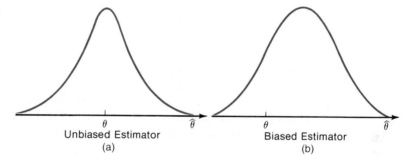

Unbiased Estimator
(a)

Biased Estimator
(b)

The second desirable property of an estimator is that the spread (measured by the variance) of the sampling distribution of the estimator be as small as possible. This ensures a high probability that an individual estimate will fall close to θ. The sampling distributions for two estimators, one with a small variance* and the other with a larger variance, are shown in Figure 8.3(a) and (b), respectively. Naturally, we would prefer the estimator with the smaller variance, the sampling distribution shown in Figure 8.3(a), because the estimates tend to lie closer to θ than in the distribution, Figure 8.3(b).

There are often several (or many) different statistics that could be used to estimate the same parameter. From among these, we would prefer the estimator with the smallest amount of bias and possessing the smallest variance. The best possible estimator is one that is unbiased and that possesses a variance smaller than that of any other unbiased estimator. This is called a minimum variance unbiased estimator (MVUE).

---

* Statisticians usually use the term *variance of an estimator* when in fact they mean the variance of the sampling distribution of the estimator. This contractive expression is used almost universally.

Figure 8.3 Comparison of estimator variability

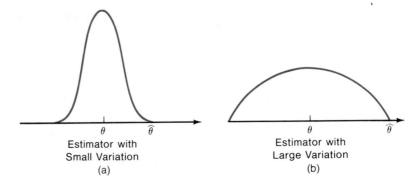

Estimator with
Small Variation
(a)

Estimator with
Large Variation
(b)

The goodness of an interval estimator is analyzed in much the same manner as is a point estimator. Samples of the same size are repeatedly drawn from the population and the interval estimate is calculated on each occasion. This process will generate a large number of intervals rather than points. A good interval estimator would successfully enclose the true value of the parameter a large fraction of the time. This fraction is called the confidence coefficient for the estimator; the estimator itself is often called a confidence interval.

The selection of a "best" estimator—the proper formula to use in calculating the estimates—involves the comparison of various methods of estimation. This is the task of the theoretical statistician and is beyond the scope of this text. Throughout the remainder of this chapter and succeeding chapters, populations and parameters of interest will be defined and the appropriate estimator indicated along with its expected value and standard deviation.

8.4
Point Estimation of a Population Mean

Practical problems very often lead to the estimation of a population mean, μ. We are concerned with the average achievement of college students in a particular university, in the average strength of a new type of steel, in the average number of deaths per capita in a given social class, and in the average demands for a new product. Conveniently, the estimation of μ serves as a very practical application of statistical inference as well as an excellent illustration of the principles of estimation discussed in Section 8.3. Many estimators are available for estimating the population mean, μ, including the sample median, the average between the largest and smallest measurements in the sample, and the sample mean, $\bar{y}$. Each would possess a sampling distribution and, depending upon the population and practical problem involved, would possess certain advantages and disadvantages. Although the sample median and the average of the sample extremes are easier to calculate, the sample mean, $\bar{y}$, is usually superior in that, for some populations, its variance is a minimum and, furthermore, regardless of the population, it is always unbiased.

In Section 7.5, we noted that the sampling distribution of the sample mean possessed three properties. Regardless of the probability distribution of the sampled population, the following facts hold true (proof omitted):

Properties of the Sample Mean

1: The expected value of $\bar{y}$ is equal to μ, the population mean.

2. The standard deviation of $\bar{y}$ is equal to*

$$\sigma_{\bar{y}} = \frac{\sigma}{\sqrt{n}}$$

3. When n is large, $\bar{y}$ will be approximately normally distributed according to the Central Limit Theorem (assuming that μ and σ are finite numbers).

Thus $\bar{y}$ is an unbiased estimator of μ. The fact that its standard deviation is proportional to the population standard deviation, σ, and inversely proportional to the square root of the sample size, n, is intuitively reasonable. Certainly, the more variable the population data, measured by σ, the more variable will be $\bar{y}$. On the other hand, more information will be available for estimating μ as n becomes large. Hence the estimates should fall closer to μ, and $\sigma_{\bar{y}}$ should decrease.

As noted in Section 7.5, the sampling distribution of $\bar{y}$ will be approximately normally distributed when the sample size n is large. If the population possesses a normal distribution, the sampling distribution of $\bar{y}$ will be normally distributed regardless of the sample size. The sampling distributions for $\bar{y}$ based on random samples of $n = 5$, $n = 20$, and $n = 80$ from a normal distribution are shown in Figure 8.4. Notice how these distributions center about μ and how the spread of the distributions decreases as n increases.

With the properties of the sampling distribution of $\bar{y}$ in mind, suppose that we draw a single sample of n measurements from a population and calculate the sample mean, $\bar{y}$. How good will this estimate of μ be; that is, how far will it deviate from the mean, μ? Although we cannot state that $\bar{y}$ will definitely lie within a specified distance of μ, Tchebysheff's Theorem states that if we were to draw many samples of the same size n from the population and compute $\bar{y}$ each time, at least three-fourths of the values of $\bar{y}$ would lie within $2\sigma_{\bar{y}}$ of the mean of the distribution of $\bar{y}$s, that is, μ.

The relative frequency distribution of the 100 values of $\bar{y}$, each calculated from a sample $n = 5$ die tosses, illustrates this concept. As noted, Section 8.3, the mean and standard deviation of the population of die tosses is $\mu = 3.5$ and

---

* When the number n of observations in a sample is large relative to the number N of observations in the population, the standard deviation of $\bar{y}$ is equal to

$$\sigma_{\bar{y}} = \frac{\sigma}{\sqrt{n}} \sqrt{\frac{N-n}{N-1}}.$$

In the following discussion, we shall assume that N is large relative to the sample size n and hence that $\sqrt{(N-n)/(N-1)}$ is approximately equal to 1. Then, $\sigma_{\bar{y}} = \sigma/\sqrt{n}$.

Figure 8.4 Sampling distributions for $\bar{y}$ based on random sampling from a normal distribution, $n = 5, 20,$ and 80

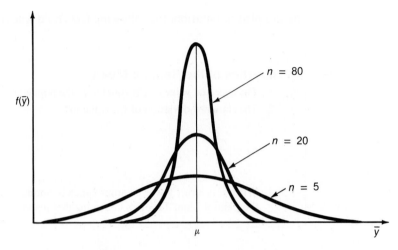

$\sigma = 1.71$. Therefore,

$$2\sigma_{\bar{y}} = \frac{2\sigma}{\sqrt{n}} = \frac{2(1.71)}{\sqrt{5}} = \frac{2(1.71)}{2.24} = 1.53.$$

Even though this sample size is small and the sampling distribution of $\bar{y}$ is not exactly normally distributed, you can see that most of the sample means fall within 1.53 of the mean, $\mu = 3.5$, in Figure 8.5.

We will assume in this chapter that the sample sizes are always large and, therefore, that the estimators that we will study possess sampling distribu-

Figure 8.5 Histogram of sample means for the die-tossing experiment, Section 7.5

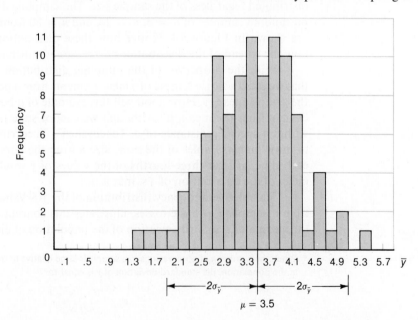

tions that are approximately normal because of the Central Limit Theorem. Consequently, if we define the difference between a particular estimate $\hat{\theta}$ and the parameter θ it estimates as the *error of estimation*, we would expect the error of estimation to be less than $1.96_{\hat{\theta}}$ with probability approximately equal to .95 (see Figure 8.6). Thus, most estimates will be less than $1.96\sigma_{\hat{\theta}}$ away from θ and this quantity provides a practical upper limit or *bound on the error of estimation*. It is possible (with probability .05) that the error of estimation will exceed this bound, but it is very unlikely.

Figure 8.6 Sampling distribution of $\bar{y}$ for large n

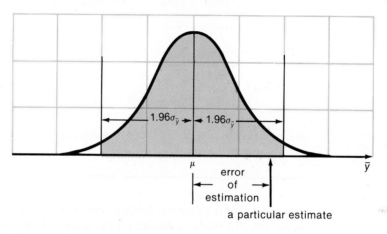

Definition
If θ is a population parameter, and $\hat{\theta}$ is an estimate of θ, then $|\hat{\theta} - \theta|$ is called the error of estimation.*

Point Estimator of a Population Mean
Estimator: $\bar{y}$.

Bound on error: $1.96\sigma_{\bar{y}} = \dfrac{1.96\sigma}{\sqrt{n}}$, where σ is the standard deviation of the sampled population and n is the sample size.

Now consider the following example of point estimation.

Example 8.1

Suppose that we wish to estimate the average daily yield of a chemical manufactured in a chemical plant. The daily yield, recorded for $n = 50$ days, produced a mean and standard

---

* The bars enclosing $\hat{\theta} - \theta$ mean "absolute value of" that is, $|\hat{\theta} - \theta|$ represents the unsigned value of the deviation.

deviation equal to

$$\bar{y} = 871 \text{ tons},$$

$$s = 21 \text{ tons}.$$

Estimate the average daily yield, μ.

Solution

The estimate of the daily yield is $\bar{y} = 871$ tons. The bound on the error of estimation is

$$1.96\sigma_{\bar{y}} = \frac{1.96\sigma}{\sqrt{n}} = \frac{1.96\sigma}{\sqrt{50}}.$$

Although σ is unknown, we may approximate its value by using s, the estimator of σ. Thus the bound on the error of estimation is approximately

$$\frac{1.96s}{\sqrt{n}} = \frac{1.96(21)}{\sqrt{50}} = 5.82.$$

We would feel fairly confident that our estimate of 871 tons is within 5.82 tons of the true average yield.

Example 8.1 deserves further comment in regard to two points. The erroneous use of 1.96σ as a bound on the error of estimation rather than $1.96\sigma_{\bar{y}}$ is common to beginners. Certainly, if we wish to discuss the distribution of $\bar{y}$, we must use its standard deviation, $\sigma_{\bar{y}}$, to describe its variability. Care must be taken not to confuse the descriptive measures of one distribution with another.

A second point of interest concerns the use of s to approximate σ. This approximation will be reasonably good when n is large, say, 30 or greater. If the sample size is not large, we can resort to a small sample estimation procedure described in Chapter 10. The choice of $n = 30$ as the division between "large" and "small" samples is arbitrary. The reasoning for its selection will become apparent in Chapter 10.

Tips on Problem Solving

1. In this and in the following sections, determine the parameter that you wish to estimate and try to attach practical significance to its value. It does have practical significance, and you need to identify it.

2. Specify the estimator (point or interval) that you plan to use to estimate the value of the parameter.

3. When evaluating the reliability of the estimator, be certain that you use the standard deviation of its sampling distribution. For example, the standard deviation of the sampling distribution of the sample mean $\bar{y}$ is $\sigma/\sqrt{n}$, not σ (σ is the standard deviation of the sampled population).

Exercises

8.1.   A random sample of ten observations is selected from a population with mean μ and variance $\sigma^2 = 4$. If the sample mean $\bar{y}$ is used to estimate μ, what is the bound on the error of estimation? Can you make a precise statement about the probability that the error of estimation will be less than the bound? Explain.

8.2.   The Environmental Protection Agency (EPA) released initial findings (*Environment News*, October 1976) on the occurrence of polychlorinated biphenyls (PCBs), a hazardous substance, in the milk of nursing mothers. An analysis of milk samples taken from 67 women in 10 states showed the presence of PCBs in the milk of 65. The sample mean for the 65 was reported to be 1.7 parts per million (ppm) and the largest reading was 10.6. To evaluate this estimate of the mean level of contamination in the sampled population, you need to know σ, the measure of variation of the contamination readings.

   a. Although the article does not give the value of σ, it is unlikely that σ would be larger than 2.7, and it probably is nearer in value to 2. Why?

   b. Assume that $\sigma = 2.5$. Use this value to assess the accuracy of the estimate, $\bar{y} = 1.7$ ppm, of the population mean, μ. Interpret your result.

8.3.   Geologists are interested in shifts and movements of the earth's surface indicated by fractures (cracks) in the earth's crust. One of the most famous large fractures is the San Andreas fault (moving fracture) in California. A geologist attempting to study the movement of the relative shifts in the earth's crust at a particular location found many fractures in the local rock structure. In an attempt to determine the mean angle of the breaks, he sampled $n = 50$ fractures and found the sample mean and standard deviation to be 39.8° and 17.2°, respectively. Estimate the mean angular direction of the fractures and place a bound on the error of estimation.

8.4.   An increase in the rate of consumer savings is frequently tied to a lack of confidence in the economy and is said to be an indicator of a recessional tendency in the economy. A random sampling of $n = 200$ savings accounts in a local community showed a mean increase in savings account values of 7.2 percent over the past 12 months with a standard deviation of 5.6 percent. Estimate the mean percent increase in savings account values over the past 12 months for depositors in the community. Place a bound on your error of estimation.

8.5.   The mean and standard deviation for the life of a random sample of 100 light bulbs were calculated be 1280 and 142 hours, respectively. Estimate the mean life of the population of light bulbs from which the sample was drawn and place bounds on the error of estimation.

8.6.   Suppose that the population mean, Exercise 8.5, was really 1285 hours, with $\sigma = 150$ hours. What is the probability that the mean of a random sample of $n = 100$ measurements would exceed 1300 hours?

8.7.   In Section 8.3 we stated that for the die-toss population, $\mu = 3.5$ and $\sigma = 1.71$.

   a. If you were to toss a die $n = 5$ times, what is the approximate probability that the mean $\bar{y}$ of the sample would fall in the interval, $2.5 \le \bar{y} \le 4.5$?

   b. If you were to toss a die $n = 10$ times, what is the approximate probability that $\bar{y}$ would fall in the interval, $2.5 \le \bar{y} \le 4.5$?

   c. You will note [part (b)] that $P(2.5 \le \bar{y} \le 4.5)$ is large. If you have access to a die, toss it $n = 10$ times and calculate $\bar{y}$. Does $\bar{y}$ fall in the interval, $2.5 \le \bar{y} \le 4.5$? Suppose that the mean of the population, $\mu = 3.5$, were unknown and that you were using your mean to estimate μ. What is your error of estimation?

8.8.   Most of the claims on a small company's medical insurance plan are in the neighborhood of $800, but a few are very large. As a consequence, the distribution of claims is highly skewed to the right and possesses a standard deviation σ equal to $2000. The first 40 claims received this month possess a mean $\bar{y}$ equal to $930. Suppose that we were to regard this group of 40 claims as a random sample from the population of all potential claims and used $\bar{y}$ to estimate the population mean μ.

a. What is the bound on the error of estimation?

b. Can you make a precise statement about the probability that the error of estimation will be less than the bound, part (a)? Explain.

8.5
Interval Estimation of a Population Mean

Constructing an interval estimate is like attempting to rope an immobile steer. In this case, the parameter that you wish to estimate corresponds to the steer and the interval to the loop formed by the cowboy's lariat. Each time you draw a sample, you construct a confidence interval for a parameter and you hope to "rope it," that is, include it in the interval. You will not be successful for every sample. The probability that an interval will enclose the estimated parameter is the *confidence coefficient*.

Definition

The probability that a confidence interval will enclose the estimated parameter is called the **confidence coefficient**.

To consider a practical example, suppose that you wish to estimate the mean number of bacteria per cubic centimeter in a polluted stream. If we were to draw ten samples, each containing $n = 30$ observations, and construct a confidence interval for the population mean, μ, for each sample, the intervals might appear as shown in Figure 8.7. The horizontal line segments represent the

Figure 8.7 Ten confidence intervals for the mean number of bacteria per cubic centimeter (each based on a sample of $n = 30$ observations)

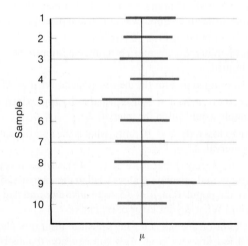

ten intervals and the vertical line represents the location of the true mean number of bacteria per cubic centimeter. Note that the parameter is fixed and that the interval location and width vary from sample to sample. Thus we speak of "the probability that the interval encloses μ" not "the probability that μ falls in the interval" because μ is fixed. The interval is random. Having grasped the concept of a confidence interval, let us now consider how to find the confidence interval for a population mean, μ, based on a random sample of n observations.

An interval estimator, or **confidence interval,** for a population mean can be obtained from the result of Section 8.4. It is possible that $\bar{y}$ might lie either above or below the population mean, although we would not expect it to deviate more than approximately $1.96\sigma_{\bar{y}}$ from μ. Hence, if we choose $(\bar{y} - 1.96\sigma_{\bar{y}})$ as the lower point of the interval, called the **lower confidence limit, LCL,** and $(\bar{y} + 1.96\sigma_{\bar{y}})$ as the upper point or **upper confidence limit, UCL,** the interval most probably will enclose the true population mean, μ. (See Figure 8.8.) In fact, if n is large and the distribution of $\bar{y}$ is approximately normal, we would expect approximately 95 percent of the intervals obtained in repeated sampling to enclose the population mean, μ.

Figure 8.8 Confidence limits for μ

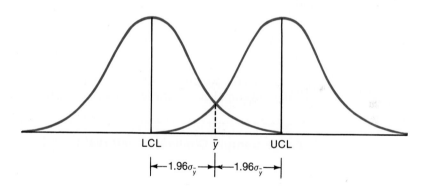

The confidence interval described is called a **large-sample confidence interval** (or confidence limits) because n must be large enough for the Central Limit Theorem to be effective and hence for the distribution of $\bar{y}$ to be approximately normal. Inasmuch as σ is usually unknown, the sample standard deviation must be used to estimate σ. As a rule of thumb, the confidence interval would be appropriate when $n = 30$ or more. This sample size will be sufficiently large to ensure (for most populations) that the normal distribution will be a very good approximation to the sampling distribution of $\bar{y}$ and that s is a good approximation to σ. Consequently, the 95 percent (confidence coefficient equals .95) confidence interval for a population mean is

$$\bar{y} \pm 1.96\sigma_{\bar{y}}$$

or

$$\text{LCL} = \bar{y} - 1.96\sigma_{\bar{y}}$$

$$= \bar{y} - 1.96\,\frac{\sigma}{\sqrt{n}}$$

$$\text{UCL} = \bar{y} + 1.96\sigma_{\bar{y}}$$

$$= 1.96\,\frac{\sigma}{\sqrt{n}}.$$

A 90 percent confidence interval for μ can be constructed in a similar manner. Recalling that .90 of the measurements in a normal distribution will fall within $z = 1.645$ standard deviations of the mean (Table 3, the Appendix), we could construct 90 percent confidence intervals by using

$$\text{LCL} = \bar{y} - 1.645\sigma_{\bar{y}} = \bar{y} - \frac{1.645\sigma}{\sqrt{n}}$$

and

$$\text{UCL} = \bar{y} + 1.645\sigma_{\bar{y}} = \bar{y} + \frac{1.645\sigma}{\sqrt{n}}.$$

In general, we can construct confidence intervals for μ corresponding to any desired confidence coefficient, say $(1 - \alpha)$, by use of the following:

Large-Sample Confidence Interval for a Population Mean, μ

$$\bar{y} \pm \frac{z_{\alpha/2}\sigma}{\sqrt{n}},$$

where $z_{\alpha/2}$ is the z value corresponding to an area, $\alpha/2$, in the upper tail of a standard normal z distribution,

n = sample size

σ = standard deviation of the sampled population.

If σ is unknown, it can be approximated by the sample standard deviation s.

Assumption: $n \geq 30$

Since we define the quantity $z_{\alpha/2}$ to be the value in the z table such that the area to the right of $z_{\alpha/2}$ is equal to $\alpha/2$ (see Figure 8.9), that is, $P(z > z_{\alpha/2}) = \alpha/2$, a

Figure 8.9 Location of $z_{\alpha/2}$

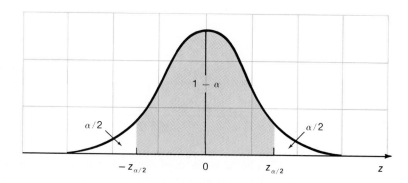

confidence coefficient equal to .95 would imply that $\alpha = .05$ and $z_{.025} = 1.96$. The value of z employed for a 90 percent confidence interval would be $z_{.05} = 1.645$.

Example 8.2

Find a 90 percent confidence interval for the population mean of Example 8.1. Recall that $\bar{y} = 871$ tons and $s = 21$ tons.

Solution

The 90 percent confidence limits would be

$$\bar{y} \pm \frac{1.645\sigma}{\sqrt{n}}.$$

Using s to estimate σ, we obtain

$$871 \pm (1.645) \frac{(21)}{\sqrt{50}}$$

or

$$871 \pm 4.89.$$

Therefore, we estimate that the average daily yield, μ, lies in the interval 866.11 to 875.89 tons. The confidence coefficient, .90, implies that in repeated sampling, 90 percent of the confidence intervals would enclose μ.

Note that the width of the confidence interval increases as the confidence coefficient increases, a result that is in agreement with our intuition. Certainly if we wish to be more confident that the interval will enclose μ, we would increase the width of the interval. Confidence limits corresponding to some of the commonly used confidence coefficients are tabulated in Table 8.2.

Table 8.2 Confidence
limits for μ

Confidence Coefficient	α	$z_{\alpha/2}$	LCL	UCL
.90	.10	1.645	$\bar{y} - 1.645\sigma/\sqrt{n}$	$\bar{y} + 1.645\sigma/\sqrt{n}$
.95	.05	1.96	$\bar{y} - 1.96\sigma/\sqrt{n}$	$\bar{y} + 1.96\sigma/\sqrt{n}$
.99	.01	2.58	$\bar{y} - 2.58\sigma/\sqrt{n}$	$\bar{y} + 2.58\sigma/\sqrt{n}$

The choice of the confidence coefficient to be used in a given situation is made by the experimenter and will depend upon the degree of confidence that the experimenter wishes to place in the estimate. As we have pointed out, the larger the confidence coefficient, the wider the interval. As a result of this freedom of choice, it has become the custom of many experimenters to use a .95 confidence coefficient, although there is no logical foundation for its popularity.

At this point you may see little distinction between point estimators and interval estimators. When we place bounds on the error of a point estimate, for all practical purposes we construct an interval estimate. Furthermore, the point estimate falls in the middle of the interval estimate when a population mean is being estimated. While this close relationship will exist for most of the parameters estimated in this text, it is not generally true. It is not obvious that the best point estimator will fall in the middle of the best interval estimator—in many cases it does not. Furthermore, it is not a foregone conclusion that the best interval estimator will even be a function of the best point estimator. Although these problems are of a theoretical nature, they are important and worth mentioning. From a practical point of view, point and interval estimation are closely related and the choice between the point and the interval estimator in an actual problem depends upon the preference of the experimenter.

Exercises

8.9.   A random sample of n measurements is selected from a population with unknown mean μ and known standard deviation, $\sigma = 10$. Calculate the width of a 95 percent confidence interval for μ when:

a. $n = 100$.   b. $n = 200$.   c. $n = 400$.

8.10.   Compare the confidence intervals in Exercise 8.9. What is the effect on the width of a confidence interval when:

a. You double the sample size?

b. You quadruple the sample size?

8.11.   Refer to Exercise 8.9 and calculate the widths of:

a. A 90 percent confidence interval for μ when $n = 100$.

b. A 99 percent confidence interval for μ when $n = 100$.

c. Compare the widths of 90, 95, and 99 percent confidence intervals for μ. What is the effect on the width of the confidence interval of increasing the confidence coefficient?

8.12. In Exercise 3.11, we commented on a Columbia University survey of the earnings of American authors, noting, in particular, that a *New York Times* (June 15, 1981) summary of the study stated that the "average" American author's writings net less than $5000 per year. The meaning of this statement is not clear, but let us suppose that the *Times* writer was trying to tell us that the mean income from writing of the 2239 authors included in the survey was less than $5000. If the standard deviation σ of the distribution of writing incomes is equal to $20,000, how accurately would the sample mean $\bar{y}$ of the 2239 authors' incomes estimate the population mean of the writing incomes of all American authors? To answer this question, assume that $\bar{y} = 5000$ and find a 95 percent confidence interval for μ.

8.13. An analysis of the vitamin C content of a random sample of 40 school lunches produced a sample mean equal to 7.6 milligrams and standard deviation equal to 3.1 milligrams. Find a 90 percent confidence interval for the mean amount of vitamin C contained in one of the school's lunches. Interpret the interval.

8.6
Estimation from Large Samples

Estimation of a population mean, Sections 8.4 and 8.5, sets the stage for the other estimation problems to be discussed in this chapter. A thread of unity runs through all, which, once it is observed, will simplify the learning process for the beginner. The following conditions will be satisfied for all estimation problems discussed in this chapter. Each point estimator of a parameter, say θ, will be unbiased. That is, the mean of the sampling distribution of the estimator will equal the parameter estimated. The standard deviation of the (sampling distribution of the) estimator will be given so that we may place a 1.96 standard deviation, $1.96\sigma_{\hat{\theta}}$, bound on the error of estimation. In each case, the point estimator will be approximately normally distributed by the Central Limit Theorem, when n is large, and the probability that the error will be less than the bound, $1.96\sigma_{\hat{\theta}}$, will be approximately .95.

The corresponding interval estimators will assume that the sample is large enough for the Central Limit Theorem to imply normality in the sampling distribution of the point estimator of θ as well as to provide a good estimate of any other unknown (for example, σ). Then the confidence intervals for any confidence coefficient, $(1 - \alpha)$, will equal

---

Large-Sample Confidence Interval for θ

$$\hat{\theta} \pm z_{\alpha/2}\sigma_{\hat{\theta}}.$$

Exercises

8.14. Some researchers think that vitamin C may be useful in reducing the buildup of cholesterol deposits on the inner walls of arteries, thereby reducing the possibility of heart attacks (*Gainesville Sun*, June 14, 1976). The cholesterol level of each of 50 persons (with higher than normal cholesterol levels) was recorded before, and then after, a one-month daily regime of 500 milligrams (mg) of vitamin C per day. The data collected for this sample showed the mean and standard deviation of the drop in cholesterol level to be $\bar{y} = 64.3$ mg per 100 milliliters (ml) and $s = 18.9$ mg per 100 ml. Estimate the mean drop per person in cholesterol level by using a 95 percent confidence interval.

8.15. Owing to a variation in laboratory techniques, impurities in materials, and other unknown factors, the results of an experiment in a chemistry laboratory will not always yield the same numerical answer. In an electrolysis experiment, a class measured the amount of copper precipitated from a saturated solution of copper sulfate over a 30-minute period. The $n = 30$ students acquired a sample mean and standard deviation equal to .145 and .0051 mole, respectively. Find a 90 percent confidence interval for the mean amount of copper precipitated from the solution over the period of time.

8.16. According to *Environment News* (September 1975), acid rain, caused by the reaction of certain air pollutants with rainwater, appears to be a growing problem in the northeastern United States. (Acid rain affects the soil and causes corrosion on exposed metal surfaces.) Pure rain falling through clean air registers a pH value (pH is a measure of acidity; 0 is acid, 14 is alkaline) of 5.7. Suppose that water samples from 40 rainfalls are analyzed for pH and that $\bar{y}$ and s are equal to 3.7 and .5, respectively. Find a 99 percent confidence interval for the mean pH in rainfalls and interpret the interval. What assumption must be made in order that the confidence interval be valid?

8.17. In a psychology study of susceptibility to perceptual illusions, 50 male subjects judged the length of an illusory figure. When each judgment was scored in terms of magnitude of the deviation from the actual length, the resultant scores had a mean and standard deviation equal to

$$\bar{y} = 81 \text{ millimeters,}$$
$$s = 12 \text{ millimeters.}$$

Find a 95 percent confidence interval for the mean magnitude of deviation, μ.

8.7
Estimating the Difference between Two Means

A problem of equal importance to the estimation of population means is the comparison of two population means. For instance, we might wish to compare the effectiveness of two teaching methods. Students would be randomly divided into two groups, the first subjected to method 1 and the second to method 2. We would then make inferences concerning the difference in average student achievement as measured by some testing procedure.

Or, we might wish to compare the average yield in a chemical plant using raw materials furnished by two suppliers, *A* and *B*. Samples of daily yield, one for each of the two suppliers, would be recorded and used to make inferences concerning the difference in mean yield.

For each of these examples we postulate two populations, the first with mean and variance μ_1 and σ_1^2, and the second with mean and variance μ_2 and σ_2^2. A random sample of n_1 measurements is drawn from population I and n_2 from population II, where the samples are assumed to have been drawn

independently of one another. Finally the estimates of the population parameters, $\bar{y}_1$, s_1^2, $\bar{y}_2$, and s_2^2, are calculated from the sample data.

The point estimator of the difference between the population means,
$(\mu_1 - \mu_2)$, is $(\bar{y}_1 - \bar{y}_2)$, the difference between the sample means. If repeated
pairs of samples of n_1 and n_2 measurements are drawn from the two populations and the estimate, $(\bar{y}_1 - \bar{y}_2)$, calculated for each pair, a distribution of
estimates will result. The sampling distribution of the estimator, $(\bar{y}_1 - \bar{y}_2)$,
will be approximately normally distributed* for large samples (say n_1 and n_2
both equal to 30 or more), with mean and standard deviation:

Mean and Standard Deviation of $(\bar{y}_1 - \bar{y}_2)$

$$E(\bar{y}_1 - \bar{y}_2) = \mu_1 - \mu_2,$$

$$\sigma_{(\bar{y}_1 - \bar{y}_2)} = \sqrt{\frac{\sigma_1^2}{n_1} + \frac{\sigma_2^2}{n_2}}$$

Although the formula for the standard deviation of $(\bar{y}_1 - \bar{y}_2)$ may appear
to be complicated, a result derived in mathematical statistics will assist in its
memorization. Certainly the variability of a difference between two independent
random variables would seem, intuitively, to be greater than the variability of
either of the two variables, since one may be extremely large at the same time
that the other is extremely small. Hence each contributes a portion of its
variability to the variability of the difference. This intuitive explanation is
supported by a theorem in mathematical statistics, which states that the
variance of either the sum or the difference of two independent random
variables is equal to the sum of their respective variances. That is,

$$\sigma_{(y_1 + y_2)}^2 = \sigma_{y_1}^2 + \sigma_{y_2}^2$$

and

$$\sigma_{(y_1 - y_2)}^2 = \sigma_{y_1}^2 + \sigma_{y_2}^2.$$

Therefore,

$$\sigma_{(\bar{y}_1 - \bar{y}_2)}^2 = \sigma_{\bar{y}_1}^2 + \sigma_{\bar{y}_2}^2 = \frac{\sigma_1^2}{n_1} + \frac{\sigma_2^2}{n_2},$$

* When n_1 and n_2 are large, the sampling distributions of $\bar{y}_1$ and $\bar{y}_2$ are approximately normally
distributed because of the Central Limit Theorem. It can be shown (proof omitted) that the
difference between two normally distributed random variables possesses a normal probability
distribution. This justifies the approximate normality of the sampling distribution of $(\bar{y}_1 - \bar{y}_2)$
when n_1 and n_2 are large.

and the standard deviation is

$$\sigma_{(\bar{y}_1 - \bar{y}_2)} = \sqrt{\frac{\sigma_1^2}{n_1} + \frac{\sigma_2^2}{n_2}}.$$

Then, when n_1 and n_2 are large, the estimator, $(\bar{y}_1 - \bar{y}_2)$, will have a sampling distribution that is approximately normal and the bound on the error of the point estimate is

$$1.96 \sqrt{\frac{\sigma_1^2}{n_1} + \frac{\sigma_2^2}{n_2}}.$$

The sample variances, s_1^2 and s_2^2, may be used to estimate σ_1^2 and σ_2^2 when these parameters are unknown. This approximation will be reasonably good when n_1 and n_2 are equal to 30 or more.

Example 8.3

A comparison of the wearing quality of two types of automobile tires was obtained by road testing samples of $n_1 = n_2 = 100$ tires for each type. The number of miles until wear-out was recorded, where wear-out was defined as a specific amount of tire wear. The test results were as follows:

$$\bar{y}_1 = 26{,}400 \text{ miles}, \qquad \bar{y}_2 = 25{,}100 \text{ miles};$$
$$s_1^2 = 1{,}440{,}000, \qquad s_2^2 = 1{,}960{,}000.$$

Estimate the difference in mean time to wear-out and place bounds on the error of estimation.

Solution

The point estimate of $(\mu_1 - \mu_2)$ is

$$(\bar{y}_1 - \bar{y}_2) = 26{,}400 - 25{,}100 = 1{,}300 \text{ miles}$$

and

$$\sigma_{(\bar{y}_1 - \bar{y}_2)} = \sqrt{\frac{\sigma_1^2}{n_1} + \frac{\sigma_2^2}{n_2}}$$
$$\approx \sqrt{\frac{s_1^2}{n_1} + \frac{s_2^2}{n_2}} = \sqrt{\frac{1{,}440{,}000}{100} + \frac{1{,}960{,}000}{100}}$$
$$= \sqrt{34{,}000} = 184 \text{ miles.}$$

We would expect the error of estimation to be less than $1.96\sigma_{(\bar{y}_1 - \bar{y}_2)}$, or 361 miles. Therefore, it would appear that tire type 1 is superior to type 2 in wearing quality when subjected to the road test.

A confidence interval for $(\mu_1 - \mu_2)$ with confidence coefficient $(1 - \alpha)$ can be obtained by using:

Large-Sample Confidence Interval for $(\mu_1 - \mu_2)$

$$(\bar{y}_1 - \bar{y}_2) \pm z_{\alpha/2} \sqrt{\frac{\sigma_1^2}{n_1} + \frac{\sigma_2^2}{n_2}}$$

Assumption: n_1 and n_2 are both greater than or equal to 30.

Example 8.4

Place a confidence interval on the difference in mean time to wear-out for the problem described in Example 8.3. Use a confidence coefficient of .99.

Solution

The confidence interval will be

$$(\bar{y}_1 - \bar{y}_2) \pm 2.58 \sqrt{\frac{\sigma_1^2}{n_1} + \frac{\sigma_2^2}{n_2}}.$$

When we use the results of Example 8.3, we find that the confidence interval is

$$1300 \pm 2.58(184).$$

Therefore, LCL = 825, UCL = 1775, and the difference in mean time to wear-out is estimated to lie between these two points. Note that the confidence interval is wider than if we had constructed the 95 percent confidence interval, $\bar{y} \pm 1.96\sigma_{(\bar{y}_1 - \bar{y}_2)}$. To gain greater confidence (.99) that the interval will enclose $(\mu_1 - \mu_2)$, we had to employ a wider confidence interval.

Exercises

8.18.   Independent random samples were selected from two populations, 1 and 2. The sample sizes, means, and variances were as shown below:

	Population	
	1	**2**
Sample Size	50	48
Sample Mean	39.4	40.7
Sample Variance	.70	.93

Find a 90 percent confidence interval for the difference in the population means and interpret your result.

8.19. Independent random samples were selected from two populations, 1 and 2. The sample sizes, means, and variances were as shown below:

	Population	
	1	2
Sample Size	100	100
Sample Mean	1.3	2.7
Sample Variance	1.12	.64

Find a 99 percent confidence interval for the difference in the population means and interpret your answer.

8.20. A small amount of the trace element selenium, from 50 to 200 micrograms per day, is considered essential to good health (*Prevention*, September, 1980). Suppose that random samples of $n_1 = n_2 = 30$ adults were selected from two regions of the United States, and a day's intake of selenium, from both liquids and solids, was recorded for each person. The mean and standard deviation of the selenium daily intakes for the 30 adults from Region 1 were $\bar{y}_1 = 167.1$ and $s_1 = 24.3$ micrograms, respectively. The corresponding statistics for the 30 adults from Region 2 were $\bar{y}_2 = 140.9$ and $s_2 = 17.6$. Find a 95 percent confidence interval for the difference in the mean selenium intake for the two regions.

8.21. A study was conducted to compare the mean number of police emergency calls per eight-hour shift in two districts of a large city. Samples of 100 eight-hour shifts were randomly selected from the police records for each of the two regions, and the number of emergency calls were recorded for each shift. The sample statistics are shown below:

	Region	
	1	2
Sample Size	100	100
Sample Mean	2.4	3.1
Sample Variance	1.44	2.64

Find a 90 percent confidence interval for the difference in the mean number of police emergency calls per shift between the two districts of the city. Interpret the interval.

8.22. An experiment was conducted to compare two diets, A and B, designed for weight reduction. Two groups of 30 overweight dieters each were randomly selected. One group was placed on diet A and the other on diet B and their weight losses were recorded over a 30-day period. The means and standard deviations of the weight loss measurements for the two groups are shown in the accompanying table. Find a 95 percent confidence interval for the difference in mean weight loss for the two diets. Interpret your confidence interval.

Diet A	Diet B
$\bar{y}_A = 21.3$	$\bar{y}_B = 13.4$
$s_A = 2.6$	$s_B = 1.9$

8.23. One method for solving the electric power shortage employs the construction of floating nuclear power plants located a few miles offshore in the ocean. Because there is concern about the possibility of a ship collision with the floating (but anchored) plant, an estimate of the density of ship traffic in the area is needed. The number of ships passing within 10 miles of the proposed power-plant location per day, recorded for $n = 60$ days during July and August, possessed sample mean and variance equal to

$$\bar{y} = 7.2$$
$$s^2 = 8.8.$$

a. Find a 95 percent confidence interval for the mean number of ships passing within 10 miles of the proposed power-plant location during a one-day time period.

b. The density of ship traffic was expected to decrease during the winter months. A sample of $n = 90$ daily recordings of ship sightings for December, January, and February gave the following mean and variance:

$$\bar{y} = 4.7$$
$$s^2 = 4.9.$$

Find a 90 percent confidence interval for the difference in mean density of ship traffic between the summer and winter months.

c. What is the population associated with your estimate, part (b)? What could be wrong with the sampling procedure, parts (a) and (b)?

8.24. The in-city gasoline mileage per gallon was computed for two economy automobiles, each for 40 tanks of gasoline. The mean and standard deviation of the mile-per-gallon readings were as given in the table. Carefully define the populations associated with the mean mile-per-gallon ratings, μ_1 and μ_2, associated with the two automobiles. Find a 90 percent confidence interval for the difference in mean miles per gallon for the two automobiles. Interpret the confidence interval.

Auto 1	Auto 2
$\bar{y}_1 = 22.3$	$\bar{y}_2 = 25.1$
$s_1 = 1.1$	$s_2 = 1.3$

8.8
Estimating the
Parameter of a
Binomial
Population

The best point estimator of the binomial parameter, p, is also the estimator that would be chosen intuitively. That is, the estimator, $\hat{p}$, would equal

$$\hat{p} = \frac{y}{n},$$

the total number of successes divided by the total number of trials. By "best" we mean that $\hat{p}$ is unbiased and possesses a minimum variance compared with other possible unbiased estimators.

We recall, Section 7.7, that the sampling distribution of $\hat{p}$ is approximately normally distributed when n is large. The mean and standard deviation of the sampling distribution of $\hat{p}$ are:

Mean and Standard Deviation of $\hat{p}$

$E(\hat{p}) = p,$

$$\sigma_{\hat{p}} = \sqrt{\frac{pq}{n}}.$$

Therefore, the bound on the error of a point estimate will be

$$1.96 \sqrt{\frac{pq}{n}},$$

and the $100(1 - \alpha)$ percent confidence interval, appropriate for large n, is:

Large-Sample Confidence Interval for p

$$\hat{p} \pm z_{\alpha/2} \sqrt{\frac{\hat{p}\hat{q}}{n}}.$$

Assumption: $\hat{p}$ is approximately normally distributed. (The same conditions given in Section 7.6 must be satisfied.)

The sample size will be considered large when we can assume that $\hat{p}$ is approximately normally distributed. The conditions necessary for the approximate normality of a binomial distribution were discussed in Section 7.6.

The only difficulty encountered in our procedure will be in calculating $\sigma_{\hat{p}}$, which involves p (and $q = 1 - p$), which is unknown. Note that we have substituted $\hat{p}$ for the parameter p in the standard deviation, $\sqrt{pq/n}$. When n is large, little error will be introduced by this substitution. As a matter of fact, the standard deviation changes only slightly as p changes. This can be observed in Table 8.3, where $\sqrt{pq}$ is recorded for several values of p. Note that $\sqrt{pq}$ changes very little as p changes, especially when p is near .5.

Table 8.3 Some calculated values of $\sqrt{pq}$

p	$\sqrt{pq}$
.5	.50
.4	.49
.3	.46
.2	.40
.1	.30

Example 8.5

A random sample of $n = 100$ voters in a community produced $y = 59$ voters in favor of candidate A. Estimate the fraction of the voting population favoring A and place a bound on the error of estimation.

Solution

The point estimate is

$$\hat{p} = \frac{y}{n} = \frac{59}{100} = .59,$$

and the bound on the error of estimation is

$$1.96\sigma_{\hat{p}} = 1.96\sqrt{\frac{pq}{n}} \approx 1.96\sqrt{\frac{(.59)(.41)}{100}} = .096.$$

A 95 percent confidence interval for p would be

$$\hat{p} \pm 1.96\sqrt{\frac{\hat{p}\hat{q}}{n}}$$

or

$$.59 \pm 1.96(.049).$$

Thus we would estimate that p lies in the interval .494 to .686 with confidence coefficient .95.

Exercises

8.25. A random sample of $n = 1000$ observations from a binomial population produced $y = 740$ successes. Find a 95 percent confidence interval for p and interpret the interval.

8.26. Suppose that the number of successes observed in $n = 1000$ trials of a binomial experiment is 950. Find a 95 percent confidence interval for p. Why is this confidence interval narrower than the confidence interval, Exercise 8.25?

8.27. A survey conducted by the President's Commission on Pension Policy in the fall of 1979 revealed that a high proportion of Americans are very pessimistic about their prospects of receiving an adequate income when they eventually retire (*Orlando Sentinel Star*, May 3, 1980). When asked if they expected their retirement to be adequate, 62.9 percent of the 6100 respondents, full-time workers 18 years old or older, indicated that in their opinion, their income upon retirement definitely would not (or probably would not) be adequate. Find a 95 percent confidence interval for the proportion of all workers, 18 or older, who believe that, when they retire, their retirement income will be inadequate. Interpret the interval.

8.28. A published report on a Gallup Youth Survey (*Gainesville Sun*, May 18, 1977) states that 1069 teenagers were questioned concerning their opinion on what they consider to be the key problems facing youth today. The key problems were topped by drug use and abuse (27%), getting along with and communicating with parents (20%), alcohol use and abuse (7%), and finding employment (6%). If the 1069 teenagers can be regarded as a random sample of all teenagers in the United States, estimate the fraction that regard drug use and abuse as the number one problem. Use a 99 percent confidence interval.

8.29. A sample of 400 human subjects produced $y = 280$ students who were classified as right-eye-dominant on the basis of a sighting task. Estimate the fraction of the entire population who are right-eye-dominant; use a 95 percent confidence interval.

8.30. A *Wall Street Journal* (July 23, 1981) report on a survey by Warwick, Welsh & Miller, a New York advertising agency, indicates that matters of taste cannot be ignored in television advertising. In a mail survey of 3440 people, 40 percent indicated that they found TV commercials to be in poor taste, 55 percent said that they avoided products whose commercials were judged to be in poor taste and, of this latter group, only 20 percent ever complained to a TV station or an advertiser about their dissatisfaction.

   a. Find a 95 percent confidence interval for the percentage of TV viewers who find TV commercials to be in poor taste.

   c. Find a 95 percent confidence interval for the percentage of TV viewers who avoid products that are promoted by TV commercials they consider to be in poor taste.

   c. Find a 95 percent confidence interval for the percentage of those who avoid products and who have complained to the TV station or the advertiser about poor taste in a TV commercial.

8.31. Osteomyelitis (bone infection) is an ailment that often fails to respond to treatment. Doctors at the Naval Regional Medical Center in Long Beach, California, have been testing the effect of a pressurized oxygen environment on patients (*Gainesville Sun*, October 21, 1977). Of 70 patients treated with the pressurized oxygen environment, all improved and 63 percent have remained free of the disease. If the treated patients represent a random sample from the population of all patients with osteomyelitis, estimate the proportion of treated patients that will remain free of the disease. Use a 95 percent confidence interval.

8.32. The *Orlando Sentinel Star* (December 1, 1977) reported on a survey they conducted to investigate the attitudes of television viewers concerning a certain national sports commentator. They state that of 380 letters received, 325 wanted the sports commentator "off the air." The remaining 55 were in favor of retaining him.

a. The validity of a confidence interval calculated for the proportion p of the newspaper's readers who favor retention of the commentator is dependent on how the survey was conducted. Suppose that the viewer responses were obtained by requesting that readers send in their opinions on the question of retention or firing. Would this method of sampling yield a valid confidence interval for p?

b. Suppose that the 380 respondents represented a random sample of the newspaper's readers. Find a 95 percent confidence interval for p and interpret it.

8.9
Estimating the Difference between Two Binomial Parameters

The fourth and final estimation problem considered in this chapter is the estimation of the difference between the parameters of two binomial populations. Assume that the two populations I and II possess parameters p_1 and p_2, respectively. Independent random samples consisting of n_1 and n_2 trials are drawn from their respective populations and the estimates $\hat{p}_1$ and $\hat{p}_2$ are calculated.

The point estimator of $p_1 - p_2$ is the corresponding difference in the estimates, $\hat{p}_1 - \hat{p}_2$. As learned in Section 7.7, this estimator has a sampling distribution that is approximately normal for large samples. The mean and standard deviation of the sampling distribution of $(\hat{p}_1 - \hat{p}_2)$ are:

Mean and Standard Deviation of $(\hat{p}_1 - \hat{p}_2)$

$$E(\hat{p}_1 - \hat{p}_2) = (p_1 - p_2),$$

$$\sigma_{(\hat{p}_1 - \hat{p}_2)} = \sqrt{\frac{p_1 q_1}{n_1} + \frac{p_2 q_2}{n_2}}. \quad *$$

Therefore, $(\hat{p}_1 - \hat{p}_2)$ is an unbiased estimator of $(p_1 - p_2)$ with a bound on the error of estimation,

$$1.96 \sqrt{\frac{p_1 q_1}{n_1} + \frac{p_2 q_2}{n_2}}.$$

Since p_1 and p_2 are unknown, we use the estimates, $\hat{p}_1$ and $\hat{p}_2$, to approximate p_1 and p_2 in this formula.

---

* The formula for the standard deviation of $(\hat{p}_1 - \hat{p}_2)$ follows from our discussion in Section 8.7. Since the variance of the difference of two independent random variables is equal to the sum of their respective variances, the variance of $(\hat{p}_1 - \hat{p}_2)$ is equal to $\sigma_{\hat{p}_1}^2 + \sigma_{\hat{p}_2}^2 = (p_1 q_1/n_1) + (p_2 q_2/n_2)$. The standard deviation of $(\hat{p}_1 - \hat{p}_2)$ is equal to the square root of this quantity.

The $100(1 - \alpha)$ percent confidence interval, appropriate when n_1 and n_2 are large, is:

Large-Sample Confidence Interval for $(p_1 - p_2)$

$$(\hat{p}_1 - \hat{p}_2) \pm z_{\alpha/2}\sqrt{\frac{\hat{p}_1\hat{q}_1}{n_1} + \frac{\hat{p}_2\hat{q}_2}{n_2}}.$$

Assumption: n_1 and n_2 are sufficiently large so that the sampling distributions of $\hat{p}_1$ and $\hat{p}_2$ are approximately normal. (See Section 7.6 for guidelines.)

Example 8.6

A manufacturer of fly sprays wished to compare two new concoctions, I and II. Two rooms of equal size, each containing 1000 flies, were employed in the experiment, one treated with fly spray I and the other treated with an equal amount of fly spray II. A total of 825 and 760 flies succumbed to sprays I and II, respectively. Estimate the difference in the rate of kill for the two sprays when used in the test environment.

Solution

The point estimate of $(p_1 - p_2)$ is

$$(\hat{p}_1 - \hat{p}_2) = .825 - .760 = .065.$$

The bound on the error of estimation is

$$1.96\sqrt{\frac{p_1q_1}{n_1} + \frac{p_2q_2}{n_2}} \approx 1.96\sqrt{\frac{(.825)(.175)}{1000} + \frac{(.76)(.24)}{1000}}$$

$$= .035.$$

The corresponding confidence interval, using confidence coefficient .95, is

$$(\hat{p}_1 - \hat{p}_2) \pm 1.96\sqrt{\frac{\hat{p}_1\hat{q}_1}{n_1} + \frac{\hat{p}_2\hat{q}_2}{n_2}}.$$

The resulting confidence interval is $.065 \pm .035$.

Hence we estimate that the difference between the rates of kill, $(p_1 - p_2)$, will fall in the interval .030 to .100. We are fairly confident of this estimate because we know that if our sampling procedure were repeated over and over again, each time generating an interval estimate, approximately 95 percent of the estimates would enclose the quantity $(p_1 - p_2)$.

Exercises

8.33.   Samples of $n_1 = 400$ and $n_2 = 400$ observations were selected from binomial populations 1 and 2, and $y_1 = 80$ and $y_2 = 110$ were observed.

   a. Find a 90 percent confidence interval for the difference $(p_1 - p_2)$ in the two population parameters. Interpret the interval.

   b. What assumptions must you make in order that the confidence interval be valid?

8.34.   According to an Associated Press–NBC News nationwide telephone poll, 65 percent of a total of 1601 adults believe that the United States should reduce immigration (*Philadelphia Inquirer*, August 17, 1981). The percentage favoring reduction in immigration varied from one social group to another and tended to be larger in those states most affected by the influx. Suppose that random samples of $n_1 = n_2 = 100$ adults selected from Florida and Texas, respectively, produced $y_1 = 72$ of the Florida respondents favoring reduction of immigration and $y_2 = 78$ from Texas.

   a. Find a 95 percent confidence interval for the difference in the proportions of adults in Florida and Texas who favor a reduction in immigration.

   b. Find a 95 percent confidence interval for the difference in the proportion of adults in Florida favoring a reduction in immigration versus the national proportion.

   c. Find a 95 percent confidence interval for the difference in the proportion of adults in Texas favoring a reduction in immigration versus the national proportion.

8.35.   According to *Environment News* (April 1975), "The continuing analysis of lead levels in the drinking water of several Boston communities has verified elevated lead levels in the water supplies of Somerville, Brighton and Beacon Hill . . . ." Preliminary results of a study carried out in 1974 found that "20 percent of 248 households tested in those communities showed levels exceeding the U.S. Public Health Service standard of 50 parts per million." In contrast, in Cambridge, which adds anticorrosives to its water, "only 5 percent of the 110 households tested showed lead levels exceeding the standard." Find a 95 percent confidence interval for the difference in the proportions of households which have lead levels exceeding the standard between the communities of Somerville, Brighton, and Beacon Hill and the community of Cambridge.

8.36.   A sampling of political candidates, 200 randomly chosen from the West and 200 from the East, was classified according to whether the candidate received backing by a national labor union and whether the candidate won. A summary of the data is shown below:

	West	East
Winners backed by union	120	142

   Find a 95 percent confidence interval for the difference in the proportions of union-backed winners between the West and the East.

8.37.   In a study of the relationship between birth order and college success, an investigator found that 126 in a sample of 180 college graduates were first-born or only children; in a sample of 100 nongraduates of comparable age and socioeconomic background the number of first-born or only children was 54. Estimate the difference in proportion of first-born or only children for the two populations from which these samples were drawn. Use a 90 percent confidence interval.

8.38.   Two different vaccines developed to prevent a disease in poultry showed 80 percent and 88 percent effectiveness in preventing the disease in samples, each containing $n = 400$ vaccinated chickens exposed to the disease. Construct a 95 percent confidence interval for the difference in proportions of chickens surviving the disease for the two types of vaccine.

8.10
Choosing the Sample Size

The design of an experiment is essentially a plan for purchasing a quantity of information, which, like any other commodity, may be acquired at varying prices depending upon the manner in which the data are obtained. Some measurements contain a large amount of information concerning the parameter of interest; others may contain little or none. Since the sole product of research is information, we want to make its purchase at minimum cost.

The sampling procedure, or experimental design as it is usually called, affects the quantity of information per measurement. This, along with the sample size, n, controls the total amount of relevant information in a sample. With few exceptions we shall be concerned with the simplest sampling situation, random sampling from a relatively large population, and will devote our attention to the selection of the sample size, n.

The researcher makes little progress in planning an experiment before encountering the problem of selecting the sample size. Indeed, perhaps one of the most frequent questions asked of the statistician is: How many measurements should be included in the sample? Unfortunately, the statistician cannot answer this question without knowing how much information the experimenter wishes to buy. Certainly, the total amount of information in the sample will affect the measure of goodness of the method of inference and must be specified by the experimenter. Referring specifically to estimation, we would like to know how accurate the experimenter wishes the estimate to be. This may be stated by specifying a bound on the error of estimation.

For instance, suppose that we wish to estimate the average daily yield of a chemical, μ (Example 8.1), and we wish the error of estimation to be less than 4 tons with a probability of .95. Since approximately 95 percent of the sample means will lie within $1.96\sigma_{\bar{y}}$ of μ in repeated sampling, we are asking that $1.96\sigma_{\bar{y}}$ equal 4 tons (see Figure 8.10). Then,

$$1.96\sigma_{\bar{y}} = 4$$

or

$$\frac{1.96\sigma}{\sqrt{n}} = 4.$$

Solving for n, we obtain

$$n = \left(\frac{1.96}{4}\right)^2 \sigma^2.$$

Figure 8.10 Approximate sampling distribution of $\bar{y}$ for large samples

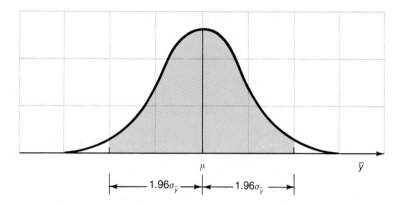

You will quickly note that we cannot obtain a numerical value for n unless the population standard deviation, σ, is known. And certainly, this is exactly what we would expect because the variability of $\bar{y}$ depends upon the variability of the population from which the sample was drawn.

Lacking an exact value for σ, we would use the best approximation available, such as an estimate, s, obtained from a previous sample or knowledge of the range in which the measurements will fall. Since the range is approximately equal to 4σ (the Empirical Rule), one-fourth of the range will provide an approximate value for σ. For our example we would use the results of Example 8.1, which provided a reasonably accurate estimate of σ equal to $s = 21$. Then,

$$n = \left(\frac{1.96}{4}\right)^2 \sigma^2$$

$$\approx \left(\frac{1.96}{4}\right)^2 (21)^2 = 105.9$$

or

$$n = 106.$$

Using a sample size $n = 106$, we would be reasonably certain (with probability approximately equal to .95) that our estimate will lie within $1.96\sigma_{\bar{y}} = 4$ tons of the true average daily yield. Actually we would expect the error of estimation to be much less than 4 tons. The probability is approximately equal to .68 that the error of estimation would be less than $\sigma_{\bar{y}} = 2$ tons.

You note that the solution $n = 106$ is a very rough approximation because we had to use an approximate value for σ in calculating the value of n. While this method of choosing the sample size is approximate for a specified desired accuracy of estimation, it is the best available and is certainly better than selecting the sample size on the basis of our intuition.

The method of choosing the sample size for all the large-sample estimation procedures discussed in preceding sections is identical to that described above.

The experimenter must specify a desired bound on the error of estimation and an associated confidence level, $(1 - \alpha)$. For example, if the parameter is θ and the desired bound is B, we would equate

$$z_{\alpha/2}\sigma_{\hat{\theta}} = B,$$

where $z_{\alpha/2}$ is the z value defined in Section 8.5; that is,

$$P(z > z_{\alpha/2}) = \frac{\alpha}{2}.$$

Procedure for Choosing the Sample Size

Let θ be the parameter to be estimated and let $\sigma_{\hat{\theta}}$ be the standard deviation of the point estimator. Then proceed as follows:

1. Choose B, the bound on the error of estimation, and a confidence coefficient $(1 - \alpha)$.
2. Solve the following equation for the sample size n:

$$z_{\alpha/2}\sigma_{\hat{\theta}} = B.$$

Note: For most estimators (all presented in this text), $\sigma_{\hat{\theta}}$ is a function of the sample size n.

We shall illustrate with examples.

Example 8.7

The reaction of an individual to a stimulus in a psychological experiment may take one of two forms, A or B. If an experimenter wishes to estimate the probability, p, that a person will react in favor of A, how many people must be included in the experiment? Assume that he will be satisfied if the error of estimation is less than .04 with probability equal to .90. Assume also that he expects p to lie somewhere in the neighborhood of .6.

Solution

For this particular example, the bound B on the error of estimation is .04. Since the confidence coefficient is $1 - \alpha = .90$, α must equal .10 and $\alpha/2 = .05$. The z value corresponding to an area equal to .05 in the upper tail of the z distribution is $z_{\alpha/2} = 1.645.$* We then require

$$1.645\sigma_{\hat{p}} = .04$$

---

* Note: It is not necessary to carry three decimal places on $z_{\alpha/2}$ for these calculations. The calculated value of n will be very approximate because we are using an approximate value for p in the calculations.

or

$$1.645 \sqrt{\frac{pq}{n}} = .04.$$

Since the variability of $\hat{p}$ is dependent upon p, which is unknown, we must use the guessed value of $p = .6$ provided by the experimenter as an approximation. Then,

$$1.645 \sqrt{\frac{(.6)(.4)}{n}} = .04$$

or

$$\sqrt{n} = \frac{1.645}{.04} \sqrt{(.6)(.4)}$$

and

$$n = 406.$$

Remember, we used an approximate value of p to obtain this value of n. Consequently, the value, $n = 406$, is not exact. It is only an approximation to the required sample size. Therefore, it is reasonable to say that a sample of approximately $n = 406$ people will allow you to estimate p with an error of estimation less than .04 with probability equal to .90.

Example 8.8

An experimenter wishes to compare the effectiveness of two methods of training industrial employees to perform a certain assembly operation. A number of employees is to be divided into two equal groups, the first receiving training method 1 and the second training method 2. Each will perform the assembly operation, and the length of assembly time will be recorded. It is expected that the measurements for both groups will have a range of approximately 8 minutes. If the estimate of the difference in mean times to assemble is desired correct to within 1 minute with probability equal to .95, how many workers must be included in each training group?

Solution

Equating $1.96\sigma_{(\bar{y}_1 - \bar{y}_2)}$ to $B = 1$ minute, we obtain

$$1.96 \sqrt{\frac{\sigma_1^2}{n_1} + \frac{\sigma_2^2}{n_2}} = 1.$$

Or, since we desire n_1 to equal n_2, we may let $n_1 = n_2 = n$ and obtain the equation

$$1.96 \sqrt{\frac{\sigma_1^2}{n} + \frac{\sigma_2^2}{n}} = 1.$$

As noted above, the variability of each method of assembly is approximately the same and hence $\sigma_1^2 = \sigma_2^2 = \sigma^2$. Since the range, equal to 8 minutes, is approximately equal to 4σ, then

$$4\sigma \approx 8$$

and

$$\sigma \approx 2.$$

Substituting this value for σ_1 and σ_2 in the above equation, we obtain

$$1.96 \sqrt{\frac{(2)^2}{n} + \frac{(2)^2}{n}} = 1;$$

or

$$1.96 \sqrt{8/n} = 1$$

and

$$\sqrt{n} = 1.96\sqrt{8}.$$

Solving, we have $n = 31$. Thus each group should contain approximately $n = 31$ members.

Exercises

8.39. Suppose that you wish to estimate a population mean based on a random sample of n observations, and that prior experience suggests that $\sigma = 15.8$. If you wish to estimate μ correct to within 3.2 with probability equal to .95, how many observations should be included in your sample?

8.40. Suppose that you wish to estimate a binomial parameter p correct to within .03 with probability equal to .95. If you suspect that p is equal to some value between .1 and .3 and you wish to be certain that your sample is large enough, how large should n be? (*Hint:* When calculating $\sigma_{\hat{p}}$, use the value of p in the interval, $.1 < p < .3$, that will give the largest sample size.)

8.41. Independent random samples of $n_1 = n_2 = n$ observations are to be selected from each of two populations, 1 and 2. If you wish to estimate the difference between the two population means correct to within .12 with probability equal to .90, how large should n_1 and n_2 be? Assume that you know that $\sigma_1^2 \approx \sigma_2^2 \approx .25$.

8.42. Independent random samples of $n_1 = n_2 = n$ observations are to be selected from each of two binomial populations, 1 and 2. If you wish to estimate the difference in the two population binomial parameters correct to within .04 with probability equal to .98, how large should n be? Assume that you have no prior information on the values of p_1 and p_2 but you wish to make certain that you have an adequate number of observations in the samples.

8.43. "More Than Half of the Women Polled Have Been Abroad" is the heading of an article in the *Gainesville Sun* (October 16, 1978). Based on a national telephone survey by Women-Poll, the survey gives the percentages of women by region, education, and income who have been abroad. The article neither gives the number of women included in the national poll nor describes the method of sampling.

a. If the sampling were random, how many women would have to be included in the poll in order to estimate the proportion of all women in the United States who have been abroad? Assume that you wish your estimate to lie within .02 of the population proportion with probability equal to .95.

b. Explain why a telephone poll might not yield a random sample. (Do you really think that Americans are so affluent that more than 50% of all women have been abroad?)

8.44. If a wildlife service wishes to estimate the mean number of days hunting per hunter for all hunters licensed in the state during a given season, with a bound on the error of estimation equal to 2 hunting days, how many hunters must be included in the survey? Assume that data collected in earlier surveys has shown σ to be approximately equal to 10.

8.45. The Federal Trade Commission samples and tests cigarettes to determine whether the nicotine and tar contents of the cigarettes agree with those claimed by the manufacturer. An *FTC News Summary* (October 29, 1976) states that for a test of True king size, filter, soft-pack cigarettes, the tar and nicotine contents were 5 and .4 milligrams (mg) per cigarette. The report does not state how many cigarettes were analyzed to obtain these figures, nor does it give a measure of the cigarette-to-cigarette variation in the measurements. Suppose that the standard deviation of the tar content measurements is approximately equal to 1 mg per cigarette. If the FTC wishes to estimate the mean tar content per cigarette correct to within .1 mg, how many cigarettes would the FTC have to analyze? (Assume that the FTC wishes the error of estimation to be less than .1 mg with probability equal to .99.)

8.46. It is desired to estimate the difference in grade-point average between two groups of college students accurate to within .2 grade point with probability approximately equal to .95. If the standard deviation of the grade-point measurements is approximately equal to .6, how many students must be included in each group? (Assume that the groups will be of equal size.)

8.47. Refer to Exercise 8.16 and the measurement of acidity in rainwater and suppose that you wish to estimate the mean pH of rainfalls in an area that suffers heavy pollution due to the discharge of smoke from a power plant. Assume you know that σ is in the neighborhood of .5 pH and that you wish your estimate to lie within .1 of μ with probability near .95. Approximately how many rainfalls must be included in your sample (one pH reading per rainfall)? Would it be valid to select all of your water specimens from a single rainfall? Explain.

8.48. Refer to Exercise 8.47. Suppose that you wish to estimate the difference between the mean acidity for rainfalls at two different locations, one in a relatively unpolluted area along the ocean and the other in an area subject to heavy air pollution. If you wish your estimate to be correct to the nearest .1 pH with probability near .90, approximately how many rainfalls (pH values) would have to be included in each sample? (Assume that the variance of the pH measurements is approximately .25 at both locations and that the samples will be of equal size.)

8.49. Refer to the comparison of the daily adult intake of selenium in two different regions of the United States, Exercise 8.20. Suppose that you wish to estimate the difference in the mean daily intake between the two regions correct to within 5 micrograms with probability equal to .90. If you plan to select an equal number of adults from the two regions (that is, $\mu_1 = \mu_2$), how large should n_1 and n_2 be?

8.50.   Refer to Exercise 8.34 and the Associated Press–NBC News nationwide telephone poll of opinions concerning the limits on immigration. Suppose that you wish to estimate the difference between the proportions of adults in Florida and Texas favoring a reduction of immigration and that you wish your estimate to lie within .08 of the true difference with probability equal to .95. If you plan to select equal sample sizes (that is, $n_1 = n_2$), how large should n_1 and n_2 be?

8.11
More on the Louis Harris Sports Poll

You will recall that our case study at the beginning of this chapter presented the results of a Louis Harris poll that was conducted to determine the favorite sport of Americans. A sample of 1091 American adults was polled and asked to indicate their favorite sport. The percentages favoring particular sports (published in the *Gainesville Sun*, May 27, 1981), shown in Table 8.1, are reproduced in Table 8.4.

Table 8.4 A sports popularity poll

Sport	Percentage of Americans Choosing Sport as Favorite
Football	36
Baseball	21
Basketball	12
Tennis	5
Auto Racing	5
Golf	4
Boxing	4
Bowling	3
Horse Racing	3
Hockey	3
Soccer	2
Track	2

How well do these percentages of the responses of 1091 adults characterize the preference percentages for the many millions of American adults? Intuitively, we would think that the larger the sample size n in relation to the number N of elements in the population, the more accurate the estimate. This is true when the number N of elements in the population is small in relation to n, but the number N of elements in the population has little effect on the accuracy of an estimate when N is ten (or more) times as large as the sample size. So, for example, a sample of 1091 adults will estimate the proportion p for a population containing $N = 5000$ adults more accurately than if the population contained 10,000 adults. But, the accuracy of estimating p based on a sample of 1091 adults will be essentially the same regardless of whether the population contains 100,000 or 100,000,000 adults.

The error of estimating the proportion p of all adults who prefer a specific sport, say football, is dependent on the sampling distribution of $\hat{p}$. Using information contained in Section 8.8, we would expect the sampling distribution of $\hat{p}$ to possess a sampling distribution that is approximately normal with mean p and standard deviation

$$\sigma_{\hat{p}} = \sqrt{\frac{pq}{n}}.$$

Since $\sigma_{\hat{p}}$ determines the spread (or variation) of the sampling distribution of $\hat{p}$, the smaller the value of $\sigma_{\hat{p}}$, the closer the estimates will lie to the mean p. Examining the formula for $\sigma_{\hat{p}}$, you can see that it depends on the sample size n and the value of p. Therefore, the sample of 1091 adults, which is quite large, should provide a reasonably accurate estimate of the proportion p of all Americans who favor a particular sport.

How accurately will the estimates obtained from the poll estimate the unknown population proportions? The answer, of course, will depend upon the sample size, the actual value of the population proportion p that is being estimated, and upon the sampling procedure. It is unlikely that the Louis Harris poll was based on a simple random sample selected from among the group of all adult Americans. (Have you ever thought how difficult it would be to select a random sample from among the totality of all adult Americans?) It is likely, however, that the sampling procedure and estimator used in the poll produced estimates that were at least as good as those that might have been obtained by random sampling. Consequently, the confidence interval of Section 8.8 will give us some notion of the reliability of the estimates.

The formula for a large-sample 95 percent confidence interval for p is:

$$\hat{p} \pm 1.96 \sqrt{\frac{\hat{p}\hat{q}}{n}},$$

where we will use $\hat{p}$ as an approximation to the value of p that appears under the radical. Since the interval will be widest when p is near .5, we will examine the 95 percent confidence interval on the proportion p of all Americans who list football as their favorite sport. From Table 8.4, you can see that the percentage of adults in the sample who favored football is 36 percent and that, therefore, $\hat{p} = .36$. Then the 95 percent confidence interval for p is,

$$\hat{p} \pm 1.96 \sqrt{\frac{\hat{p}\hat{q}}{n}} = .36 \pm 1.96 \sqrt{\frac{(.36)(.64)}{1091}}$$

or

$$.36 \pm .028.$$

Thus, we estimate the proportion of all Americans who list football as their favorite sport to fall in the interval .332 to .388; that is, 32.2 percent to 38.8 percent.

We could construct confidence intervals for the proportions corresponding to any other sport in a similar manner. Since all of the other sample percentages (and proportions) are smaller than the sample percentage (and proportion) for football, the half-width of any other confidence interval would be smaller than the 2.89 percent shown above. This indicates that the sample percentages (or proportions) are accurate estimates of their corresponding population parameters.

The preceding statements assume that the poll elicited correct responses from the 1091 respondents and that the sample was actually selected according to the designed sampling procedure. Thus, the ultimate validity of the results of a sample survey depend upon the satisfaction of any assumptions that you have made regarding the nature of the sampled population and the sampling procedure.

8.12 Summary

Chapter 8 presented the basic concepts of statistical estimation and demonstrated how these concepts can be applied to the solution of some practical problems.

We learned that estimators are rules (usually formulas) that tell us how to calculate a parameter estimate based on sample data. Point estimators produce a single number (point) that estimates the value of a population parameter. The properties of a point estimator are contained in its sampling distribution. Thus, we prefer a point estimator that is unbiased (that is, the mean of its sampling distribution is equal to the estimated parameter) and that possesses a small, preferably a minimum, variance.

The reliability of a point estimator is usually measured by a 1.96 standard deviation (that is, standard deviation of the sampling distribution of the estimator) bound on the error of estimation. When we use sample data to calculate a particular estimate, the probability that the error of estimation will be less than the bound is approximately .95.

An interval estimator uses the sample data to calculate two points, a confidence interval, that we hope will enclose the estimated parameter. Since we will want to know the probability that the interval will enclose the parameter, we need to know the sampling distribution of the statistic used to calculate the interval.

Four estimators, a sample mean, proportion, and the differences between pairs of these statistics, were used to estimate their population equivalents and to demonstrate the concepts of estimation developed in this chapter. These estimators were chosen for a particular reason. Generally, they are "good" estimators for the respective population parameters in a wide variety of applications. Fortunately, they all possess, for large samples, sampling distributions that are approximately normal. This enabled us to use the exact same procedure to construct confidence intervals for the four population parameters, μ, p, $(\mu_1 - \mu_2)$, and $(p_1 - p_2)$. Thus, we were able to demonstrate an important role that the Central Limit Theorem (Chapter 7) plays in statistical inference.

Tips on Problem Solving

In solving the exercises in this chapter, you will be required to answer a practical question of interest to a business-person, a professional person, a scientist, or a layperson. To find the answer to the question, you will need to make an inference about one or more population parameters. Consequently, the first step in solving a problem is in deciding on the objective of the exercise. What parameters do you wish to make an inference about? Answering the following two questions will help you reach a decision.

1. What *type of data* is involved? This will help you decide the type of parameters about which you will wish to make inferences, binomial proportions (p's) or population means (μ's): Check to see if the data are of the yes-no (two-possibility) variety. If they are, the data are probably binomial and you will be interested in proportions. If not, the data probably represent measurements on one or more quantitative random variables and you will be interested in means. To aid you, look for key words such as "proportions," "fractions," etc., which indicate binomial data. Binomial data often (but not exclusively) evolve from a "sample survey."

2. Do I wish to make an inference about a *single parameter*, p or μ, or about the *difference between two parameters*, $(p_1 - p_2)$ or $(\mu_1 - \mu_2)$? This is an easy question to answer. Check on the number of samples involved. One sample implies an inference about a single parameter; two samples imply a comparison of two parameters. The answers to questions 1 and 2 identify the parameter.

After identifying the parameter(s) involved in the exercise, you must identify the exercise objective. It will be one of these two:

1. Choosing the sample size required to estimate a parameter with a specified bound on the error of estimation.
2. Estimating a parameter (or difference between two parameters).

The objective will be very clear if it is 1 because the question will ask for or direct you to find the "sample size." Objective 2 will be clear because the exercise will specifically direct you to estimate a parameter (or the difference between two parameters).

To summarize these tips, your thought process should follow the decision tree shown in Figure 8.11.

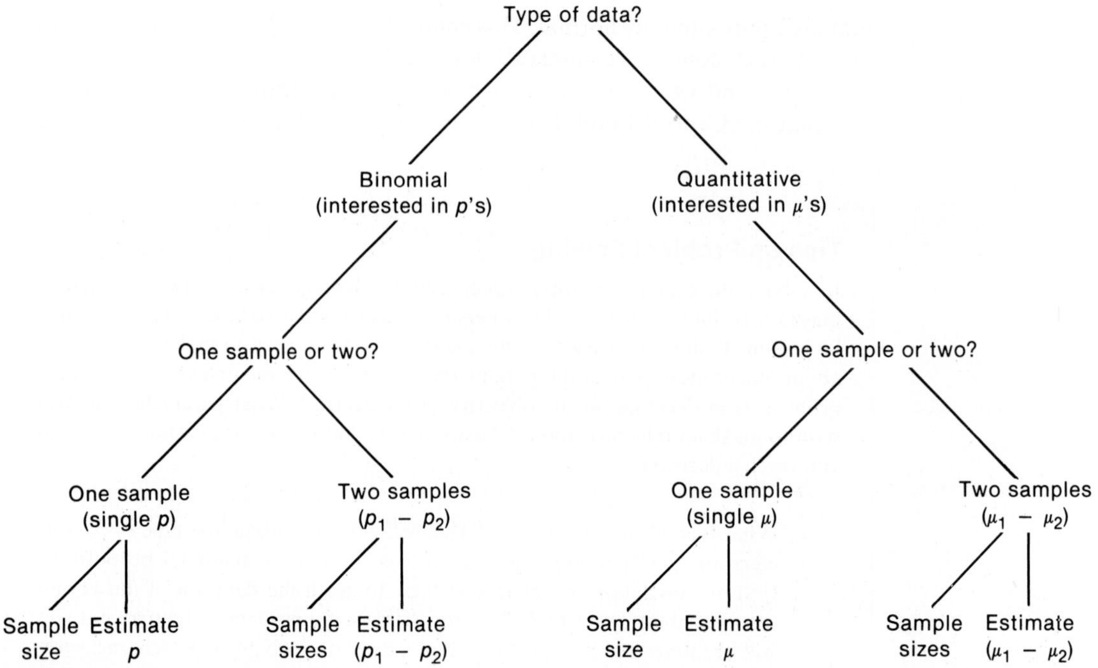

Figure 8.11

References   Dixon, W.J., and M.B. Brown, eds. *BMDP Biomedical Computer Programs, P Series.* Berkeley, Calif.: University of California, 1979.

Dixon, W.J., and F.J. Massey, Jr., *Introduction to Statistical Analysis*, 3rd ed. New York McGraw-Hill Book Company, 1969.

Freund, J.E., *Mathematical Statistics*, 2nd ed. Englewood Cliffs, N.J.: Prentice-Hall, Inc., 1971.

Helwig, J.P. and K.A. Council, eds. *SAS User's Guide.* Cary, N.C.: SAS Institute, Inc., P.O. Box 8000, 1979.

Hogg, R.V., and A.T. Craig, *Introduction to Mathematical Statistics*, 4th ed. New York: Macmillan Publishing Co., Inc., 1978.

Mendenhall, W., R. Scheaffer, and D. Wackerly, *Mathematical Statistics with Applications*, 2nd ed. Boston: Duxbury Press, 1981.

Neter, J., W. Wasserman, and G.A. Whitmore, *Applied Statistics.* Boston: Allyn and Bacon, Inc., 1978.

Nie, N., C.H. Hull, J.G. Jenkins, K. Steinbrenner, and D.H. Bent, *Statistical Package for the Social Sciences*, 2nd ed. New York: McGraw-Hill Book Company, 1979.

Scheaffer, R., W. Mendenhall, and L. Ott, *Elementary Survey Sampling*, 2nd ed. Boston: Duxbury Press, 1979.

Walpole, R.E., *Introduction to Statistics*, 2nd ed. New York: Macmillan Publishing Co., Inc., 1974.

Supplementary Exercises

[Starred (*) exercises are optional.]

8.51.   A random sample of $n = 64$ observations possessed a mean, $\bar{y} = 29.1$ and a standard deviation, $s = 3.9$. Give the point estimate for the population mean μ and give a bound on the error of estimation.

8.52.   Refer to Exercise 8.51 and find a 90 percent confidence interval for μ.

8.53.   Refer to Exercise 8.51. How many observations would be required if you wished to estimate μ with a bound on the error of estimation equal to .5 with probability equal to .95?

8.54.   Independent random samples of $n_1 = 50$ and $n_2 = 60$ observations were selected from populations 1 and 2, respectively. The sample sizes and computed sample statistics are shown below:

	Population	
	1	2
Sample Size	50	60
Sample Mean	100.4	96.2
Sample Standard Deviation	.8	1.3

Find a 90 percent confidence interval for the difference in population means and interpret the interval.

8.55.   Refer to Exercise 8.54. Suppose that you wish to estimate $(\mu_1 - \mu_2)$ correct to within .2 with probability equal to .95. If you plan to use equal sample sizes, how large should n_1 and n_2 be?

8.56.   A random sample of $n = 500$ observations from a binomial population produced $y = 240$ successes.

a.   Find a point estimate for p and place a bound on your error of estimation.

b.   Find a 90 percent confidence interval for p.

8.57.   Refer to Exercise 8.56. How large a sample would be required if you wished to estimate p correct to within .025 with probability equal to .90?

8.58.   Independent random samples of $n_1 = 40$ and $n_2 = 80$ observations were selected from binomial populations 1 and 2, respectively. The number of successes in the two samples were $y_1 = 17$ and $y_2 = 23$. Find a 90 percent confidence interval for the difference between the two binomial population proportions.

8.59.   Refer to Exercise 8.58. Suppose that you wish to estimate $(p_1 - p_2)$ correct to within .06 with probability equal to .90 and that you plan to use equal sample sizes, that is, $n_1 = n_2$. How large should n_1 and n_2 be?

8.60.   State the Central Limit Theorem. Of what value is the Central Limit Theorem in statistical inference?

8.61.   An experiment was conducted to estimate the effect of smoking on the blood pressure of a group of 25 college-age cigarette smokers. The difference for each participant was obtained by taking the difference in the blood-pressure readings at the time of graduation and again 5 years later. The sample mean increase, measured in millimeters of mercury, was $\bar{y} = 9.7$. The sample standard deviation was $s = 5.8$. Estimate the mean increase in blood pressure that one would expect for cigarette smokers over the time span indicated by the experiment. Place a bound on the error of estimation. Describe the population associated with the mean that you have estimated.

8.62. Using a confidence coefficient equal to .90, place a confidence interval on the mean increase in blood pressure for Exercise 8.61.

8.63. A random sample of 400 television tubes was tested and 40 tubes were found to be defective. With confidence coefficient equal to .90, estimate the interval within which the true fraction defective lies.

8.64. From each of two normal populations with identical means and with standard deviations of 6.40 and 7.20, independent random samples of 64 observations are drawn. Find the probability that the difference between the means of the samples exceeds .60 in absolute value.

8.65. If it is assumed that the heights of men are normally distributed with a standard deviation of 2.5 inches, how large a sample should be taken in order to be fairly sure (probability .95) that the sample mean does not differ from the true mean (population mean) by more than .50 in absolute value?

8.66. In measuring reaction time a psychologist estimates that the standard deviation is .05 second. How large a sample of measurements must he take in order to be 90 percent confident that the error of his estimate will not exceed .01 second?

8.67. In an experiment 320 out of 400 seeds germinate. Find a confidence interval for the true fraction of germinating seeds; use a confidence coefficient of .98.

8.68. To estimate the proportion of unemployed workers in Panama, an economist selected at random 400 persons from the working class. Of these, 25 were unemployed.

a. Estimate the true proportion of unemployed workers and place bounds on the error of estimation.

b. How many persons must be sampled to reduce the bound on error to .02?

8.69. A random sample of 36 cigarettes of a certain brand were tested for nicotine content. The sample gave a mean of 22 and a standard deviation of 4 milligrams. Find a 98 percent confidence interval for μ, the true mean nicotine content of the brand.

8.70. How large a sample is necessary to estimate a binomial parameter p to within .01 of its true value with probability .95? Assume the value of p is approximately .5.

8.71. A dean of men wishes to estimate the average cost of the freshman year at a particular college correct to within $500.00 with a probability of .95. If a random sample of freshmen is to be selected and requested to keep financial data, how many must be included in the sample? Assume that the dean knows only that the range of expenditure will vary from approximately $4800 to $13,000.

8.72. An experimenter wishes to estimate the difference, $(p_1 - p_2)$, for two binomial populations to within .1 of the true difference. Both p_1 and p_2 are expected to assume values between .3 and .7. If samples of equal size are to be selected from the populations to estimate $(p_1 - p_2)$, how large should they be?

8.73. An experimenter fed different rations to two groups of 100 chicks each. Assume that all factors other than rations are the same for both groups. A record of mortality for each group is as follows:

Chicks	Ration A	Ration B
n	100	100
Number died	13	6

Construct a 98 percent confidence interval for the true difference in mortality rates for the two rations.

8.74. It is desired to estimate the mean hourly yield for a process manufacturing an antibiotic. The process is observed for 100 hourly periods chosen at random, with the following results:

$$\bar{y} = 34 \text{ oz/hr}; \qquad s = 3.$$

Estimate the mean hourly yield for the process using a 95 percent confidence interval.

8.75. A quality-control engineer wants to estimate the fraction of defectives in a large lot of light bulbs. From previous experience, he feels that the actual fraction of defectives should be somewhere around .2. How large a sample should he take if he wants to estimate the true fraction to within .01 using a 95 percent confidence interval?

8.76. Samples of 400 radio tubes were selected from each of two production lines, A and B. The numbers of defectives in the samples were:

Line	Number of Defectives
A	40
B	80

Estimate the difference in the actual fractions of defectives for the two lines with a confidence coefficient of .90.

* 8.77. Refer to Exercise 8.63. Suppose that ten samples of $n = 400$ television tubes were tested and a confidence interval constructed for p for each of the ten samples. What is the probability that exactly one of the intervals will not enclose the true value of p? At least one?

8.78. The results of a Gallup Poll (*Orlando Sentinel Star*, December 2, 1976) showed that among eight physical afflictions, including cancer, heart disease, and blindness, 58 percent of those polled named cancer as the most feared affliction. If this estimate was based on a random sample of 2000 adults, how accurate is the estimate? Find a 90 percent confidence interval for the proportion of American adults who rate cancer as the most feared affliction.

9. Large-Sample Tests of Hypotheses

Chapter Objectives

General Objectives

The concept of a statistical test of an hypothesis was introduced in Chapter 6 and encountered again in examples and exercises in Chapter 7. In this chapter we will review the logic of this method of statistical inference and utilize the sampling distributions of Chapter 7 to present a large-sample statistical test procedure.

Specific Objectives

1. To review the logic of a statistical test of an hypothesis. *Sections 9.1, 9.2*

2. To present a large-sample statistical test that is based upon the sampling distributions of Chapter 7. *Sections 9.3*

3. To apply the large-sample statistical test, Section 9.3, to tests of hypotheses about population means and proportions. *Sections 9.4, 9.5, 9.6, 9.7*

4. To introduce the notion of an observed significance level (*p* value) as an alternative way of presenting statistical test results. *Sections 9.8*

5. To develop an understanding of the applications of statistical tests of hypotheses. *Sections 9.4, 9.5, 9.6, 9.7, 9.8, 9.9, 9.10*

Case Study

Counting the Big Bass

If you are an avid fisherman or fisherwoman, you know that there are many places in the United States that are noted for bass fishing. One of these is in central Florida. Lake Tohopekaliga, or simply Lake Toho, is one of the Kissimmee chain of lakes that is located west of Orlando, Florida. Approximately 12-miles long, 2- to 4-miles wide, and 5-feet deep, the lake covers over 21,000 acres. The fishing statistics concerning this lake, which are typical of the area, are impressive.

An article in *Bassmaster Magazine** states that netting and shocking surveys by Florida biologists in Lake Toho indicate that there are 455 pounds of fish per acre, about half of them gamefish. Largemouth bass make up 10 percent of the total gamefish population, and there is one bass over 10 pounds for every 5 acres of lake. These weights and numbers of fish per acre are undoubtedly based on samplings of randomly selected areas within the lake,

and the reported statistics are estimates of actual mean weights and mean numbers of fish per acre.

We learned in Chapter 8 how to estimate population means. Now we must learn how to test hypotheses about them. How can we decide whether these "fish stories" are really true? Is the mean weight of fish per acre as large as 455 pounds, and is the mean number of 10-pound bass per 5-acre plot as large as 1? In this chapter we will learn how to test hypotheses; that is, make decisions about the values of population means and proportions when the sample sizes are large. We will explain in Section 9.9 how this technique can be used to test hypotheses about the claimed mean number of 10-pound bass per 5-acre plot, and, in the process, will explain how to test hypotheses about population means when sample sizes are moderately small but the population variances are known.

* Gresham, Morris, "Go to Toho for Big Bass," *Bassmaster Magazine*, 11, no. 1 (January 1978).

9.1
Testing Hypotheses about Population Parameters

The answers to some practical questions require that we estimate the values of population parameters; others require that we test hypotheses about them. For example, if a pharmaceutical company were fermenting a vat of antibiotic, it would want to sample the potency of specimens of the antibiotic and use them to estimate the mean potency μ of the antibiotic in the vat. In contrast, suppose that there is no concern that the potency of the antibiotic will be too high; the company is only concerned that the mean potency exceed some government-specified minimum in order that the vat be declared acceptable for sale. In this case, the company would not wish to estimate the mean potency. Rather, it would want to show that the sample of potency measurements provides sufficient evidence to indicate that the mean potency of the antibiotic in the vat exceeds the minimum specified by the government. Thus, it would wish to test the null hypothesis,

$$H_0: \mu = \text{minimum allowable potency,}$$

against the alternative hypothesis,

$$H_a: \mu > \text{minimum allowable potency.}$$

(Throughout our discussion, a null hypothesis is denoted by the symbol H_0, and the alternative hypothesis by the symbol H_a.)

Thus, the pharmaceutical company would hope to show support for the alternative hypothesis (that the vat of antibiotic meets acceptable standards) by showing that the data provide sufficient evidence to reject the null hypothesis.

You will recall (Chapters 6 and 7) that we needed to know the sampling (probability) distribution of the test statistic in order to determine values of the test statistic that favored rejection of the null hypothesis and acceptance of the alternative hypothesis. These values of the test statistic were called the *rejection region for the test*.

In this chapter we will review the basic concepts of a test of an hypothesis and demonstrate the concepts with some very useful large-sample statistical tests of the values of a population mean, a population proportion, the difference between a pair of population means, and the difference between two proportions. We will employ the four point estimators discussed in Chapter 8, $\bar{y}$, $(\bar{y}_1 - \bar{y}_2)$, $\hat{p}$, and $(\hat{p}_1 - \hat{p}_2)$, as test statistics and, in doing so, will obtain a unity in these four statistical tests. All four test statistics will, for large samples, possess sampling distributions that are approximately normal. Then, as you will subsequently see, the tests possess a similarity that you cannot ignore!

9.2
A Statistical Test of an Hypothesis

The basic reasoning employed in a statistical test of an hypothesis was outlined in Section 6.6 in connection with the test of the effectiveness of a cold vaccine. In this section we will summarize the basic points involved and refer you to Section 6.6 for an intuitive presentation of the subject.

The objective of a statistical test is to test an hypothesis concerning the values of one or more population parameters. We will generally have a theory, a research hypothesis, about the parameter(s) that we wish to support. For example, we might wish to show that the mean life μ_1 of a new type of automobile tire is greater than the mean life μ_2 of a competitor's tire. Support for this research hypothesis, called the *alternative hypothesis* by statisticians, is obtained by showing (using the sample data as evidence) that the converse of the research hypothesis, the *null hypothesis*, is false. Thus support for one theory is obtained by showing lack of support for its converse, in a sense a "proof" by contradiction. For the tire example, the converse of the research (alternative) hypothesis $\mu_1 > \mu_2$ is $\mu_1 < \mu_2$. If we can show that the sample data support rejection of the null hypothesis, $\mu_1 = \mu_2$ (the means are equal), in favor of the alternative hypothesis, $\mu_1 > \mu_2$, we have achieved our research objective.* Although it is common to speak of testing a null hypothesis, keep in mind that the research objective is always to show support for the alternative hypothesis, if support is warranted.

The decision to reject or accept the null hypothesis is based upon information contained in a sample drawn from the population of interest. The sample values are used to compute a single number, corresponding to a point on a line, which operates as a decision maker and which is called the **test statistic.** The entire set of values that the test statistic may assume is divided into two sets or regions, one corresponding to the **rejection region** and the other to the acceptance region (see Figure 9.1). If the test statistic computed from a particular sample assumes a value in the rejection region, the null hypothesis is rejected and you decide in favor of the alternative hypothesis. If the test statistic falls in the acceptance region, the null hypothesis is accepted, that is, you decide in favor of the null hypothesis. Thus, a statistical test involves four elements:

Elements of a Statistical Test

1. Null hypothesis.
2. Alternative hypothesis.
3. Test statistic.
4. Rejection region.

Note that the specification of these four elements defines a particular test and that changing one or more creates a new test.

Figure 9.1 Possible values for the test statistic, y

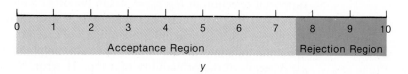

Acceptance Region   Rejection Region

y

The decision that you make when you test an hypothesis is subject to two types of errors, errors that are prevalent in any two-choice decision problem. We may reject the null hypothesis when, in fact, it is true, or we may accept H_0 when it is false and some *alternative hypothesis* is true. These errors are called the type I and type II errors, respectively, for the statistical test. The two states for the null hypothesis (that is, true or false) along with the two decisions that the experimenter may make are indicated in the two-way table, Table 9.1.

---

* Note that if the test rejects the null hypothesis, $\mu_1 = \mu_2$, in favor of the alternative, $\mu_1 > \mu_2$, then it should (our test will) be even more likely to reject a null hypothesis for values of μ_1 and μ_2 such that $\mu_1 < \mu_2$ (since this is even more contradictory to the proposition, $\mu_1 > \mu_2$, than the proposition that $\mu_1 = \mu_2$). For this reason, we usually say that we are testing the null hypothesis, $\mu_1 = \mu_2$.

Table 9.1

Decision Table

	Null Hypothesis	
Decision	True	False
Reject H_0	Type I error	Correct decision
Accept H_0	Correct decision	Type II error

The occurrences of the type I and type II errors are indicated in the appropriate cells.

The goodness of a statistical test of an hypothesis is measured by the probabilities of making a type I or a type II error, denoted by the symbols α and β, respectively. These probabilities, calculated for the elementary statistical tests presented in the exercises in Chapter 6, illustrate the basic relationship among α, β, and the sample size, n. Since α is the probability that the test statistic will fall in the rejection region, assuming H_0 to be true, an increase in the size of the rejection region will increase α and, at the same time, decrease β for a fixed sample size. Reducing the size of the rejection region will decrease α and increase β. If the sample size, n, is increased, more information will be available upon which to base the decision and both α and β will decrease.

The probability of making a type II error, β, varies depending upon the true value of the population parameter. For instance, suppose that we wish to test the null hypothesis that the binomial parameter, p, is equal to $p_0 = .4$. (We shall use a subscript 0 to indicate the parameter value specified in the null hypothesis, H_0.) Furthermore, suppose that H_0 is false and that p is really equal to an alternative value, say p_α. What will be more easily detected, a $p_a = .4001$ or a $p_a = 1.0$? Certainly, if p is really equal to 1.0, every single trial will result in a success and the sample results will produce strong evidence to support a rejection of $H_0 : p_0 = .4$. On the other hand, $p_a = .4001$ lies so close to $p_0 = .4$ that it would be extremely difficult to detect without a very large sample. In other words, the probability β of accepting H_0 will vary depending upon the difference between the true value of p and the hypothesized value, p_0. A graph of the probability of a type II error, β, as a function of the true value of the parameter is called the **operating characteristic curve** for the statistical test. Note that the operating characteristic curves for the lot acceptance sampling plans, Chapter 6, were really graphs expressing β as a function of p.

Since the rejection region is specified and remains constant for a given test, α will also remain constant and, as in lot acceptance sampling, the operating characteristic curve will describe the characteristics of the statistical test. An increase in the sample size, n, will decrease β and reduce its value for all alternative values of the parameter tested. Thus there is an operating characteristic curve corresponding to each sample size. This property of the operating characteristic curve was illustrated in the exercises in Chapter 6.

Ideally, experimenters will have in mind some values, α and β, that measure the risks of the respective errors they are willing to tolerate. They will also have in mind some deviation from the hypothesized value of the parameter, which they consider of *practical* importance and which they wish to detect. The rejection region for the test will then be located in accordance with the specified value of α. Finally, they will choose the sample size necessary to achieve an acceptable value of β for the specified deviation that they wish to detect. This could be done by consulting the operating characteristic curves, corresponding to various sample sizes, for the chosen test.

In conclusion, keep in mind that "accepting" a particular hypothesis means deciding in its favor. Regardless of the outcome of a test, you are never *certain* that the hypothesis you "accept" is true. There is always a risk of being wrong (measured by α and β). Consequently, you never "accept" H_0 if β is unknown or its value is unacceptable to you. When this situation occurs, you should withhold judgment and collect more data.

9.3
A Large-Sample Statistical Test

We stated in Section 9.1 that large-sample tests of hypotheses concerning population means and proportions are very similar. In fact, when viewed in a certain light, they are equivalent. Seeing this equivalency will make it easier to understand the tests and examples described in the following sections.

The key to the equivalency lies in the fact that all of the point estimators discussed in Chapter 8 are unbiased and possess sampling distributions that, for large samples, are approximately normal. Therefore, we can use the point estimators as test statistics to test hypotheses about the respective parameters.

To illustrate, suppose that we use the symbol θ to denote one of the four parameters, μ, $(\mu_1 - \mu_2)$, p, or $(p_1 - p_2)$, that we estimated in Chapter 8, and let $\hat{\theta}$ denote the corresponding unbiased point estimator, that is, $\hat{\mu} = \bar{y}$, etc. Then, when the null hypothesis is true (that is, θ equals some hypothesized value, say θ_0) the sampling distribution of $\hat{\theta}$ will be (for large samples) approximately normal with mean θ_0, as shown in Figure 9.2.

Suppose that, from a *practical* point of view, we are primarily concerned with the rejection of H_0 when θ is greater than θ_0. Then the alternative hypothesis would be $H_a: \theta > \theta_0$ and we would reject the null hypothesis when $\hat{\theta}$ is

Figure 9.2 Distribution of $\hat{\theta}$ when H_0 is true

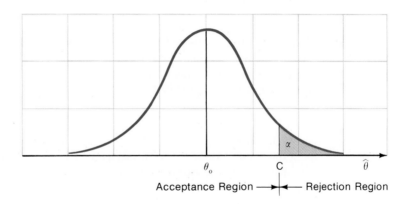

Acceptance Region $\longrightarrow$ $\mid$ $\longleftarrow$ Rejection Region

too large. Letting $\sigma_{\hat{\theta}}$ represent the standard deviation of the sampling distribution of $\hat{\theta}$, "too large" means too many standard deviations, $\sigma_{\hat{\theta}}$, away from θ_0. The rejection region for the test is shown in Figure 9.2.

Definition

The value, C, of the test statistic that separates the rejection and acceptance regions is called the critical value of the test statistic.

The probability of rejecting the null hypothesis, assuming it to be true, would equal the area under the normal curve lying above the rejection region. Thus if we desire $\alpha = .05$, we would reject when $\hat{\theta}$ is more than $1.645\sigma_{\hat{\theta}}$ to the right of θ_0. A test rejecting in one tail of the distribution of the test statistic is called a one-tailed statistical test.

Definition

A one-tailed statistical test is one that locates the rejection region in only one tail of the sampling distribution of the test statistic. To detect $\theta > \theta_0$ place the rejection region in the upper tail of the distribution of $\hat{\theta}$. To detect $\theta < \theta_0$ place the rejection region in the lower tail of the distribution of $\hat{\theta}$.

If we wish to detect departures *either* greater than or less than θ_0, the alternative hypothesis would be

$$H_a : \theta \neq \theta_0;$$

that is,

$$\theta > \theta_0$$

or

$$\theta < \theta_0.$$

The probability of a type I error, α, would be equally divided between the two tails of the normal distribution and we would reject H_0 for values of $\hat{\theta}$ greater than some critical value, $\theta_0 + C$, or less than $\theta_0 - C$ (see Figure 9.3). This is called a two-tailed statistical test.

Definition

A two-tailed statistical test is one that locates the rejection region in both tails of the sampling distribution of the test statistic. Two-tailed tests are used to detect either $\theta > \theta_0$ or $\theta < \theta_0$.

Figure 9.3 The rejection region for a two-tailed test

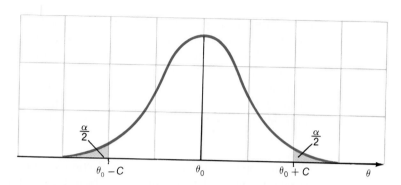

The calculation of β for the one-tailed statistical test described above can be facilitated by considering Figure 9.4. When H_0 is false and $\theta = \theta_a$, the test statistic, $\hat{\theta}$, will be normally distributed about a mean θ_a, rather than θ_0. The distribution of $\hat{\theta}$, assuming $\theta = \theta_a$, is shown by the black curve. The hypothesized distribution of $\hat{\theta}$, shown by the colored curve, locates the rejection region and the critical value of $\hat{\theta}$, C. Since β is the probability of accepting H_0, given $\theta = \theta_a$, β would equal the area under the black curve located above the acceptance region. This area, which is shaded, could be easily calculated using the methods described in Chapter 7.

Figure 9.4 Distribution of $\hat{\theta}$ when H_0 is false and $\theta = \theta_a$

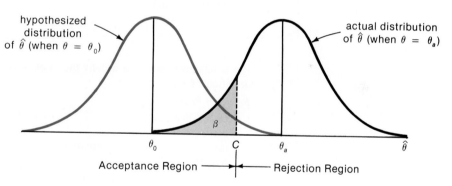

The mechanics of testing are simplified by using z as a test statistic, as noted in Example 7.11.

Large-Sample Test Statistic

$$z = \frac{\hat{\theta} - \theta_0}{\sigma_{\hat{\theta}}}.$$

Note that z is simply the deviation of a normally distributed random variable, $\hat{\theta}$, from θ_0, expressed in units of $\sigma_{\hat{\theta}}$. Thus for a two-tailed test with $\alpha = .05$ we would reject H_0 when $z > 1.96$ or $z < -1.96$.

As we have previously stated, the method of inference used in a given situation will often depend upon the preference of the experimenter. Some people wish to express an inference as an estimate; others prefer to test an hypothesis concerning the parameter of interest. The following section will demonstrate the use of the z test in testing an hypothesis concerning a population mean and, at the same time, will illustrate the close relationship between the statistical test and the large-sample confidence intervals discussed in Chapter 8.

9.4
Testing an Hypothesis about a Population Mean

We can apply the large-sample test, Section 9.3, to test an hypothesis about a population mean. The parameter θ to be tested is μ, the point estimator $\hat{\theta}$ is the sample mean $\bar{y}$, and the standard deviation $\sigma_{\hat{\theta}}$ of the sampling distribution of $\bar{y}$ is $\sigma/\sqrt{n}$. A summary of the test is shown below:

Large-Sample Statistical Test for μ

1. *Null Hypothesis:* $H_0: \mu = \mu_0$

2. *Alternative Hypothesis:*

   One-Tailed Test   *Two-Tailed Test*

   $H_a: \mu > \mu_0$ $H_a: \mu \neq \mu_0$
   (or, $H_a: \mu < \mu_0$)

3. *Test Statistic:* $z = \dfrac{\bar{y} - \mu_0}{\sigma_{\bar{y}}} = \dfrac{\bar{y} - \mu_0}{\sigma/\sqrt{n}}$

   If σ is unknown (which is usually the case), substitute the sample standard deviation s for σ.

4. *Rejection Region:*

   One-Tailed Test   *Two-Tailed Test*

   $z > z_\alpha$ $z > z_{\alpha/2}$ or $z < -z_{\alpha/2}$
   (or, $z < -z_\alpha$ when the alternative
   hypothesis is $H_a: \mu < \mu_0$)

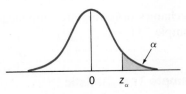
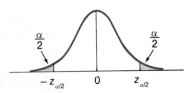

Assumptions: The n observations in the sample were randomly selected from the population and n is large, say $n \geq 30$.

Example 9.1

Refer to Example 8.1, Section 8.4. Test the hypothesis that the average daily yield of the chemical is $\mu = 880$ tons per day against the alternative that μ is either greater or less than

880 tons per day. The sample (Example 8.1), based upon $n = 50$ measurements, yielded $\bar{y} = 871$ and $s = 21$ tons.

Solution

The point estimate for μ is $\bar{y}$. Therefore, the test statistic is

$$z = \frac{\bar{y} - \mu_0}{\sigma_{\bar{y}}} = \frac{\bar{y} - \mu_0}{\sigma/\sqrt{n}}.$$

Using s to approximate σ, we obtain

$$z = \frac{871 - 880}{21/\sqrt{50}} = -3.03.$$

For $\alpha = .05$, the rejection region is $z > 1.96$ or $z < -1.96$. (See Figure 9.5.) Since the calculated value of z falls in the rejection region, we reject the hypothesis that $\mu = 880$ tons. (In fact, it appears that the mean yield is less than 880 tons per day.) The probability of rejecting H_0, assuming it to be true, is only $\alpha = .05$. Hence we are reasonably confident that our decision is correct.

Figure 9.5 Location of the rejection region in Example 9.1

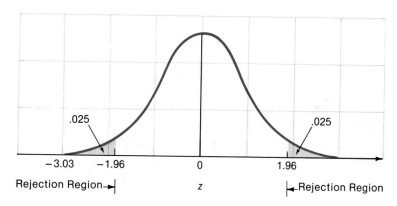

The statistical test based upon a normally distributed test statistic, with given α, and the $(1 - \alpha)$ confidence interval, Section 8.5, are clearly related. The interval $\bar{y} \pm 1.96\sigma/\sqrt{n}$, or approximately 871 ± 5.82, is constructed such that, in repeated sampling, $(1 - \alpha)$ of the intervals will enclose μ. Noting that $\mu = 880$ does not fall in the interval, we would be inclined to reject $\mu = 880$ as a likely value and conclude that the mean daily yield was, indeed, less.

The following example will demonstrate the calculation of β for the statistical test, Example 9.1.

Example 9.2

Referring to Example 9.1, calculate the probability, β, of accepting H_0 when μ is actually equal to 870 tons.

Solution

The acceptance region for the test, Example 9.1, is located in the interval $\mu_0 \pm 1.96\sigma_{\bar{y}}$. Substituting numerical values, we obtain

$$880 \pm 1.96(21/\sqrt{50})$$

or

874.18 to 885.82.

The probability of accepting H_0, given $\mu = 870$, is equal to the area under the sampling distribution for the test statistic, $\bar{y}$, above the interval 874.18 to 885.82. Since $\bar{y}$ will be normally distributed with mean equal to 870 and $\sigma_{\bar{y}} = 21/\sqrt{50} = 2.97$, β is equal to the area under the normal curve (Figure 9.6) located to the right of 874.18 (because the area to the right of 885.82 is negligible). Calculating the z value corresponding to 874.18, we obtain

$$z = \frac{\bar{y} - \mu}{\sigma/\sqrt{n}} = \frac{874.18 - 870}{21/\sqrt{50}} = 1.41.$$

We see from Table 3, the Appendix, that the area between $z = 0$ and $z = 1.41$ is .4207. Therefore,

$$\beta = .5 - .4207 = .0793.$$

Thus the probability of accepting H_0, given that μ is really equal to 870, is .0793 or, approximately, 8 chances in 100.

Figure 9.6 Calculating β in Example 9.2

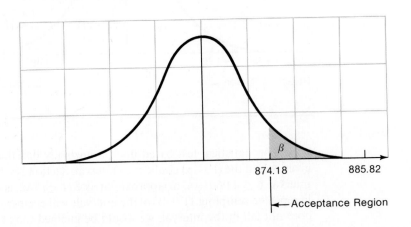

Acceptance Region

Exercises

9.1. A random sample of $n = 40$ observations from a population produced a mean, $\bar{y} = 30.9$, and standard deviation equal to 2.4. Suppose that you wish to show that the population mean μ exceeds 30.

a. Give the alternative hypothesis for the test.

b. Give the null hypothesis for the test.

c. If you wish your probability of (erroneously) deciding that $\mu > 30$, when in fact $\mu = 30$, to equal .05, what is the value of α for the test?

d. Glance at the data and use your intuition to decide whether the sample mean, $\bar{y} = 30.9$, implies that $\mu > 30$. (Do not conduct the statistical test.) Now test the null hypothesis. Do the data provide sufficient evidence to indicate that $\mu > 30$? Test using $\alpha = .05$.

9.2. Refer to Exercise 9.1. Suppose that you wish to show that the sample data support the hypothesis that the mean of the population is less than 30.0. Give the null and alternative hypotheses for this test. Would this test be a one- or a two-tailed test? Explain.

9.3. Refer to Exercises 9.1 and 9.2. Suppose that you wish to detect a value of μ that differs from 30; that is, a value of μ either greater or less than 30. State the null and alternative hypotheses for the test. Would the alternative hypothesis imply a one- or a two-tailed test?

9.4. High airline occupancy rates on scheduled flights are essential to profitability. Suppose that a scheduled flight must average at least 60 percent occupancy in order to be profitable and that an examination of the occupancy rates for 120 10:00 A.M. flights from Atlanta to Dallas showed a mean occupancy rate per flight of 58 percent and a standard deviation of 11 percent.

a. If μ is the mean occupancy per flight and if the company wishes to determine whether or not this scheduled flight is unprofitable, give the alternative and the null hypotheses for the test.

b. Does the alternative hypothesis, part (a), imply a one- or a two-tailed test? Explain.

c. Do the occupancy data for the 120 flights suggest that this scheduled flight is unprofitable? Test using $\alpha = .10$.

9.5. A drug manufacturer claimed that the mean potency of one of its antibiotics was 80 percent. A random sample of $n = 100$ capsules was tested and produced a sample mean of $\bar{y} = 79.7$ percent, with a standard deviation $s = .8$ percent. Do the data present sufficient evidence to refute the manufacturer's claim? Let $\alpha = .05$.

a. State the null hypothesis to be tested.

b. State the alternative hypothesis.

c. Conduct a statistical test of the null hypothesis and state your conclusion.

9.6. The pH factor is a measure of the acidity or alkalinity of water. A reading of 7.0 is neutral; values in excess of 7.0 indicate alkalinity; those below 7.0 imply acidity. Loren Hill, in *Bassmaster Magazine* (September/October 1980), states that the best chance of catching bass occurs when the pH of the water is in the range of 7.5 to 7.9. Suppose that you suspect that acid rain is lowering the pH of your favorite fishing spot and you wish to determine whether the pH is less than 7.5.

a. State the alternative and null hypotheses that you would choose for a statistical test.

b. Would the alternative hypothesis, part (a), imply a one- or a two-tailed test? Explain.

c. Suppose that a random sample of 30 water specimens gave pH readings with $\bar{y} = 7.3$ and $s = .2$. Just glancing at the data, do you think that the difference, $\bar{y} - 7.5 = -.2$ is large enough to indicate that the mean pH of the water samples is less than 7.5? (Do not conduct the test.)

d. Now conduct a statistical test of the null hypothesis, part (a), and state your conclusions. Test using $\alpha = .05$. Compare your statistically based decision with your intuitive decision, part (c).

9.7. An article in *U.S. News & World Report* (September 28, 1981) states that approximately 21.3 million workers, more than one-fifth of the United States work force, work unorthodox schedules. More than 9.3 million are involved in flexitime (the worker schedules his or her own work hours) or compressed work weeks. A company that was contemplating the installation of a flexitime schedule estimated that it needed a minimum mean of 7 hours per day per assembly worker in order to operate effectively. Each of a random sample of 80 of the company's assemblers was asked to submit a tentative flexitime schedule. If the mean of the number of hours per day for Monday was 6.7 hours and the standard deviation was 2.7 hours, do the data provide sufficient evidence to indicate that the mean number of hours worked per day on Mondays, for all of the company's assemblers, will be less than 7 hours? Test using $\alpha = .05$.

9.5
Testing an Hypothesis about the Difference between Two Population Means

A test of an hypothesis about the difference $(\mu_1 - \mu_2)$ between two population means, μ_1 and μ_2, based on independent random samples of n_1 and n_2 observations, respectively, is summarized below:

Large-Sample Statistical Test for $(\mu_1 - \mu_2)$

1. *Null Hypothesis:* $H_0:(\mu_1 - \mu_2) = D_0$, where D_0 is some specified difference that you wish to test. For many tests, you will wish to hypothesize that there is no difference between μ_1 and μ_2, i.e., $D_0 = 0$.

2. *Alternative Hypothesis:*

   One-Tailed Test
   $H_a:(\mu_1 - \mu_2) > D_0$
   (or, $H_a:(\mu_1 - \mu_2) < D_0$)

   Two-Tailed Test
   $H_a:(\mu_1 - \mu_2) \neq D_0$

3. *Test Statistic:* $z = \dfrac{(\bar{y}_1 - \bar{y}_2) - D_0}{\sigma_{(\bar{y}_1 - \bar{y}_2)}} = \dfrac{(\bar{y}_1 - \bar{y}_2) - D_0}{\sqrt{\dfrac{\sigma_1^2}{n_1} + \dfrac{\sigma_2^2}{n_2}}}$

   If σ_1^2 and σ_2^2 are unknown (which is usually the case), substitute the sample variances, s_1^2 and s_2^2, for σ_1^2 and σ_2^2, respectively.

4. *Rejection Region:*

   One-Tailed Test
   $z > z_\alpha$
   (or $z < -z_\alpha$ when the alternative hypothesis is $H_a:(\mu_1 - \mu_2) < D_0$)

   Two-Tailed Test
   $z > z_{\alpha/2}$ or $z < -z_{\alpha/2}$

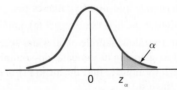

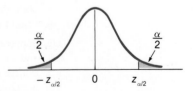

Assumptions: The samples were randomly and independently selected from the two populations and $n_1 \geq 30$ and $n_2 \geq 30$.

Example 9.3

A university investigation, conducted to determine whether car ownership affects academic achievement, was based upon two random samples of 100 male students, each drawn from student body. The grade-point average for the $n_1 = 100$ non–car owners possessed an average and variance equal to $\bar{y}_1 = 2.70$ and $s_1^2 = .36$ as opposed to a $\bar{y}_2 = 2.54$ and $s_2^2 = .40$ for the $n_2 = 100$ car owners. Do the data present sufficient evidence to indicate a difference in the mean achievement between car owners and non–car owners? Test using $\alpha = .10$.

Solution

Since we wish to detect a difference, if it exists, between the mean academic achievement for non–car owners, μ_1, and car owners, μ_2, we will wish to test the null hypothesis that there is no difference between the means, against the alternative hypothesis that $(\mu_1 - \mu_2) \neq 0$; that is,

$$H_0 : \mu_1 - \mu_2 = D_0 = 0 \quad \text{and} \quad H_a : \mu_1 - \mu_2 \neq 0.$$

Substituting into the formula for the test statistic, we obtain

$$z = \frac{(\bar{y}_1 - \bar{y}_2) - D_0}{\sqrt{\dfrac{\sigma_1^2}{n_1} + \dfrac{\sigma_2^2}{n_2}}} = \frac{2.70 - 2.54}{\sqrt{\dfrac{.36}{100} + \dfrac{.40}{100}}} = 1.84.$$

Using a two-tailed test with $\alpha = .10$, we will place $\alpha/2 = .05$ in each tail of the z distribution and reject H_0 if $z > 1.645$ or $z < -1.645$. (See Figure 9.7.) Since $z = 1.84$ exceeds $z_{\alpha/2} = 1.645$, it falls in the rejection region. Therefore, we would reject the null hypothesis that there is no difference in the average academic achievement of car owners versus non–car owners. The chance of rejecting H_0, assuming H_0 true, is only $\alpha = .10$, and hence we would be inclined to think that we have made a reasonably good decision.

Figure 9.7 Location of the rejection region in Example 9.3

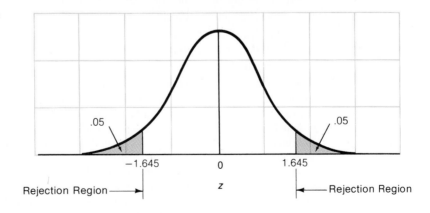

Exercises

9.8. Independent random samples of $n_1 = 100$ and $n_2 = 100$ were selected from populations 1 and 2, respectively. The population parameters and the sample means and variances are shown below:

	Population	
	1	2
Population Mean	μ_1	μ_2
Population Variance	σ_1^2	σ_2^2
Sample Size	100	100
Sample Mean	18.7	16.4
Sample Variance	35.1	58.2

a. If your research objective is to show that μ_1 is larger than μ_2, state the alternative and the null hypotheses that you would choose for a statistical test.

b. Is the test, part (a), a one- or a two-tailed test?

c. Give the test statistic that you would use for the test, parts (a) and (b), and give the rejection region for $\alpha = .10$.

d. Look at the data. Based on your intuition, do you think that the data provide sufficient evidence to indicate that μ_1 is larger than μ_2? (We will employ a statistical test to reach this decision in part (e).)

e. Conduct the test and draw your conclusions. Do the data present sufficient evidence to indicate that $\mu_1 > \mu_2$?

9.9.   Refer to Exercise 9.8. Explain the practical conditions that would motivate you to run a one-tailed z test with the rejection region in the lower tail of the z distribution. Give the alternative and null hypotheses for this test.

9.10.   Refer to Exercise 9.8.

a. Explain the practical conditions that would motivate you to run a two-tailed z test.

b. Give the alternative and null hypotheses.

c. Use the data, Exercise 9.8, to conduct the test. Do the data provide sufficient evidence to reject H_0 and accept H_a? Test using $\alpha = .05$.

d. What practical conclusions can be drawn from the test, part (c)?

9.11.   Suppose that you wish to detect a difference between μ_1 and μ_2 (either $\mu_1 > \mu_2$ or $\mu_1 < \mu_2$) and that, instead of running a two-tailed test using $\alpha = .10$, you employ the following test procedure. You wait until you have collected the sample data and have calculated $\bar{y}_1$ and $\bar{y}_2$. If $\bar{y}_1$ is larger than $\bar{y}_2$, you choose the alternative hypothesis, $H_a: \mu_1 > \mu_2$, and run a one-tailed test placing $\alpha_1 = .10$ in the upper tail of the z distribution. If, on the other hand, $\bar{y}_2$ is larger than $\bar{y}_1$, you reverse the procedure and run a one-tailed test, placing $\alpha_2 = .10$ in the lower tail of the z distribution. If you use this procedure and if μ_1 actually equals μ_2, what is the probability, α, that you will conclude that μ_1 is not equal to μ_2 (i.e., what is the probability, α, that you will incorrectly reject H_0 when H_0 is true)? This exercise demonstrates why statistical tests should be formulated *prior* to observing the data.

9.12.   An experiment was planned to compare the mean time (in days) required to recover from a common cold for persons given a daily dose of 4 grams of vitamin C versus those who were not given a vitamin supplement. Suppose that 35 adults were randomly selected for each treatment category and that the mean recovery times and standard deviations for the two groups were as follows:

	Treatment	
	No Vitamin Supplement	4 mg Vitamin C
Sample Size	35	35
Sample Mean	6.9	5.8
Sample Standard Deviation	2.9	1.2
Population Mean	μ_1	μ_2

a. Suppose your research objective is to show that the use of vitamin C reduces the mean time required to recover from a common cold and its complications. Give the null and alternative hypotheses for the test.

b. Does your alternative hypothesis, part (a), imply a one- or a two-tailed test? Explain.

c. Examine the data and use your intuition to decide whether the data provide sufficient evidence to indicate that vitamin C reduces the mean time to recover from a common cold. (Do not base your answer on a statistical test.)

d. Now, conduct the statistical test of the null hypothesis, part (a), and state your conclusion. Test using $\alpha = .05$. Compare your answer to your intuitive conclusion, part (c).

9.13. Studies of the habits of the white-tailed deer indicate that they live and feed within very limited ranges (150 to 205 acres).* To determine whether there was a difference in the ranges of deer located in two different geographical areas, 40 deer were caught, tagged, and fitted with small radio transmitters. Several months later, the deer were tracked and identified and the distance y from the release point was recorded. The mean and standard deviation of the distances from the release point were as follows:

	Location	
	1	2
Sample Size	40	40
Sample Mean	2980 ft	3205 ft
Sample Standard Deviation	1140 ft	963 ft
Population Mean	μ_1	μ_2

a. If you have no preconceived reason for believing one population mean to be larger than another, what would you choose for your alternative hypothesis? Your null hypothesis?

b. Would your alternative hypothesis, part (a), imply a one- or a two-tailed test? Explain.

c. Do the data provide sufficient evidence to indicate that the mean distances differ for the two geographical locations? Test using $\alpha = .10$.

9.14. *Psychology Today* (April 1981) reports on a study by environmental psychologist Karen Frank and two colleagues, that was conducted to determine whether it is more difficult to make friends in a large city than in a small town. Two groups of graduate students were

* Dickey, Charles, "A Strategy for Big Bucks," *Field and Stream*, October 1980.

used for the study, 45 new arrivals at a private university in New York City and 47 at a prestigious university located in a town of 31,000 in an upstate rural area. The number of friends (more than just acquaintances) made by each student was recorded for each student at the end of two months and then again at the end of seven months. The seven-month sample means are shown below.*

	Location	
	New York City	**Upstate Rural Small Town**
Sample Size	45	47
Sample Mean (7 months)	5.1	5.3
Population Mean (7 months)	μ_1	μ_2

a. If your theory is that it is more difficult to make friends in a city than in a small town, state the alternative hypothesis that you would use for a statistical test. State your null hypothesis.

b. Suppose that $\sigma_1 = 2.2$ and $\sigma_2 = 2.3$ for the populations of "numbers of friends" made by new graduate students after seven months at a new location. Do the data provide sufficient evidence to indicate that the mean number of friends acquired after seven months differs between the two locations? Test using $\alpha = .05$.

9.15.   Do soldiers who re-enlist enjoy greater job satisfaction than those who do not re-enlist or do they re-enlist because of other factors, such as re-enlistment bonuses, a lack of opportunity in civilian life, etc.? A study by Chisholm, Gauntner, and Munzenrider[†] of the attitudes of a sample of soldiers stationed in the United States in 1976 addresses this and other questions relating to army life. Each soldier included in the study completed a job satisfaction questionnaire and received a job satisfaction score and, in addition, indicated his/her re-enlistment intentions. The sample sizes, means, and standard deviations for the two groups are shown below:

	Intention	
	To Re-enlist	**Not to Re-enlist**
Sample Size	30	297
Sample Mean	136.9	108.8
Sample Standard Deviation	29.8	31.3
Population Mean	μ_1	μ_2

* Reprinted from *Psychology Today Magazine.* Copyright © 1981 Ziff-Davis Publishing Company.
† Chisholm, R.F., D.E. Gauntner and R.F. Munzenrider, "Pre-enlistment Expections/Perceptions of Army Life, Satisfaction and Re-enlistment of Volunteers," *Journal of Political and Military Sociology*, 8 (1980), pp. 31–42.

a. Suppose that you had a preconceived theory that the job satisfaction mean score μ_1 for soldiers intending to re-enlist is higher than the mean score μ_2 for soldiers who do not intend to re-enlist. Based on this theory, what alternative hypothesis would you choose for a statistical test of the null hypothesis, $H_0: \mu_1 = \mu_2$?

b. Does your alternative hypothesis, part (a), imply a one-tailed or a two-tailed statistical test? Locate the rejection region for the test.

c. Conduct the test and state your conclusions.

9.6
Testing an Hypothesis about a Population Proportion

The large-sample statistical test for a population proportion p can be summarized as follows:

Large-Sample Test for a Population Proportion p

1. *Null Hypothesis:* $H_0: p = p_0$

2. *Alternative Hypothesis:*

   One-Tailed Test   Two-Tailed Test

   $H_a: p > p_0$ $H_a: p \neq p_0$

   (or, $H_a: p < p_0$)

3. *Test Statistic:* $z = \dfrac{\hat{p} - p_0}{\sigma_{\hat{p}}} = \dfrac{\hat{p} - p_0}{\sqrt{\dfrac{p_0 q_0}{n}}}$ where $\hat{p} = \dfrac{y}{n}$

4. *Rejection Region:*

   One-Tailed Test   Two-Tailed Test

   $z > z_\alpha$ $z > z_{\alpha/2}$ or $z < -z_{\alpha/2}$

   (or $z < -z_\alpha$ when the alternative
   hypothesis is $H_a: p < p_0$)

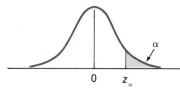 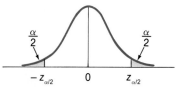

Assumptions: The sampling satisfies the assumptions of a binomial experiment (Section 6.1) and n is large enough so that the sampling distribution of y (and, consequently, of $\hat{p}$) will be approximately normally distributed. These conditions are given in Section 7.6.

Example 9.4

It is known that approximately one in ten smokers favor cigarette brand A. After a promotional campaign in a given sales region, a sample of 200 cigarette smokers were interviewed to determine the effectiveness of the campaign. The result of the survey showed that a total of 26 people expressed a preference for brand A. Do these data present sufficient evidence to indicate an increase in the acceptance of brand A in the region? (Note that, for all practical purposes, this problem is identical to the cold-serum problem given in Example 7.11.)

Solution

It is assumed that the sample satisfies the requirements of a binomial experiment. The question posed may be answered by testing the hypothesis

$$H_0 : p = .10$$

against the alternative

$$H_a : p > .10.$$

A one-tailed statistical test would be utilized because we are primarily concerned with detecting a value of p greater than .10. For this situation it can be shown that the probability of a type II error, β, is minimized by placing the entire rejection region in the upper tail of the distribution of the test statistic.

The point estimator of p is $\hat{p} = y/n$ and the test statistic would be

$$z = \frac{\hat{p} - p_0}{\sigma_{\hat{p}}} = \frac{\hat{p} - p_0}{\sqrt{p_0 q_0/n}}.$$

Or, multiplying numerator and denominator by n, we obtain

$$z = \frac{y - np_0}{\sqrt{np_0 q_0}},$$

which is the test statistic used in Example 7.13. Note that the two test statistics are equivalent.

Once again we require a value of p so that $\sigma_{\hat{p}} = \sqrt{pq/n}$, appearing in the denominator of z, may be calculated. Since we have hypothesized that $p = p_0$, it would seem reasonable to use p_0 as an approximation for p. Note that this differs from the estimation procedure, where, lacking knowledge of p, we chose $\hat{p}$ as the best approximation. This apparent inconsistency will have negligible effect on the inference, whether it is the result of a test or of estimation, when n is large.

Choosing $\alpha = .05$, we would reject H_0 when $z > 1.645$. (See Figure 9.8.) Substituting the numerical values into the test statistic, we obtain

$$z = \frac{\hat{p} - p_0}{\sqrt{p_0 q_0/n}} = \frac{.13 - .10}{\sqrt{\dfrac{(.10)(.90)}{200}}} = 1.41.$$

The calculated value, $z = 1.41$, does not fall in the rejection region, and hence we *do not reject H_0*.

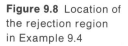

Figure 9.8 Location of the rejection region in Example 9.4

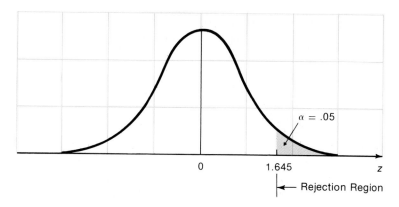

$\alpha = .05$

0   1.645   z

Rejection Region

Do we accept H_0? No, not until we have stated an alternative value of p that is larger than $p_0 = .10$ and considered to be of *practical* significance. The probability of a type II error, β, should be calculated for this alternative. If β is sufficiently small, we would accept H_0 and would do so with the risk of an erroneous decision fully known.

Tips on Problem Solving

When testing an hypothesis concerning p, use p_0 (not $\hat{p}$) to calculate $\sigma_{\hat{p}}$ in the denominator of the z statistic. The reason for this is that the rejection region is determined by the distribution of $\hat{p}$ when the null hypothesis is true, namely, when $p = p_0$.

Examples 9.1 and 9.4 illustrate an important point. **If the data present sufficient evidence to reject H_0, the probability of an erroneous conclusion, α, is known in advance because α is used in locating the rejection region. Since α is usually small, we are fairly certain that we have made a correct decision. On the other hand, if the data present insufficient evidence to reject H_0, the conclusions are not so obvious.** Ideally, following the statistical test procedure outlined in Section 9.4, we would have specified a practically significant alternative, p_a, in advance and chosen n such that β would be small. Unfortunately, many experiments are not conducted in this ideal manner. Someone chooses a sample size and the experimenter or statistician is left to evaluate the evidence.

The calculation of β is not too difficult for the statistical test procedure outlined in this section but may be extremely difficult, if not beyond the capability of the beginner, in other test situations. A much simpler procedure is to *not reject H_0*, rather than to accept it; then estimate using a confidence interval. The interval will give you a range of possible values for p.

Exercises

9.16.   A random sample of $n = 2000$ observations from a binomial population produced $y = 740$ successes.

    a. If your research hypothesis is that p is less than .4, what should you choose for your alternative hypothesis? Your null hypothesis?

    b. Does your alternative hypothesis, part (a), imply a one- or a two-tailed statistical test? Explain.

    c. Do the data provide sufficient evidence to indicate that p is less than .4? Test using $\alpha = .10$.

9.17.   A check cashing service has found that approximately 5 percent of all checks submitted to the service for cashing are bad. After instituting a check verification system to reduce its losses, the service found that only 45 checks were bad in a total of 1124 cashed.

    a. If you wish to conduct a statistical test to determine whether the check verification system reduces the probability that a bad check will be cashed, what should you choose for the alternative hypothesis? The null hypothesis?

    b. Does your alternative hypothesis, part (a), imply a one- or a two-tailed test? Explain.

    c. Noting the data, what does your intuition tell you? Do you think that the check verification system is effective in reducing the proportion of bad checks that were cashed?

    d. Conduct a statistical test of the null hypothesis, part (a), and state your conclusions. Test using $\alpha = .05$. Do the test conclusions agree with your intuition, part (c)?

9.18.   The *U.S. News and World Report* (September 1, 1980) states that a new drug extracted from a fungus, cyclosporin A, increases the success rate in organ transplant operations. According to the article, 22 patients receiving kidney transplants were treated with the new drug. Results of the operations showed an 86 percent success rate in comparison with a success rate of 60 percent that had been obtained in the past using a standard treatment.

    a. For this experiment, the researcher wishes to show that the success rate using cyclosporin A exceeds the standard 60 percent rate of success. State the null and alternative hypotheses that you would use for testing an hypothesis about the success rate p.

    b. Is the sample size, $n = 22$, large enough so that the sampling distribution of the number y of successful operations is normally distributed? Explain.

    c. Does use of cyclosporin A increase the probability of a successful transplant? Conduct the test and state your conclusions. Test using $\alpha = .10$.

9.19.   If you are a bass fisherman, you may wish to heed the advice of Doug Hannon of Odessa, Florida.* After 48 months of keeping records on his personal bass fishing experiences, Hannon, in his statistics, indicates that two-thirds of all big bass are caught on the three days before and after the new and the full moon (a total of 12 days) and the other one-third are caught during the remaining 18 days of the month. To test Hannon's theory, suppose that you monitored the catches of Florida professional guides over a 12-month period of time and that, of 320 large bass (5 pounds or over) caught, 228 were caught during the three-day period on either side of the new or the full moon.

    a. If you wish to show support for Hannon's theory, what should you choose for your alternative hypothesis? Your null hypothesis?

    b. Explain whether (and why) the alternative hypothesis, part (a), implies a one- or a two-tailed statistical test. Locate the rejection region for the test.

* Carter, W. Horace, "A Unique Guide to Trophy Bass," *Bassmaster Annual*, 1978.

c. Just glancing at the data, does it appear that the proportion of all large bass catches in Central Florida during the three-day new and full moon intervals is as large or larger than two-thirds? (We will conduct a statistical test in part (d).) Or, do you think that catching as many as 228 of 320 large bass during the time intervals is just due to chance?

d. Conduct a statistical test of the null hypothesis, part (a), and state your conclusions. Test using $\alpha = .05$.

9.20.   Human skin can be removed from a dead person, stored in a skin bank, and then grafted onto burn victims. Only skin from flat areas, such as the chest, back, and thighs, is acceptable and, because persons with infections, cancer, etc. must be excluded, only approximately 7 percent of all potential donors are acceptable. The *New York Times* (June 14, 1981) notes that a burn research center at one large city hospital screens approximately 8000 deaths annually and, from these, gets only 80 donors, or, 1 percent. Do the data provide sufficient evidence to indicate that the burn research center is selecting their donors from a population containing a lower percentage of acceptables than found in the population at large (7 percent)?

9.21.   An editorial appearing in the *Journal of the American Medical Association* (June 1977) comments on the findings of a study group that recently toured mainland China. The group reported that Chinese doctors claimed 90 percent overall effectiveness of acupuncture as an anesthesia for surgery. Of 48 patients the Americans personally observed, only 73 percent experienced satisfactory relief with acupuncture. Are the Americans' observations inconsistent with the claims of the Chinese physicians?

9.22.   The *Orlando Sentinel Star* (October 6, 1981), reporting on a September, 1981, AP–NBC news poll, states that "even among those who personally believe abortion is wrong, more than 40 percent say it still should be legal." The poll of 1601 adults in a nationwide "scientific random sampling" showed that 44 percent of the respondents viewed abortion as not wrong, 49 percent viewed it as wrong, and 7 percent had no opinion. Forty percent of the 49 percent who regarded abortion as wrong still believed that it should be legal. The article concludes, "the results [of the poll] are subject to an error margin of 3 percentage points either way because of chance variations. That is, if one could have talked this past week to all Americans with telephones, there is only 1 chance in 20 that the findings would vary by more than 3 percentage points."

a. Do you agree that, based on 1601 respondents, the margin of error for the reported percentages should be less than 3 percent with probability equal to .95?

b. If the sample included only those adults who have telephones, do you think that the 1601 respondents represent a random sample from the population of all United States adults?

9.7
Testing an Hypothesis about the Difference between Two Population Proportions

The large-sample statistical test for the difference between two population proportions can be summarized as follows:

A Large-Sample Statistical Test for ($p_1 - p_2$)

1. *Null Hypothesis:* $H_0:(p_1 - p_2) = D_0$, where D_0 is some specified difference that you wish to test. For many tests, you will wish to hypothesize that there is no difference between p_1 and p_2, i.e., $D_0 = 0$.

2. *Alternative Hypothesis:*

One-Tailed Test	Two-Tailed Test
$H_a:(p_1 - p_2) > D_0$	$H_a:(p_1 - p_2) \neq D_0$
(or $H_a:(p_1 - p_2) < D_0$)	

3. *Test Statistic:* $z = \dfrac{(\hat{p}_1 - \hat{p}_2) - D_0}{\sigma_{\hat{p}_1 - \hat{p}_2}} = \dfrac{(\hat{p}_1 - \hat{p}_2) - D_0}{\sqrt{\dfrac{p_1 q_1}{n_1} + \dfrac{p_2 q_2}{n_2}}}$

where $\hat{p}_1 = y_1/n_1$ and $\hat{p}_2 = y_2/n_2$. Since p_1 and p_2 are unknown, we will need to approximate their values in order to calculate the standard deviation of $(\hat{p}_1 - \hat{p}_2)$ that appears in the denominator of the z statistic. Approximations are available for two cases:

Case I: If we hypothesize that p_1 equals p_2, that is,

$$H_0:p_1 = p_2$$

or, equivalently, that

$$(p_1 - p_2) = 0,$$

then $p_1 = p_2 = p$ and the best estimate of p is obtained by pooling the data from both samples. Thus, if y_1 and y_2 are the numbers of successes obtained from the two samples, then

$$\hat{p} = \frac{y_1 + y_2}{n_1 + n_2}.$$

The test statistic would be

$$z = \frac{(\hat{p}_1 - \hat{p}_2) - 0}{\sqrt{\dfrac{\hat{p}\hat{q}}{n_1} + \dfrac{\hat{p}\hat{q}}{n_2}}}$$

or

$$z = \frac{\hat{p}_1 - \hat{p}_2}{\sqrt{\hat{p}\hat{q}\left(\dfrac{1}{n_1} + \dfrac{1}{n_2}\right)}}$$

Case II: On the other hand, if we hypothesize that D_0 is *not equal* to zero, that is

$$H_0:(p_1 - p_2) = D_0,$$

where $D_0 \neq 0$, then the best estimates of p_1 and p_2 are $\hat{p}_1$ and $\hat{p}_2$, respectively. The test statistic would be

$$z = \frac{(\hat{p}_1 - \hat{p}_2) - D_0}{\sqrt{\dfrac{\hat{p}_1\hat{q}_1}{n_1} + \dfrac{\hat{p}_2\hat{q}_2}{n_2}}}$$

4. *Rejection Region:*

One-Tailed Test	*Two-Tailed Test*
$z > z_\alpha$	$z > z_{\alpha/2}$ or $z < -z_{\alpha/2}$

(or $z < -z_\alpha$ when the alternative hypothesis is $H_a{:}(p_1 - p_2) < D_0$)

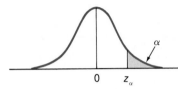
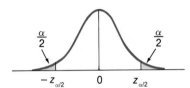

Assumptions: Samples were selected in a random and independent manner from two binomial populations and n_1 and n_2 are large enough so that the sampling distributions of y_1 and y_2 (and, therefore, of $\hat{p}_1$ and $\hat{p}_2$) will be approximately normally distributed. These conditions are given in Section 7.6.

Example 9.5

The records of a hospital show that 52 men in a sample of 1000 men versus 23 women in a sample of 1000 women were admitted because of heart disease. Do these data present sufficient evidence to indicate a higher rate of heart disease among men admitted to the hospital?

Solution

We shall assume that the number of patients admitted for heart disease will follow approximately a binomial probability distribution for both men and women with parameters p_1 and p_2, respectively. Then, since we wish to determine whether $p_1 > p_2$, we will test the null hypothesis that $p_1 = p_2$; that is, $H_0{:}(p_1 - p_2) = 0$ against the alternative hypothesis, $H_a{:}p_1 > p_2$ or, equivalently, that $(p_1 - p_2) > 0$. To conduct this test, we will use the z test statistic and approximate the value of $\sigma_{(\hat{p}_1 - \hat{p}_2)}$ using the pooled estimate of p described in Case I. Since H_a implies a one-tailed test, we will reject H_0 only for large values of z. Thus, for $\alpha = .05$, we will reject H_0 if $z > 1.645$ (see Figure 9.9).

The pooled estimate of p required for $\sigma_{(\hat{p}_1 - \hat{p}_2)}$ is

$$\hat{p} = \frac{y_1 + y_2}{n_1 + n_2} = \frac{52 + 23}{1000 + 1000} = .0375.$$

Figure 9.9 Location of the rejection region in Example 9.5

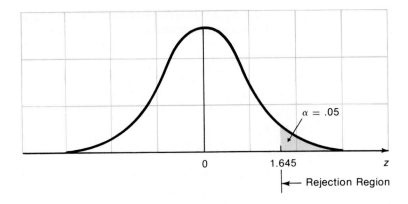

The test statistic is

$$z = \frac{\hat{p}_1 - \hat{p}_2}{\sqrt{\hat{p}\hat{q}\left(\dfrac{1}{n_1} + \dfrac{1}{n_2}\right)}} = \frac{.052 - .023}{\sqrt{(.0375)(.9625)\left(\dfrac{1}{1000} + \dfrac{1}{1000}\right)}}$$

or

$$z = 3.41.$$

Since the computed value of z falls in the rejection region, we reject the hypothesis that $p_1 = p_2$ and conclude that the data present sufficient evidence to indicate that the percentage of men entering the hospital because of heart disease is higher than that of women. (*Note:* This does not imply that the *incidence* of heart disease is higher in men. Perhaps fewer women enter the hospital when afflicted with the disease!)

Exercises

9.23. Independent random samples of $n_1 = 120$ and $n_2 = 120$ observations were randomly selected from binomial populations 1 and 2, respectively. The number of successes in the samples and the population parameters are shown below.

	Population	
	1	**2**
Sample Size	120	120
Number of Successes	69	76
Binomial Parameter	p_1	p_2

    a. Suppose that you have no preconceived theory concerning which parameter, p_1 or p_2, is the larger, and that you only wish to detect a difference between the two parameters, if it exists. What should you choose for the alternative hypothesis for a statistical test? The null hypothesis?

    b. Does your alternative hypothesis, part (a), imply a one- or a two-tailed test?

    c. Conduct the test and state your conclusions. Test using $\alpha = .05$.

9.24.   Refer to Exercise 9.23. Suppose that for practical reasons, you know that p_1 cannot be larger than p_2.

    a. Given this knowledge, what should you choose as the alternative hypothesis for your statistical test? Your null hypothesis?

    b. Will your alternative hypothesis, part (a), imply a one- or a two-tailed test? Explain.

    c. Conduct the test and state your conclusions. Test using $\alpha = .05$.

9.25.   "Paralytic shellfish poisoning, PSP, remains a major problem in California."* Shellfish containing the denoflagellate, Gonyaulax catenella, a one-celled organism, are most often found in the cool waters of coastal northern California from May through October, but they are also found in decreasing levels as you proceed south along the California coast. The reaction from eating these shellfish varies from a mild case of nausea, stomach cramps, or what appears to be a 24-hour stomach flu, to death within a few hours. Of 400 cases reported between 1927 and 1980, 30 were fatal. In contrast, in 1980 alone there were 100 reported cases and three were fatal. There were undoubtedly many more cases of PSP during these time periods, undiagnosed and unreported, but, do these data imply a difference in death rate p_1 for reported cases, between the 1927–1980 period, and the death rate p_2 for the 12 months of 1980?

    a. Suppose that you only wish to detect a difference between p_1 and p_2. State the alternative and null hypotheses that you would use for a statistical test.

    b. Based on a visual examination of the data, do you believe that p_1 differs from p_2? (We will conduct the statistical test in part (c).)

    c. Conduct the statistical test using $\alpha = .05$. Do the data present sufficient evidence to indicate a difference in the fatality rates for reported cases between the two time periods?

9.26.   A surprising event occurred during the 1976 presidential race. The Gallup and the Harris polls published conflicting results. The Gallup Poll of 1078 registered voters showed Ford beating Humphrey by 51 to 39 percent. The Harris Poll (taken slightly later) of 950 people gave Humphrey a 52 to 41 margin over Ford. If you were to compare the percentages in the two samples favoring Ford and if the two samples were randomly selected from the same population, what is the probability that the sample percentages favoring Ford would differ by as much as 10 percent (51 versus 41 percent)? Do the data provide sufficient evidence to indicate that the two samples were selected from different populations? Or could the disparity in the conclusions of the pollsters be due to the methods of selecting the samples?

9.27.   To curb the spread of the swine flu, President Ford ordered a massive vaccination program in 1976. Shortly thereafter, numerous cases of the paralyzing Guillain-Barre syndrome were reported. In December 1976, fearful that the vaccine was causing the syndrome, the vaccination program was suspended to allow for an analysis of the data. Of 220 million Americans, approximately 50 million received swine flu shots. Of 383 persons who contracted the Guillain-Barre syndrome, 202 received the swine flu shots. Do you think these

* Hudgins, Shirley, "New Threats from PSP," *Sea Grant Today*, May/June 1981.

data suggest a dependency between persons who received the swine flu shot and the contraction of the Guillain-Barre syndrome? Test the null hypothesis that the probability of contracting the Guillain-Barre syndrome is the same for the vaccinated portion of the general public as for the unvaccinated portion. Test using $\alpha = .05$.

9.8
Another Way to Report the Results of Statistical Tests: *p* Values

The probability α of making a type I error is often called the significance level of the statistical test, a term that originated in the following way. The probability of the observed value of the test statistic, or some value even more contradictory to the null hypothesis, measures, in a sense, the weight of evidence favoring rejection of H_0. Some experimenters report test results as being significant (we would reject H_0) at the 5 percent significance level but not at the 1 percent level. This means that we would reject H_0 if α were .05 but not if α were .01.

The smallest value of α for which test results are statistically significant is often called the *p* value, or the **observed significance level**, for the test. Some statistical computer programs compute *p* values for statistical tests correct to four or five decimal places. But if you are using statistical tables to determine a *p* value, you will only be able to approximate its value. This is because most statistical tables give the critical values of test statistics only for large differential values of α (for example, .01, .025, .05, .10, etc.). Consequently, the *p* value reported by most experimenters is the largest tabulated value of α for which the test remains statistically significant. For example, if a test result is statistically significant for $\alpha = .10$, but not for $\alpha = .05$, then the *p* value for the test would be given as $p = .10$ or, more precisely, as $p < .10$.

Many scientific journals require researchers to report the *p* values associated with statistical tests because these values provide a reader with *more information* than simply stating that a null hypothesis is or is not to be rejected for some value of α chosen by the experimenter. In a sense, it allows the reader of published research to evaluate the extent to which the data disagree with the null hypothesis. Particularly, it enables each reader to choose his or her own personal value for α and then to decide whether or not the data lead to rejection of the null hypothesis.

The procedure for finding the *p* value for a test will be illustrated by the following examples.

Example 9.6

Find the *p* value for the statistical test, Example 9.1. Interpret your results.

Solution

Example 9.1 presents a test of the null hypothesis, $H_0: \mu = 880$, against the alternative hypothesis, $H_a: \mu \neq 880$. The value of the test statistic, computed from the sample data, was $z = -3.03$. Therefore, the *p* value for this two-tailed test is the probability that $z \leq -3.03$ or $z \geq 3.03$ (the shaded areas in Figure 9.10).

From Table 3, the Appendix, you can see that the tabulated area under the normal curve between $z = 0$ and $z = 3.03$ is .4988 and the area to the right of $z = 3.03$ is

Figure 9.10 Locating the p value for the test, Example 9.1

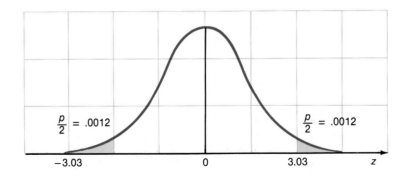

$.5 - .4988 = .0012$. Then, since this was a two-tailed test, the area corresponding to the z values, $z > 3.03$ or $z < -3.03$, is $2(.0012) = .0024$. Consequently, we would report the p value for the test as $p = .0024$.

Example 9.7

Find the p value for the statistical test, Example 9.4. Interpret your results.

Solution

Example 9.4 presented a one-tailed test of the null hypothesis, $H_0:p = .10$, against the alternative hypothesis, $H_a:p > .10$, and the observed value of the test statistic was $z = 1.41$. Therefore, the p value for the test is the probability of observing a value of the z statistic larger than 1.41. This value is the area under the normal curve to the right of $z = 1.41$ (the shaded area in Figure 9.11).

Figure 9.11 Finding the p value for the test, Example 9.4

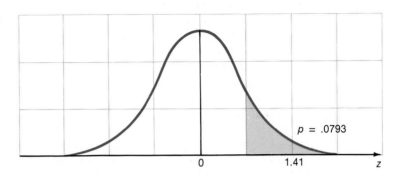

From Table 3, the Appendix, the area under the normal curve between $z = 0$ and $z = 1.41$ is .4207. Therefore, the area under the normal curve to the right of $z = 1.41$, the p value for the test, is $p = .5 - .4207 = .0793$.

Advocating that a researcher report the p value for a test and leave its interpretation to a reader does not violate the traditional statistical test procedure described in the preceding sections. It simply leaves the decision of whether or not to reject the null hypothesis (with the potential for a type I or type II error) to the reader. Thus, it shifts the responsibility for choosing the value of α, and possibly the problem of evaluating the probability, β, of making a type II error, to the reader.

Exercises

9.28.   Suppose that you tested the null hypothesis, $H_0 : \mu = 15$, against the alternative hypothesis, $H_a : \mu > 15$. For a random sample of $n = 38$ observations, $\bar{y} = 15.7$ and $s = 2.4$,

a. Give the observed significance level for the test.

b. If you wish to conduct your test using $\alpha = .05$, what would be your test conclusions?

9.29.   Find the observed significance level for the test, Exercise 9.8. Compare it with the value of α given for the test and state your conclusions.

9.30.   Find the observed significance level for the test, Exercise 9.16. Compare it with the value of α given for the test and state your conclusions.

9.31.   Find the observed significance level for the test, Exercise 9.23. Compare it with the value of α given for the test and state your conclusions.

9.32.   A serious problem facing the United States is the depletion in the quality and quantity of our fresh water supply. *Bioscience* (July/August 1981), in "Features and News," reports that the water table is dropping in many areas of the United States. To illustrate, the rate of depletion of a reservoir near Plainview, Texas, from 1936 to 1973, has averaged 76 centimeters (approximately 2.5 feet) per year and the rate of depletion has increased since 1973. The water level in one well in the Houston area dropped at an average rate of 2.4 meters (8 feet) per year from 1939 to 1973.

a. Suppose that you wish to determine whether the mean water table in an area has dropped more than 3 feet in a given year. If you plan to sample and to conduct a statistical test, what should you choose for your alternative hypothesis? Your null hypothesis?

b. Given your alternative hypothesis, part (a), will this be a one- or a two-tailed test? Explain.

c. Suppose that the drop in the water level over a one-year period was measured for 37 wells in the region and that the mean and standard deviation of the drops in water levels were $\bar{y} = 3.2$ feet and $s = 6.4$ inches. Find the observed significance level for the test.

d. If you wish α to be equal to .05, would you reject the null hypothesis? What conclusions would you derive from the test?

9.33.   How would you like to live to be 200 years old? For centuries, humankind has sought the key to the mystery of aging. What causes it? How can it be slowed? Recent studies focus on biomarkers, physical or biological changes that occur at a predictable time in a person's life. The theory is that if ways can be found to delay the occurrence of these biomarkers, human life can be extended. A key biomarker, according to scientists, is forced vital capacity (FVC), the volume of air that you can expel after taking a deep breath. A study of 5209 men and women aged 30 to 62 showed that FVC declined on the average 3.8 deciliters per decade for men and 3.1 deciliters per decade for women.* Suppose that you wished to determine

* Boddé, Tineke, "Biomarkers of Aging: Key to a Younger Life?" *Bioscience*, 31, 8 (1981), pp. 566–567. Copyright © 1981 by the American Institute of Biological Sciences.

whether a physical fitness program for men and women, aged 50 to 60, would delay aging and that you measured the FVC for 30 men and 30 women at the beginning and end of the 50–to–60–year age interval and recorded the drop in FVC for each person. The data are shown below:

	Men	Women
Sample Size	30	30
Sample Average Drop in FVC	3.6	2.7
Sample Standard Deviation	1.1	1.2
Population Mean Drop in FVC	μ_1	μ_2

a. Do the data provide sufficient evidence to indicate that the decrease in the mean FVC over the decade for the men on the physical fitness program is less than 3.8 deciliters? Find the observed significance level for the test.

b. Refer to part (a). If you choose $\alpha = .05$, do the data support the contention that the mean decrease in FVC is less than 3.8 deciliters?

c. Test to determine whether the FVC drop for women on the physical fitness program was less than 3.1 deciliters for the decade. Find the observed significance level for the test.

d. Refer to part (c). If you choose $\alpha = .05$, do the data support the contention that the mean decrease in FVC is less than 3.1 deciliters?

9.9
Testing the Veracity of a Fish Story

Tales of the mean number of 10-pound bass per 5 acres of Lake Toho can be checked by statistical sampling and a statistical test of an hypothesis concerning the mean number μ of 10-pound fish per 5-acre plot. Particularly, if we wish to determine whether the data support the biologists' claims, we would want to show that μ is larger than 1.0 10-pound bass per 5 acres. Therefore, we would choose as the alternative hypothesis, $H_a: \mu > 1.0$, and would run a one- (upper) tailed test of the null hypothesis, $H_0: \mu = 1.0$.

To conduct this test, we need to sample portions of the lake. Sampling fish or wildlife (that is, counting the number of fish or animals per acre) is not a simple matter. For example, you cannot simply select a 5-acre plot, pour the lake water contained therein through a sieve, and count the fish. Methods for counting the approximate number of fish within smaller plots by netting or shocking are possible, but most estimates of the size of fish or animal populations are obtained by tagging methods. A known number of fish, say 100, of a certain type are caught, marked, and then released. After the marked fish have had time to disperse themselves throughout the lake, some fish (a sample) are caught. Then, based on the number of marked fish in the sample, methods are available to estimate or test hypotheses about the size of the fish population. A description of estimation (and test) methods based on tagging procedures can be found in Scheaffer, Mendenhall, and Ott (1979).

For our purposes, we will suppose that we have a method for counting the number y of 10-pound bass per 5-acre plot of lake and that we have the counts for a random sample of $n = 12$ plots. Can we employ the z test of Section 9.4 to test the null hypothesis, $H_0:\mu = 1.0$, against the alternative hypothesis, $H_a:\mu > 1.0$? At first glance, we would say "no" because the sample size $n = 12$ violates the requirement that n be greater than or equal to 30. But further examination indicates that our sampling situation provides an exception to the rule, an exception that we may find to be useful in the future.

The primary reason for requiring that the sample size n be as large as 30 is that, for most practical situations, the standard deviation σ of the sampled population will be unknown and must be approximated (in the z statistic) by the sample standard deviation s. A relatively large sample size, $n \geq 30$, is needed in order that the value of s provide a good estimate of σ. However, *when the value of σ is known in advance* (which, we will explain, is true for our example), this value of σ can be substituted directly into the formula for the z statistic and the standard normal z test will be valid for samples of modest size, even for sample sizes as small as $n = 12$.*

For our particular sampling situation, the sampled population consists of a number of counts, one corresponding to each 5-acre plot in the lake. This count, y, the number of 10-pound bass per plot, is a discrete random variable that can assume values 0, 1, 2, 3, . . . , and which, according to our null hypothesis, possesses a mean $\mu = 1$. When certain assumptions are true, namely that the 10-pound bass are randomly distributed throughout the lake, it can be shown (proof omitted) that y, the number of 10-pound bass per 5-acre plot, possesses a probability distribution given by the equation,

$$p(y) = \frac{\mu^y e^{-\mu}}{y!}$$

where

y = number of 10-pound bass per 5-acre plot

μ = mean number of 10-pound bass per 5-acre plot

$e = 2.718 \ldots$

This probability distribution, the Poisson probability distribution, was employed in the Thyroid Disease and Three Mile Island Case Study of Chapter 5.

To refresh your memory, we stated in Section 5.6 that the Poisson probability distribution is a good approximation to the probability distribution of the number of counts of events that occur in a specified unit of area, volume, time, etc. (See Optional Exercise 5.44 and some of the other optional exercises

---

* The sample sum possesses a Poisson sampling distribution. This distribution tends to normality as the mean of the Poisson distribution increases, a fact which (along with the Central Limit Theorem) justifies the use of the z test for our data. This test would not be appropriate for small samples.

in Chapter 5 for applications.) Unlike the binomial probability distribution which possesses two parameters, n and p, the Poisson probability distribution contains only one, the parameter μ. This parameter, μ, is the mean of the Poisson distribution, and it can be shown (proof omitted) that $\sigma = \sqrt{\mu}$.

How do we know whether a sample size, $n = 12$, will be large enough that $\bar{y}$ will be approximately normally distributed? Just as the number y of 10-pound bass in a 5-acre plot is a Poisson random variable, so also is the number of 10-pound bass in a 1-acre plot, in one 10-acre plot, or in one plot of 60 acres. The only difference between these different Poisson random variables is that their means (and hence their standard deviations) differ. For this reason, a Poisson random variable y can be viewed as a sum (that is, the number of 10-pound bass in a 60-acre plot is equal to the sum of the 60 counts of the numbers of bass in the sixty 1-acre plots) and therefore y, like the binomial random variable, will be approximately normally distributed by the Central Limit Theorem when the sample size n is large. This condition will be adequately satisfied (proof omitted) when the mean μ is $\mu \geq 2\sigma$ or, since $\sigma = \sqrt{\mu}$, when μ is greater than or equal to 4.

To consider whether our situation satisfies this condition, note that we have data for $n = 12$ 5-acre plots or the equivalent of 60 acres. The null hypothesis, $\mu = 1$ 10-pound bass per 5 acres, is equivalent to $\mu = 12$ 10-pound bass per 60 acres. Therefore, since the data available in our sample correspond to 60 acres and $\mu = 12$ exceeds 4, the sampling distribution of the total number of 10-pound bass in the 60 acres will be approximately normally distributed; that is, the approximation will be adequate to employ a large-sample z to test an hypothesis about μ.

It does not make any difference whether we test $H_0:\mu = 1$ per 5-acre plot or $H_0:\mu = 12$ per 60-acre plot, the tests are equivalent and the calculated values of the z statistic will be identical. If we test

$$H_0:\mu = 1 \text{ per 5-acre plot}$$

against

$$H_a:\mu > 1,$$

then $\sigma = \sqrt{\mu} = \sqrt{1} = 1$ and the test statistic is,

$$z = \frac{\bar{y} - \mu_0}{\sigma/\sqrt{n}} = \frac{\bar{y} - 1}{1/\sqrt{12}}.$$

Since we wish to detect only values of μ larger than 1.0, (that is, $H_a:\mu > 1.0$) we will run a one-tailed statistical test and locate the rejection region in the upper tail of the z distribution. The rejection region for $\alpha = .05$ is shown in Figure 9.12.

Figure 9.12 The rejection region for one-tailed test, $H_0: \mu = 1.0, H_a: \mu > 1.0$

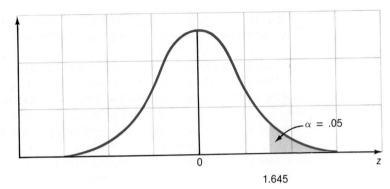

Now, suppose that a total of sixteen 10-pound bass were observed in the $n = 12$ 5-acre plots. Then the mean number of 10-pound bass in the $n = 12$ 5-acre plots is,

$$\bar{y} = \frac{\sum\limits_{i=1}^{n} y_i}{n} = \frac{16}{12} = 1.33$$

and the test statistic is,

$$z = \frac{\bar{y} - \mu_0}{\sigma/\sqrt{n}} = \frac{1.33 - 1.0}{1/\sqrt{12}} = 1.14.$$

Since this value of z is less than $z_{.05} = 1.645$, there is no evidence to indicate that the mean number of 10-pound bass per 5-acre plot μ is greater than 1.0.

This test conclusion may clash with your intuition. It might seem that a sample mean as large as 1.33 would indicate that $\mu > 1.0$. But, putting intuition aside, we must measure the deviation between the observed value of $\bar{y}$, 1.33, and the hypothesized value of μ, $\mu = 1.0$, in units of the standard deviation, $\sigma/\sqrt{n} = 1/\sqrt{12}$, of the sampling distribution of $\bar{y}$. On this yardstick, $\bar{y}$ is not very far away from μ ($1.14\sigma_{\bar{y}}$) and there is no evidence to support the contention that the mean number of 10-pound bass per 5-acre plot is larger than 1.0.

If you reexamine our analyses of the hyperthyroid births near Three Mile Island, Section 5.6, you will see that we were, in effect, testing the null hypothesis that the mean number of hyperthyroid births (during a specified period of time) per three-county area was equal to $H_0: \mu = 3$ against the alternative hypothesis, $H_a: \mu > 3$. We did not employ a large-sample standard normal z test to test the hypothesis for two reasons. We had not discussed tests of hypotheses and the concept of normality at that stage of our studies and, even if we had, the standard normal z test would not have been appropriate because the mean value of the total number of counts, $\mu = 3$, is not large enough (μ must be larger than 4) in order that the number y of hyperthyroid births be approximately normally distributed. Fortunately, for that example, the difference between $y = 11$, the observed number of hyperthyroid deaths, and the hypothesized mean, $\mu = 3$, is large enough so that there was little question that $\mu > 3$. The observed significance level for the hypothesized births test, $P\{y \geq 11\}$, could be calculated using the Poisson distribution with $\mu = 3$. It would be very small.

9.10
Some Comments on the Theory of Tests of Hypotheses

As outlined in Section 9.2, the theory of a statistical test of an hypothesis is indeed a very clear-cut procedure, enabling the experimenter to either reject or accept the null hypothesis with measured risks α and β. Unfortunately, as we noted, the theoretical framework does not suffice for all practical situations.

The crux of the theory requires that we be able to specify a meaningful alternative hypothesis that permits the calculation of the probability β of a type II error for all alternative values of the parameter(s). This indeed can be done for many statistical tests, including the large-sample test discussed in Section 9.3, although the calculation of β for various alternatives and sample sizes may, in some cases, be difficult. On the other hand, in some test situations it is extremely difficult to clearly specify alternatives to H_0 that have practical significance. This may occur when we wish to test an hypothesis concerning the values of a set of parameters, a situation that we will encounter in Chapter 13 in analyzing enumerative data.

The obstacle that we mention does not invalidate the use of statistical tests. Rather, it urges caution in drawing conclusions when insufficient evidence is available to reject the null hypothesis. The difficulty of specifying meaningful alternatives to the null hypothesis, together with the difficulty in calculating and tabulating β for other than the simplest statistical tests, justifies skirting this issue in an introductory text. Hence, we can adopt one of two procedures. We can present the p value associated with a statistical test and leave the interpretation to the reader. Or, we can agree to adopt the procedure described in Example 9.4 when tabulated values of β (the operating characteristic curve) are unavailable for the test. When the test statistic falls in the acceptance region, we will "not reject" rather than "accept" the null hypothesis. Further conclusions may be made by calculating an interval estimate for the parameter or by consulting one of several published statistical handbooks for tabulated values of β. We will not be too surprised to learn that these tabulations are inaccessible, if not completely unavailable, for some of the more complicated statistical tests.

Finally, we might comment on the choice between a one- or a two-tailed test for a given situation. We emphasize that this choice is dictated by the practical aspects of the problem and will depend on the alternative value of the parameter, say θ, that the experimenter is trying to detect. If we were to sustain a large financial loss if θ were greater than θ_0, but not if it were less, we would concentrate our attention on the detection of values of θ greater than θ_0. Hence we would reject in the upper tail of the distribution for the test statistics previously discussed. On the other hand, if we are equally interested in detecting values of θ that are either less than or greater than θ_0, we would employ a two-tailed test.

9.11 Summary

In this chapter we have reviewed the basic concepts of a test of an hypothesis and have demonstrated the procedure with four large-sample tests. Considering that all four tests employ a standard normal z as a test statistic, they may be viewed as equivalent.

The key to a statistical test is the research hypothesis about a population parameter, call it θ. For some practical reason, you wish to show that θ is larger than some value, for example 50, smaller than some value, say 50, or, you may wish to show that θ is either larger or smaller than some value. This hypothesis is called the alternative hypothesis, H_a.

To gain evidence to support H_a, we define a null hypothesis. This hypothesis is defined in such a way that if we decide to reject it, we are left with only one alternative; that is, support of the research (or alternative) hypothesis. For example, if we wished to show that a population mean was larger than 50, the alternative hypothesis would be $H_a: \mu > 50$ and the null hypothesis would be $H_0: \mu = 50$.

To conduct a test, we need a decision maker (that is, a test statistic) that is calculated from the sample data. Knowing its sampling distribution, we can then decide which values (the rejection region) of the test statistic favor acceptance of the alternative hypothesis and which values (the nonrejection region) favor the null hypothesis. For each of the four tests presented in this chapter, we used a point estimator (from Chapter 8) to form a standard normal z test statistic. Values of z contradictory to the null hypothesis were those in the tail (one-tailed test) or tails (two-tailed test) of the z distribution.

Equally important as the test conclusion is a measure of its reliability. This measure is the probability of making an incorrect decision, α, if we reject H_0 and accept the alternative hypothesis, and β, if we accept the null hypothesis. You can present the results of a statistical test either by stating the test conclusions (that is, the decision) and the accompanying values of α and β, or by presenting the observed significance level for a test. The observed significance level for the test is the probability of observing a value of the test statistic at at least as contradictory to the null hypothesis (and as supportive of the alternative hypothesis) as the one calculated from the sample data. This latter procedure enables a reader of the test results to choose his or her own value of α.

In addition to the test statistics presented in this chapter, many other test statistics possess sampling distributions that are approximately normal when the sample sizes are large (you will encounter several in Chapter 16). But when the sample sizes are small, the sampling distributions of most test statistics are non-normal. One of these non-normal sampling distributions, the t distribution will be employed in Chapter 10 to obtain confidence intervals and tests of hypotheses for a single population mean and the difference between two population means.

References

Dixon, W.J., and M.B. Brown, eds. *BMDP Biomedical Computer Programs, P Series*. Berkeley, Calif.: University of California, 1979.

Dixon, W.J., and F.J. Massey, Jr., *Introduction to Statistical Analysis*, 3rd ed. New York: McGraw-Hill Book Company, 1969.

Freund, J.E., *Mathematical Statistics*, 2nd ed. Englewood Cliffs, N.J.: Prentice-Hall, Inc., 1971.

Helwig, J.P., and K.A. Council, eds. *SAS User's Guide*. Cary, N.C.: SAS Institute, Inc., P.O. Box 8000, 1979.

Hogg, R.V., and A.T. Craig, *Introduction to Mathematical Statistics*, 4th ed. New York: Macmillan Publishing Co., Inc., 1978.

Mendenhall, W., R. Scheaffer, and D. Wackerly, *Mathematical Statistics with Applications*, 2nd ed. Boston: Duxbury Press, 1981.

Neter, J., W. Wasserman, and G.A. Whitmore, *Applied Statistics*. Boston: Allyn and Bacon, Inc., 1978.

Nie, N., C.H. Hull, J.G. Jenkins, K. Steinbrenner, and D.H. Bent, *Statistical Package for the Social Sciences*, 2nd ed. New York: McGraw-Hill Book Company, 1979.

Scheaffer, R., W. Mendenhall, and L. Ott, *Elementary Survey Sampling*, 2nd ed. Boston: Duxbury Press, 1979.

Walpole, R.E., *Introduction to Statistics*, 2nd ed. New York: Macmillan Publishing Co., Inc., 1974.

Supplementary Exercises

[Starred (*) exercises are optional.]

9.34. Define α and β for a statistical test of an hypothesis.

9.35. What is the observed significance level of a test?

9.36. The daily wages in a particular industry are normally distributed with a mean of $23.20 and a standard deviation of $4.50. Suppose that a company in this industry employing 40 workers pays these workers $21.20 on the average. Based on this sample mean, could these workers be viewed as a random sample from among all workers in the industry?

a. Find the observed significance level of the test.

b. If you planned to conduct your test using $\alpha = .01$, what would be your test conclusions?

9.37. Refer to Exercise 8.16 and the collection of water samples to estimate the mean acidity (in pH) of rainfalls in the northeastern United States. As noted, the pH for pure rain falling through clean air is approximately 5.7. The sample of $n = 40$ rainfalls produced pH readings with $\bar{v} = 3.7$ and $s = .5$. Do the data provide sufficient evidence to indicate that the mean pH for rainfalls is more acidic $(H_a:\mu < 5.7$ pH$)$ than pure rainwater? Test using $\alpha = .05$. Note that this inference is appropriate only for the area in which the rainwater specimens were collected.

9.38.   A manufacturer of automatic washers provides a particular model in one of three colors, A, B, or C. Of the first 1000 washers sold, it is noticed that 400 were of color A. Would you conclude that more than one-third of all customers have a preference for color A?

a. Find the observed significance level of the test.

b. If you planned to conduct your test using $\alpha = .05$, what would be your test conclusions?

9.39   What conditions must be met in order that the z test may be used to test an hypothesis concerning a population mean, μ?

9.40.   According to the *Washington Post* (December 7, 1976), federal scientists in Oklahoma "have been able to forecast tornadoes in Oklahoma with an accuracy that is better than than 90 percent." Suppose that the scientists tracked 274 storms that did not result in tornadoes and 37 that did. Of these 311 storms, they correctly forecast the type of storm in 273 cases. Would this disagree with the claimed forecasting ability of the federal scientists?

9.41.   In a study to assess various effects of using a female model in automobile advertising, each of 100 male subjects was shown photographs of two automobiles matched for price, color, and size but of different makes. One of the automobiles was shown with a female model to 50 of the subjects (group A) and both automobiles were shown without the model to the other 50 subjects (group B). In group A, the automobile shown with the model was judged as more expensive by 37 subjects, while in group B the same automobile was judged as the more expensive by 23 subjects. Do these results indicate that using a female model influences the perceived expensiveness of an automobile? Use a one-tailed test with $\alpha = .05$.

9.42.   Random samples of 200 bolts manufactured by machine A and 200 bolts manufactured by machine B showed 16 and 8 defective bolts, respectively. Do these data present sufficient evidence to suggest a difference in the performance of the machines? Use $\alpha = .05$.

9.43.   Suppose that the true fraction p in favor of the death penalty is the same for Democrats as it is for Republicans. Independent random samples are selected, one consisting of 800 Republicans and the other of 800 Democrats. Find the probability that the sample fraction of Republicans favoring the death penalty exceeds that of the Democrats by more than .03 if $p = .5$. What is this probability if $p = .1$?

9.44.   A social scientist believes that the fraction p_1 of Republicans in favor of the death penalty is greater than the fraction p_2 of Democrats in favor of the death penalty. She acquired independent random samples of 200 Republicans and 200 Democrats, respectively, and found 46 Republicans and 34 Democrats favoring the death penalty. Do these data support the social scientist's belief?

a. Find the observed significance level of the test.

b. If you planned to conduct your test using $\alpha = .05$, what would be your test conclusions?

* 9.45.   Refer to Exercise 9.44. Some thought should have been given to designing a test for which β is tolerably low when p_1 exceeds p_2 by an important amount. For example, find a common sample size n for a test with $\alpha = .05$ and $\beta \leq .20$ when in fact p_1 exceeds p_2 by .1. [*Hint:* The maximum value of $p(1 - p)$ is .25.]

9.46.   In comparing the mean weight loss for two diets the following sample data were obtained:

	Diet I	Diet II
Sample size, n	40	40
Sample mean, $\bar{y}$	10 lb	8 lb
Sample variance, s^2	4.3	5.7

Do the data provide sufficient evidence to indicate that diet I produces a greater mean weight loss than diet II? Use $\alpha = .05$.

9.47.   A test of the breaking strengths of two different types of cables was conducted using samples of $n_1 = n_2 = 100$ pieces of each type of cable.

Cable I	Cable II
$\bar{y}_1 = 1925$	$\bar{y}_2 = 1905$
$s_1 = 40$	$s_2 = 30$

Do the data provide sufficient evidence to indicate a difference between the mean breaking strengths of the two cables? Use $\alpha = .10$.

9.48.   The braking ability was compared for two types of 1982 automobiles. Random samples of 64 automobiles were tested for each type. The recorded measurement was the distance (in feet) required to stop when the brakes were applied at 40 miles per hour. The computed sample means and variances were:

$$\bar{y}_1 = 118, \qquad \bar{y}_2 = 109;$$
$$s_1^2 = 102, \qquad s_2^2 = 87.$$

Do the data provide sufficient evidence to indicate a difference in the mean stopping distance for the two types of automobiles?

9.49.   A fruit grower wishes to test a new spray which a manufacturer claims will *reduce* the loss due to damage by a certain insect. To test the claim, the grower sprays 200 trees with the new spray and 200 trees with the standard spray. The following data were recorded:

	New Spray	Standard Spray
Mean yield per tree, $\bar{y}$ (lb)	240	227
Variance, s^2	980	820

a. Do the data provide sufficient evidence to conclude that the new spray is better than the old? (Use $\alpha = .05$.)

b. Set up a 95 percent confidence interval for the difference in mean yields for the two sprays.

* 9.50.   Refer to Example 9.2. Use the procedure described in Example 9.2 to calculate β for several alternative values of μ. (For example, $\mu = 873, 875,$ and 877.) Use the three computed values of β along with the value computed in Example 9.2 to construct an operating characteristic curve for the statistical test. The resulting graph will be similar to that shown in the accompanying figure.

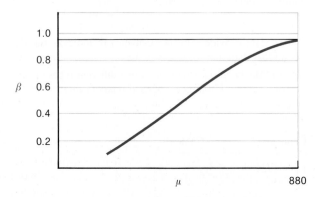

* 9.51.   Repeat the procedure described in the preceding exercise for a sample size $n = 25$ (as opposed to $n = 50$ used in Exercise 9.50) and compare the two operating characteristic curves.

9.52.   A biologist hypothesizes that high concentrations of actinomysin D inhibit RNA synthesis in cells and hence the production of proteins as well. An experiment conducted to test this theory compared the RNA synthesis in cells treated with two concentrations of actinomysin D, .6 and .7 microgram per milliliter, respectively. Cells treated with the lower concentration (.6) of actinomysin D showed that 55 of 70 developed normally, whereas only 23 of 70 appeared to develop normally for the higher concentration (.7). Do these data provide sufficient evidence to indicate a difference in the rate of normal RNA synthesis for cells exposed to the two different concentrations of actinomysin D?

   a. Find the observed significance level of the test.

   b. If you planned to conduct your test using $\alpha = .10$, what would be your test conclusions?

9.53.   The 1979–1980 national statistics on admissions to medical colleges, classified according to undergraduate major programs, are shown in the accompanying table.*

Major	Number of Admissions	Percentage of Total of Major Applications Admitted
Biology	6139	44.2
Chemistry	2273	53.5
Zoology	886	43.7
Premed	658	50.3
Biochemistry	859	59.2
Interdisciplinary	86	62.8

* Data from *Medical School Admission Requirements 1979–1980, United States and Canada,* Association of American Medical Colleges, 1981.

a. Suppose that prior to viewing this data, you decided to compare the chances of admission between majors in biology and chemistry. Do the data provide sufficient evidence to indicate a difference in the probabilities of admission for these two majors? Explain.

b. Find a 95 percent confidence interval for the difference in the probabilities of admission between biology majors and biochemistry majors. Interpret the confidence interval.

(*Note:* In Chapter 13 we shall address the general question of whether the data provide sufficient evidence to indicate a difference in the probabilities of admission among the six major program categories.)

9.54. Because of our infatuation with expensive automobiles, the Bureau of Labor Statistics claims that the American family has reached the point of spending proportionately more money on transportation than on food. In fact, it notes that the mean number of cars in 1972 increased to 1.3 over a value of 1.0 a decade earlier (*Orlando Sentinel Star*, May 15, 1977). In deciding where to place its emphasis in advertising, the market research department for a major automobile manufacturer wished to compare the mean number of automobiles per family in two regions of the United States. Suppose that a preliminary study of the number of cars per family for $n = 200$ families from each of the two regions gave the means and variances for the two samples as shown in the table.

	Area 1	Area 2
Sample size	200	200
Sample mean	1.30	1.37
Sample variance	.53	.64

a. Note that a small increase in the mean number of automobiles per family can represent a very large number of automobiles for a region. Do the data provide sufficient evidence to indicate a difference in the mean number of automobiles per family for the two regions?

b. Picture the data associated with either of the two populations. What values will y assume? Imagine the probability distributions for these two populations. Will their nature violate the conditions necessary in order that the test, part (a), be valid? Explain.

c. Find a 95 percent confidence interval for the mean number of automobiles per family for region 2. Interpret the interval.

d. Find a 90 percent confidence interval for the difference in the mean number of automobiles for the two regions. Interpret the interval.

9.55. "Schizophrenia, a disease marked by withdrawal from reality, confusion, flat and often inappropriate emotions and, often, hallucinations," may be linked to a long-standing viral disease (*Orlando Sentinel Star*, October 21, 1977). Researchers have found that a higher proportion of schizophrenics are born during months in which viral infections are prevalent. Particularly, they found that of a total of 53,584 schizophrenics born between 1920 and 1955, 6.9 percent more were born in the peak months of March and April than in the other months. Suppose that we interpret this statement to mean that during any of the other two-month periods, the observed percentage of schizophrenic births was 16.48 and that during the March–April period, the percentage was 6.9 percent larger, or 17.61

percent. Is the percentage of schizophrenic births observed by the researchers during the March–April period statistically significant? That is, do the data provide sufficient evidence to indicate that the proportion of schizophrenics born during March–April is higher than might be expected if schizophrenia and birth date are unrelated? Test using $\alpha = .05$. (*Hint:* If the proportion p of schizophrenics born during any two-month period is independent of the months included in the pair, then $p = 1/6$.)

10. Inference from Small Samples

Chapter Objectives

The basic concepts of statistical estimation and tests of hypotheses were presented in Chapters 8 and 9 along with summaries of the methodologies for some large-sample estimation and test procedures. Large-sample estimation and test procedures for population means and proportions were used to illustrate concepts as well as to give you some useful tools for solving some practical problems. Because all these techniques rely on the Central Limit Theorem to justify the normality of the estimators and test statistics, they apply only when the sample sizes are large. Consequently, the objective of Chapter 10 is to supplement the results of Chapters 8 and 9 by presenting small-sample statistical test and estimation procedures for population means and variances. These techniques differ substantially from those of Chapters 8 and 9 because they require that the relative frequency distributions of the sampled populations be approximately normal.

Specific Objectives

1. To present the Student's t statistic and identify the relationship between it and the z statistic of Chapter 9. *Section 10.2*

2. To show the relationship between the normal distribution and a Student's t distribution and to explain how to use the table values for the Student's t statistic. *Section 10.2*

3. To show how the t statistic is used to construct small-sample confidence intervals and tests of hypotheses for population means and the difference between two population means. *Sections 10.3,10.4,10.6*

4. To give our first example of an experiment designed to reduce the cost of experimentation, the paired-difference experiment. We shall show how the techniques of Section 10.3 can be used to analyze these data. *Section 10.5*

5. To show how to estimate and test hypotheses concerning population variances and to give examples and exercises that indicate the applications of these techniques to the solution of certain practical problems. *Sections 10.7,10.8*

6. To emphasize the assumptions you must make in order for the estimation and test procedures described in this chapter to be valid. *Section 10.9*

Case Study

Just a Pinch between the Cheek and the Gum

Those interested in controlling and reducing stress should not listen too closely to the advice of the pro-football running backs Walt Garrison and Earl Campbell. If we are to believe a report in the *Orlando Sentinel Star* (November 18, 1981), just a pinch (of "smokeless tobacco," i.e., snuff) between the cheek and the gum can be harmful to your health.

According to the *Sentinel Star*, researchers at Texas Lutheran College found that "oral tobacco increased the work of the heart and drove up blood pressure in humans." These conclusions were based on research reported by William G. Squires, an associate professor of biology, at the American Heart Association's 54th Scientific Sessions.* The experimental results were based on a study of 20 Texas Lutheran male athletes, ten who used oral tobacco and ten who did not. After obtaining initial electrocardiogram and blood pressure readings, the researchers gave both tobacco users and non-users "a good healthy man-size dip" (2.5 grams) of tobacco. After 20 minutes, the average heart rate of the tobacco dippers had "increased from 69 to 88 beats per minute" and the "average blood pressure increased from 118 over 72 to 126

over 78." The article further notes that oral tobacco use might pose a health hazard to some persons, particularly those suffering from hypertension.

Do these statistics really indicate that the use of smokeless tobacco produces a significant increase in a person's heart rate and blood pressure? The data provided in the study are based upon independent random samples, $n_1 = 10$ who did not use smokeless tobacco prior to the experiment, and $n_2 = 10$ who did. Thus, the researchers wish to detect a change in mean heart rate and mean blood pressure produced by the use of smokeless tobacco for both regular users and non-users. Regardless of the comparisons that the experimenters might wish to make, it is clear that the sample sizes are too small (less than 30) to use the methods of Chapters 8 and 9 and that new methodology, appropriate for small samples, is needed to analyze the data.

In this chapter we will present some methods for estimating and testing hypotheses about population means when the sample sizes are small. We will discuss the Texas Lutheran College's research on smokeless tobacco in Section 10.6.

* Squires, William G., et al., "Hemodynamics affects of oral tobacco in experimental animals and young adults," *Circulation Supp.*, October 1981, IV, 186 (Abstract). By permission of the American Heart Association, Inc., and the author.

10.1
Introduction

Large-sample methods for making inferences concerning population means and the difference between two means were discussed with examples in Chapters 8 and 9. Frequently cost, available time, and other factors limit the size of the sample that may be acquired. When this occurs, large-sample procedures are inadequate and other tests and estimation procedures must be employed. In this chapter we shall study several small-sample inferential procedures that are closely related to the large-sample methods presented in Chapters 8 and 9. Specifically, we shall consider methods for estimating and testing hypotheses concerning population means, the difference between two means, a population variance, and a comparison of two population variances. Small-sample tests and confidence intervals for binomial parameters will be omitted from our discussion.*

10.2
Student's *t*
Distribution

We introduce our topic by considering the following problem. A very costly experiment has been conducted to evaluate a new process for producing synthetic diamonds. Six diamonds have been generated by the new process with recorded weights .46, .61, .52, .48, .57, and .54 karat.

A study of the process costs indicates that the average weight of the diamonds must be greater than .5 karat in order that the process be operated at a profitable level. Do the six diamond-weight measurements present sufficient evidence to indicate that the average weight of the diamonds produced by the process is in excess of .5 karat?

Recall that, according to the Central Limit Theorem,

$$z = \frac{\bar{y} - \mu}{\sigma/\sqrt{n}}$$

possesses approximately a normal distribution in repeated sampling when n is large. For $\alpha = .05$, we would employ a one-tailed statistical test and reject H_0 when $z > 1.645$. This, of course, assumes that σ is known or that a good estimate, s, is available and is based upon a reasonably large sample (we have suggested $n \geq 30$). Unfortunately, this latter requirement will not be satisfied for the $n = 6$ diamond-weight measurements. How, then, may we test the hypothesis that $\mu = .5$ against the alternative that $\mu > .5$ when we have a small sample?

The problem that we pose is not new; rather, it is one that received serious attention by statisticians and experimenters at the turn of the century. If a sample standard deviation, s, were substituted for σ in z, would the resulting quantity possess approximately a standardized normal distribution in repeated sampling? More specifically, would the rejection region $z > 1.645$ be appropriate; that is, would approximately 5 percent of the values of the test statistic, computed in repeated sampling, exceed 1.645? The answers to these questions, not unlike many of the problems encountered in the sciences, may be resolved by experimentation. That is, we could draw a small sample, say $n = 6$ measurements, and compute the value of the test statistic. Then we would repeat this process over and over again a very large number of times and construct a frequency distribution for the computed values of the test statistic. The general

---

* A small-sample test for the binomial parameter p was presented in Chapter 6.

shape of the distribution and the location of the rejection region would then be evident.

The distribution of the test statistic

$$t = \frac{\bar{y} - \mu}{s/\sqrt{n}}$$

for samples drawn from a normally distributed population was discovered by W.S. Gosset and published in 1908 under the pen name of Student. He referred to the quantity under study as t and it has ever since been known as Student's t. We omit the complicated mathematical expression for the density function for t but describe some of its characteristics.

The distribution of the test statistic

$$t = \frac{\bar{y} - \mu}{s/\sqrt{n}}$$

in repeated sampling is, like z, mound-shaped and perfectly symmetrical about $t = 0$. Unlike z, it is much more variable, tailing rapidly out to the right and left, a phenomenon that may readily be explained. The variability of z in repeated sampling is due solely to $\bar{y}$; the other quantities appearing in z (n and σ) are nonrandom. On the other hand, the variability of t is contributed by *two* random quantities, $\bar{y}$ and s, which can be shown to be independent of one another. Thus when $\bar{y}$ is very large, s may be very small, and vice versa. As a result, t will be more variable than z in repeated sampling (see Figure 10.1). Finally, as we might surmise, the variability of t decreases as n increases because the estimate s of σ will be based upon more and more information. When n is infinitely large, the t and z distributions will be identical. Thus Gosset discovered that the distribution of t depended upon the sample size, n.

The divisor of the sum of squares of deviations, $(n - 1)$, that appears in the formula for s^2 is called the number of degrees of freedom associated with s^2. The origin of the term "degrees of freedom" is linked to the statistical theory underlying the probability distribution of s^2. We shall not pursue this point

Figure 10.1 Standard normal, z, and the t distribution based on $n = 6$ measurements (5 d.f.)

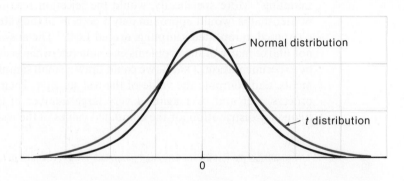

further except to note that one may say that the test statistic t is based upon a sample of n measurements or that it possesses $(n - 1)$ degrees of freedom (d.f.).

The critical values of t that separate the rejection and acceptance regions for the statistical test are presented in Table 4, the Appendix. Table 4 is partially reproduced in Table 10.1. The tabulated value, t_α, records the value of t such that an area, α, lies to its right as shown in Figure 10.2. The degrees of freedom associated with t, d.f., are shown in the first and last columns of the table (see Table 10.1), and the t_α, corresponding to various values of α, appear in the top row. Thus, if we wish to find the value of t, such that 5 percent of the area lies to its right, we would use the column marked $t_{.05}$. The critical value of t for our example, found in the $t_{.05}$ column opposite d.f. $= (n - 1) = (6 - 1) = 5$, is $t = 2.015$ (shaded in Table 10.1). Thus we would reject $H_0 : \mu = .5$ in favor of $H_a : \mu > .5$ when $t > 2.015$.

Note that the critical value of t will always be larger than the corresponding critical value of z for a specified α. For example, where $\alpha = .05$, the critical value

Table 10.1 Format of the Student's *t* table, Table 4, the Appendix

d.f.	$t_{.100}$	$t_{.050}$	$t_{.025}$	$t_{.010}$	$t_{.005}$	d.f.
1	3.078	6.314	12.706	31.821	63.657	1
2	1.886	2.920	4.303	6.965	9.925	2
3	1.638	2.353	3.182	4.541	5.841	3
4	1.533	2.132	2.776	3.747	4.604	4
5	1.476	2.015	2.571	3.365	4.032	5
6	1.440	1.943	2.447	3.143	3.707	6
7	1.415	1.895	2.365	2.998	3.499	7
8	1.397	1.860	2.306	2.896	3.355	8
9	1.383	1.833	2.262	2.821	3.250	9
⋮	⋮	⋮	⋮	⋮	⋮	⋮
26	1.315	1.706	2.056	2.479	2.779	26
27	1.314	1.703	2.052	2.473	2.771	27
28	1.313	1.701	2.048	2.467	2.763	28
29	1.311	1.699	2.045	2.462	2.756	29
inf.	1.282	1.645	1.960	2.326	2.576	inf.

Figure 10.2 Tabulated values of Student's *t*

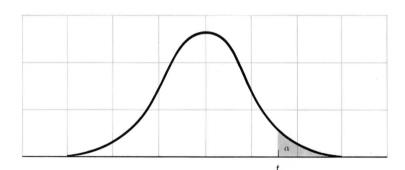

of t for $n = 2$ (d.f. $= 1$) is $t = 6.314$, which is very large when compared with the corresponding $z_{.05} = 1.645$. Proceeding down the $t_{.05}$ column, we note that the critical value of t decreases, reflecting the effect of a larger sample size (more degrees of freedom) on the estimation of σ. Finally, when n is infinitely large, the critical value of t will equal $z_{.05} = 1.645$.

The reason for choosing $n = 30$ (an arbitrary choice) as the dividing line between large and small samples is apparent. For $n = 30$ (d.f. $= 29$), the critical value of $t_{.05} = 1.699$ is numerically quite close to $z_{.05} = 1.645$. For a two-tailed test based upon $n = 30$ measurements and $\alpha = .05$, we would place .025 in each tail of the t distribution and reject $H_0 : \mu = \mu_0$ when $t > 2.045$ or $t < -2.045$. Note that this is very close to the $z_{.025} = 1.96$ employed in the z test.

It is important to note that the Student's t and corresponding tabulated critical values are based upon the assumption that the sampled population possesses a normal probability distribution. This indeed is a very restrictive assumption because, in many sampling situations, the properties of the population will be completely unknown and may well be nonnormal. If this were to seriously affect the distribution of the t statistic, the application of the t test would be very limited. Fortunately, this point is of little consequence, as it can be shown that the distribution of the t statistic possesses nearly the same shape as the theoretical t distribution for populations that are nonnormal but possess a mound-shaped probability distribution. This property of the t statistic and the common occurrence of mound-shaped distributions of data in nature enhance the value of Student's t for use in statistical inference.

Having discussed the origin of Student's t and the tabulated critical values, Table 4, the Appendix, we now return to the problem of making an inference about the mean diamond weight based upon our sample of $n = 6$ measurements. Prior to considering the solution, you may wish to test your built-in inference-making equipment by glancing at the six measurements and arriving at a conclusion concerning the significance of the data.

10.3
Small-Sample Inferences Concerning a Population Mean

The statistical test of an hypothesis concerning a population mean may be stated as follows:

Small-Sample Test of an Hypothesis Concerning a Population Mean

1. *Null Hypothesis:* $H_0 : \mu = \mu_0$
2. *Alternative Hypothesis:*

One-Tailed Test	Two-Tailed Test
$H_a : \mu > \mu_0$	$H_a : \mu \neq \mu_0$
(or, $H_a : \mu < \mu_0$)	

3. *Test Statistic:* $t = \dfrac{\bar{y} - \mu_0}{s/\sqrt{n}}$

---

4. *Rejection Region*:

One-Tailed Test	Two-Tailed Test
$t > t_\alpha$	$t > t_{\alpha/2}$ or $t < -t_{\alpha/2}$

(or, $t < -t_\alpha$ when the alternative hypothesis is $H_a: \mu < \mu_0$)

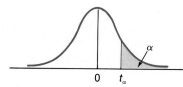
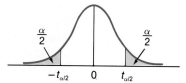

The critical values of t, t_α and $t_{\alpha/2}$, are based upon $(n - 1)$ degrees of freedom. These tabulated values can be found in Table 4, the Appendix.

Assumption: The sample has been randomly selected from a normally distributed population.

To apply this test to the diamond-weight data, we must first calculate the sample mean, $\bar{y}$, and standard deviation, s. This latter quantity is calculated using the formula of Section 3.7.

You can verify that the mean and standard deviation for the six diamond weights are .53 and .0559, respectively.

Recall that we wish to test the null hypothesis that the mean diamond weight is .5 against the alternative hypothesis that it is greater than .5. Then the elements of the test as defined above are:

$H_0: \mu = .5$.

$H_a: \mu > .5$.

Test Statistic: $t = \dfrac{\bar{y} - \mu_0}{s/\sqrt{n}} = \dfrac{.53 - .5}{.0559/\sqrt{6}} = 1.31$.

Rejection Region: The rejection region for $\alpha = .05$ and $(n - 1) = (6 - 1) = 5$ degrees of freedom is $t > 2.015$. Noting that the calculated value of the test statistic does not fall in the rejection region, we do not reject H_0. This implies that the data do not present sufficient evidence to indicate that the mean diamond weight exceeds .5 karat.

The calculation of the probability of a type II error, β, for the t test is very difficult and is beyond the scope of this text. Therefore, we shall avoid this problem and obtain an interval estimate for μ, as noted in Section 8.5.

We recall that the large-sample confidence interval for μ is

$$\bar{y} \pm \frac{z_{\alpha/2}\sigma}{\sqrt{n}},$$

where $z_{\alpha/2} = 1.96$ for a confidence coefficient equal to .95. This result assumes that σ is known and simply involves a measurement of $1.96\sigma_{\bar{y}}$ (or approximately $2\sigma_{\bar{y}}$) on either side of $\bar{y}$ in conformity with the Empirical Rule. When σ is unknown and must be estimated by a small-sample standard deviation, s, the large-sample confidence interval will not enclose μ 95 percent of the time in repeated sampling. But this situation can be easily corrected; that is, we can make the confidence coefficient equal .95, if we substitute $t_{.025}$ for $z_{.025}$ in the formula for the large-sample 95 percent confidence interval. Although we omit the derivation, the corresponding small-sample confidence interval for μ, with confidence coefficient $(1 - \alpha)$, will be:

Small-Sample Confidence Interval for μ

$$\bar{y} \pm \frac{t_{\alpha/2}s}{\sqrt{n}},$$

where $s/\sqrt{n}$ is the *estimated* standard deviation of $\bar{y}$.

Assumption: The sampled population is approximately normally distributed.

For our example,

$$\bar{y} \pm \frac{t_{\alpha/2}s}{\sqrt{n}} = .53 \pm 2.571 \frac{.0559}{\sqrt{6}} \qquad \text{or} \qquad .53 \pm .059.$$

The interval estimate for μ is therefore .471 to .589 with confidence coefficient equal to .95. If the experimenter wishes to detect a small increase in mean diamond weight in excess of .5 karat, the width of the interval must be reduced by obtaining more diamond-weight measurements. This will decrease both $1/\sqrt{n}$ and $t_{\alpha/2}$ and thereby decrease the width of the interval. Or, looking at it from the standpoint of a statistical test of an hypothesis, more information will be available upon which to base a decision and the probability of making a type II error will decrease.

Example 10.1   A manufacturer of gunpowder has developed a new powder that is designed to produce a muzzle velocity equal to 3000 feet per second. Eight shells are loaded with the charge and the muzzle velocities measured. The resulting velocities are shown in Table 10.2. Do

Table 10.2

Muzzle Velocity (ft/sec)	
3005	2995
2925	3005
2935	2935
2965	2905

the data present sufficient evidence to indicate that the average velocity differs from 3000 feet per second?

Solution

Testing the null hypothesis that $\mu = 3000$ feet per second against the alternative that μ is either greater than or less than 3000 feet per second will result in a two-tailed statistical test. Thus,

$$H_0 : \mu = 3000,$$

$$H_a : \mu \neq 3000.$$

Using $\alpha = .05$ and placing .025 in each tail of the t distribution, we find that the critical value of t for $n = 8$ measurements [or $(n - 1) = 7$ d.f.] is $t = 2.365$. Hence we will reject H_0 if $t > 2.365$ or $t < -2.365$. (Recall that the t distribution is symmetrical about $t = 0$.) The sample mean and standard deviation for the recorded data are

$$\bar{y} = 2958.75 \qquad \text{and} \qquad s = 39.26.$$

Then

$$t = \frac{\bar{y} - \mu_0}{s/\sqrt{n}} = \frac{2958.75 - 3000}{39.26/\sqrt{8}} = -2.97.$$

Since the observed value of t falls in the rejection region, we will reject H_0 and conclude that the average velocity differs from 3000 feet per second. Furthermore, we will be reasonably confident that we have made the correct decision. When we use our procedure, we should erroneously reject H_0 only $\alpha = .05$ of the time in repeated applications of the statistical test.

Example 10.2

Find a 95 percent confidence interval for the mean, Example 10.1.

Solution

Substituting $t_{.025} = 2.365$, $s = 39.26$, and $n = 8$ into the formula

$$\bar{y} \pm \frac{t_{\alpha/2} s}{\sqrt{n}},$$

we obtain

$$2958.75 \pm (2.365) \frac{39.26}{\sqrt{8}}$$

or

$$2958.75 \pm 32.83.$$

Thus, we estimate the average muzzle velocity to lie in the interval 2926 to 2992 feet per second. A more accurate interval estimate (a shorter interval) can be obtained by increasing the sample size.

Example 10.3

If you planned to report the results of the statistical test, Example 10.1, what p value would you report?

Solution

The p value for this test is the probability of observing a value of the t statistic at least as contradictory to the null hypothesis as the one observed for this set of data, namely $t = -2.97$. Since this is a two-tailed test, the p value is the probability that $t \le -2.97$ or $t \ge 2.97$.

Unlike the table of areas under the normal curve (Table 3, the Appendix), Table 4, the Appendix, does not give the areas corresponding to various values of t. Rather, it gives the values of t corresponding to upper-tail areas equal to .10, .05, .025, .010, and .005. Consequently, we can only approximate the upper-tail area that corresponds to the probability that $t > 2.97$. Since the t statistic for this test is based on seven degrees of freedom, we consult the d.f. $= 7$ row of Table 4 and find that 2.97 falls between $t_{.025} = 2.365$ and $t_{.010} = 2.998$. Since the observed value of t was 2.97, we reject H_0 if $t \le -2.365$ with $\alpha = .05$ (α is double the upper-tail area because the test is two-tailed), but cannot reject H_0 for $\alpha = .02$. Therefore, the p value for the test would be reported as $p < .05$.

Since most researchers would be using a t table similar or identical to Table 4, the reader of a published research report would realize that the exact p value for the test was probably less than $p = .05$, a value between .05 and .01.

Exercises

10.1. A random sample of $n = 7$ observations from a normal population produced a mean, $\bar{y} = 6.4$, and standard deviation, $s = .29$.

   a. If you wish to use the data to determine whether the population mean μ exceeds 6.0, state the null and alternative hypotheses that you would use for your statistical test.

   b. Refer to part (a) and find the rejection region for the test if $\alpha = .10$.

   c. Conduct the test, part (a), and state your conclusions.

10.2. Suppose that your objective was to estimate the mean, Exercise 10.1. Find a 90 percent confidence interval for μ.

10.3.   A random sample of $n = 5$ observations from a normal population produced the data 1.0, 1.2, 1.0, .6, and .8.

a. Glance at the data and use your intuition to decide whether you think that the population mean μ is less than 1.0.

b. Conduct a statistical test to determine whether μ is less than 1.0. Test using $\alpha = .05$. State your conclusions and compare them to your answer, part (a).

10.4.   Scholastic Aptitude Test scores have been falling slowly since the inception of the tests. Originally, a score of 500 was intended to be "average," but the mean scores have fallen to approximately 424 for the verbal test, and 466 for the mathematics test. A sampling of the test scores of 20 seniors at an urban high school produced the following mean scores and standard deviations:

	Verbal	**Mathematics**
Sample Mean	412	446
Sample Standard Deviation	57	69

a. Do the data provide sufficient evidence to indicate that the mean verbal SAT score for the urban high school seniors is less than the 1981 national mean of 424? Test using $\alpha = .05$.

b. Find the approximate observed significance level for the test, part (a), and interpret its value.

c. Do the data provide sufficient evidence to indicate that the mean mathematics SAT score for the urban high school seniors is less than the 1981 national mean of 466? Test using $\alpha = .05$.

d. Find the observed significance level for the test, part (c), and interpret its value.

10.5.   Noise limits, established by the EPA for heavy trucks, were set at a level of 83 decibels in 1978 (*Environment News*, October 1976). If a random sample of six heavy trucks in 1977 produced the following noise levels (in decibels),*

85.4, 86.8, 86.1, 85.3, 84.8, 86.0,

how far from the 1978 limit was the mean noise level in 1977? Express your estimate as a 95 percent confidence interval. Interpret your findings.

10.6.   The tremendous growth of the Florida lobster (called spiny lobster) industry over the past 20 years has made it the state's second most valuable fishery industry. A recent declaration by the Bahamian government that prohibits U.S. lobstermen from fishing on the Bahamian portion of the continental shelf is expected to produce a dramatic reduction in the landings in pounds per lobster per trap. According to the records, the mean landings per trap is 30.31 pounds. A random sampling of 20 lobster traps since the Bahamian fishing

---

* Decibels are measured on a logarithmic scale. An increase of 3 decibels represents a doubling of the actual noise energy.

restriction went into effect gave the following results (in pounds):

17.4	18.9	39.6	34.4	19.6
33.7	37.2	43.4	41.7	27.5
24.1	39.6	12.2	25.5	22.1
29.3	21.1	23.8	43.2	24.4

Do these landings provide sufficient evidence to support the contention that the mean landings per trap has decreased since imposition of the Bahamian restrictions? Test using $\alpha = .05$.

10.7.   Refer to Exercise 10.6 and find a 90 percent confidence interval for the mean landings per lobster trap (in pounds) and interpret your result.

10.8.   PCBs (polychlorinated biphenyls), used in the manufacture of large electrical power transformers and capacitors, are known to be hazardous substances which have serious health effects. *Environment News* (October 1976) reports that the federal and state governments are investigating PCB contamination in the Acushnet River in the New Bedford, Massachusetts, area. As part of their investigation, they found that analyses of three clam specimens showed PCB levels of 21, 23, and 53 parts per million (ppm) (the FDA tolerance level for PCBs is 5 ppm). Use these three clam PCB concentration levels to estimate the mean concentration in clams in the Acushnet River area. Use a 90 percent confidence interval. Interpret your results. What assumptions must be made in order that your inference be valid?

10.9.   Measurements of water intake, obtained from a sample of 17 rats that had been injected with a sodium chloride solution, produced a mean and standard deviation of 31.0 and 6.2 cubic centimeters, respectively. Given that the average water intake for noninjected rats observed over a comparable period of time is 22.0 cubic centimeters, do the data indicate that injected rats drink more water than noninjected rats? Test at the 5 percent level of significance. Find a 90 percent confidence interval for the mean water intake for injected rats.

10.10.   Organic chemists often purify organic compounds by a method known as fractional crystallization. An experimenter desired to prepare and purify 4.85 grams of aniline. Ten 4.85-gram quantities of aniline were individually prepared and purified to acetanilide. The following dry yields were recorded:

3.85	3.80
3.88	3.85
3.90	3.36
3.62	4.01
3.72	3.83

Estimate the mean number of grams of acetanilide that could be recovered from an initial amount of 4.85 grams of aniline. Use a 95 percent confidence interval.

10.11.   Refer to Exercise 10.10. Approximately how many 4.85-gram specimens of aniline would be required if you wished to estimate the mean number of grams of acetanilide correct to within .06 grams with probability equal to .95?

10.12.   An experimenter, interested in determining the mean thickness of the cortex of the sea urchin egg, employed an experimental procedure developed by Sakai. The thickness of the cortex was measured for $n = 10$ sea urchin eggs. The following measurements were obtained:

4.5	5.2
6.1	2.6
3.2	3.7
3.9	4.6
4.7	4.1

Estimate the mean thickness of the cortex using a 95 percent confidence interval.

10.13 Industrial wastes and sewage dumped into our rivers and streams absorb oxygen and thereby reduce the amount of dissolved oxygen available for fish and other forms of aquatic life. One state agency requires a minimum of 5 parts per million of dissolved oxygen in order that the oxygen content be sufficient to support aquatic life. Six water specimens taken from a river at a specific location during the low-water season (July) gave readings of 4.9, 5.1, 4.9, 5.0, 5.0, and 4.7 parts per million of dissolved oxygen. Do the data provide sufficient evidence to indicate that the dissolved oxygen content is less than 5 parts per million? Test using $\alpha = .05$.

10.14. If you are renting an apartment and you think your rent is too high, consider that part of the rent you are paying is caused by the high interest rate on borrowed money. What will be the prime rate of interest next September 1? A random sample of $n = 12$ economic forecasters produced a mean, $\bar{y} = 16.7\%$, and standard deviation, $s = 2.1\%$, If the forecasters' predictions are "unbiased;" that is, if the mean of the population of forecasts of all economic forecasters will equal the actual interest rate next fall, find a 90 percent confidence interval for the September 1 prime interest rate.

10.4
Small-Sample Inferences Concerning the Difference between Two Means

The physical setting for the problem that we consider is identical to that discussed in Section 8.7. Independent random samples of n_1 and n_2 measurements, respectively, are drawn from two populations that possess means and variances μ_1, σ_1^2 and μ_2, σ_2^2. Our objective is to make inferences concerning the difference between the two population means, $(\mu_1 - \mu_2)$.

The following small-sample methods for testing hypotheses and placing a confidence interval on the difference between two means are, like the case for a single mean, founded upon assumptions regarding the probability distributions of the sampled populations. Specifically, we shall assume that both populations possess a normal probability distribution and, also, that the population variances, σ_1^2 and σ_2^2, are equal. In other words, we assume that the variability of the measurements in the two populations is the same and can be measured by a common variance, which we will designate as σ^2 (that is, $\sigma_1^2 = \sigma_2^2 = \sigma^2$).

The point estimator of $\mu_1 - \mu_2$, $(\bar{y}_1 - \bar{y}_2)$, the difference between the sample means, was discussed in Section 8.7, where it was observed to be unbiased and to possess a standard deviation,

$$\sigma_{(\bar{y}_1 - \bar{y}_2)} = \sqrt{\frac{\sigma_1^2}{n_1} + \frac{\sigma_2^2}{n_2}},$$

in repeated sampling. This result was used in placing bounds on the error of estimation, the construction of a large-sample confidence interval, and the z test statistic,

$$z = \frac{(\bar{y}_1 - \bar{y}_2) - D_0}{\sqrt{\dfrac{\sigma_1^2}{n_1} + \dfrac{\sigma_2^2}{n_2}}},$$

for testing an hypothesis, $H_0: \mu_1 - \mu_2 = D_0$ (i.e., D_0 is the hypothesized difference between μ_1 and μ_2). Utilizing the assumption that $\sigma_1^2 = \sigma_2^2 = \sigma^2$, the z test statistic could be simplified as follows:

$$z = \frac{(\bar{y}_1 - \bar{y}_2) - D_0}{\sqrt{\dfrac{\sigma^2}{n_1} + \dfrac{\sigma^2}{n_2}}} = \frac{(\bar{y}_1 - \bar{y}_2) - D_0}{\sigma \sqrt{\dfrac{1}{n_1} + \dfrac{1}{n_2}}}.$$

For small-sample tests of the hypothesis $H_0: (\mu_1 - \mu_2) = D_0$, it would seem reasonable to use the test statistic

$$t = \frac{(\bar{y}_1 - \bar{y}_2) - D_0}{s \sqrt{\dfrac{1}{n_1} + \dfrac{1}{n_2}}};$$

that is, we would substitute a sample standard deviation s for σ. Surprisingly enough, this test statistic will possess a Student's t distribution in repeated sampling when the stated assumptions are satisfied, a fact that can be proved mathematically or verified by experimental sampling from two normal populations.

The estimate, s, to be used in the t statistic could be either s_1 or s_2, the standard deviations for the two samples, although the use of either would be wasteful since both estimate σ. Since we wish to obtain the best estimate available, it would seem reasonable to use an estimator that would pool the information from both samples. This estimator, utilizing the sums of squares of the deviations about the mean for both samples, is

---

Pooled Estimator of σ^2

$$s^2 = \frac{\displaystyle\sum_{i=1}^{n_1} (y_i - \bar{y}_1)^2 + \sum_{i=1}^{n_2} (y_i - \bar{y}_2)^2}{n_1 + n_2 - 2}.$$

Note that the pooled estimator may also be written as

$$s^2 = \frac{(n_1 - 1)s_1^2 + (n_2 - 1)s_2^2}{n_1 + n_2 - 2},$$

because

$$s_1^2 = \frac{\sum\limits_{i=1}^{n_1} (y_i - \bar{y}_1)^2}{n_1 - 1}$$

and

$$s_2^2 = \frac{\sum\limits_{i=1}^{n_2} (y_i - \bar{y}_2)^2}{n_2 - 1}.$$

As in the case for the single sample, the denominator in the formula for s^2, $(n_1 + n_2 - 2)$, is called the "number of degrees of freedom" associated with s^2. It can be proved either mathematically or experimentally that the expected value of the pooled estimator, s^2, is equal to σ^2 and hence s^2 is an unbiased estimator of the common population variance. Finally, we note that the divisors of the sums of squares of deviations in s_1^2 and s_2^2, $(n_1 - 1)$ and $(n_2 - 1)$, respectively, are the numbers of degrees of freedom associated with these two independent estimators of σ^2. It is interesting to note that an estimator that uses the pooled information from both samples would possess $(n_1 - 1) + (n_2 - 1)$ or $(n_1 + n_2 - 2)$ degrees of freedom.

Summarizing, the small-sample statistical test for the difference between two means is as follows.

Test of an Hypothesis Concerning the Difference between Two Means

1. *Null Hypothesis:* $H_0:(\mu_1 - \mu_2) = D_0$ where D_0 is some specified difference that you wish to test. For many tests, you will wish to hypothesize that there is no difference between μ_1 and μ_2; that is, $D_0 = 0$.

2. *Alternative Hypothesis:*

    One-Tailed Test   *Two-Tailed Test*
    $H_a:(\mu_1 - \mu_2) > D_0$ $H_a:(\mu_1 - \mu_2) \neq D_0$
    (or, $H_a:(\mu_1 - \mu_2) < D_0$)

3. *Test Statistic:* $t = \dfrac{(\bar{y}_1 - \bar{y}_2) - D_0}{s\sqrt{\dfrac{1}{n_1} + \dfrac{1}{n_2}}}$

    where $s^2 = \dfrac{\sum\limits_{i=1}^{n_1} (y_i - \bar{y}_1)^2 + \sum\limits_{i=1}^{n_2} (y_i - \bar{y}_2)^2}{n_1 + n_2 - 2}$

4. *Rejection Region:*

One-Tailed Test	Two-Tailed Test
$t > t_\alpha$	$t > t_{\alpha/2}$ or $t < -t_{\alpha/2}$
(or, $t < -t_\alpha$ when the alternative hypothesis is $H_a:(\mu_1 - \mu_2) < D_0$)	

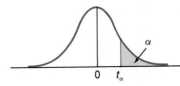   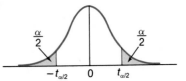

The critical values of t, t_α and $t_{\alpha/2}$, will be based upon $(n_1 + n_2 - 2)$ degrees of freedom. The tabulated values can be found in Table 4, the Appendix.

Assumptions: The samples were randomly and independently selected from normally distributed populations. The variances of the populations, σ_1^2 and σ_2^2, are equal.

Example 10.4

An assembly operation in a manufacturing plant requires approximately a one-month training period for a new employee to reach maximum efficiency. A new method of training was suggested and a test was conducted to compare the new method with the standard procedure. Two groups of nine new employees were trained for a period of three weeks, one group using the new method and the other following standard training procedure. The length of time in minutes required for each employee to assemble the device was recorded at the end of the three-week period. These measurements appear in Table 10.3. Do the data present sufficient evidence to indicate that the mean time to assemble at the end of a three-week training period is less for the new training procedure?

Table 10.3

Standard Procedure	New Procedure
32	35
37	31
35	29
28	25
41	34
44	40
35	27
31	32
34	31

Solution

Let μ_1 and μ_2 equal the mean time to assemble for the standard and the new assembly procedures, respectively. Also, assume that the variability in mean time to assemble is essentially a function of individual differences and that the variability for the two populations of measurements will be approximately equal.

The sample means and sums of squares of deviations are

$$\bar{y}_1 = 35.22,$$

$$\sum_{i=1}^{9} (y_i - \bar{y}_1)^2 = 195.56;$$

$$\bar{y}_2 = 31.56,$$

$$\sum_{i=1}^{9} (y_i - \bar{y}_2)^2 = 160.22.$$

Then the pooled estimate of the common variance is

$$s^2 = \frac{\sum_{i=1}^{9} (y_i - \bar{y}_1)^2 + \sum_{i=1}^{9} (y_i - \bar{y}_2)^2}{n_1 + n_2 - 2}$$

$$= \frac{195.56 + 160.22}{9 + 9 - 2}$$

$$= 22.24,$$

and the standard deviation is $s = 4.72$.

The null hypothesis to be tested is

$$H_0 : (\mu_1 - \mu_2) = 0.$$

Suppose that we are concerned only with detecting whether the new method reduces the assembly time, and therefore that the alternative hypothesis is

$$H_a : (\mu_1 - \mu_2) > 0.$$

This would imply that we should use a one-tailed statistical test and that the rejection region for the test will be located in the upper tail of the t distribution. Referring to Table 4, the Appendix, we note that the critical value of t for $\alpha = .05$ and $(n_1 + n_2 - 2) = 16$ degrees of freedom is 1.746. Therefore, we will reject H_0 when $t > 1.746$.

The calculated value of the test statistic is

$$t = \frac{\bar{y}_1 - \bar{y}_2}{s \sqrt{\frac{1}{n_1} + \frac{1}{n_2}}} = \frac{35.22 - 31.56}{4.72 \sqrt{\frac{1}{9} + \frac{1}{9}}}$$

$$= 1.64.$$

Comparing this with the critical value, $t = 1.746$, we note that the calculated value does not fall in the rejection region. Therefore, we must conclude that there is insufficient evidence to indicate that the new method of training is superior at the .05 level of significance. As we shall show in Example 10.6, the variability of the data is sufficient to mask a difference in μ_1 and μ_2, if in fact it exists. Consequently, the plant manager may wish to increase the number of employees in each sample and repeat the test.

Example 10.5

Find the p value that would be reported for the statistical test, Example 10.4.

Solution

The observed value of t for this one-tailed test was $t = 1.64$. Therefore, the p value for the test would be the probability that $t > 1.64$. Since we cannot obtain this probability from Table 4, the Appendix, we would report the p value for the test as the smallest tabulated value for α that leads to the rejection of H_0. Consulting the row in Table 4 corresponding to 16 degrees of freedom, we see that we would reject H_0 for this one-tailed test if t were greater than $t_{.10} = 1.337$ but not for $t_{.05} = 1.746$. Therefore, the p value for the test would be reported as $p < .10$. Because of the limitations of most t tables, a reader of this reported value would realize that the exact p value for the test was less than or equal to .10 but larger than .05.

The small-sample confidence interval for $(\mu_1 - \mu_2)$ is based upon the same assumptions as was the statistical test procedure. This confidence interval, with confidence coefficient $(1 - \alpha)$, is given by the formula:

Small-Sample Confidence Interval for $(\mu_1 - \mu_2)$

$$(\bar{y}_1 - \bar{y}_2) \pm t_{\alpha/2} s \sqrt{\frac{1}{n_1} + \frac{1}{n_2}},$$

where s is obtained from the pooled estimate of σ^2.

Assumptions: The samples were randomly and independently selected from normally distributed populations. The variances of the populations, σ_1^2 and σ_2^2, are equal.

Note the similarity in the procedures for constructing the confidence intervals for a single mean, Section 10.3, and the difference between two means. In both cases, the interval is constructed by using the appropriate point estimator and then adding and subtracting an amount equal to $t_{\alpha/2}$ times the *estimated* standard deviation of the point estimator.

Example 10.6

Find an interval estimate for $(\mu_1 - \mu_2)$, Example 10.4, using a confidence coefficient equal to .95.

Solution

Substituting into the formula

$$(\bar{y}_1 - \bar{y}_2) \pm t_{\alpha/2} s \sqrt{\frac{1}{n_1} + \frac{1}{n_2}},$$

we find the interval estimate (or 95 percent confidence interval) to be

$$(35.22 - 31.56) \pm (2.120)(4.72)\sqrt{\frac{1}{9} + \frac{1}{9}}$$

or

$$3.66 \pm 4.72.$$

Thus, we estimate the difference in mean time to assemble, $(\mu_1 - \mu_2)$, to fall in the interval -1.06 to 8.38. Note that the interval width is considerable and that it would seem advisable to increase the size of the samples and reestimate.

Before concluding our discussion we should comment on the two assumptions upon which our inferential procedures are based. Moderate departures from the assumption that the populations possess a normal probability distribution do not seriously affect the distribution of the test statistic and the confidence coefficient for the corresponding confidence interval. On the other hand, the population variances should be nearly equal in order that the aforementioned procedures be valid. A procedure will be presented in Section 10.8 for testing an hypothesis concerning the equality of two population variances.

If there is reason to believe that the population variances are unequal or that the normality assumptions have been violated, you can test for a shift in location of two population distributions using the nonparametric Mann–Whitney U test of Chapter 16. This test procedure, which requires fewer assumptions concerning the nature of the population probability distributions, is almost as sensitive in detecting a difference in population means when the conditions necessary for the t test are satisfied. It may be more sensitive when the assumptions are not satisfied.

Exercises

10.15. Independent random samples of $n_1 = 12$ and $n_2 = 16$ observations were selected from two normal populations with equal variances. The sample means and variances are shown below:

	Population	
	1	**2**
Sample Size	12	16
Sample Mean	34.7	32.1
Sample Variance	4.9	6.1

a. Suppose that you wish to detect a difference between the population means. State the null and alternative hypotheses that you would use for the test.

b. Find the rejection region for the test, part (a), for $\alpha = .10$.

c. Find the value of the test statistic.

d. Find the approximate observed significance level for the test.

e. Conduct the test and state your conclusions.

10.16.   Refer to Exercise 10.15. Find a 90 percent confidence interval for $(\mu_1 - \mu_2)$.

10.17.   As *Environment Midwest* (December 1976) notes, air pollution spread by air currents is becoming an increasing problem in rural as well as urban areas, particularly in the east. The data shown in the accompanying table presume to show the maximum amount of ozone recorded at each of eight Ohio air-pollution monitoring stations as well as the number of violations of the maximum allowable limit at each location recorded over the period, June 14 to August 31, 1974.

City	Designation	Maximum Concentration (ppm)	Days Exceeding Standard (%)	Number of Violations (Total)
Canton	urban	.14	44	148
Cincinnati	urban	.18	44	54
Cleveland	urban	.14	26	51
Columbus	urban	.15	27	113
Dayton	urban	.13	35	114
McConnelsville	rural	.16	56	239
Wilmington	rural	.18	58	259
Wooster	rural	.17	55	262

a. Do the data provide sufficient evidence to indicate that there is a difference in the mean number of violations between urban and rural areas? Test using $\alpha = .05$.

b. Find the approximate observed significance level for the test and interpret its value.

c. What must you assume in order to make your inference, part (a)?

10.18.   The length of time to recovery was recorded for patients randomly assigned and subjected to two different surgical procedures. The data, recorded in days, are as follows:

Procedure I	Procedure II
$n_1 = 21$	$n_2 = 23$
$\bar{y}_1 = 7.3$	$\bar{y}_2 = 8.9$
$s_1^2 = 1.23$	$s_2^2 = 1.49$

a. Do the data present sufficient evidence to indicate a difference in mean recovery time for the two surgical procedures? Test using $\alpha = .05$.

b. Find the approximate observed significance level for the test and interpret its value.

10.19. Refer to Exercise 10.13 where we measured the dissolved oxygen content in river water to determine whether a stream possessed sufficient oxygen to support aquatic life. A pollution control inspector suspected that a river community was releasing amounts of semitreated sewage into a river. To check his theory, he drew five randomly selected specimens of river water at a location above the town and another five below. The dissolved oxygen readings, in parts per million, are as follows:

Above Town	4.8	5.2	5.0	4.9	5.1
Below Town	5.0	4.7	4.9	4.8	4.9

a. Do the data provide sufficient evidence to indicate that the mean oxygen content between locations above and below the town is less than the mean oxygen content above? Test using $\alpha = .05$.

b. Suppose that you prefer estimation as a method of inference. Estimate the difference in mean dissolved oxygen content between locations above and below the town. Use a 95 percent confidence interval.

10.20. An experiment was conducted to compare the mean lengths of time required for the bodily absorption of two drugs, A and B. Ten people were randomly selected and assigned to each drug treatment. Each person received an oral dosage of the assigned drug and the length of time (in minutes) for the drug to reach a specified level in the blood was recorded. The means and variances for the two samples are as follows:

Drug A	Drug B
$\bar{y}_1 = 27.2$	$\bar{y}_2 = 33.5$
$s_1^2 = 16.36$	$s_2^2 = 18.92$

a. Do the data provide sufficient evidence to indicate a difference in mean times to absorption for the two drugs? Test using $\alpha = .10$.

b. Find the approximate observed significance level for the test and interpret its value.

10.21. Refer to Exercise 10.20. Find a 95 percent confidence interval for the difference in mean times to absorption.

10.22. Refer to Exercise 10.21. Suppose that you wished to estimate the difference in mean times to absorption correct to within 1 minute with probability approximately equal to .95.

a. Approximately how large a sample would be required for each drug (assume that the sample sizes will be equal)?

b. To conduct the experiment using the sample sizes of part (a) will require a large amount of time and money. Can anything be done to reduce the sample sizes and still achieve the 1-minute bound on the error of estimation?

10.23. In an article concerning the adverse effects of alcohol abuse on offspring during pregnancy, Ouellette* and colleagues comment on the mean age of the heavy drinkers (group A) versus the mean age of the moderate drinkers or abstinents (group B). The age data for the females included in the experiment are shown in the accompanying table. The authors utilize a t test to test the null hypothesis, $\mu_1 - \mu_2$, against the alternative that $\mu_1 \neq \mu_2$.

Group A Heavy Drinkers	Group B Abstinent or Moderate Drinkers
$n_A = 58$	$n_B = 575$
$\bar{y}_A = 25.7$ years	$\bar{y}_B = 22.8$ years
$s_A = 5.9$ years	$s_B = 5.5$ years

a. Is it necessary to use the t test for the analysis of these data? Explain.

b. The authors state that "($P < .001$ by t test)" or, equivalently, that the results are statistically significant at the .001 level. This means that if t_0 is the computed (observed) value of the t statistic, $P(t > t_0 \text{ or } t < -t_0) < .001$. Do you agree with this statement? Check their calculations.

10.24. A survey by *Industrial Research and Development* magazine found that salaries for research and development personnel rose 9.6 percent in 1981 to an average of $31,221 per year (*Orlando Sentinel Star*, April 26, 1981). The mean salaries in specific areas of application tended to be higher or lower than this overall mean. If a random sample of the salaries of 20 physicists possessed a mean, $\bar{y} = \$33,120$, and standard deviation, $s = \$2,140$, would you conclude that physicists are paid, on the average, substantially more than the overall mean of $31,221 per year? Test using $\alpha = .05$.

10.25. The survey, Exercise 10.24, found that the mean salary for chemists was substantially lower than the mean salary for physicists. Suppose that a random sample of the salaries of 20 chemists possessed a mean salary, $\bar{y} = \$30,000$, and standard deviation, $s = \$1,760$. Use these statistics and the statistics of Exercise 10.24 to find a 90 percent confidence interval for the mean annual difference in physicists' and chemists' salaries.

10.5
A Paired-Difference Test

A manufacturer wished to compare the wearing qualities of two different types of automobile tires, A and B. To make the comparison, a tire of type A and one of type B were randomly assigned and mounted on the rear wheels of each of five automobiles. The automobiles were then operated for a specified number of miles and the amount of wear was recorded for each tire. These measurements appear in Table 10.4. Do the data present sufficient evidence to indicate a difference in the average wear for the two tire types?

---

* Ouellette, E.M., H.L. Rosett, N.P. Rosman, and L. Weiner, "Adverse Effects on Offspring of Maternal Alcohol Abuse During Pregnancy," *New England J. of Med.*, *297* (September 8, 1977), pp. 528–530.

Table 10.4

Automobile	Tire A	Tire B
1	10.6	10.2
2	9.8	9.4
3	12.3	11.8
4	9.7	9.1
5	8.8	8.3
	$\bar{y}_1 = 10.24$	$\bar{y}_2 = 9.76$

Analyzing the data, we note that the difference between the two sample means is $(\bar{y}_1 - \bar{y}_2) = .48$, a rather small quantity, considering the variability of the data and the small number of measurements involved. At first glance it would seem that there is little evidence to indicate a difference between the population means, a conjecture that we may check by the method outlined in Section 10.4.

The pooled estimate of the common variance, σ^2, is

$$s^2 = \frac{\sum_{i=1}^{n_1} (y_i - \bar{y}_1)^2 + \sum_{i=1}^{n_2} (y_i - \bar{y}_2)^2}{n_1 + n_2 - 2} = \frac{6.932 + 7.052}{5 + 5 - 2} = 1.748,$$

and

$$s = 1.32.$$

The calculated value of t used to test the hypothesis that $\mu_1 = \mu_2$ is

$$t = \frac{\bar{y}_1 - \bar{y}_2}{s\sqrt{\frac{1}{n_1} + \frac{1}{n_2}}} = \frac{10.24 - 9.76}{1.32\sqrt{\frac{1}{5} + \frac{1}{5}}} = .57,$$

a value that is not nearly large enough to reject the hypothesis that $\mu_1 = \mu_2$.

The corresponding 95 percent confidence interval is

$$(\bar{y}_1 - \bar{y}_2) \pm t_{\alpha/2}s\sqrt{\frac{1}{n_1} + \frac{1}{n_2}} = (10.24 - 9.76) \pm (2.306)(1.32)\sqrt{\frac{1}{5} + \frac{1}{5}}$$

or -1.45 to 2.41. Note that the interval is quite wide, considering the small difference between the sample means.

A second glance at the data reveals a marked inconsistency with this conclusion. We note that the wear measurement for the type A is larger than the corresponding value for type B for *each* of the five automobiles. These differences, recorded as $d = A - B$, are as follows:

Automobile	$d = A - B$
1	.4
2	.4
3	.5
4	.6
5	.5
	$\bar{d} = .48$

Suppose that we were to use y, the number of times that A is larger than B, as a test statistic, as was done in Exercise 6.51. Then the probability that A would be larger than B on a given automobile, assuming no difference between the wearing quality of the tires, would be $p = 1/2$, and y would be a binomial random variable. If the null hypothesis were true, the expected value of y would be $\mu = np = 5(1/2) = 2.5$.

If we choose the most extreme values of y, $y = 0$ and $y = 5$, as the rejection region for a two-tailed test, then $\alpha = p(0) + p(5) = 2(1/2)^5 = 1/16$. We would then reject $H_0 : (\mu_1 = \mu_2)$ with a probability of type 1 error equal to $\alpha = 1/16$. Certainly this is evidence to indicate that a difference exists in the mean wear of the two tire types.

You will note that we have employed two different statistical tests to test the same hypothesis. Is it not peculiar that the t test, which utilizes more information (the actual sample measurements) than the binomial test, fails to supply sufficient evidence for rejection of the hypothesis $\mu_1 = \mu_2$?

The explanation of this seeming inconsistency is quite simple. The t test described in Section 10.4 is *not* the proper statistical test to be used for our example. The statistical test procedure, Section 10.4, required that the two samples be *independent* and random. Certainly, the independence requirement was violated by the manner in which the experiment was conducted. The (pair of) measurements, an A and B, for a particular automobile are definitely related. A glance at the data will show that the readings are of approximately the same magnitude for a particular automobile but vary from one automobile to another. This, of course, is exactly what we might expect. Tire wear, in a large part, is determined by driver habits, the balance of the wheels, and the road surface. Since each automobile had a different driver, we might expect a large amount of variability in the data from one automobile to another.

The familiarity that we have gained with interval estimation has shown that the width of the large- and small-sample confidence intervals will depend upon the magnitude of the standard deviation of the point estimator of the parameter. The smaller its value, the better the estimate and the more likely that the test statistic will reject the null hypothesis if it is, in fact, false. Knowledge of this phenomenon was utilized in *designing* the tire wear experiment.

The experimenter would realize that the wear measurements would vary greatly from auto to auto and that this variability could not be separated from the data if the tires were assigned to the 10 wheels in a *random* manner. (A random assignment of the tires would have implied that the data be analyzed according to the procedure of Section 10.4.) Instead, a comparison of the wear between the tire types A and B made on each automobile resulted in the five difference measurements. This design eliminates the effect of the car-to-car variability and yields more information on the mean difference in the wearing quality for the two tire types.

The proper analysis of the data would utilize the five difference measurements to test the hypothesis that the average difference μ_d is equal to zero, or, equivalently, to test the null hypothesis $H_0:\mu_d = \mu_1 - \mu_2 = 0$ against the alternative hypothesis, $\mu_d = \mu_1 - \mu_2 \neq 0$.

Paired-Difference Test for $(\mu_1 - \mu_2) = \mu_d$

1. *Null Hypothesis:* $H_0:\mu_d = 0$

2. *Alternative Hypothesis:*

One-Tailed Test	Two-Tailed Test
$H_a:\mu_d > 0$	$H_a:\mu_d \neq 0$
(or, $H_a:\mu_d < 0$)	

3. *Test Statistic:* $\quad t = \dfrac{\bar{d} - 0}{s_d/\sqrt{n}} = \dfrac{\bar{d}}{s_d/\sqrt{n}} \quad$ where

   n = number of paired differences

$$s_d = \sqrt{\frac{\sum\limits_{i=1}^{n} (d_i - \bar{d})^2}{n - 1}}$$

4. *Rejection Region:*

One-Tailed Test	Two-Tailed Test
$t > t_\alpha$	$t > t_{\alpha/2} \quad$ or $\quad t < -t_{\alpha/2}$
(or, $t < -t_\alpha$ when the alternative hypothesis is $H_a:\mu_d < 0$)	

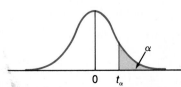

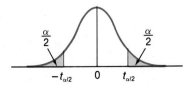

The critical values of t, t_α and $t_{\alpha/2}$, are based upon $(n - 1)$ degrees of freedom. These tabulated values are given in Table 4, the Appendix.

Assumptions: The n paired observations are randomly selected from normally distributed populations.

Example 10.7

Do the data, Table 10.4, provide sufficient evidence to indicate a difference in mean wear for tire types A and B? Test using $\alpha = .05$.

Solution

You can verify that the average and standard deviation of the five difference measurements are

$$\bar{d} = .48,$$

$$s_d = .0837.$$

Then,

$$H_0: \mu_d = 0 \quad \text{and} \quad H_a: \mu_d \neq 0$$

and

$$t = \frac{\bar{d} - 0}{s_d/\sqrt{n}} = \frac{.48}{.0837/\sqrt{5}} = 12.8.$$

The critical value of t for a two-tailed statistical test, $\alpha = .05$ and four degrees of freedom, is 2.776. Certainly, the observed value of $t = 12.8$ is extremely large and highly significant. Hence we would conclude that there is a difference in the mean amount of wear for tire types A and B.

$(1 - \alpha)$ 100 Percent Small-Sample Confidence Interval for $(\mu_1 - \mu_2) = \mu_d$, Based on a Paired-Difference Experiment

$$\bar{d} \pm t_{\alpha/2} \frac{s_d}{\sqrt{n}},$$

where $n =$ number of paired differences and

$$s_d = \sqrt{\frac{\sum\limits_{i=1}^{n} (d_i - \bar{d})^2}{n - 1}}.$$

> *Assumptions:* The n paired observations are randomly selected from normally distributed populations.

Example 10.8

Find a 95 percent confidence interval for $(\mu_1 - \mu_2) = \mu_d$.

Solution

A 95 percent confidence interval for the difference between the mean wear would be

$$\bar{d} \pm \frac{t_{\alpha/2} s_d}{\sqrt{n}} = .48 \pm (2.776)\frac{.0837}{\sqrt{5}}$$

or $.48 \pm .10$.

The statistical design of the tire experiment represents a simple example of a **randomized block design** and the resulting statistical test is often called a paired-difference test. Note that the pairing occurred when the experiment was planned and *not* after the data were collected. Comparisons of tire wear were made within relatively homogeneous blocks (automobiles) with the tire types randomly assigned to the two automobile wheels.

An indication of the gain in the amount of information obtained by blocking the tire experiment may be observed by comparing the calculated confidence interval for the unpaired (and incorrect) analysis with the interval obtained for the paired-difference analysis. The confidence interval for $(\mu_1 - \mu_2)$ that might have been calculated, had the tires been randomly assigned to the 10 wheels (unpaired), is unknown but likely would have been of the same magnitude as the interval -1.45 to 2.41 calculated by analyzing the observed data in an unpaired manner. Pairing the tire types on the automobiles (blocking) and the resulting analysis of the differences produced the interval estimate .38 to .58. Note the difference in the width of the intervals indicating the very sizable increase in information obtained by blocking in this experiment.

While blocking proved to be very beneficial in the tire experiment, this may not always be the case. We observe that the degrees of freedom available for estimating σ^2 are less for the paired than for the corresponding unpaired experiment. If there were actually no difference between the blocks, the reduction in the degrees of freedom would produce a moderate increase in the $t_{\alpha/2}$ employed in the confidence interval and hence increase the width of the interval. This, of course, did not occur in the tire experiment because the large reduction in the standard deviation of $\bar{d}$ more than compensated for the loss in degrees of freedom.

Before concluding, we want to reemphasize a point. Once you have used a paired design for an experiment, you no longer have the option of using the unpaired analysis of Section 10.4. The assumptions upon which that test are based have been violated. Your only alternative is to use the correct method of analysis, the paired-difference test (and associated confidence interval) of this section.

Exercises

10.26. A paired-difference experiment was conducted to compare the means of two populations. The data are shown below:

			Pairs		
Population	1	2	3	4	5
1	1.3	1.6	1.1	1.4	1.7
2	1.2	1.5	1.1	1.2	1.8

a. Do the data provide sufficient evidence to indicate that μ_1 differs from μ_2? Test using $\alpha = .05$.

b. Find the approximate observed significance level for the test and interpret its value.

c. Find a 95 percent confidence interval for $(\mu_1 - \mu_2)$. Compare your interpretation of the confidence interval with your test results, part (a).

d. What assumptions must you make in order that your inferences be valid?

10.27. Persons submitting computing jobs to a computer center are usually required to estimate the amount of computer time required to complete the job. This time is measured in CPU's, the amount of time that a job will occupy a portion of the computer's central processing unit's memory. A computer center decided to perform a comparison of the estimated versus actual CPU times for a particular customer. The corresponding times were available for 11 jobs. The sample data are shown below:

CPU Time minutes						Job Number					
	1	2	3	4	5	6	7	8	9	10	11
Estimated	.50	1.40	.95	.45	.75	1.20	1.60	2.6	1.30	.85	.60
Actual	.46	1.52	.99	.53	.71	1.31	1.49	2.9	1.41	.83	.74

a. Why would you expect these pairs of data to be correlated?

b. Do the data provide sufficient evidence to indicate that, *on the average*, the customer tends to underestimate CPU time required for computing jobs? Test using $\alpha = .10$.

c. Find the observed significance level for the test and interpret its value.

d. Find a 90 percent confidence interval for the difference in mean estimated CPU time versus mean actual CPU time.

10.28. Two computers are often compared by running a collection of different "benchmark" programs and recording the difference in CPU time required to complete the same program. Six benchmark programs, run on two computers, produced the following CPU times (in minutes):

Computer	Benchmark Program					
	1	2	3	4	5	6
1	1.12	1.73	1.04	1.86	1.47	2.10
2	1.15	1.72	1.10	1.87	1.46	2.15

a. Do the data provide sufficient evidence to indicate a difference in mean CPU times required for the two computers to complete a job? Test using $\alpha = .05$.

b. Find the approximate observed significance level for the test and interpret its value.

c. Find a 95 percent confidence interval for the difference in mean CPU time required for the two computers to complete a job.

10.29. The earth's temperature (which affects seed germination, crop survival in bad weather, and many other aspects of agricultural production) can be measured using either ground-based sensors or infrared sensing devices mounted in aircraft or space satellites. Ground-based sensing is tedious, requiring many replications to obtain an accurate estimate of ground temperature. On the other hand, airplane or satellite sensoring of infrared waves appears to introduce a bias in the temperature readings. To determine the bias, readings were obtained at six different locations using both ground- and air-based temperature sensors. The readings, in degrees Celsius, were as follows:

Location	Ground	Air
1	46.9	47.3
2	45.4	48.1
3	36.3	37.9
4	31.0	32.7
5	24.7	26.2

a. Do the data present sufficient evidence to indicate a bias in the air-collected temperature readings? Explain. (Use $\alpha = .05$.)

b. Estimate the difference in mean temperature between ground- and air-based sensors using a 95 percent confidence interval.

10.30. Why use paired observations to estimate the difference between two population means in preference to estimation based on independent random samples selected from the two populations? Is a paired experiment always preferable? Explain.

10.31. Refer to Exercise 10.29. How many paired observations would be required to estimate the difference in mean temperature between ground- and air-based sensors correct to within .2 degree Celsius with probability approximately equal to .95?

10.32. In response to a complaint that a particular tax assessor (A) was biased, an experiment was conducted to compare the assessor named in the complaint with another tax assessor (B) from the same office. Eight properties were selected and each was assessed by both assessors. The assessments (in thousands) are shown in the table.

Property	Assessor A	Assessor B
1	36.3	35.1
2	48.4	46.8
3	40.2	37.3
4	54.7	50.6
5	28.7	29.1
6	42.8	41.0
7	36.1	35.3
8	39.0	39.1

a. Do the data provide sufficient evidence to indicate that assessor A tends to give higher assessments than assessor B? Test with $\alpha = .05$.

b. Estimate the difference in mean assessments for the two assessors.

c. What assumptions must you make in order that the inferences, (a) and (b), are valid?

d. Suppose the assessor A had been compared with a more stable standard, say the average, $\bar{y}$, of the assessments given by four assessors selected from the tax office. Thus each property would be assessed by A and also by each of the four other assessors and $y_A - \bar{y}$ would be calculated. If the test, part (a), is valid, could you use the paired-difference t test to test the hypothesis that the bias, the mean difference between A's assessments and the mean of the assessments of the four assessors, is equal to 0? Explain.

 10.33.   Seven obese persons were placed on a diet for one month and the weights, at the beginning and at the end of the month, were recorded. They are shown in the table.

Subjects	Weights Initial	Final
1	310	263
2	295	251
3	287	249
4	305	259
5	270	233
6	323	267
7	277	242
8	299	265

Estimate the mean weight loss for obese persons when placed on the diet for a one-month period. Use a 95 percent confidence interval and interpret your results. What assumptions must you make in order that your inference be valid?

10.6
More on Smokeless Tobacco

You will recall that the Texas Lutheran College study of smokeless tobacco was based on an independent random sampling of the heart rates and blood pressures of ten athletes who used smokeless tobacco and ten who did not. The objective of the experiment was to determine whether tobacco usage caused an increase in the heart rate and blood pressure of a user. The data for the experiment, obtained from Professor Squires, are shown in Table 10.5.*

Table 10.5 Heart rates and diastolic and systolic blood pressures before and after the use of smokeless tobacco

Subject	Group	HRT1	HRT2	DYS1	DYS2	SYS1	SYS2
1	1	81	105	64	80	127	150
2	1	81	91	79	79	110	125
3	1	68	87	78	81	116	142
4	1	61	86	70	83	124	142
5	1	67	82	72	85	118	134
6	1	74	78	70	71	110	119
7	1	75	87	77	84	125	141
8	1	64	94	70	77	118	130
9	1	70	93	78	74	124	127
10	1	60	90	69	75	116	126
11	2	69	73	72	76	113	126
12	2	65	73	64	72	114	124
13	2	62	71	83	90	136	136
14	2	82	98	78	87	118	147
15	2	71	91	69	80	120	140
16	2	77	87	65	69	112	120
17	2	53	57	74	78	113	120
18	2	69	95	70	79	110	123
19	2	62	90	70	76	117	131
20	2	70	100	70	76	120	130

Each row of Table 10.5 corresponds to a particular athlete. The first column gives the number assigned to identify the athlete. The second column identifies the athlete as a non-user (Group 1) or a user (Group 2) of smokeless tobacco. Columns 3 and 4 give the heart rate at the beginning of the experiment (HRT1) and the heart rate after holding the tobacco for a 20-minute period (HRT2). Corresponding before and after readings for the diastolic blood pressures (DYS1 and DYS2) and the systolic blood pressures (SYS1 and SYS2) are shown in columns 5, 6, 7, and 8, respectively.

Does the 20-minute use of 2.5 grams of smokeless tobacco increase the heart rate, and the diastolic and systolic blood pressures? To answer this question for any one of these three variables, we must examine the paired differences in the variable reading before and after tobacco use, $d = y_{after} - y_{before}$, for each of the 20 athletes. You can see at a glance, based on the number of positive differences, that there appears to be sufficient evidence to indicate

* Personal communication.

that all three variables—heart rate, and diastolic and systolic blood pressures—increase after the use of smokeless tobacco, and this is true regardless of whether a person was a user or non-user of tobacco prior to the experiment. Confining our attention to the non-user group (group 1), you can see that the heart rate increased for all ten of ten subjects after the use of the smokeless tobacco. Eight of the ten experienced an increase in diastolic blood pressure, subject #2 experienced no change, and ten of ten recorded an increase in systolic blood pressure. An examination of the paired differences for users of tobacco (group 2) shows similar results. Assuming that the dip of tobacco had no effect on heart rate or blood pressure, the probability that an individual subject would experience a positive difference (an increase) for one of the variables is .5, and the probability of observing nine or ten positive differences in ten subjects is practically nil. Thus, using the logic employed in our cold vaccine test, Section 6.6, there is ample evidence to indicate that the use of smokeless tobacco produces an increase in heart rate and blood pressure for populations of individuals of whom the athlete subjects are representative.

We can reach the same conclusions using the more powerful paired-difference t test and can also use the t statistic to estimate the mean change in a variable after use of the smokeless tobacco. We will confine our attention to an analysis of the change in heart rates for non-users.

The before and after use of smokeless tobacco and their respective differences are shown in Table 10.6 for the ten non-user subjects.

Table 10.6 Heart rates before and after using smokeless tobacco for the non-user group

Subject	Heart Rate Before	Heart Rate After	Difference d	d^2
1	81	105	24	576
2	81	91	10	100
3	68	87	19	361
4	61	86	25	625
5	67	82	15	225
6	74	78	4	16
7	75	87	12	144
8	64	94	30	900
9	70	93	23	529
10	60	90	30	900
		Totals	192	4376

We will first calculate $\bar{d}$, s_d^2, and s_d:

$$\bar{d} = \frac{\sum_{i=1}^{10} d_i}{n} = \frac{192}{10} = 19.2$$

$$\sum_{i=1}^{n}(d_i - \bar{d})^2 = \sum_{i=1}^{n} d_i^2 - \frac{\left(\sum_{i=1}^{n} d_i\right)^2}{n} = 4376 - \frac{(192)^2}{10} = 689.6$$

$$s_d^2 = \frac{\sum_{i=1}^{n}(d_i - \bar{d})^2}{n-1} = \frac{689.6}{9} = 76.6222$$

$$s_d = 8.75.$$

To test the null hypothesis, $H_0: \mu_d = 0$, against the alternative hypothesis, $H_a: \mu_d > 0$, we employ the test statistic,

$$t = \frac{\bar{d} - \mu_d}{s_d/\sqrt{n}},$$

and, for $\alpha = .05$ and d.f. $= 9$, reject H_0 if the calculated value of t is $t > 1.833$. See Figure 10.3.

Figure 10.3 Rejection region to detect change in mean heart rate

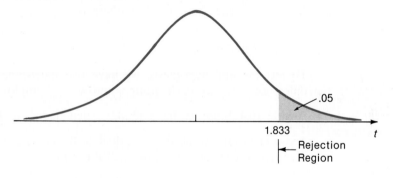

For the data, Table 10.6,

$$t = \frac{19.2 - 0}{8.75/\sqrt{10}} = 6.94.$$

This very large value of t lies in the rejection region and provides ample evidence to indicate that the 20-minute use of smokeless tobacco will produce an increase in the mean heart rate for non-users. Similar tests could be conducted and would verify that the heart rates and blood pressures increase after the use of the tobacco for both users and non-users.

We can also estimate the mean change in heart rate and blood pressure using confidence intervals. For example, a 95 percent confidence interval for the mean change in heart rate for non-users is

$$\bar{d} \pm t_{.025} s_d/\sqrt{n}.$$

Substituting into this formula,

$$19.2 \pm (2.262)\frac{(8.75)}{\sqrt{10}} = 19.2 \pm 6.26,$$

or, we estimate that the interval, 12.94 to 25.46, encloses the mean change in heart rate for athletes who, prior to the experiment, had been non-users of smokeless tobacco.

Do the data indicate a difference in the mean change in heart rate (or blood pressure) between the non-user and user groups? For example, we might suspect that users of tobacco are conditioned to its effects and that, although their heart rates (or blood pressures) might increase, the mean increase would be less than the mean increase for non-users. To test this theory, we would wish to test:

$$H_0 : \mu_{dU} - \mu_{dN} = 0,$$

where μ_{dU} and μ_{dN} are the mean changes in heart rates for users and non-users, respectively. The alternative hypothesis is:

$$H_a : \mu_{dU} < \mu_{dN}.$$

To test the null hypothesis, we have two independent random samples of differences—ten for each group—and we will employ the unpaired t test of Section 10.4. Since we have already calculated $\bar{d}_N$ and $\sum_{i=1}^{n} (d_i - \bar{d}_N)^2$ for the non-user group, we need only calculate the corresponding quantities for the user group. The data are shown in Table 10.7.

Table 10.7 Heart rates before and after using smokeless tobacco for the user group

Subject	Heart Rate Before	Heart Rate After	Difference d	d^2
11	69	73	4	16
12	65	73	8	64
13	62	71	9	81
14	82	98	16	256
15	71	91	20	400
16	77	87	10	100
17	53	57	4	16
18	69	95	26	676
19	62	90	28	784
20	70	100	30	900
		Totals	155	3293

We first calculate,

$$\bar{d}_U = \frac{\sum_{i=1}^{n} d_i}{n} = \frac{155}{10} = 15.5$$

and

$$\sum_{i=1}^{n} (d_i - \bar{d}_U)^2 = \sum_{i=1}^{n} d_i^2 - \frac{\left(\sum_{i=1}^{n} d_i\right)^2}{n} = 3293 - \frac{(155)^2}{10} = 890.5.$$

Then the pooled estimate of σ^2 is,

$$s^2 = \frac{\sum_{i=1}^{n_1} (d_i - \bar{d}_U)^2 + \sum_{i=1}^{n_2} (d_i - \bar{d}_N)^2}{n_1 + n_2 - 2}$$

$$= \frac{890.5 + 689.6}{10 + 10 - 2} = 87.78$$

and

$$s = 9.37.$$

Since we wish to detect only $\mu_{dU} < \mu_{dN}$, we will employ a one-tailed t test, with $\alpha = .05$ and d.f. $= 18$, and reject H_0 if the calculated value of t is $t < -1.734$ (see Figure 10.4).

The calculated value of t is,

$$t = \frac{(\bar{d}_U - \bar{d}_N) - 0}{s\sqrt{\frac{1}{n_1} + \frac{1}{n_2}}} = \frac{15.5 - 19.2}{9.37\sqrt{\frac{1}{10} + \frac{1}{10}}} = -.88.$$

Figure 10.4 Rejection region for comparing changes in mean heart rates for users and non-users of smokeless tobacco

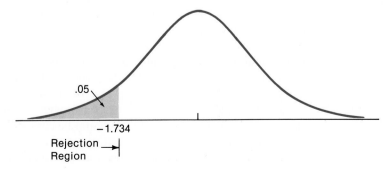

.05

−1.734

Rejection Region

Since this value of t does not fall in the rejection region, there is insufficient evidence to indicate a difference in the mean change in heart rate between groups 1 and 2.

In our preceding discussion, we analyzed only the heart rate data for the two groups of subjects. Similar analyses could be conducted for the blood pressure data. The mean changes in blood pressure and heart rate for the two groups of subjects are summarized in Table 10.8. We have indicated that there is ample evidence to conclude that the mean heart rate and blood pressure increases for both groups of subjects, and have analyzed the heart rate data, performing tests and constructing confidence intervals using the t statistic. Similar tests could be performed, and confidence intervals could be constructed, to make comparisons of blood pressure, before and after the use of smokeless tobacco, and to compare the changes between groups. We leave this as an exercise for the student.

Table 10.8 A summary of mean changes in heart rate and blood pressure

	Heart Rate	Diastolic Blood Pressure	Systolic Blood Pressure
Group 1 (Non-Users)	19.2	6.2	14.8
Group 2 (Users)	15.5	6.8	12.4

Is the use of smokeless tobacco harmful to your health? The answer depends upon a second question—whether the magnitudes of the changes in heart rate and blood pressure, indicated by the Squires study, could increase the risk of heart malfunctions. This question is one that can only be answered by the medical profession or by additional experimentation.

10.7
Inferences Concerning a Population Variance

We have seen in the preceding sections that an estimate of the population variance, σ^2, is fundamental to procedures for making inferences about population means. Moreover, there are many practical situations where σ^2 is the primary objective of an experimental investigation; thus it may assume a position of far greater importance than that of the population mean.

Scientific measuring instruments must provide unbiased readings with a very small error of measurement. An aircraft altimeter that measured the correct altitude on the *average* would be of little value if the standard deviation of the error of measurement were 5000 feet. Indeed, bias in a measuring instrument can often be corrected, but the precision of the instrument, measured by the standard deviation of the error of measurement, is usually a function of the design of the instrument itself and cannot be controlled.

Machined parts in a manufacturing process must be produced with minimum variability in order to reduce out-of-size and hence defective products.

And, in general, it is desirable to maintain a minimum variance in the measurements of the quality characteristics of an industrial product in order to achieve process control and therefore minimize the percentage of poor-quality product.

The sample variance,

$$s^2 = \frac{\sum_{i=1}^{n} (y_i - \bar{y})^2}{n-1},$$

is an unbiased estimator of the population variance, σ^2. Thus the distribution of sample variances generated by repeated sampling will have a probability distribution that commences at $s^2 = 0$ (since s^2 cannot be negative) with a mean equal to σ^2. Unlike the distribution of $\bar{y}$, the sampling distribution of s^2 is nonsymmetrical, the exact form being dependent upon the probability distribution of the population.

For the methodology that follows we will assume that the sample is drawn from a normal population and that s^2 is based upon a random sample of n measurements. Or, using the terminology of Section 10.2, we would say that s^2 possesses $(n - 1)$ degrees of freedom.

The next and obvious step would be to consider the distribution of s^2 in repeated sampling from a specified normal distribution—one with a specific mean and variance—and to tabulate the critical values of s^2 for some of the commonly used tail areas. If this is done, we shall find that the distribution of s^2 is independent of the population mean, μ, but possesses a distribution for each sample size and each value of σ^2. This task would be quite laborious, but fortunately it may be simplified by *standardizing*, as was done by using z in the normal tables.

The quantity

$$\chi^2 = \frac{(n-1)s^2}{\sigma^2},$$

called a **chi-square variable** by statisticians, admirably suits our purposes. Its distribution in repeated sampling is called, as we might suspect, a **chi-square probability distribution.** The equation of the density function for the chi-square distribution is well known to statisticians who have tabulated critical values corresponding to various tail areas of the distribution. These values are presented in Table 5, the Appendix.

The shape of the chi-square distribution, like that of the t distribution, will vary with the sample size, or, equivalently, with the degrees of freedom associated with s^2. Thus Table 5, the Appendix, is constructed in exactly the same manner as the t table, with the degrees of freedom shown in the first and last column. A partial reproduction of Table 5, the Appendix, is shown in Table 10.9. The symbol χ_α^2 indicates that the tabulated χ^2 value is such that an area, α, lies

Table 10.9 Format of the chi-square table, Table 5, the Appendix

d.f.	$\chi^2_{0.995}$	$\cdots$	$\chi^2_{0.950}$	$\chi^2_{0.900}$	$\chi^2_{0.100}$	$\chi^2_{0.050}$	$\cdots$	$\chi^2_{0.005}$	d.f.
1	0.0000393		0.0039321	0.0157908	2.70554	3.84146		7.87944	1
2	0.0100251		0.102587	0.210720	4.60517	5.99147		10.5966	2
3	0.0717212		0.351846	0.584375	6.25139	7.81473		12.8381	3
4	0.206990		0.710721	1.063623	7.77944	9.48773		14.8602	4
5	0.411740		1.145476	1.61031	9.23635	11.0705		16.7496	5
6	0.675727		1.63539	2.20413	10.6446	12.5916		18.5476	6
$\vdots$	$\vdots$		$\vdots$	$\vdots$	$\vdots$	$\vdots$		$\vdots$	$\vdots$
15	4.60094		7.26094	8.54675	22.3072	24.9958		32.8013	15
16	5.14224		7.96164	9.31223	23.5418	26.2962		34.2672	16
17	5.69724		8.67176	10.0852	24.7690	27.5871		35.7185	17
18	6.26481		9.39046	10.8649	25.9894	28.8693		37.1564	18
19	6.84398		10.1170	11.6509	27.2036	30.1435		38.5822	19
$\vdots$	$\vdots$		$\vdots$	$\vdots$	$\vdots$	$\vdots$		$\vdots$	$\vdots$

to its right. (See Figure 10.5.) Stated in probabilistic terms,

$$P(\chi^2 > \chi^2_\alpha) = \alpha.$$

Thus, 99 percent of the area under the χ^2 distribution would lie to the right of $x^2_{.99}$. We note that the extreme values of χ^2 must be tabulated for both the lower and upper tail of the distribution because it is nonsymmetrical.

Figure 10.5 A chi-square distribution

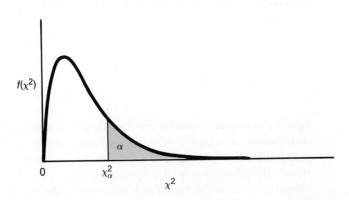

You can check your ability to use the table by verifying the following statements. The probability that χ^2, based upon $n = 16$ measurements (d.f. = 15), will exceed 24.9958 is .05. For a sample of $n = 6$ measurements (d.f. = 5), 95 percent of the area under the χ^2 distribution will lie to the right of $\chi^2 = 1.145476$. These values of χ^2 are shaded in Table 10.9. The statistical test of a null hypothesis concerning a population variance,

$$H_0\!:\!\sigma^2 = \sigma_0^2,$$

will employ the test statistic

$$\chi^2 = \frac{(n-1)s^2}{\sigma_0^2}.$$

If σ^2 is really greater than the hypothesized value, σ_0^2, then the test statistic will tend to be large and will likely fall toward the upper tail of the distribution. If $\sigma^2 < \sigma_0^2$, the test statistic will tend to be small and will likely fall toward the lower tail of the χ^2 distribution. As in other statistical tests, we may use either a one- or two-tailed statistical test, depending upon the alternative hypothesis that we choose.

Test of an Hypothesis Concerning a Population Variance

1. *Null Hypothesis:* $H_0\!:\!\sigma^2 = \sigma_0^2$.

2. *Alternative Hypothesis:*

One-Tailed Test	*Two-Tailed Test*
$H_a\!:\!\sigma^2 > \sigma_0^2$	$H_a\!:\!\sigma^2 \neq \sigma_0^2$
(or, $H_a\!:\!\sigma^2 < \sigma_0^2$)	

3. *Test Statistic:* $\chi^2 = \dfrac{(n-1)s^2}{\sigma_0^2}.$

4. *Rejection Region:*

   One-Tailed Test

   $\chi^2 > \chi_\alpha^2$
   (or, $\chi^2 < \chi_{(1-\alpha)}^2$ when the alternative hypothesis is $H_a\!:\!\sigma^2 < \sigma_0^2$) where χ_α^2 and $\chi_{(1-\alpha)}^2$ are, respectively, the upper- and lower-tail values of χ^2 that place α in the tail areas.

   Two-Tailed Test

   $\chi^2 > \chi_{\alpha/2}^2$ or $\chi^2 < \chi_{(1-\alpha/2)}^2$ where $\chi_{\alpha/2}^2$ and $\chi_{(1-\alpha/2)}^2$ are, respectively, the upper- and lower-tail values of χ^2 that place $\alpha/2$ in the tail areas.

The critical values of χ^2 are based upon $(n-1)$ degrees of freedom. These tabulated values are given in Table 5, the Appendix.

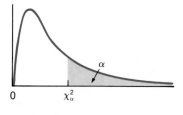
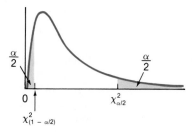

Assumption: The sample has been randomly selected from a normal population.

Example 10.9

A cement manufacturer claimed that concrete prepared from his product would possess a relatively stable compressive strength and that the strength measured in kilograms per square centimeter would lie within a range of 40 kilograms per square centimeter. A sample of $n = 10$ measurements produced a mean and variance equal to, respectively,

$$\bar{y} = 312,$$
$$s^2 = 195.$$

Solution

Do these data present sufficient evidence to reject the manufacturer's claim?

As stated, the manufacturer claimed that the range of the strength measurements would equal 40 kilograms per square centimeter. We will suppose that he meant that the measurements would lie within this range 95 percent of the time and, therefore, that the range would equal approximately 4σ and that $\sigma = 10$. We would then wish to test the null hypothesis

$$H_0 : \sigma^2 = (10)^2 = 100$$

against the alternative

$$H_a : \sigma^2 > 100.$$

The alternative hypothesis would require a one-tailed statistical test with the entire rejection region located in the upper tail of the χ^2 distribution. The critical value of χ^2 for $\alpha = .05$ and $(n - 1) = 9$ degrees of freedom is $\chi^2 = 16.9190$, which implies that we will reject H_0 if the test statistic exceeds this value.

Calculating, we obtain

$$\chi^2 = \frac{(n-1)s^2}{\sigma_0^2} = \frac{1755}{100} = 17.55.$$

Since the value of the test statistic falls in the rejection region, we conclude that the null hypothesis is false and that the range of concrete-strength measurements will exceed the manufacturer's claim.

Example 10.10

Find the approximate observed significance level for the test, Example 10.9.

Solution

Examining the row corresponding to 9 degrees of freedom in Table 5, the Appendix, you will see that he observed value of chi-square, $\chi^2 = 17.55$, is larger than the tabulated value, $\chi^2_{.05} = 16.9190$, and less than $\chi^2_{.025} = 19.0228$. Therefore, the observed significance level (p value) for the test lies between .05 and .025 and we would report the observed significance level for the test as $p < .05$. This tells us that we would reject the null hypothesis for any value of α equal to or greater than .05.

A confidence interval for σ^2 with a $(1 - \alpha)$ confidence coefficient can be shown to be:

$(1 - \alpha)$ 100 Percent Confidence Interval for σ^2

$$\frac{(n - 1)s^2}{\chi^2_{\alpha/2}} < \sigma^2 < \frac{(n - 1)s^2}{\chi^2_{(1 - \alpha/2)}},$$

where $\chi^2_{\alpha/2}$ and $\chi^2_{(1 - \alpha/2)}$ are the upper and lower χ^2 values which would locate one-half of α in each tail of the chi-square distribution.

Assumption: The sample has been randomly selected from a normal population.

Example 10.11

Find a 90 percent confidence interval for σ^2, Example 10.9.

Solution

The first step would be to obtain the tabulated values of $\chi^2_{.95}$ and $\chi^2_{.05}$ corresponding to $(n - 1) = 9$ degrees of freedom.

$$\chi^2_{(1 - \alpha/2)} = \chi^2_{.95} = 3.32511,$$
$$\chi^2_{\alpha/2} = \chi^2_{.05} = 16.9190.$$

Substituting these values and $s^2 = 195$ into the formula for the confidence interval,

$$\frac{(n - 1)s^2}{\chi^2_{\alpha/2}} < \sigma^2 < \frac{(n - 1)}{\chi^2_{(1 - \alpha/2)}}.$$

The interval estimate for σ^2 would be

$$\frac{9(195)}{16.9190} < \sigma^2 < \frac{9(195)}{3.32511}$$

or

$$103.73 < \sigma^2 < 527.80.$$

Example 10.12

An experimenter was convinced that his measuring equipment possessed a variability measured by a standard deviation, $\sigma = 2$. During an experiment he recorded the measurements 4.1, 5.2, 10.2. Do these data disagree with his assumption? Test the hypothesis, $H_0: \sigma = 2$ or $\sigma^2 = 4$. Then place a 90 percent confidence interval on σ^2.

Solution

The calculated sample variance is $s^2 = 10.57$. Since we wish to detect $\sigma^2 > 4$ as well as $\sigma^2 < 4$, we should employ a two-tailed test. When we use $\alpha = .10$ and place .05 in each tail, we will reject H_0 when $\chi^2 > 5.99147$ or $\chi^2 < .102587$.

The calculated value of the test statistic is

$$\chi^2 = \frac{(n-1)s^2}{\sigma_0^2} = \frac{2(10.57)}{4} = 5.29.$$

Since the test statistic does not fall in the rejection region, the data do not provide sufficient evidence to reject the null hypothesis, $H_0 : \sigma^2 = 4$.

The corresponding 90 percent confidence interval is

$$\frac{(n-1)s^2}{\chi_{\alpha/2}^2} < \sigma^2 < \frac{(n-1)s^2}{\chi_{(1-\alpha/2)}^2}.$$

The values of $\chi_{(1-\alpha/2)}^2$ and $\chi_{\alpha/2}^2$ are

$$\chi_{(1-\alpha/2)}^2 = \chi_{.95}^2 = .102587,$$

$$\chi_{\alpha/2}^2 = \chi_{.05}^2 = 5.99147.$$

Substituting these values into the formula for the interval estimate, we obtain

$$\frac{2(10.57)}{5.99147} < \sigma^2 < \frac{2(10.57)}{.102587}$$

or

$$3.53 < \sigma^2 < 206.07.$$

Thus we estimate the population variance to fall in the interval, 3.53 to 206.07. This very wide confidence interval indicates how little information on the population variance is obtained in a sample of only three measurements. Consequently, it is not surprising that there was insufficient evidence to reject the null hypothesis, $\sigma^2 = 4$. To obtain more information on σ^2, the experimenter needs to increase the sample size.

Exercises

10.34.   A random sample of 22 observations from a normal population possessed a variance equal to 37.3.

   a. Do these data provide sufficient evidence to indicate that the population variance exceeds 30.0? Test using $\alpha = .05$.

   b. Find the approximate observed significance level for the test and interpret its value.

10.35.   Refer to the comparison of the estimated versus actual CPU times for computer jobs submitted by the customer, Exercise 10.27.

CPU Time	Job Number										
(minutes)	1	2	3	4	5	6	7	8	9	10	11
Estimated	.50	1.40	.95	.45	.75	1.20	1.60	2.6	1.30	.85	.60
Actual	.46	1.52	.99	.53	.71	1.31	1.49	2.9	1.41	.83	.74

Find a 90 percent confidence interval for the variance of the differences between estimated and actual CPU times.

10.36. Refer to Exercise 10.35. Suppose that the computer center expects customers to estimate CPU time correct to within .2 minutes most (say 95 percent) of the time.

a. Do the data provide sufficient evidence to indicate that the customer's error in estimating job CPU time exceeds the computer center's expectation? Test using $\alpha = .10$.

b. Find the approximate observed significance level for the test and interpret its value.

10.37. In Exercise 10.5 we mentioned that the EPA set a maximum noise level for heavy trucks in 1978 at 83 decibels. How this limit will be applied will greatly affect the industry and the public. One way to apply the limit would be to require all trucks to conform to the noise limit. A second but less satisfactory method would be to require the truck fleet mean noise level to be less than the limit. If the latter were the rule, variation in the noise level from truck to truck would be important because a large value of σ^2 would imply many trucks exceeding the limit, even if the mean fleet level was 83 decibels. The data for the six trucks, Exercise 10.5, in decibels, were

85.4, 86.8, 86.1, 85.3, 84.8, 86.0.

Use these data to construct a 90 percent confidence interval for σ^2, the variance of the truck noise emission readings. Interpret your results.

10.38. A precision instrument is guaranteed to read accurately to within 2 units. A sample of four instrument readings on the same object yielded the measurements 353, 351, 351, and 355. Test the null hypothesis that $\sigma = .7$ against the alternative $\sigma > .7$. Conduct the test at the $\alpha = .05$ level of significance.

10.39. Find a 90 percent confidence interval for the population variance in Exercise 10.38.

10.40. In order to properly treat patients, drugs prescribed by physicians must possess a potency that is accurately defined. Consequently, the distribution of potency values for shipments of a drug must not only possess a mean value as specified on the drug's container, but the variation in potency must be small. Otherwise, pharmacists would be distributing drug prescriptions that could be harmfully potent or possess a low potency that would be ineffective. A drug manufacturer claims that his drug is marketed with a potency of $5 \pm .1$ milligram per cubic centimeter. A random sample of four containers gave potency readings equal to 4.94, 5.09, 5.03, and 4.90 mg/cc.

a. Do the data present sufficient evidence to indicate that the mean potency differs from 5 mg/cc?

b. Do the data present sufficient evidence to indicate that the variation in potency differs from the error limits specified by the manufacturer? [Hint: It is sometimes difficult to determine exactly what is meant by limits on potency as specified by a manufacturer. Since he implies that the potency values will fall in the interval $5 \pm .1$ mg/cc with very high probability—the implication is "always"—let us assume that the range .2(.49 to .51) represents 6σ, as suggested by the Empirical Rule. Note that letting the range equal 6σ rather than 4σ places a stringent interpretation on the manufacturer's claim. We want the potency to fall in the interval $5.0 \pm .1$ with very high probability.]

10.41.   Refer to Exercise 10.40. Testing of an additional 60 randomly selected containers of the drug gave a sample mean and variance equal to 5.04 and .0063 (for the total of $n = 64$ containers). Using a 95 percent confidence interval, estimate the variance of the manufacturer's potency measurements.

10.42.   A manufacturer of hard safety hats for construction workers is concerned about the mean and the variation of the forces helmets transmit to wearers when subjected to a standard external force. The manufacturer desires the mean force transmitted by helmets to be 800 pounds (or less), well under the legal 1000-pound limit, and σ to be less than 40. A random sample of $n = 40$ helmets was tested and the sample mean and variance were found to be equal to 825 pounds and 2350 (pounds)^2, respectively.

   a. If $\mu = 800$ and $\sigma = 40$, is it likely that any helmet, subjected to the standard external force, will transmit a force to a wearer in excess of 1000 pounds? Explain.

   b. Do the data provide sufficient evidence to indicate that when subjected to the standard external force, the mean force transmitted by the helmets exceeds 800 pounds?

   c. Do the data provide sufficient evidence to indicate that σ exceeds 40?

10.8
Comparing Two Population Variances

The need for statistical methods to compare two population variances is readily apparent from the discussion in Section 10.7. We may frequently wish to compare the precision of one measuring device with that of another, the stability of one manufacturing process with that of another, or even the variability in the grading procedure of one college professor with that of another.

Intuitively, we might compare two population variances, σ_1^2 and σ_2^2, using the ratio of the sample variances s_1^2/s_2^2. If s_1^2/s_2^2 is nearly equal to 1, we would find little evidence to indicate that σ_1^2 and σ_2^2 are unequal. On the other hand, a very large or very small value for s_1^2/s_2^2 would provide evidence of a difference in the population variances.

How large or small must s_1^2/s_2^2 be in order that sufficient evidence exists to reject the null hypothesis,

$$H_0: \sigma_1^2 = \sigma_2^2?$$

The answer to this question may be acquired by studying the distribution of s_1^2/s_2^2 in repeated sampling.

When independent random samples are drawn from two normal populations with equal variances, that is, $\sigma_1^2 = \sigma_2^2$, then s_1^2/s_2^2 possesses a probability distribution in repeated sampling that is known to statisticians as an **F distribution.** We need not concern ourselves with the equation of the density function for F except to state that, as we might surmise, it is reasonably complex. For our purposes it will suffice to accept the fact that the distribution is well known and that critical values have been tabulated. These appear in Tables 6, 7, 8, 9, and 10, the Appendix.

The shape of the F distribution is nonsymmetrical and will depend upon the number of degrees of freedom associated with s_1^2 and s_2^2. We will represent these quantities as v_1 and v_2, respectively. This fact complicates the tabulation of critical values for the F distribution and necessitates the construction of a table for each value that we may choose for a tail area, α. Thus, Tables 6, 7, 8, 9, and 10, the Appendix, present critical values corresponding to $\alpha = .10, .05, .025, .01,$ and $.005$, respectively. A typical F probability distribution ($v_1 = 10$ and $v_2 = 10$) is shown in Figure 10.6.

Figure 10.6 An F distribution with $v_1 = 10$ and $v_2 = 10$

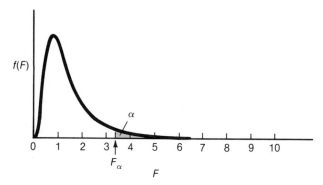

A partial reproduction of Table 7, the Appendix, is shown in Table 10.10. Table 7 records the value $F_{.05}$ such that the probability that F will exceed $F_{.05}$ is $.05$. Another way of saying this is that 5 percent of the area under the F distribution lies to the right of $F_{.05}$. The degrees of freedom for s_1^2, v_1, are

Table 10.10 Format of the F table, Table 7, the Appendix, $\alpha = .05$

v_2 (d.f.)	1	2	3	4	5	6	7	8	9	$\cdots$	60	120	∞	v_2 (d.f.)
1	161.4	199.5	215.7	224.6	230.2	234.0	236.8	238.9	240.5		252.2	253.3	254.3	1
2	18.51	19.00	19.16	19.25	19.30	19.33	19.35	19.37	19.38		19.48	19.49	19.50	2
3	10.13	9.55	9.28	9.12	9.01	8.94	8.89	8.85	8.81		8.57	8.55	8.53	3
4	7.71	6.94	6.59	6.39	6.26	6.16	6.09	6.04	6.00		5.69	5.66	5.63	4
5	6.61	5.79	5.41	5.19	5.05	4.95	4.88	4.82	4.77		4.43	4.40	4.36	5
6	5.99	5.14	4.76	4.53	4.39	4.28	4.21	4.15	4.10		3.74	3.70	3.67	6
7	5.59	4.74	4.35	4.12	3.97	3.87	3.79	3.73	3.68		3.30	3.27	3.23	7
8	5.32	4.46	4.07	3.84	3.69	3.58	3.50	3.44	3.39		3.01	2.97	2.93	8
9	5.12	4.26	3.86	3.63	3.48	3.37	3.29	3.23	3.18		2.79	2.75	2.71	9
$\vdots$	$\vdots$	$\vdots$	$\vdots$	$\vdots$	$\vdots$	$\vdots$	$\vdots$	$\vdots$	$\vdots$		$\vdots$	$\vdots$	$\vdots$	$\vdots$
15	4.54	3.68	3.29	3.06	2.90	2.79	2.71	2.64	2.59		2.16	2.11	2.07	15
16	4.49	3.63	3.24	3.01	2.85	2.74	2.66	2.59	2.54		2.11	2.06	2.01	16
17	4.45	3.59	3.20	2.96	2.81	2.70	2.61	2.55	2.49		2.06	2.01	1.96	17
18	4.41	3.55	3.16	2.93	2.77	2.66	2.58	2.51	2.46		2.02	1.97	1.92	18
19	4.38	3.52	3.13	2.90	2.74	2.63	2.54	2.48	2.42		1.98	1.93	1.88	19
$\vdots$	$\vdots$	$\vdots$	$\vdots$	$\vdots$	$\vdots$	$\vdots$	$\vdots$	$\vdots$	$\vdots$		$\vdots$	$\vdots$	$\vdots$	$\vdots$

indicated across the top of the table, and the degrees of freedom for s_2^2, v_2, appear in the first column on the left.

Referring to Table 10.10 we note that $F_{.05}$ for sample sizes $n_1 = 7$ and $n_2 = 10$ (that is, $v_1 = 6$, $v_2 = 9$) is 3.37. Likewise, the critical value, $F_{.05}$, for sample sizes $n_1 = 9$ and $n_2 = 16$ ($v_1 = 8$, $v_2 = 15$) is 2.64. These values of F are shaded in Table 10.10.

In a similar manner, the critical values for a tail area, $\alpha = .01$, are presented in Table 9, Appendix. Thus

$$P(F > F_{.01}) = .01.$$

The statistical test of the null hypothesis,

$$H_0 : \sigma_1^2 = \sigma_2^2,$$

utilizes the test statistic,

$$F = \frac{s_1^2}{s_2^2}.$$

When the alternative hypothesis implies a one-tailed test, that is,

$$H_a : \sigma_1^2 > \sigma_2^2,$$

we may use the tables directly. However, when the alternative hypothesis requires a two-tailed test,

$$H_a : \sigma_1^2 \neq \sigma_2^2,$$

we note that the rejection region will be divided between the lower and upper tails of the F distribution and that tables of critical values for the lower tail are conspicuously missing. The reason for their absence is explained as follows: We are at liberty to identify either of the two populations as population I. If the population with the larger sample variance is designated as population II, then $s_2^2 > s_1^2$ and we will be concerned with rejection in the lower tail of the F distribution. Since the identification of the populations was arbitrary, we may avoid this difficulty by designating the population with the larger sample variance as population I. In other words, always place the larger sample variance in the numerator of

$$F = \frac{s_1^2}{s_2^2}$$

and designate that population as I. Then, since the area in the right-hand tail will represent only $\alpha/2$, we double this value to obtain the correct value for the probability of a type I error, α. Hence, if we use Table 7, the Appendix, for a two-tailed test, the probability of a type I error will be $\alpha = .10$.

Test of an Hypothesis Concerning the Equality of Two Population Variances

1. *Null Hypothesis:* $H_0: \sigma_1^2 = \sigma_2^2$.

2. *Alternative Hypothesis:*

One-Tailed Test	Two-Tailed Test
$H_a: \sigma_1^2 > \sigma_2^2$	$H_a: \sigma_1^2 \neq \sigma_2^2$
(or, $H_a: \sigma_2^2 > \sigma_1^2$)	

3. *Test Statistic:*

One-Tailed Test	Two-Tailed Test
$F = \dfrac{s_1^2}{s_2^2}$	$F = \dfrac{s_1^2}{s_2^2}$

$$\left(\text{or, } F = \frac{s_2^2}{s_1^2} \text{ for } H_a: \sigma_2^2 > \sigma_1^2 \right)$$ where s_1^2 is the larger sample variance.

4. *Rejection Region:*

One-Tailed Test	Two-Tailed Test
$F > F_\alpha$	$F > F_{\alpha/2}$

When $F = s_1^2/s_2^2$, the critical values of F, F_α and $F_{\alpha/2}$, are based on $v_1 = n_1 - 1$ and $v_2 = n_2 - 1$ degrees of freedom. These tabulated values, for $\alpha = .10, .05, .025, .01$, and $.005$, can be found in Tables 6, 7, 8, 9, and 10, the Appendix, respectively.

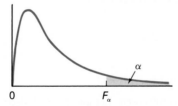
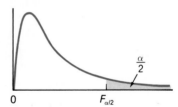

Assumptions: The samples were randomly and independently selected from normally distributed populations.

We illustrate with examples.

Example 10.13   Two samples consisting of 10 and 8 measurements each were observed to possess sample variances equal to $s_1^2 = 7.14$ and $s_2^2 = 3.21$, respectively. Do the sample variances present sufficient evidence to indicate that the population variances are unequal?

Solution   Assume that the populations possess probability distributions that are reasonably mound-shaped and hence will satisfy, for all practical purposes, the assumption that the populations are normal.

We wish to test the null hypothesis,

$$H_0: \sigma_1^2 = \sigma_2^2,$$

against the alternative,

$$H_a: \sigma_1^2 \neq \sigma_2^2.$$

Using Table 7, the Appendix, and doubling the tail area, we will reject H_0 when $F > 3.68$ with $\alpha = .10$.

The calculated value of the test statistic is

$$F = \frac{s_1^2}{s_2^2} = \frac{7.14}{3.21} = 2.22.$$

Noting that the test statistic does not fall in the rejection region, we do not reject $H_0: \sigma_1^2 = \sigma_2^2$. Thus, there is insufficient evidence to indicate a difference in the population variances.

___

The confidence interval, with confidence coefficient $(1 - \alpha)$, for the ratio of two population variances can be shown to equal:

A Confidence Interval for σ_1^2/σ_2^2

$$\frac{s_1^2}{s_2^2} \frac{1}{F_{v_1, v_2}} < \frac{\sigma_1^2}{\sigma_2^2} < \frac{s_1^2}{s_2^2} F_{v_2, v_1},$$

where $v_1 = n_1 - 1$ and $v_2 = n_2 - 1$.

Assumptions: The samples were randomly and independently selected from normally distributed populations.

---

where F_{v_1, v_2} is the tabulated critical value of F corresponding to v_1 and v_2 degrees of freedom in the numerator and denominator of F, respectively. Similar to the two-tailed test, the α will be double the tabulated value. Thus F values extracted from Tables 7 and 8 will be appropriate for confidence coefficients equal to .90 and .95, respectively.

Example 10.14   Refer to Example 10.13 and find a 90 percent confidence interval for σ_1^2/σ_2^2.

Solution   The 90 percent confidence interval for σ_1^2/σ_2^2, Example 10.13, is

$$\frac{s_1^2}{s_2^2} \frac{1}{F_{v_1, v_2}} < \frac{\sigma_1^2}{\sigma_2^2} < \frac{s_1^2}{s_2^2} F_{v_2, v_1},$$

where

$$s_1^2 = 7.14, \qquad\qquad s_2^2 = 3.21,$$

$$v_1 = (n_1 - 1) = 9, \qquad v_2 = (n_2 - 1) = 7,$$

$$F_{v_1,v_2} = F_{9,7} = 3.68, \qquad F_{v_2,v_1} = F_{7,9} = 3.29.$$

Substituting these values into the formula for the confidence interval we obtain

$$\left(\frac{7.14}{3.21}\right)\frac{1}{3.68} < \frac{\sigma_1^2}{\sigma_2^2} < \frac{(7.14)(3.29)}{3.21}$$

or

$$.60 < \frac{\sigma_1^2}{\sigma_2^2} < 7.32.$$

The calculated interval estimate, .60 to 7.32, is observed to include 1.0, the value hypothesized in H_0. This indicates that it is quite possible that $\sigma_1^2 = \sigma_2^2$ and therefore agrees with our test conclusions. Do not reject $H_0 : \sigma_1^2 = \sigma_2^2$.

Example 10.15

The variability in the amount of impurities present in a batch of chemical used for a particular process depends upon the length of time the process is in operation. A manufacturer using two production lines, 1 and 2, has made a slight adjustment to process 2, hoping to reduce the variability as well as the average amount of impurities in the chemical. Samples of $n_1 = 25$ and $n_2 = 25$ measurements from the two batches yield means and variances as follows:

$$\bar{y}_1 = 3.2, \quad s_1^2 = 1.04;$$

$$\bar{y}_2 = 3.0, \quad s_2^2 = .51.$$

Do the data present sufficient evidence to indicate that the process variability is less for process 2? Test the null hypothesis, $H_0 : \sigma_1^2 = \sigma_2^2$.

Solution

The practical implications of Example 10.15 are illustrated in Figure 10.7. We believe that the mean levels of impurities in the two production lines are nearly equal (in fact, that they may be equal) but that there is a possibility that the variation in the level of impurities is substantially less for line 2. Then distributions of impurity measurements for the two production lines would have nearly the same mean level but they would differ in their variation. A large variance for the level of impurities increases the probability of producing shipments of chemical with an unacceptably high level of impurities. Consequently, we hope to show that the process change in line 2 has made σ_2^2 less than σ_1^2.

Testing the null hypothesis,

$$H_0 : \sigma_1^2 = \sigma_2^2,$$

Figure 10.7 Distributions of impurity measurements for two production lines

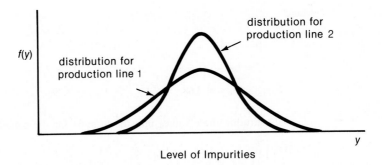

against the alternative,

$$H_a : \sigma_1^2 > \sigma_2^2,$$

at an $\alpha = .05$ significance level, we will reject H_0 when F is greater than $F_{.05} = 1.98$; that is, we shall employ a one-tailed statistical test.

We readily observe that the calculated value of the test statistic,

$$F = \frac{s_1^2}{s_2^2} = \frac{1.04}{.51} = 2.04,$$

falls in the rejection region, and hence we conclude that the variability of process 2 is less than that for process 1.

The 90 percent confidence interval for the ratio σ_1^2/σ_2^2 is

$$\frac{s_1^2}{s_2^2} \frac{1}{F_{v_1, v_2}} < \frac{\sigma_1^2}{\sigma_2^2} < \frac{s_1^2}{s_2^2} F_{v_2, v_1}$$

$$\frac{1.04}{(.51)(1.98)} < \frac{\sigma_1^2}{\sigma_2^2} < \frac{(1.04)(1.98)}{.51}$$

or

$$1.03 < \frac{\sigma_1^2}{\sigma_2^2} < 4.04.$$

Thus we estimate the reduction in the variance of the amount of impurities to be as large as 4.04 to 1 or as small as 1.03 to 1. The actual reduction would likely be somewhere between these two extremes. This would suggest that the adjustment is quite effective in reducing the variation in the amount of impurities in the chemical.

Exercises

10.43. Independent random samples from two normal populations produced the following variances:

Population	Sample Size	Sample Variance
1	25	72
2	18	54

a. Do the data provide sufficient evidence to indicate that σ_1^2 differs from σ_2^2? Test using $\alpha = .05$.

b. Find the approximate observed significance level for the test and interpret its value.

10.44. The stability of measurements of the characteristics of a manufactured product is important in maintaining product quality. In fact, it is sometimes better to possess small variation in the measured value of some important characteristic of a product and have the process mean slightly off target than to suffer wide variation with a mean value that perfectly fits requirements. The latter situation may produce a higher percentage of defective product than the former. A manufacturer of light bulbs suspected that one of his production lines was producing bulbs with a high variation in length of life. To test this theory, he compared the lengths of life of $n = 50$ bulbs randomly sampled from the suspect line and $n = 50$ from a line that seemed to be "in control." The sample means and variances for the two samples were as follows:

"Suspect Line"	Line "in Control"
$\bar{y}_1 = 1520$ $s_1^2 = 92,000$	$\bar{y}_2 = 1476$ $s_2^2 = 37,000$

a. Do the data provide sufficient evidence to indicate that bulbs produced by the "suspect line" possess a larger variance in length of life than those produced by the line that is assumed to be in control? Use $\alpha = .05$.

b. Find the approximate observed significance level for the test and interpret its value.

10.45. Use the method of Section 10.8 to obtain a 90 percent confidence interval for the variance ratio, Exercise 10.44.

10.46. A personnel manager, planning to use a Student's t test to compare the mean number of monthly absences for two categories of employees, noticed a possible difficulty. The variation in the numbers of absences per month seemed to differ for the two groups. As a check, the personnel officer randomly selected five months and counted the number of absences for each group. The data are shown in the table.

Category A	Category B
20	37
14	29
19	51
22	40
25	26

a. About which assumption necessary for use of the t test was the personnel officer concerned?

b. Do the data provide sufficient evidence to indicate that the variances differ for the populations of absences for the two employee categories? Test with $\alpha = .10$ and interpret the results of the test.

10.47. A pharmaceutical manufacturer purchases a particular material from two different suppliers. The mean level of impurities in the raw material is approximately the same for both suppliers but the manufacturer is concerned about the variability of the impurities from shipment to shipment. If the level of impurities tends to vary excessively for one source of supply, it could affect the quality of the pharmaceutical product. To compare the variation in percentage impurities for the two suppliers, the manufacturer selects ten shipments from each of the two suppliers and measures the percentage of impurities in the raw material for each shipment. The sample means and variances are shown in the table.

Supplier A	Supplier B
$\bar{y}_1 = 1.89$	$\bar{y}_2 = 1.85$
$s_1^2 = .273$	$s_2^2 = .094$
$n_1 = 10$	$n_2 = 10$

a. Do the data provide sufficient evidence to indicate a difference in the variability of the shipment impurity levels for the two suppliers? Test using $\alpha = .10$. Based on the results of your test, what recommendation would you make to the pharmaceutical manufacturer?

b. Find a 90 percent confidence interval for σ_B^2 and interpret your results.

10.48. The blood cholesterol values for randomly selected patients were compared for two diets, one low fat and the other normal. The sample means, variances, and sample sizes are as follows:

Low Fat	Normal
$\bar{y}_1 = 170$	$\bar{y}_2 = 196$
$s_1^2 = 198$	$s_2^2 = 435$
$n_1 = 19$	$n_2 = 24$

a. Do the data provide sufficient evidence to indicate a difference in variation for the populations of patients from which the two samples were drawn? Use $\alpha = .10$.

b. Why is the answer to part (a) of importance if we wish to determine whether the low-fat diet is effective in lowering blood cholesterol?

10.9
Assumptions

As noted earlier, the tests and confidence intervals based on the Student's t, the chi-square, and the F statistic require that the data satisfy specific assumptions in order that the error probabilities (for the tests) and the confidence coefficients (for the confidence intervals) be equal to the values that we have specified. For example, if the assumptions are violated by selecting a sample from a nonnormal population, and the data are used to construct a 95 percent confidence interval for μ, the actual confidence coefficient might (unknown to us) be equal to .85 instead of .95. The assumptions are summarized next for your convenience.

Assumptions

1. For all tests and confidence intervals described in this chapter, it is assumed that samples are randomly selected from normally distributed populations.

2. When two samples are selected, we assume that they are selected in an independent manner except in the case of the paired-difference experiment.

3. For tests or confidence intervals concerning the difference between two population means, μ_1 and μ_2, based on independent random samples, we assume that $\sigma_1^2 = \sigma_2^2$.

In a practical sampling situation, you never know everything about the probability distribution of the sampled population. If you did, there would be no need for sampling or statistics. Second, it is highly unlikely that a population would possess, exactly, the characteristics described above. Consequently, to be useful, the inferential methods described in this chapter must give good inferences when moderate departures from the assumptions are present. For example, if the population possesses a mound-shaped distribution that is nearly normal, we would like a 95 percent confidence interval constructed for μ to be one with a confidence coefficient close to .95. Similarly, if we conduct a t test of the null hypothesis, $\mu_1 = \mu_2$, based on independent random samples from normal populations where σ_1^2 and σ_2^2 are not exactly equal, we want the probability of incorrectly rejecting the null hypothesis, α, to be approximately equal to the value we used in locating the rejection region.

A statistical method that is insensitive to departures from the assumptions upon which the method is based is said to be **robust.** The t tests are quite robust to moderate departures from normality. In contrast, the chi-square and F tests are sensitive to departures from normality. The t test for comparing two means is moderately robust to departures from the assumption $\sigma_1^2 = \sigma_2^2$ when $n_1 = n_2$. However, the test becomes sensitive to departures from this assumption as n_1 becomes large relative to n_2 (or vice versa).

If you are concerned that your data do not satisfy the assumptions prescribed for one of the statistical methods described in this chapter, you may be able to use a nonparametric statistical method to make your inference. These methods, which require few or no assumptions about the nature of the population probability distributions, are particularly useful for testing hypotheses and some nonparametric methods have been developed for estimating population parameters. Tests of hypotheses concerning the location of a population distribution or a test for the equivalence of two population distributions are presented in Chapter 16. If you can select relatively large samples, you can usually use one of the large-sample estimation or the test procedures of Chapters 8 and 9.

10.10
Summary

It is important to note that the t, χ^2, and F statistics employed in the small-sample statistical methods discussed in the preceding sections are based upon the assumption that the sampled populations possess a normal probability distribution. This requirement will be adequately satisfied for many types of experimental measurements.

You will observe the very close relationship connecting Student's t and the z statistic and, therefore, the similarity of the methods for testing hypotheses and the construction of confidence intervals. The χ^2 and F statistics employed in making inferences concerning population variances do not, of course, follow this pattern, but the reasoning employed in the construction of the statistical tests and confidence intervals is identical for all the methods we have presented.

Tips on Problem Solving

To help you decide whether the techniques of this chapter are appropriate for the solution of a problem, ask yourself the following questions:

1. Does the problem imply that an inference should be made about a population mean or the difference between two means? Are the samples small, say $n <$ 30? If the answers to both questions are "yes," you may be able to use one of the methods of Sections 10.3, 10.4, or 10.5. If the sample sizes are large, $n \geq 30$, you can use the methods of Chapters 8 and 9. In practice, you would also need to verify that the assumptions underlying each procedure are satisfied. Are the population distributions nearly normal and have the sampling procedures conformed to those prescribed for the statistical method?

2. When comparing population means, were the observations from the two populations selected in a paired manner? If they were, you must use the paired-difference analysis of Section 10.5. If the samples were selected independently and in a random manner, use the methods of Section 10.4.

3. Is data variation the primary objective of the problem? If it is, you may be required to make an inference about a population variance, σ^2 (Section 10.7), or to compare two population variances, σ_1^2 and σ_2^2 (Section 10.8).

   (*Note:* The tips on problem solving following Section 8.12 will also be helpful in solving problems in this chapter.)

References

Dixon, W.J., and M.B. Brown, eds. *BMDP Biomedical Computer Programs, P Series.* Berkeley, Calif.: University of California, 1979.

Helwig, J.P., and K.A. Council, eds. *SAS User's Guide.* Cary, N.C.: SAS Institute, Inc., P.O. Box 8000, 1979.

Nie, N., C.H. Hull, J.G. Jenkins, K. Steinbrenner, and D.H. Bent, *Statistical Package for the Social Sciences*, 2nd ed. New York: McGraw-Hill Book Company, 1979.

Snedecor, G.W., and W.G. Cochran, *Statistical Methods*, 7th ed. Ames, Iowa: The Iowa State University Press, 1980.

Steel, R.G.D., and J.H. Torrie, *Principles and Procedures of Statistics*. New York: McGraw-Hill Book Company, 1960.

Supplementary Exercises

10.49. What assumptions are made when Student's t test is employed to test an hypothesis concerning a population mean?

10.50. A manufacturer can tolerate a small amount (.05 milligrams per liter) of impurities in a raw material needed for manufacturing its product. Because the laboratory test for the impurities is subject to experimental error, the manufacturer tests each batch ten times. Assume that the mean value of the experimental error is 0 and hence that the mean value of the ten test readings is an unbiased estimate of the true amount of the impurities in the batch. For a particular batch of the raw material, the mean of the ten test readings is .058 milligrams per liter (mg/l), with a standard deviation of .012 mg/l. Do the data provide sufficient evidence to indicate that the amount of impurities in the batch exceeds .05 mg/l? Find the p value for the test and interpret its value.

10.51. The main stem growth measured for a sample of 17 four-year-old red pine trees produced a mean and standard deviation equal to 11.3 and 3.4 inches, respectively. Find a 90 percent confidence interval for the mean growth of a population of four-year-old red pine trees subjected to similar environmental conditions.

10.52. In a general chemistry experiment, it is desired to determine the amount (in milliliters) of sodium (NaOH) solution needed to neutralize 1 gram of a specified acid. This will be an exact amount, but when run in the laboratory, variation will occur as the result of experimental error. Three titrations are made using phenolphthalein as an indicator of the neutrality of the solution (pH equals 7 for a neutral solution). The three volumes of sodium hydroxide required to attain a pH of 7 in each of the three titrations are as follows: 82.10, 75.75, and 75.44 milliliters. Use a 90 percent confidence interval to estimate the mean number of milliliters required to neutralize 1 gram of the acid.

10.53. What assumptions are made about the populations from which independent random samples are obtained when utilizing the t distribution in making small-sample inferences concerning the difference in population means?

10.54. A production plant has two extremely complex fabricating systems, with one being twice the age of the other. Both systems are checked, lubricated, and maintained once every two weeks. The number of finished products fabricated daily by each of the systems is recorded for 30 working days. The results are given in the table. Do these data present sufficient evidence to conclude that the variability in daily production warrants increased maintenance of the older fabricating system? Use a 5 percent level of significance.

New System	Old System
$\bar{y}_1 = 246$	$\bar{y}_2 = 240$
$s_1 = 15.6$	$s_2 = 28.2$

10.55. Four sets of identical twins (pairs A, B, C, and D) were selected at random from a population of identical twins. One child was taken at random from each pair to form an "experimental group." These four children were sent to school. The other four children were kept at home as a control group. At the end of the school year the following IQ scores were obtained:

Pair	Experimental Group	Control Group
A	110	111
B	125	120
C	139	128
D	142	135

Does this evidence justify the conclusion that lack of school experience has a depressing effect on IQ scores? Use $\alpha = .10$.

10.56. A comparison of reaction times for two different stimuli in a psychological word-association experiment produced the following results when applied to a random sample of 16 people:

Stimulus	Reaction Time (sec)							
1	1	3	2	1	2	1	3	2
2	4	2	3	3	1	2	3	3

Do the data present sufficient evidence to indicate a difference in mean reaction time for the two stimuli? Test at the $\alpha = .05$ level of significance.

10.57. Refer to Exercise 10.56. Suppose that the word-association experiment had been conducted using eight people as blocks and making a comparison of reaction time within each person; that is, each person would be subjected to both stimuli in a random order. The data for the experiment are as follows:

Person	Reaction Time (sec)	
	Stimulus 1	Stimulus 2
1	3	4
2	1	2
3	1	3
4	2	1
5	1	2
6	2	3
7	3	3
8	1	3

Do the data present sufficient evidence to indicate a difference in mean reaction time for the two stimuli? Test at the $\alpha = .05$ level of significance.

10.58.   Obtain a 90 percent confidence interval for $(\mu_1 - \mu_2)$ in Exercise 10.56.

10.59.   Obtain a 95 percent confidence interval for $(\mu_1 - \mu_2)$ in Exercise 10.57.

10.60.   Analyze the data in Exercise 10.57 as though the experiment had been conducted in an unpaired manner. Calculate a 95 percent confidence interval for $(\mu_1 - \mu_2)$ and compare with the answer to Exercise 10.59. Does it appear that blocking increased the amount of information available in the experiment?

10.61.   The following data give readings in foot-pounds of the impact strength on two kinds of packaging material. Determine whether there is evidence of a difference in mean strength between the two kinds of material. Test at the $\alpha = .10$ level of significance.

A	B
1.25	.89
1.16	1.01
1.33	.97
1.15	.95
1.23	.94
1.20	1.02
1.32	.98
1.28	1.00
1.21	.98
$\sum y_i = 11.13$	$\sum y_i = 8.80$
$\bar{y}_1 = 1.237$	$\bar{y}_2 = .978$
$\sum y_i^2 = 13.7973$	$\sum y_i^2 = 8.6240$

10.62.   Would the amount of information extracted from the data in Exercise 10.61 be increased by pairing successive observations and analyzing the differences? Calculate 90 percent confidence intervals for $(\mu_1 - \mu_2)$ for the two methods of analysis (unpaired and paired) and compare the widths of the intervals.

10.63.   When should one employ a paired-difference analysis in making inferences concerning the difference between two means?

10.64.   An experiment was conducted to compare the density of cakes prepared from two different cake mixes, A and B. Six cake pans received batter A and six received batter B. Expecting a variation in oven temperature, the experimenter placed an A and a B side by side at six different locations within the oven. The six paired observation are as follows:

Mix	Density (oz/in.^{3})					
A	.135	.102	.098	.141	.131	.144
B	.129	.120	.112	.152	.135	.163

Do the data present sufficient evidence to indicate a difference in the average density for cakes prepared using the two types of batter? Test at the $\alpha = .05$ level of significance.

10.65.   Place a 95 percent confidence interval on the difference between the average densities for the two mixes in Exercise 10.64.

10.66.   The 1977 interim emission standards for hydrocarbons (HC) and carbon monoxide (CO) in automobile exhaust systems is 1.5 grams per mile (HC) and 15 grams per mile (CO). Analysis of the exhaust for a random sample of six automobiles of a particular 1977 model gave the following (HC) readings:

1.37, 1.44, 1.28, 1.51, 1.39, 1.32.

a. Do the tests on the six automobiles provide sufficient evidence to indicate that the mean fleet automobile output of HC for this model is less than the 1.5 grams per mile limit? Test using $\alpha = .05$.

b. Find a 90 percent confidence interval for the mean HC output for the fleet.

10.67.   Under what assumptions may the F distribution be used in making inferences about the ratio of population variances?

10.68.   A dairy is in the market for a new bottle-filling machine and is considering models A and B manufactured by company X and company Y, respectively. If ruggedness, cost, and convenience are comparable in the two models, the deciding factor is the variability of fills (the model producing fills with the smaller variance being preferred). Let σ_1^2 and σ_2^2 be the fill variances for models A and B, respectively, and consider various tests of the null hypothesis $H_0: \sigma_1^2 = \sigma_2^2$. Obtaining samples of fills from the two machines and utilizing the test statistic s_1^2/s_2^2, one could set up as the rejection regions an upper-tail area, a lower-tail area, or a two-tailed area of the F distribution depending on his point of view. Which type of rejection region would be most favored by the following persons:

a. The manager of the dairy? Why?

b. A salesman for company X? Why?

c. A salesman for company Y? Why?

10.69.   The closing prices of two common stocks were recorded for a period of 15 days. The means and variances are

$$\bar{y}_1 = 40.33, \qquad \bar{y}_2 = 42.54,$$

$$s_1^2 = 1.54, \qquad s_2^2 = 2.96.$$

Do these data present sufficient evidence to indicate a difference in variability of the closing prices of the two stocks for the populations associated with the two samples? Give the p value for the test and interpret its value.

10.70.   Place a 90 percent confidence interval on the ratio of the two population variances in Exercise 10.69.

10.71.   Refer to Exercise 10.68. Wishing to demonstrate that the variability of fills is less for model A than for model B, a salesman for company X acquired a sample of 30 fills from a machine of model A and a sample of 10 fills from a machine of model B. The sample variances were $s_1^2 = .027$ and $s_2^2 = .065$, respectively. Does this result provide statistical support at the .05 level of significance for the salesman's claim?

10.72. A chemical manufacturer claims that the purity of his product never varies more than 2 percent. Five batches were tested and given purity readings of 98.2, 97.1, 98.9, 97.7, and 97.9 percent. Do the data provide sufficient evidence to contradict the manufacturer's claim? (*Hint:* To be generous, let a range of 2 percent equal 4σ.)

10.73. Refer to Exercise 10.72. Find a 90 percent confidence interval for σ^2.

10.74. A cannery prints "weight 16 ounces" on its label. The quality-control supervisor selects nine cans at random and weighs them. He finds $\bar{y} = 15.7$ and $s = .5$. Do the data present sufficient evidence to indicate that the mean weight is less than that claimed on the label? (Use $\alpha = .05$.)

10.75. A psychologist wishes to verify that a certain drug increases the reaction time to a given stimulus. The following reaction times in tenths of a second were recorded before and after injection of the drug for each of four subjects:

	Reaction Time	
Subject	Before	After
1	7	13
2	2	3
3	12	18
4	12	13

Test at the 5 percent level of significance to determine whether the drug significantly increases reaction time.

10.76. PCBs are found not only in the milk of pregnant women (Exercise 8.2) and in clams (Exercise 10.8); concentrations of PCBs have been found to be dangerously high in some game birds along the Coosa River in Georgia (*Environment News*, January 1977). The FDA considers concentrations higher than 5 parts per million (ppm) dangerous for human consumption. Six wild ducks tested in the Coosa area (four woodcocks and two mallards) showed the following PCB concentrations:

28, 26, 11, 7.8, 11.5, 11.5.

If these ducks can be regarded as a random sample of ducks from the Coosa area, estimate the mean concentration per duck using a 90 percent confidence interval. Interpret the interval. What assumptions must you make in order that your test be valid?

10.77. How much combustion efficiency should a homeowner expect from an oil furnace? The EPA (*Environment News*, January 1977) states that 80 percent or above is excellent, 75 to 79 is good, 70 to 74 is fair, and below 70 percent is poor. A home heating contractor who sells two makes of oil heaters (call them A and B) decided to compare their mean efficiencies. An analysis was made of the efficiencies for eight heaters of type A and six of type B. The efficiency ratings, in percentages, for the 14 heaters are shown in the table.

Type *A*	Type *B*
72	78
78	76
73	81
69	74
75	82
74	75
69	
75	

a. Do the data provide sufficient evidence to indicate a difference in mean efficiencies for the two makes of home heaters? Find the p value for the test and interpret its value.

b. Find a 90 percent confidence interval for $(\mu_A - \mu_B)$ and interpret the result.

10.78.   At a time when energy conservation is so important, some scientists think we should give closer scrutiny to the cost (in energy) of producing various forms of food. One recent study compares the mean amount of oil required to produce one acre of different types of crops. For example, suppose that we wish to compare the mean amount of oil required to produce one acre of corn versus one acre of cauliflower. The readings in barrels of oil per acre, based on 20-acre plots, seven for each crop, are shown in the table. Use these data to find a 90 percent confidence interval for the difference in the mean amount of oil required to produce these two crops.

Corn	Cauliflower
5.6	15.9
7.1	13.4
4.5	17.6
6.0	16.8
7.9	15.8
4.8	16.3
5.7	17.1

10.79.   The effect of alcohol consumption on the body appears to be much greater at high altitudes than at sea level. To test this theory, a scientist randomly selects 12 subjects and randomly divides them into two groups of six each. One group is transported to an altitude of 12,000 feet and each subject ingests a drink containing 100 cc of alcohol. The second group of six receives the same drink at sea level. After two hours, the amount of alcohol in the blood (grams per 100 cc) for each subject is measured. The data are shown in the table. Do the data provide sufficient evidence to support the theory that retention of alcohol in the blood is greater at high altitude? Test with $\alpha = .10$.

Sea Level	12,000 feet
.07	.13
.10	.17
.09	.15
.12	.14
.09	.10
.13	.14

10.80. In an earlier exercise, we mentioned the Bureau of Labor Statistic's surveys of consumer spending habits, particularly focusing on the estimate of the mean number of automobiles per family in the United States. Suppose that you were to run a very small survey for your community. You randomly sample $n = 20$ households and for each you record the number of automobiles per household. Suppose that $\bar{y} = 1.75$ and $s = .55$. Is it valid to use the t statistic to form a 90 percent confidence interval for the mean number of automobiles per household for the community? Explain.

10.81. A cigarette manufacturer claimed that its cigarettes contained no more than 25 milligrams of nicotine. A sample of 16 cigarettes yielded a mean and standard deviation equal to 26.4 and 2, respectively. Do the data provide sufficient evidence to refute the manufacturer's claim? (Use $\alpha = .10$.)

10.82. Refer to Exercise 10.81. Estimate the mean nicotine content of this brand of cigarette using a 90 percent confidence interval.

10.83. An experiment is conducted to compare two new automobile designs. Twenty people are randomly selected and each person is asked to rate each design on a scale of 1 (poor) to 10 (excellent). The resulting ratings will be used to test the null hypothesis that the mean level of approval is the same for both designs against the alternative hypothesis that one of the automobile designs is preferred. Would these data satisfy the assumptions required for the Student's t test of Section 10.4? Explain.

10.84. The following data were collected on lost-time accidents (the figures given are mean man-hours lost per month over a period of 1 year) both before and after an industrial safety program was put into effect. Data were recorded for six industrial plants. Do the data provide sufficient evidence to indicate whether the safety program was effective in reducing lost-time accidents? (Use $\alpha = .10$.)

	Plant Number					
	1	2	3	4	5	6
Before program	38	64	42	70	58	30
After program	31	58	43	65	52	29

10.85. To compare the demand for two different entrees, the manager of a cafeteria recorded the number of purchases for each entree on seven consecutive days. The data are shown in the table. Do the data provide sufficient evidence to indicate a greater mean demand for one of the entrees?

Day	A	B
Monday	420	391
Tuesday	374	343
Wednesday	434	469
Thursday	395	412
Friday	637	538
Saturday	594	521
Sunday	679	625

10.86.   The EPA limit on the allowable discharge of suspended solids into rivers and streams is 60 milligrams per liter (mg per l) per day. A study of water samples selected from the discharge at a phosphate mine shows that over a long period of time, the mean daily discharge of suspended solids is 48 mg per l but the day-to-day discharge readings are very variable. State inspectors measured the discharge rates of suspended solids for $n = 20$ days and found $s^2 = 39$ $(mg/l)^2$. Find a 90 percent confidence interval for σ^2. Interpret your results.

11. Linear Regression and Correlation

Chapter Objectives

General Objective

We presented methods for making inferences about population means based on large random samples in Chapters 8 and 9 and small random samples in Chapter 10. The object of this chapter is to extend this methodology to consider the case in which the mean value of y is related to another variable, call it x. By making simultaneous observations on y and the x variable, we can use information contained in the x measurements to estimate the mean value of y and to predict particular values of y for preassigned values of x. This chapter will be devoted to the case where y is a linear function of one predictor variable x. The general case, where y is related to one or more predictor variables, say $x_1, x_2, \ldots, x_k$, will be discussed in Chapter 12.

Specific Objectives

1. To give practical illustrations of the types of problems that can be solved using the techniques of linear regression and correlation. *Sections 11.1 through 11.10*

2. To distinguish between deterministic and probabilistic models and to identify their advantages and limitations, and to present a linear probabilistic model for relating a reponse y to a single independent variable x. *Section 11.2*

3. To explain how the method of least squares can be used to fit a linear probabilistic model to data. *Section 11.3*

4. To provide a method for determining whether x contributes information for the prediction of y. *Sections 11.4, 11.5*

5. To present a confidence interval for estimating the mean value of y for a given value of x. *Section 11.6*

6. To give a prediction interval for predicting a particular value of y for a given value of x. *Section 11.7*

7. To present measures of the strength of the linear relationship between y and x—the simple linear coefficient of correlation and the coefficient of determination. *Section 11.8*

8. To familiarize you with a typical computer printout for an interesting application of regression analysis. *Section 11.9*

9. To emphasize the assumptions you must make in order for the estimation and test procedures described in this chapter to be valid. *Section 11.11*

Case Study

Does It Pay to Save?

Does it pay to save? If you had or were to acquire some extra money, should you put it aside for a rainy day (just in case Social Security is not with us in the future)? What is the effect of government taxation on your incentive to put your savings in a bank, a savings and loan association, a money fund, real estate, or some other investment? This question is addressed by Dr. Srully Blotnick, a practicing psychologist and columnist for *Forbes* magazine.*

The essence of Dr. Blotnick's article is that we are inclined to save and invest if we are allowed to keep some of the rewards. (It *does* make sense!) But, the greater the amount of government taxation on the return (interest, dividends, capital gains, etc.) from our savings and/or investments, the less we tend to save and invest. (Why should we?) Data suggest that if saving and investment produce low return, people spend their money on entertainment and material goods, and if they have large amounts available for investment, they attempt to place their funds in tax shelters (tax-exempt investments). Thus, according to one theory, heavy taxation decreases capital formation, the life-blood needed for investment in research and the modernization of industry, for the revitalization of our economy, and ultimately, for our own individual economic welfare.

To illustrate, it is not uncommon for the combined personal income of a young college-educated husband and wife to equal $25,000 to as much as $50,000 per year. If they have two children, with normal medical, mortgage, and other expenses, and if their joint income totals $50,000 per year, they are required to pay approximately 62% of this income in taxes if they live in Sweden, 26% in the United States, 25% in Italy, 24% in Japan, 20% in Germany, and only 11% in France (prior to the 1981 change to a socialist government). Given the distribution of taxed percentages of earned income for most countries in the Western world, the United States falls near the middle.

In contrast to the taxation on *earned* income, and in comparison with other countries, the United States imposes a tax rate on *invested* income that is among the highest. Suppose that you received the $50,000 annual income as interest, dividends, or capital gains on investment; that is, income from *invested* savings. Dr. Blotnick provides a table that shows, for each of eight countries, the personal savings rate, that is, the percent of total national earned income that is saved, and the approximate[†] percentage of investment income that must be paid in taxes. If theory is correct, countries that show low taxation of invested income should show a high percentage of savings of earned income and, if the taxation rate is high, the savings rate should be low. The data are shown in Table 11.1.

An examination of Table 11.1 motivates some comments and questions. In the preceding chapters, each measurement resulted from one observation

---

* "Psychology and Investing," *Forbes*, May 25, 1981. Reprinted by permission. Copyright 1981, Forbes Inc.

† The percentage investment income tax liability shown in the table was based on an investment income of $49,000.

Table 11.1 A comparison of personal savings rates and taxation rates on investment income

Country	Personal Savings Rate %	Investment Income Tax Liability %
Italy	23.1	6.4
Japan	21.5	14.4
France	17.2	7.3
W. Germany	14.5	11.8
United Kingdom	12.2	32.5
Canada	10.3	30.0
Sweden	9.1	52.7
United States	6.3	33.5

Source: New York Stock Exchange with assistance of Price Waterhouse

on a single experimental unit. The data shown in Table 11.1 are very different. Two observations, the savings rate y and the taxation rate x, are made on each country (the experimental unit). Note also, for the data shown, the taxation rate of invested income for the United States is nearly the highest and the savings rate is the lowest. Is the savings rate y of a country related to the taxation rate x of invested income? Can we, knowing the taxation rate x, predict the percentage of national income that will be saved?

Chapter 11 presents an extension of the methodology of Chapters 1 through 10. In the earlier chapters we made a single measurement y on each experimental unit, and we were interested in the mean, standard deviation, and other features of the relative frequency distribution for the population of y measurements. In this chapter, we will take pairs of observations (x, y) on each experimental unit. We will want to determine whether y and x are related and whether knowledge of the value of x will enable us to obtain a better estimate of the mean value of y or a better prediction of some future value of y. In Section 11.12 the techniques we develop will be applied to the savings rate and taxation rate data of this Case Study.

11.1
Introduction

An estimation problem of more than casual interest to high school seniors, freshmen entering college, their parents, and a university administration concerns the expected academic achievements of a particular student after he or she has enrolled in a university. For example, we might wish to estimate a student's grade-point average at the end of the freshman year *before* the student has been accepted or enrolled in the university. At first glance this would seem to be a difficult task.

The statistical approach to this problem is, in many respects, a formalization of the procedure we might follow intuitively. If data were available giving the high school academic grades, psychological and sociological information, as well as the grades attained at the end of the college freshman year for a large number of students, we might categorize the students into groups possessing similar characteristics. Certainly, highly motivated students who have had a high rank in their high school class, have graduated from a high school with known superior academic standards, and so on, should achieve, on the average, a high grade-point average at the end of the college freshman year. On the other hand, students who lack proper motivation, who achieved only moderate success in high school, would not be expected, on the average, to do as well. Carrying this line of thought to the ultimate and idealistic extreme, we would expect the grade-point average of a student to be a *function* of the many variables that define the characteristics, psychological and physical, of the individual as well as those that define the environment, academic and social, to which he will be exposed. Ideally, we would like to possess a mathematical equation that would relate a student's grade-point average to all these independent variables so that it could be used for prediction.

You will observe that the problem we have defined is of a very general nature. We are interested in a random variable, y, that is related to a number of independent variables, $x_1, x_2, x_3, \ldots$. The variable, y, for our example, would be the student's grade-point average, and the independent variables might be

$x_1 = $ rank in high school class,

$x_2 = $ score on a mathematics achievement test,

$x_3 = $ score on a verbal achievement test,

and so on. The ultimate objective would be to measure $x_1, x_2, x_3, \ldots$ for a particular student, substitute these values into the prediction equation, and thereby predict the student's grade-point average. To accomplish this end, we must first locate the related variables $x_1, x_2, x_3, \ldots$ and obtain a measure of the strength of their relationship to y. Then we must construct a good prediction equation that will express y as a function of the selected independent variables.

Practical examples of our prediction problem are very numerous in business, industry, and the sciences. The stockbroker wishes to predict stock market behavior as a function of a number of "key indices" which are observable and serve as the predictor variables, $x_1, x_2, x_3, \ldots$. The manager of a manufacturing plant would like to relate yield of a chemical to a number of process variables. The manager would then use the prediction equation to find settings for the controllable process variables that would provide the maximum yield of the chemical. The personnel director of a corporation, like the admissions director of a university, wishes to test and measure individual characteristics so that the corporation may hire the person best suited for a particular job. The biologist would like to relate body characteristics to the amounts of various glandular secretions. The political scientist may wish to relate success in a political campaign to the characteristics of a candidate, the nature of the opposition, and various campaign issues and promotional techniques. Certainly, all these prediction problems are, in many respects, one and the same.

In this chapter we shall be primarily concerned with the *reasoning* involved in acquiring a prediction equation based upon one or more predictor variables. Thus we will restrict our attention to the simple problem of predicting y as a *linear* **function** of a *single* variable and observe that the solution for the multivariable problem, for example, predicting student grade-point average, will consist of a generalization of our technique. We shall show you how to fit a simple linear model to a set of data, a process called a **regression analysis,** and shall show you how to use the model for estimation and prediction. The methodology for finding the multivariable predictor, called a **multiple regression analysis,** will be discussed in Chapter 12.

11.2
A Simple Linear Probabilistic Model

We will introduce our topic by considering the problem of predicting a student's final grade in a college freshman calculus course based upon his score on a mathematics achievement test administered prior to college entrance. As noted in Section 11.1, we wish to determine whether the achievement test is really worthwhile—whether the achievement test score is related to a student's grade in calculus—and, in addition, we wish to obtain an equation that may be useful for prediction purposes. The evidence, presented in Table 11.2, represents a sample of the achievement test scores and calculus grades for ten college freshmen. Hopefully, the ten students represent a random sample drawn from the population of freshmen who have already entered the university or will do so in the immediate future.

Our initial approach to the analysis of the data of Table 11.2 would be to plot the data as points on a graph, representing a student's calculus grade

Table 11.2 Mathematics achievement test scores and final calculus grades for college freshmen

Student	Mathematics Achievement Test Score	Final Calculus Grade
1	39	65
2	43	78
3	21	52
4	64	82
5	57	92
6	47	89
7	28	73
8	75	98
9	34	56
10	52	75

as y and the corresponding achievement test score as x. The graph is shown in Figure 11.1. You will quickly observe that y appears to increase as x increases. Do you think this arrangement of the points could have occurred due to chance even if x and y were unrelated?

Figure 11.1 Plot of the data in Table 11.2

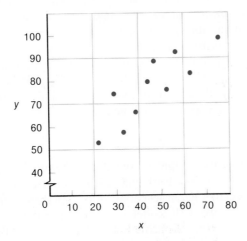

One method of obtaining a prediction equation relating y to x would be to place a ruler on the graph and move it about until it seems to pass through the points and provide what we might regard as the "best fit" to the data. Indeed, if we were to draw a line through the points, it would appear that our prediction problem was solved. Certainly, we may now use the graph to predict a student's calculus grade as a function of his score on the mathematics achievement test. Furthermore, we note that we have chosen a *mathematical model* that expresses the supposed functional relation between y and x.

You should recall several facts concerning the graphing of mathematical functions. First, the mathematical equation of a straight line is

$$y = \beta_0 + \beta_1 x,$$

where β_0 is the y intercept, the value of y when $x = 0$, and β_1 is the slope of the line, the change in y for a one-unit change in x (see Figure 11.2). Second, the line that we may graph corresponding to any linear equation is unique. Each equation will correspond to only one line and vice versa. Thus, when we draw a line through the points, we have automatically chosen a mathematical equation

$$y = \beta_0 + \beta_1 x,$$

where β_0 and β_1 have unique numerical values.

Figure 11.2 The *y* intercept and slope for a line

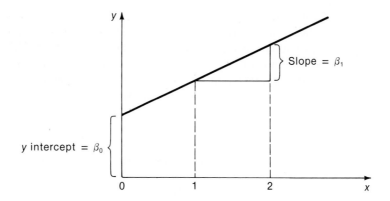

Equation of a Straight Line

$$y = \beta_0 + \beta_1 x$$

$\beta_0 = y$ intercept

$\quad = $ value of y when $x = 0$

$\beta_1 = $ slope of the line

$\quad = $ change in y for a one unit increase in x

The linear model,

$$y = \beta_0 + \beta_1 x,$$

is said to be a **deterministic mathematical model** because when a value of x is substituted into the equation the value of y is determined and no allowance is made for error. Fitting a straight line through a set of points by eye produces a deterministic model. Many other examples of deterministic mathematical models may be found by leafing through the pages of elementary chemistry, physics, or engineering textbooks.

Deterministic models are quite suitable for explaining physical phenomena and predicting when the error of prediction is negligible for practical purposes. Thus, Newton's Law, which expresses the relation between the force, F, imparted by a moving body with mass, m, and acceleration, a, given by the deterministic model

$$F = ma,$$

predicts force with very little error for most practical applications. "Very little" is, of course, a relative concept. An error of .1 inch in forming an I-beam for a bridge is extremely small but would be impossibly large in the manufacture of parts for a wristwatch. Thus, in many physical situations the error of prediction cannot be ignored. Indeed, consistent with our stated philosophy, we would be hesitant to place much confidence in a prediction unaccompanied by a measure of its goodness. For this reason, a visual choice of a line to relate the calculus grade and achievement test score would be of limited utility.

In contrast to the deterministic model, we might employ a **probabilistic mathematical model** to explain a physical phenomenon. As we might suspect, probabilistic mathematical models contain one or more random elements with specified probability distributions. For our example, we shall relate the calculus score to the achievement test score by the equation

$$y = \beta_0 + \beta_1 x + \epsilon,$$

where ϵ is assumed to be a random error variable with expected value equal to zero and variance equal to σ^2. In addition, we shall assume that any pair of random errors, ϵ_i and ϵ_j, corresponding to two observations, y_i and y_j, are independent. In other words, we assume that the *average or expected value* of y is linearly related to x and that observed values of y will deviate above and below this line by a random amount, ϵ. Furthermore, we have assumed that the distribution of errors about the line will be identically the same, regardless of the value of x, and that any pair of errors will be independent of one another. The assumed line, giving the expected value of y for a given value of x, is indicated in Figure 11.3. The probability distribution of the random error, ϵ, is shown for several values of x.

In Section 11.3, we shall consider the problem of finding the prediction equation, or regression line, as it is commonly known in statistics.

Figure 11.3 Linear probabilistic model

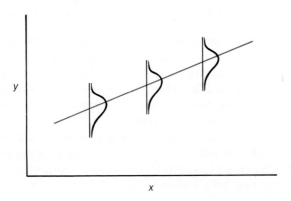

Exercises

11.1.   Graph the line corresponding to the equation $y = 3x + 2$ by graphing the points corresponding to $x = 0, 1,$ and 2. Give the y intercept and slope for the line.

11.2.   Graph the line corresponding to the equation $y = -3x + 2$ by graphing the points corresponding to $x = 0, 1,$ and 2. Give the y intercept and slope for the line. Give the similarities and differences between this line and the line of Exercise 11.1.

11.3.   Graph the line corresponding to the equation, $3y = -x + 4$.

11.4.   How is the line, $3y = x + 4$, related to the line of Exercise 11.3?

11.5.   Give the equation and graph for a line with a y intercept equal to 1 and a slope equal to -2.

11.6.   Give the equation and graph for a line with a y intercept equal to -1 and a slope equal to 2.

11.7.   What is the difference between deterministic and probabilistic mathematical models?

11.3
The Method of
Least Squares

The statistical procedure for finding the "best-fitting" straight line for a set of points would seem, in many respects, a formalization of the procedure employed when we fit a line by eye. For instance, when we visually fit a line to a set of data, we move the ruler until we think that we have minimized the deviations of the points from the prospective line. If we denote the predicted value of y obtained from the fitted line as $\hat{y}$, the prediction equation will be

$$\hat{y} = \hat{\beta}_0 + \hat{\beta}_1 x,$$

where $\hat{\beta}_0$ and $\hat{\beta}_1$ represent estimates of the parameters, β_0 and β_1. This line for the data of Table 11.2 is shown in Figure 11.4. The vertical lines drawn from the prediction line to each point represent the deviations of the points from the predicted value of y. Thus the deviation of the ith point is

$$y_i - \hat{y}_i, \qquad \text{where} \qquad \hat{y}_i = \hat{\beta}_0 + \hat{\beta}_1 x_i.$$

Having decided that in some manner or other we will attempt to minimize the deviations of the points in choosing the best-fitting line, we must now define what we mean by "best." That is, we wish to define a criterion for "best fit" that will seem intuitively reasonable, which is objective, and which under certain conditions will give the best prediction of y for a given value of x.

Figure 11.4 Linear prediction equation

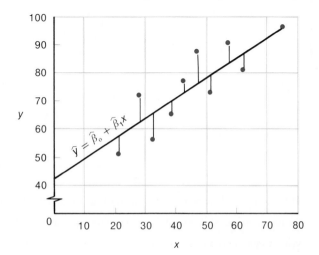

We shall employ a criterion of goodness that is known as the **principle of least squares** and which may be stated as follows. **Choose as the "best-fitting" line the line that minimizes the sum of squares of the deviations of the observed values of y from those predicted.*** Expressed mathematically, we wish to choose values for $\hat{\beta}_0$ and $\hat{\beta}_1$ that minimize

$$SSE = \sum_{i=1}^{n} (y_i - \hat{y}_i)^2.$$

The symbol **SSE** represents the sum of squares of deviations or, as commonly called, the **sum of squares for error.**

Substituting for $\hat{y}_i$ in SSE, we obtain

$$SSE = \sum_{i=1}^{n} [y_i - (\hat{\beta}_0 + \hat{\beta}_1 x_i)]^2.$$

The method for finding the numerical values of $\hat{\beta}_0$ and $\hat{\beta}_1$ that minimize SSE utilizes the differential calculus and hence is beyond the scope of this text. We simply state that the least-squares solutions for $\hat{\beta}_0$ and $\hat{\beta}_1$ are given by the following formulas:

Least-Squares Estimators of β_0 and β_1

$$\hat{\beta}_1 = \frac{SS_{xy}}{SS_x} \quad \text{and} \quad \hat{\beta}_0 = \bar{y} - \hat{\beta}_1 \bar{x},$$

where

$$SS_x = \sum_{i=1}^{n} (x_i - \bar{x})^2 = \sum_{i=1}^{n} x_i^2 - \frac{\left(\sum_{i=1}^{n} x_i\right)^2}{n}$$

and

$$SS_{xy} = \sum_{i=1}^{n} (x_i - \bar{x})(y_i - \bar{y}) = \sum_{i=1}^{n} x_i y_i - \frac{\left(\sum_{i=1}^{n} x_i\right)\left(\sum_{i=1}^{n} y_i\right)}{n}.$$

---

* The deviations of the points about the least-squares line satisfy another property. The sum of the deviations equals 0, i.e., $\sum_{i=1}^{n} (y_i - \hat{y}_i) = 0$ (proof omitted). Many other lines can be found that satisfy this property.

Note that SS_x (sum of squares for x) is computed using the familiar shortcut formula for calculating sums of squares of deviations that was presented in Chapter 3. SS_{xy} is computed using a very similar formula and hence should be easy to remember. Once $\hat{\beta}_0$ and $\hat{\beta}_1$ have been computed, substitute their values into the equation of a line to obtain the least-squares prediction equation,

$$\hat{y} = \hat{\beta}_0 + \hat{\beta}_1 x.$$

There is one important point to note here. Rounding errors can greatly affect the answer you obtain in calculating SS_x and SS_{xy}. If you must round a number, it is recommended that you carry six or more significant figures in the calculations. (Note also that in working exercises, rounding errors might cause some slight discrepancies between your answers and the answers given in the back of the text.)

Example 11.1

Obtain the least-squares prediction line for the data in Table 11.2.

Solution

The calculation of $\hat{\beta}_0$ and $\hat{\beta}_1$ for the data in Table 11.2 is simplified by the use of Table 11.3.
Substituting the appropriate sums from Table 11.3 into the least-squares equations, we obtain

$$SS_x = \sum_{i=1}^{n} x_i^2 - \frac{\left(\sum_{i=1}^{n} x_i\right)^2}{n} = 23{,}634 - \frac{(460)^2}{10}$$
$$= 2474$$

$$SS_{xy} = \sum_{i=1}^{n} x_i y_i - \frac{\left(\sum_{i=1}^{n} x_i\right)\left(\sum_{i=1}^{n} y_i\right)}{n} = 36{,}854 - \frac{(460)(760)}{10}$$
$$= 1894$$

$$\bar{y} = \frac{\sum_{i=1}^{n} y_i}{n} = \frac{760}{10} = 76$$

and

$$\bar{x} = \frac{\sum_{i=1}^{n} x_i}{n} = \frac{460}{10} = 46.$$

Hence

$$\hat{\beta}_1 = \frac{SS_{xy}}{SS_x} = \frac{1894}{2474} = .765562 \approx .77$$

Table 11.3 Calculations for the data, Table 11.2

	y_i	x_i	x_i^2	$x_i y_i$	y_i^2
	65	39	1521	2535	4225
	78	43	1849	3354	6084
	52	21	441	1092	2704
	82	64	4096	5248	6724
	92	57	3249	5244	8464
	89	47	2209	4183	7921
	73	28	784	2044	5329
	98	75	5625	7350	9604
	56	34	1156	1904	3136
	75	52	2704	3900	5625
Sum	760	460	23,634	36.854	59,816

and

$$\hat{\beta}_0 = \bar{y} - \hat{\beta}_1 \bar{x} = 76 - (.765562)(46) = 40.7841 \approx 40.78.$$

Then, according to the principle of least squares, the best-fitting straight line relating the calculus grade to the achievement test score is

$$\hat{y} = \hat{\beta}_0 + \hat{\beta}_1 x$$

or

$$\hat{y} = 40.78 + .77x.$$

The graph of this equation is shown in Figure 11.4. Note that the y intercept, 40.78, is the value of $\hat{y}$ when $x = 0$. The slope of the line, .77, gives the estimated change in y for a one-unit increase in x.

We may now predict y for a given value of x by referring to Figure 11.4 or by substituting into the prediction equation. For example, if a student scored $x = 50$ on the achievement test, the student's predicted calculus grade would be

$$\hat{y} = \hat{\beta}_0 + \hat{\beta}_1 x = 40.78 + (.77)(50) = 79.28.$$

How accurate will this prediction be? To answer this question, we will want to place a bound upon our error of prediction. We shall consider this and related problems in succeeding sections.

Tips on Problem Solving

1. Be careful of rounding errors. Carry at least six significant figures in computing sums of squares of deviations or the sums of squares of cross products of deviations.
2. Always plot the data points and graph your least-squares line. If the line does not provide a reasonable fit to the data points, you may have made an error in your calculations.

Exercises

11.8. Given five points whose coordinates are:

x	-2	-1	0	1	2
y	5	3	2	2	1

a. Find the least-squares line for the data.

b. As a rough check on the calculations in part (a), plot the five points and graph the line. Does the line appear to provide a good fit to the data points?

11.9. Given five points whose coordinates are:

x	-5	-3	-1	1	3	5
y	0	0	1	2	2	3

a. Find the least-squares line for the data.

b. As a rough check on the calculations in part (a), plot the five points and graph the line. Does the line appear to provide a good fit to the data points?

11.10. Given the points whose coordinates are:

x	1	2	3	4	5	6
y	4.7	3.9	3.7	3.0	2.6	1.9

a. Find the least-squares line for the data.

b. As a rough check on the calculations in part (a), plot the six points and graph the line. Does the line appear to provide a good fit to the data points?

11.11. The murder rate per 100,000 people increased in a city over a six-year period as follows:

Year, x	1	2	3	4	5	6
Murder Rate, y	8	11	16	19	25	29

a. Find the least-squares line relating y to x.

b. As a check on your calculations, plot the six points and graph the line.

11.12. The EPA 1980 49-state (all except California) combined mileage rating and engine volume are shown below for ten standard-transmission, four-cylinder, gasoline-fueled, subcompact cars. The engine sizes are in total cubic inches of cylinder volume.

Car	Cylinder Volume x	mpg (combined) y
VW Rabbit	97	24
Datsun 210	85	29
Chevette	98	26
Dodge Omni	105	24
Mazda 626	120	24
Oldsmobile Starfire	151	22
Mercury Capri	140	23
Toyota Celica	134	23
Datsun 810	146	21

a. Plot the data points on graph paper.

b. Find the least-squares line for the data.

c. Graph the least-squares line to see how well it fits the data.

d. Use the least-squares line to estimate the mean miles per gallon for a subcompact automobile which has 125 cubic inches of engine volume. (*Note:* We shall find a confidence interval for this mean in Section 11.6.)

11.13. Cucumbers are usually preserved by fermenting them in a low-salt brine (6 to 9 percent sodium chloride) and then storing them in a high-salt brine until they are used by processors to produce various types of pickles. The high-salt brine is needed to retard softening of the pickles and to prevent freezing when stored outside in northern climates. Data showing the reduction in firmness of pickles stored over time in a low-salt brine (2 to 3 percent) are shown below:*

	Weeks in Storage at 72 °F				
	0	4	14	32	52
Firmness in pounds y	19.8	16.5	12.8	8.1	7.5

* Buescher, R.W., J.M. Hudson, J.R. Adams, and D.H. Wallace, "Calcium Makes It Possible to Store Cucumber Pickles in Low-Salt Brine," *Arkansas Farm Research*, 30, no. 4 (July/August 1981).

a. Fit a least-squares line to the data.

b. As a check on your calculations, plot the five data points and graph the line. Does the line appear to provide a good fit to the data points?

c. Use the least-squares line to estimate the mean firmness of pickles stored for 20 weeks. (*Note:* We will find a confidence interval for this mean in Section 11.6.)

11.4
Calculating s^2, an Estimator of σ^2

Recall that we constructed a probabilistic model for y in Section 11.2,

$$y = \beta_0 + \beta_1 x + \epsilon,$$

where ϵ is a random error with mean value equal to 0 and variance equal to σ^2. Thus each observed value of y is subject to a random error, ϵ, which will enter into the computations of $\hat{\beta}_0$ and $\hat{\beta}_1$ and will introduce errors in these estimates. Further, if we use the least-squares line,

$$\hat{y} = \hat{\beta}_0 + \hat{\beta}_1 x,$$

to predict some future value of y, the random errors will affect the error of prediction. Consequently, the variability of the random errors, measured by σ^2, plays an important role when estimating or predicting using the least-squares line.

    The first step toward acquiring a bound on a prediction error requires that we estimate σ^2, the variance of the random error, ϵ. For this purpose it would seem reasonable to use SSE, the sum of squares of deviations (sum of squares for error) about the predicted line. Indeed, it can be shown that

$$\hat{\sigma}^2 = s^2 = \frac{\text{SSE}}{n - 2}$$

provides a good estimator for σ^2, one that is unbiased and that is based upon $(n - 2)$ degrees of freedom.

Estimator of σ^2

$$\hat{\sigma}^2 = s^2 = \frac{\text{SSE}}{n - 2}$$

where n = number of data points and

$$\text{SSE} = \sum_{i=1}^{n} (y_i - \hat{y}_i)^2.$$

The sum of squares of deviations, SSE, may be calculated directly by using the prediction equation to calculate $\hat{y}$ for each point, then calculating the deviations $(y_i - \hat{y}_i)$, and finally calculating

$$\text{SSE} = \sum_{i=1}^{n} (y_i - \hat{y}_i)^2.$$

This tends to be a very tedious procedure and is rather poor from a computational point of view because the numerous subtractions tend to introduce computational rounding errors. An easier and computationally better procedure is to use the following formula:

Method for Computing SSE

$$\text{SSE} = \text{SS}_y - \hat{\beta}_1 \text{SS}_{xy},$$

where

$$\text{SS}_y = \sum_{i=1}^{n} (y_i - \bar{y})^2 = \sum_{i=1}^{n} y_i^2 - \frac{\left(\sum_{i=1}^{n} y_i\right)^2}{n}$$

and

$$\text{SS}_{xy} = \sum_{i=1}^{n} x_i y_i - \frac{\left(\sum_{i=1}^{n} x_i\right)\left(\sum_{i=1}^{n} y_i\right)}{n}.$$

(Note that SS_{xy} was used in the calculation of $\hat{\beta}_1$ and hence has already been computed.)

Example 11.2

Calculate an estimate of σ^2 for the data of Table 11.2.

Solution

SS_{xy} and $\hat{\beta}_1$ were computed in Example 11.1. We found $\text{SS}_{xy} = 1894$ and $\hat{\beta}_1 = .765562$. To find SS_y, we need the values of $\sum_{i=1}^{n} y_i$ and $\sum_{i=1}^{n} y_i^2$ given in Table 11.3. Substituting these values into the formula for SS_y, we have

$$SS_y = \sum_{i=1}^{n} y_i^2 - \frac{\left(\sum_{i=1}^{n} y_i\right)^2}{n}$$

$$= 59{,}816 - \frac{(760)^2}{10} = 2056.$$

Then,

$$SSE = SS_y - \hat{\beta}_1 SS_{xy}$$
$$= 2056 - (.765562)(1894)$$
$$= 606.03.$$

Then, since the number of data points is $n = 10$,

$$s^2 = \frac{SSE}{n-2} = \frac{606.03}{8} = 75.754.$$

How can you interpret these values of SSE and s^2? Turn to Figure 11.4 and note the deviations of the $n = 10$ points from the least-squares line (shown as the vertical line segments between the points and the line). SSE = 606.03 is equal to the sum of squares of the numerical values of these deviations. This quantity is then used to calculate $s^2 = 75.754$ and $s = \sqrt{75.754} = 8.70$, estimates of σ^2 and σ.

The practical interpretation that can be given to s ultimately rests on the meaning of σ. Since σ measures the spread of the y values about the line of means $E(y) = \beta_0 + \beta_1 x$ (see Figure 11.3), we would expect (from the Empirical Rule) approximately 95 percent of the y values to fall within 2σ of that line. Since we do not know σ, $2s$ provides an approximate value for the half width of this interval. Now return to Figure 11.4 and note the location of the data points about the least-squares line. Since we used the $n = 10$ data points to fit the least-squares line, you would not be too surprised to find that most of the points fall within $2s = 2(8.7) = 17.40$ of the line. If you check Figure 11.4, you will see that all ten points fall within $2s$ of the least-squares line. (You will find that, in general, most of the data points used to fit the least-squares line will fall within $2s$ of the line. This provides you with a rough check for your calculated value of s.)

But s will play a much more important role in this chapter than the application described. As mentioned at the beginning of this section, the less the variability of the y values about the line of means (i.e., the smaller the value of σ), the closer the least-squares line will be to the line of means. Consequently, s will play an important role in evaluating the goodness of all of the inferential methods described in this chapter.

Tips on Problem Solving

1. To reduce rounding error, always carry at least six significant figures when calculating SS_y and SSE. You can round when you obtain the answer for SSE if you desire.

2. As a check on your calculated value of s, remember that s measures the spread of the points about the least-squares line. Therefore, you would expect (by the Empirical Rule) most of the points to fall within $2s$ of the least-squares line. For example, if the points appear to fall in a band roughly equal to 4 units in width on the scale of the y variable and if your calculated value of s is 10, your value of s is too large. You have made an error. For example, perhaps you forgot to divide SSE by $(n - 2)$.

Exercises

11.14. Calculate SSE and s^2 for the data, Exercise 11.8.

11.15. Calculate SSE and s^2 for the data, Exercise 11.9.

11.16. Calculate SSE and s^2 for the data, Exercise 11.10.

11.17. Calculate SSE and s^2 for the data, Exercise 11.11.

11.18. Calculate SSE and s^2 for the data, Exercise 11.12.

11.19. Calculate SSE and s^2 for the data, Exercise 11.13.

11.5

Inferences Concerning the the Slope of the Line, β_1

The initial inference desired in studying the relationship between y and x concerns the existence of the relationship. Does x contribute information for the prediction of y? That is, do the data present sufficient evidence to indicate that y increases (or decreases) linearly as x increases over the region of observation? Or, is it quite probable that the points would fall on the graph in a manner similar to that observed in Figure 11.1 when y and x are completely unrelated?

The practical question we pose concerns the value of β_1, which is the average change in y for a one-unit increase in x. Stating that y does not increase (or decrease) linearly as x increases is equivalent to saying that $\beta_1 = 0$. Thus, we would wish to test an hypothesis that $\beta_1 = 0$ against the alternative that $\beta_1 \neq 0$. As we might suspect, the estimator, $\hat{\beta}_1$, is extremely useful in constructing a test statistic to test this hypothesis. Therefore, we wish to examine the distribution of estimates, $\hat{\beta}_1$, that would be obtained when samples, each containing n points, are repeatedly drawn from the population of interest. If we assume that the random error, ϵ, is normally distributed, in addition to the previously stated assumptions, it can be shown that both $\hat{\beta}_0$ and $\hat{\beta}_1$ will be normally distributed in repeated sampling and that the expected value and variance of $\hat{\beta}_1$ will be

$$E(\hat{\beta}_1) = \beta_1,$$

$$\sigma_{\hat{\beta}_1}^2 = \frac{\sigma^2}{SS_x}.$$

Thus, $\hat{\beta}_1$ is an unbiased estimator of β_1, we know its standard deviation, and hence we can construct a z statistic in the manner described in Section 9.3. Then,

$$z = \frac{\hat{\beta}_1 - \beta_1}{\sigma_{\hat{\beta}_1}} = \frac{\hat{\beta}_1 - \beta_1}{\sigma/\sqrt{SS_x}}$$

would possess a standardized normal distribution in repeated sampling. Since the actual value of σ^2 is unknown, we would wish to obtain the estimated standard deviation of $\hat{\beta}_1$, which is $s/\sqrt{SS_x}$. Substituting s for σ in z, we obtain, as in Chapter 10, a test statistic,

$$t = \frac{\hat{\beta}_1 - \beta_1}{s/\sqrt{SS_x}} = \frac{\hat{\beta}_1 - \beta_1}{s}\sqrt{SS_x},$$

which can be shown to follow a Student's t distribution in repeated sampling with $(n - 2)$ degrees of freedom. Note that the number of degrees of freedom associated with s^2 determines the number of degrees of freedom associated with t. Thus we observe that the test of an hypothesis that β_1 equals some particular numerical value, say $\beta_{1,0}$, is the familiar t test encountered in Chapter 10.

Test of an Hypothesis Concerning the Slope of a Line

1. *Null Hypothesis:* $H_0: \beta_1 = \beta_{1,0}$
2. *Alternative Hypothesis:*

One-Tailed Test	*Two-Tailed Test*
$H_a: \beta_1 > \beta_{1,0}$	$H_a: \beta_1 \neq \beta_{1,0}$
(or, $\beta_1 < \beta_{1,0}$)	

3. *Test Statistic:* $t = \dfrac{\hat{\beta}_1 - \beta_{1,0}}{s}\sqrt{SS_x}$

   where $SS_x = \displaystyle\sum_{i=1}^{n}(x_i - \bar{x})^2 = \sum_{i=1}^{n} x_i^2 - \frac{\left(\displaystyle\sum_{i=1}^{n} x_i\right)^2}{n}.$

When the assumptions noted below are satisfied, the test statistics will possess a Student's t distribution with $(n - 2)$ degrees of freedom.

4. *Rejection Region:*

One-Tailed Test	*Two-Tailed Test*
$t > t_\alpha$	$t > t_{\alpha/2}$ or $t < -t_{\alpha/2}$
(or, $t < -t_\alpha$ when the	
alternative hypothesis is	
$H_a:\beta_1 < \beta_{1,0}$)	

The values of t_α and $t_{\alpha/2}$ are given in Table 4, the Appendix. Use the values of t corresponding to $(n - 2)$ degrees of freedom.

Assumptions: The assumptions are those associated with the linear probabilistic model, Section 11.2. They are summarized and discussed in Section 11.11.

Example 11.3

Use the data in Table 11.2 to determine whether a linear relationship exists between a freshman's mathematics achievement test score, x, and his or her final calculus grade, y.

Solution

We wish to test the null hypothesis

$$H_0:\beta_1 = 0 \quad \text{against} \quad H_a:\beta_1 \neq 0$$

for the calculus grade–achievement test score data in Table 11.2. The test statistic will be

$$t = \frac{\hat{\beta}_1 - 0}{s}\sqrt{SS_x}$$

and, if we choose $\alpha = .05$, we will reject H_0 when $t > 2.306$ or $t < -2.306$. The critical value of t is obtained from the t table using $(n - 2) = 8$ degrees of freedom. Substituting into the test statistic, we obtain

$$t = \frac{\hat{\beta}_1}{s}\sqrt{SS_x} = \frac{.765562}{8.70}\sqrt{2474}$$

or

$$t = 4.377.$$

Observing that the test statistic exceeds the critical value of t, we will reject the null hypothesis, $\beta_1 = 0$, and conclude that there is evidence to indicate that the calculus final grade is linearly related to the achievement test score.

Once we have decided that x and y are linearly related, we would be interested in examining this relationship in detail. If x increases by one unit, what is the predicted change in y and how much confidence can be placed in the estimate? In other words, we require an estimate of the slope β_1. You will not be surprised to observe a continuity in the procedures of Chapters 10 and 11. That is, the confidence interval for β_1, with confidence coefficient $(1 - \alpha)$, can be shown to be

$$\hat{\beta}_1 \pm t_{\alpha/2}(\text{estimated } \sigma_{\hat{\beta}_1})$$

or:

A 100$(1 - \alpha)$% Confidence Interval for β_1

$$\hat{\beta}_1 \pm \frac{t_{\alpha/2}s}{\sqrt{SS_x}},$$

where $t_{\alpha/2}$ is based on $(n - 2)$ degrees of freedom and

$$SS_x = \sum_{i=1}^{n} (x_i - \bar{x})^2 = \sum_{i=1}^{n} x_i^2 - \frac{\left(\sum_{i=1}^{n} x_i \right)^2}{n}$$

Example 11.4   Find a 95 percent confidence interval for β_1 based upon the data in Table 11.2.

The 95 percent confidence interval for β_1, based upon the data in Table 11.2, is

$$\hat{\beta}_1 \pm \frac{t_{.025}s}{\sqrt{SS_x}}.$$

Substituting, we obtain

$$.77 \pm \frac{(2.306)(8.70)}{\sqrt{2474}}$$

or

.77 $\pm$ .40.

Several points concerning the interpretation of our results deserve particular attention. As we have noted, β_1 is the slope of the assumed line over the region of observation and indicates the *linear* change in y for a one-unit change in x. Even though we do not reject the null hypothesis, $\beta_1 = 0$, x and y may be related. In the first place, we must be concerned with the probability of committing a type II error, that is, accepting the null hypothesis that the slope β_1 equals 0 when this hypothesis is false. Second, it is possible that x and y might be *perfectly* related in a curvilinear but not in a linear manner. For example, Figure 11.5 depicts a curvilinear relationship between y and x over the domain of x, $a \leq x \leq f$. We note that a straight line would provide a good predictor of y if fitted over a small interval in the x domain, say $b \leq x \leq c$. The resulting line would be line 1. On the other hand, if we attempt to fit a line over the region $c \leq x \leq d$, β_1 will equal zero and the best fit to the data will be the horizontal line 2. This would occur even though *all* the points fell perfectly on the curve and y and x possessed a functional relation as defined in Section 11.2. Thus, we must take care in drawing conclusions if we do not find evidence to indicate that β_1 differs from zero. Perhaps we have chosen the wrong type of probabilistic model for the physical situation.

Figure 11.5 Curvilinear relation

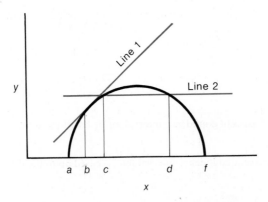

Note that these comments contain a second implication. If the data provide values of x in an interval $b \leq x \leq c$, then the calculated prediction equation is appropriate only over this region. You can see that extrapolation in predicting y for values of x outside of the region, $b \leq x \leq c$, for the situation indicated in Figure 11.5 would result in a serious prediction error.

Finally, if the data present sufficient evidence to indicate that β_1 differs from zero, we do not conclude that the true relationship between y and x is linear. Undoubtedly, y is a function of a number of variables, which demonstrate their existence to a greater or lesser degree in terms of the random error, ϵ, which appears in the model. This, of course, is why we have been obliged to use a probabilistic model in the first place. Large errors of prediction imply either curvatures in the true relation between y and x, the presence of other important variables that do not appear in the model, or, as most often is the case, both. All we can say is that we have evidence to indicate that y changes as x changes and that we may obtain a better prediction of y using x and the linear predictor than simply using $\bar{y}$ and ignoring x. Note that this *does not* imply a *causal* relationship between x and y. A third variable may have caused the change in both x and y, producing the relationship that we have observed.

Exercises

11.20. Do the data, Exercise 11.8, present sufficient evidence to indicate that y and x are linearly related? (Test the hypothesis that $\beta_1 = 0$; use $\alpha = .05$.)

11.21. Find a 90 percent confidence interval for the slope of the line in Exercise 11.8. Interpret this interval estimate.

11.22. Do the data, Exercise 11.9, present sufficient evidence to indicate that y and x are linearly related? (Test the hypothesis that $\beta_1 = 0$; use $\alpha = .05$.)

11.23. Find a 95 percent confidence interval for the slope of the line in Exercise 11.9. Interpret this interval estimate.

11.24. Do the data, Exercise 11.10, present sufficient evidence to indicate that y and x are linearly related? (Test the hypothesis that $\beta_1 = 0$; use $\alpha = .05$.)

11.25. Find a 90 percent confidence interval for the slope of the line in Exercise 11.10. Give a practical interpretation to this interval estimate.

11.26. Do the data, Exercise 11.11, present sufficient evidence to indicate that y and x are linearly related? (Test the hypothesis that $\beta_1 = 0$; use $\alpha = .05$.)

11.27. Find a 95 percent confidence interval for the slope of the line in Exercise 11.11. Give a practical interpretation to this interval estimate.

11.28. Do the data, Exercise 11.12, present sufficient evidence to indicate that y and x are linearly related? (Test the hypothesis that $\beta_1 = 0$; use $\alpha = .05$.)

11.29. Find a 90 percent confidence interval for the slope of the line in Exercise 11.12. Interpret this interval estimate.

11.30. Do the data, Exercise 11.13, present sufficient evidence to indicate that y and x are linearly related? (Test the hypothesis that $\beta_1 = 0$; use $\alpha = .05$.)

11.31. Find a 95 percent confidence interval for the slope of the line in Exercise 11.13. Interpret this interval estimate.

11.32. A study was conducted to determine the effects of sleep deprivation on subjects' ability to solve simple problems. The amount of sleep deprivation varied over 8, 12, 16, 20, and 24 hours without sleep. A total of ten subjects participated in the study, two at each sleep-deprivation level. After his specified sleep-deprivation period, each subject was administered a set of simple addition problems and the number of errors recorded. The following results were obtained:

Number of Errors, y	8, 6	6, 10	8, 14	14, 12	16, 12
Number of Hours Without Sleep, x	8	12	16	20	24

a. Find the least-squares line appropriate to these data.

b. Plot the points and graph the least-squares line as a check on your calculations.

c. Calculate s^2.

11.33. Do the data in Exercise 11.32 present sufficient evidence to indicate that number of errors is linearly related to number of hours without sleep? (Test using $\alpha = .05$.) Would you expect the relation between y and x to be linear if x were varied over a wider range (say $x = 4$ to $x = 48$)?

11.34. Find a 95 percent confidence interval for the slope of the line in Exercise 11.32. Give a practical interpretation to this interval estimate.

∞ 11.35. Writing in *World Tennis*, Hedrick, Mikic, and Ramnath* present data resulting from a series of physical tests on 26 popular tennis racquets. Among the measurements are the racquet head area, x, and a "sweet spot index," y. The "sweet spot" in a racquet is the region in the racquet face that will deliver a solid impact to the ball. The larger the sweet spot, and presumably the larger the sweet spot index, the greater the likelihood that a player will solidly hit the ball. Some tennis players believe that racquets that possess large face areas have larger sweet spots. To examine this theory, we have randomly selected 12 racquets from among the 26. The face area, x, and sweet spot index, y are shown for each.

	Aldila Cannon	Dunlop Black Max	Le Coq Sportif TXF	Prince Graphite	Slazenger Phantom II	Wilson Advantage	Davis Classic III	Rossignol R-40	Yonex R-1	Spalding Natural	Snauwaert Graphite Composite	Wilson Ultra PWS
x Head Area (sq. in.)	75.15	88.4	72.6	108.8	81.8	65.5	69.3	71.9	90.2	67.8	67.1	71.4
y Sweet Spot Index	5.65	5.0	4.75	5.75	5.5	5.35	4.45	3.75	5.5	7.5	3.75	8.0

a. Plot the data points. From a visual examination of the data, does it appear that the racquet face area x helps predict the sweet spot index y?

b. Find the least-squares line for the data.

c. Do the data provide sufficient evidence to indicate that racquet face area x contributes information for predicting the sweet spot index y? Test using $\alpha = .05$. (*Note:* The result

* Hedrick, Karl, Bora Mikic, and Ram Ramnath, "Table of Racquet Performance Data," *World Tennis*, December 1979. Copyright 1979 CBS Publications, the Consumer Publishing Division of CBS Inc.

of this test will neither support nor refute the theory that the size of the sweet spot is related to the face area of a racquet. We do not know how the sweet spot index was computed or whether it is closely related to the size of a racquet's sweet spot.)

11.36. A marketing research experiment was conducted to study the relationship between the length of time necessary for a buyer to reach a decision and the number of alternative package designs of a product presented. Brand names were eliminated from the packages to reduce the effects of brand preferences. The buyers made their selections using the manufacturer's product descriptions on the packages as the only buying guide. The length of time necessary to reach a decision is recorded for 15 participants in the marketing research study.

Length of Decision Time, y (sec)	5, 8, 8, 7, 9	7, 9, 8, 9, 10	10, 11, 10, 12, 9
Number of Alternatives, x	2	3	4

a. Find the least-squares line appropriate for these data.

b. Plot the points and graph the line as a check on your calculations.

c. Calculate s^2.

d. Do the data present sufficient evidence to indicate that the length of decision time is linearly related to the number of alternative package designs? (Test at the $\alpha = .05$ level of significance.)

e. Find the approximate observed significance level for the test and interpret its value.

11.37. Is the per capita consumption of cheese growing in the United States? A cheese importer would tell you that it depends upon the type. The data shown in the table give the per capita consumption of two types of cheese, Swiss and the combination of the Dutch cheese, Edam and Gouda, for the period 1965 to 1976.*

		Per Capita Consumption (Pounds)	
Year	$x =$ Year $-$ 1970	Swiss	Edam and Gouda
1965	-5	.73	.07
1966	-4	.80	.10
1967	-3	.81	.10
1968	-2	.93	.15
1969	-1	.85	.10
1970	0	.90	.11
1971	1	.95	.10
1972	2	1.08	.11
1973	3	1.08	.12
1974	4	1.21	.11
1975	5	1.12	.11
1976	6	1.28	.11

* *Dairy Situation*, Economic Research Service, U.S. Department of Agriculture, September 1977.

a. Let y represent the per capita consumption of Swiss cheese. Find a least-squares line appropriate for the data.

b. Plot the points and graph the line as a check on your calculations.

c. Calculate s^2.

d. Do the data provide sufficient increase to indicate that the mean annual change β_1 in the per capita consumption of Swiss cheese differs from 0? Test using $\alpha = .10$.

e. Find a 90 percent confidence interval for the mean annual change in the consumption rate. Interpret this interval estimate.

f. If you were a Swiss cheese importer, how much of an increase would you expect in the mean per capita consumption of Swiss cheese in 1978 over 1977?

11.38.   Refer to Exercise 11.37 and perform the same type of data analysis for the per capita consumption measurements for the Dutch (Edam and Gouda) cheese.

11.6
Estimating the Expected Value of y for a Given Value of x

In Chapters 8 and 10 we studied methods for estimating a population mean, μ, and encountered numerous practical applications of these methods in the examples and exercises. Now let us consider a generalization of this problem.

Estimating the mean value of y for a given value of x [that is, estimating $E(y|x)$] can be a very important practical problem. If a corporation's profit, y, is linearly related to advertising expenditures, x, the corporation would wish to estimate the mean profit for a given expenditure, x. Similarly, a research physician might wish to estimate the mean response of a human to a specific drug dosage, x, and an educator might wish to know the mean grade expected of calculus students who acquired a mathematics achievement test score of $x = 50$. Let us see how our least-squares prediction equation can be employed to obtain these estimates.

Assume that x and y are linearly related according to the probabilistic model defined in Section 11.2 and therefore that $E(y|x) = \beta_0 + \beta_1 x$ represents the expected value of y for a given value of x. Since the fitted line

$$\hat{y} = \hat{\beta}_0 + \hat{\beta}_1 x$$

attempts to estimate the true linear relation (that is, we estimate β_0 and β_1), then $\hat{y}$ would be used to estimate the *expected* value of y as well as a *particular* value of y for a given value of x. If would seem quite reasonable to assume that the errors of estimation and prediction would differ for these two cases. In this situation, we consider the estimation of the expected value of y for a given value of x.

Observe that two lines are drawn in Figure 11.6. The first line represents the line of means for the true relationship,

$$E(y|x) = \beta_0 + \beta_1 x,$$

and the second is the fitted prediction equation,

$$\hat{y} = \hat{\beta}_0 + \hat{\beta}_1 x.$$

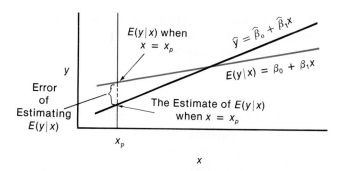

Figure 11.6 Estimating $E(y|x)$ when $x = x_p$

You can see that the error of estimating the expected value of y when $x = x_p$ will be the deviation between the two lines above the point x_p and that this error will increase as we move to the end points of the interval over which x has been measured. It can be shown that the predicted value,

$$\hat{y} = \hat{\beta}_0 + \hat{\beta}_1 x,$$

is an unbiased estimator of $E(y|x)$, that is, $E(\hat{y}) = \beta_0 + \beta_1 x$, and that it will be normally distributed with variance

$$\sigma_{\hat{y}}^2 = \sigma^2 \left[\frac{1}{n} + \frac{(x_p - \bar{x})^2}{SS_x} \right].$$

The corresponding estimated variance of $\hat{y}$ would use s^2 to replace σ^2 in the above expression.

The results outlined above may be used to test an hypothesis concerning the mean or expected value of y for a given value of x, say x_p. (This, of course, would also enable us to test an hypothesis concerning the y intercept, β_0, which is the special case where $x_p = 0$.) The null hypothesis would be

$$H_0: E(y|x = x_p) = E_0,$$

where E_0 is the hypothesized numerical value of $E(y)$ when $x = x_p$. Once again, it can be shown that the quantity

$$t = \frac{\hat{y} - E_0}{\text{estimated } \sigma_{\hat{y}}}$$

$$= \frac{\hat{y} - E_0}{s \sqrt{\frac{1}{n} + \frac{(x_p - \bar{x})^2}{SS_x}}}$$

follows a Student's t distribution in repeated sampling with $(n - 2)$ degrees of freedom. Thus the statistical test is conducted in exactly the same manner as the other t tests previously discussed.

A Test Concerning the Expected Value of y when x = x_p

1. *Null Hypothesis:* $H_0: E(y|x = x_p) = E_0$.

2. *Alternative Hypothesis:*

    One-Tailed Test

    $H_a: E(y|x = x_p) > E_0$
    (or, $E(y|x = x_p) < E_0$)

    Two-Tailed Test

    $H_a: E(y|x = x_p) \neq E_0$

3. *Test Statistic:* $t = \dfrac{\hat{y} - E_0}{s\sqrt{\dfrac{1}{n} + \dfrac{(x_p - \bar{x})^2}{SS_x}}}$

   where $SS_x = \displaystyle\sum_{i=1}^{n} (x_i - \bar{x})^2 = \sum_{i=1}^{n} x_i^2 - \dfrac{\left(\displaystyle\sum_{i=1}^{n} x_i\right)^2}{n}$

4. *Rejection Region:*

    One-Tailed Test

    $t > t_\alpha$
    (or, $t < -t_\alpha$ when the
    alternative hypothesis is
    $H_a: E(y|x = x_p) < E_0$)

    Two-Tailed Test

    $t > t_{\alpha/2}$ or $t < -t_{\alpha/2}$

The values of t_α and $t_{\alpha/2}$ are given in Table 4, the Appendix. Use the value of t corresponding to $(n - 2)$ degrees of freedom

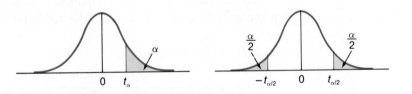

Assumptions: The assumptions are those associated with the linear probabilistic model, Section 11.2. They are summarized and discussed in Section 11.11.

The corresponding confidence interval, with confidence coefficient $(1 - \alpha)$, for the expected value of y, given $x = x_p$, is:

A 100(1 $-$ α)% Confidence Interval for $E(y|x)$ when $x = x_p$

$$\hat{y} \pm t_{\alpha/2}s\sqrt{\frac{1}{n} + \frac{(x_p - \bar{x})^2}{SS_x}}$$

where $t_{\alpha/2}$ is based on $(n - 2)$ degrees of freedom and

$$SS_x = \sum_{i=1}^{n}(x_i - \bar{x})^2 = \sum_{i=1}^{n}x_i^2 - \frac{\left(\sum\limits_{i=1}^{n}x_i\right)^2}{n}.$$

Example 11.5

Find a 95 percent confidence interval for the expected value of y, the final calculus grade, given that the mathematics achievement test score is $x = 50$.

Solution

To estimate the mean calculus grade for students whose achievement test score was $x_p = 50$, we would use

$$\hat{y} = \hat{\beta}_0 + \hat{\beta}_1 x_p$$

to calculate $\hat{y}$, the estimate of $E(y|x = 50)$. Then,

$$\hat{y} = 40.78 + (.77)(50) = 79.28.$$

The formula for the 95 percent confidence interval would be

$$\hat{y} \pm t_{.025}s\sqrt{\frac{1}{n} + \frac{(x_p - \bar{x})^2}{SS_x}}.$$

Substituting into this expression, we find the 95 percent confidence interval for the expected (mean) calculus grade, given an achievement test score of 50, is

$$79.28 \pm (2.306)(8.70)\sqrt{\frac{1}{10} + \frac{(50 - 46)^2}{2474}}$$

or

$$79.28 \pm 6.55.$$

Thus we estimate that the mean calculus grade for the population of students acquiring mathematics achievement test scores of $x = 50$ will fall in the interval 79.28 ± 6.55, or 72.73 to 85.83.

Exercises

11.39.   Refer to Exercise 11.8. Estimate the expected value of y when $x = 1$ using a 90 percent confidence interval.

11.40.   Refer to Exercise 11.9. Estimate the expected value of y when $x = -1$ using a 90 percent confidence interval.

11.41.   Refer to Exercise 11.12. Find a 90 percent confidence interval for the mean number of miles per gallon (EPA combined) that you would expect to obtain from a subcompact with a 125 cubic-inch engine. Interpret the interval.

11.42.   Refer to Exercise 11.32. Estimate the mean number of errors corresponding to $x = 20$ hours of sleep deprivation. Use a 95 percent confidence interval.

11.43.   The table shown below gives the catches in Peruvian anchovy (in millions of metric tons) and the prices of fish meal (in current dollars per ton) for the years 1965 to 1978.*

	1965	1966	1967	1968	1969	1970	1971	1972	1973	1974	1975	1976	1977	1978
Price, Fish Meal, y	190	160	134	129	172	197	167	239	542	372	245	376	454	410
Anchovy Catch, x	7.23	8.53	9.82	10.26	8.96	12.27	10.28	4.45	1.78	4.0	3.3	4.3	0.8	0.5

a. Find the least-squares line appropriate for these data.

b. Plot the points and graph the line as a check on your calculations.

c. Calculate s^2.

d. Do the data present sufficient evidence to indicate that the size of the anchovy catch, x, contributes information for the prediction of the price, y, of fish meal? Test using $\alpha = .10$.

e. Find the approximate observed significance level for the test, part (d), and interpret its value.

f. Find a 90 percent confidence interval for the mean price per ton of fish meal when the anchovy catch is 5 million metric tons.

11.44.   In manufacturing an antibiotic, the yield is a function of time. Data collected show that a process yielded the following pounds of antibiotic for the time periods shown:

Time, x (days)	1	2	3	4	5	6
Yield, y	23	31	40	46	52	63

a. For various reasons, it is convenient to schedule production using a 4-day cycle. Estimate the mean yield for the amount of antibiotic produced over a 4-day period. Use a 95 percent confidence interval.

b. In a practical situation, the yield for a zero length of time must be 0. Explain why the least-squares line does not go through the origin. Should it?

* Bardach, John E. and Regina M. Santerre, "Climate and the Fish in the Sea." *BioScience*, 31, no. 3 (March 1981), pp. 206 ff. Copyright © 1981 by the American Institute of Biological Sciences. Reprinted with permission of the copyright holder.

11.45. If you try to rent an apartment or buy a house, you will find that real estate representatives establish apartment rents and house prices on the basis of the square footage of the heated floor space. The data in the table give the square footages and sales prices of $n = 12$ houses randomly selected from those sold in a small city.

| Square Feet | Price |
x	y
1460	$58,700
2108	79,300
1743	71,400
1499	61,100
1864	72,400
2391	84,900
1977	75,400
1610	67,000
1530	62,400
1759	68,200
1821	74,300
2216	81,700

a. Estimate the mean increase in the price for an increase of one square foot for houses sold in the city. Use a 90 percent confidence interval. Interpret your estimate.

b. Suppose that you are a real estate salesperson and you desire an estimate of the mean sales price of houses with a total of 2000 squares feet of heated space. Use a 95 percent confidence interval and interpret your estimate.

c. Calculate the price per square foot for each house and then calculate the sample mean. Why is this estimate of the mean cost per square foot not equal to the answer in part (a)? Should it be? Explain.

11.7
Predicting a Particular Value of y for a Given Value of x

Although the expected value of y for a particular value of x is of interest for our example, Table 11.2, we are primarily interested in *using* the prediction equation, $\hat{y} = \hat{\beta}_0 + \hat{\beta}_1 x$, based upon our observed data to predict the final calculus grade for some prospective student selected from the population of interest. That is, we want to use the prediction equation obtained for the ten measurements, Table 11.2, to predict the final calculus grade for a new student selected from the population. If the student's achievement test score was x_p, we intuitively see that the error of prediction (the deviation between $\hat{y}$ and the actual grade, y, that the student will obtain) is composed of two elements. Since the student's grade will equal

$$y = \beta_0 + \beta_1 x_p + \epsilon,$$

$(y - \hat{y})$ will equal the deviation between $\hat{y}$ and the expected value of y, described in Section 11.6 (and shown in Figure 11.6), *plus* the random amount ϵ which represents the deviation of the student's grade from the expected value (see Figure 11.7). **Thus the variability in the error for predicting a single value of y will exceed the variability for estimating the expected value of y.**

Figure 11.7 Error in predicting a particular value of y

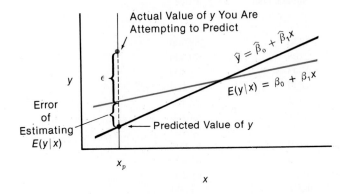

It can be shown that the variance of the error of predicting a particular value of y when $x = x_p$, that is, $(y - \hat{y})$, is

$$\sigma_{error}^2 = \sigma^2 \left[1 + \frac{1}{n} + \frac{(x_p - \bar{x})^2}{SS_x} \right].$$

When n is very large, the second and third terms in the brackets will become small and the variance of the prediction error will approach σ^2. These results may be used to construct the following prediction interval for y, given $x = x_p$. The confidence coefficient for the prediction interval is $(1 - \alpha)$.

A 100$(1 - \alpha)$% Prediction Interval for y when $x = x_p$

$$\hat{y} \pm t_{\alpha/2} s \sqrt{1 + \frac{1}{n} + \frac{(x_p - \bar{x})^2}{SS_x}}$$

where $t_{\alpha/2}$ is based on $(n - 2)$ degrees of freedom and

$$SS_x = \sum_{i=1}^{n} (x_i - \bar{x})^2 = \sum_{i=1}^{n} x_i^2 - \frac{\left(\sum_{i=1}^{n} x_i \right)^2}{n}.$$

Example 11.6

Refer to Example 11.1 and predict the final calculus grade for some new student who scored $x = 50$ on the mathematics achievement test.

Solution

The predicted value of y would be

$$\hat{y} = \hat{\beta}_0 + \hat{\beta}_1 x_p$$

or

$$\hat{y} = 40.78 + (.77)(50) = 79.28,$$

and the 95 percent prediction interval for the final calculus grade would be

$$79.28 \pm (2.306)(8.70)\sqrt{1 + \frac{1}{10} + \frac{(50 - 46)^2}{2474}}$$

or

$$79.28 \pm 21.10.$$

Note that in a practical situation we would likely possess the grades and achievement test scores for many more than the $n = 10$ students indicated in Table 11.2 and that this would reduce somewhat the width of the prediction interval. In fact, when n is very large, the prediction interval approaches $\hat{y} \pm z_{\alpha/2}s$ or, for a 95 percent prediction interval, $\hat{y} \pm 1.96s$.

Again, note the distinction between the confidence interval for $E(y|x)$ discussed in Section 11.6 and the prediction interval presented in this section. $E(y|x)$ is a mean, a parameter of a population of y values, and y is a random variable that oscillates in a random manner about $E(y|x)$. The mean value of y when $x = 50$ is vastly different from some value of y, chosen at random from the set of all y values for which $x = 50$. To make this distinction when making inferences, **we always** *estimate* **the value of a parameter and** *predict* **the value of a random variable.** As noted in our earlier discussion and as shown in Figures 11.6 and 11.7, the error of predicting y is different from the error of estimating $E(y|x)$. This is evident in the difference in widths of the two prediction and confidence intervals.

A graph of the confidence interval for $E(y|x)$ and the prediction interval for a particular value of y for the data in Table 11.2 is shown in Figure 11.8. The plot of the confidence interval is shown by solid lines, the prediction interval is identified by dashed lines. Note how the widths of the intervals

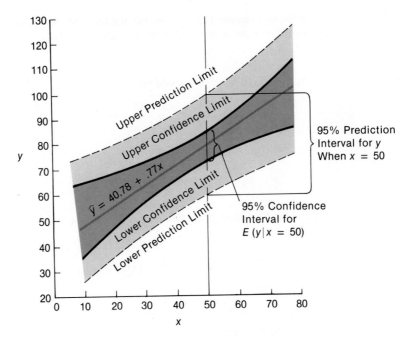

Figure 11.8 Confidence intervals for $E(y|x)$ and prediction intervals for y based on data in Table 11.2

increase as you move to the right or left of $\bar{x} = 46$. Particularly, see the confidence interval and prediction interval for $x = 50$ which were calculated in Examples 11.5 and 11.6.

Exercises

11.46.   Refer to Exercise 11.8. Find a 90 percent prediction interval for some value of y to be observed in the future when $x = 1$.

11.47.   Refer to Exercise 11.9. Find a 90 percent prediction interval for some value of y to be observed in the future when $x = -1$.

11.48.   Refer to Exercise 11.44. Suppose that the production process is operated for a 4-day period at some particular time in the future. Find a 95 percent prediction interval for the yield. Compare with the confidence interval on the mean yield for $x = 4$ days.

11.49.   Refer to Exercise 11.37. Find a 95 percent prediction interval for the per capita Swiss cheese consumption in the United States for the year 1977. Interpret your prediction interval. Are there any qualifications that you would place on your prediction? Explain.

11.50.   Refer to Exercise 11.45. Suppose that a house containing 1780 square feet of heated floor space is offered for sale. Give a 90 percent prediction interval for the price at which the house will sell. Interpret this prediction.

11.51.   In an effort to increase the production in the semiarid regions of the world, it is necessary that we be able to monitor the amount of moisture in the soil. Two methods are available. One utilizes mathematical formulas, temperatures, rainfalls, and evaporation rates to arrive at a computed value of moisture content. The other relies on direct measurement of the weight of soil before and after evaporation of the soil moisture. A comparison of the theoretically computed moisture content (in percent of dry weight of soil) with the measured moisture content is shown below for nine soil samples randomly selected from a semiarid region.

Computed Soil Moisture, x	10.2	13.6	9.7	7.3	13.1	15.6	9.0	10.4	12.1
Measured Soil Moisture, y	11.8	15.4	9.2	8.0	13.8	16.1	8.4	10.2	10.5

a. Fit a least-squares line to the data.

b. If the computed soil moisture reading at some future time was 10.0, find a 90 percent prediction interval for the actual measured soil moisture.

11.52. Most sophomore physics students are required to conduct an experiment verifying Hooke's Law. Hooke's Law states that when a force is applied to a body that is long in comparison to its cross-sectional area, the change y in its length is proportional to the force x; that is,

$$y = \beta_1 x,$$

where β_1 is a constant of proportionality. The results of an actual physics student's laboratory experiment are shown in the table. Six lengths of steel wire, .34 millimeters (mm) in diameter and 2 meters (m) long, were used to obtain the six force-length change measurements.

Force x (kg)	Change in Length y (mm)
29.4	4.25
39.2	5.25
49.0	6.50
58.8	7.85
68.6	8.75
78.4	10.00

a. Fit the model, $y = \beta_0 + \beta_1 x + \epsilon$, to the data using the method of least squares.

b. Plot the points and graph the line as a check on your calculations.

c. Find a 95 percent confidence interval for the slope of the line.

d. According to Hooke's Law, the line should pass through the point $(0, 0)$; that is, β_0 should equal 0. Test the hypothesis that $E(y) = 0$ when $x = 0$.

e. Predict the elongation of a 2-m length of wire when a force of 55 kg is applied. Use a 95 percent prediction interval.

11.53. Colin W. Clark, in an article titled "Bioeconomics of the Ocean,"* discusses the problem of overexploitation of ocean resources. In the article, Clark presents data on the Peruvian

---

* Clark, Colin W., "Bioeconomics of the Ocean," *BioScience*, 31, no. 3 (March 1981), pp. 231–237. Copyright © 1981 by the American Institute of Biological Sciences. Reprinted with the permission of the copyright holder.

anchoveta fishery, a fishery "that developed in the 1960's into the world's largest single fishery." Data showing the growth and the eventual depletion of the anchoveta catch are shown in the accompanying table.

Year	Number of Boats	Number of Fishing Days	Fishing Effort (Number of Boats) × (Number of Fishing Days) (in units of 10,000 boat-days)	Catch (million tons)
1959	414	294	12.2	1.91
1960	667	279	18.6	
1961	756	298	22.5	4.58
1962	1069	294	31.4	6.27
1963	1655	269	44.5	6.42
1964	1744	297	51.8	8.86
1965	1623	265	43.0	7.23
1966	1650	190	31.4	8.53
1967	1569	170	26.7	9.82
1968	1490	167	24.9	10.26
1969	1455	162	23.6	8.96
1970	1499	180	27.0	12.27
1971	1473	89	13.1	10.28
1972	1399	89	12.5	4.45
1973	1256	27	3.39	1.78
1974				4.00
1975				3.30
1976				4.30
1977				.80
1978				.50

The increase in the size of the fishing fleet from 1959 to 1973 can be seen in the second column of the table. The reduction in fishing days and the rise and fall in the catch over the same period of time is shown in the third and fifth columns. The column marked "Fishing Effort," the product of the number of boats and the number of fishing days, is a measure of the amount of effort (money, time, etc.) expended to obtain the catch.

a. Find the least-squares prediction equation relating catch y and the fishing effort x. (Delete years in which data are missing.)

b. Do the data provide sufficient evidence to indicate that, using the straight-line linear model, x contributes information for the prediction of y? Test using $\alpha = .10$.

c. Plot the data points (x, y). Does the distribution of plotted points appear to agree with your conclusions, part (b)?

d. Find a 90 percent confidence interval for the mean catch when the fishing effort is equal to 300,000 boat-days.

e. Find a 90 percent prediction interval for the catch when the fishing effort is equal to 300,000 boat-days.

11.8
A Coefficient
of Correlation

It is sometimes desirable to obtain an indicator of the strength of the linear relationship between two variables, y and x, which will be independent of their respective scales of measurement. We will call this a measure of the **linear correlation between y and x.**

The measure of linear correlation commonly used in statistics is called the **Pearson product moment coefficient of correlation.** This quantity, denoted by the symbol r, is defined as follows:

Pearson Product Moment Coefficient of Correlation

$$r = \frac{SS_{xy}}{\sqrt{SS_x SS_y}}$$

We shall show you how to compute the Pearson product moment coefficient of correlation for the grade-point data, Table 11.2, and then we shall explain how it measures the strength of the relationship between y and x.

Example 11.7

Calculate the coefficient of correlation for the calculus grade–achievement test score data in Table 11.2.

Solution

The coefficient of correlation for the calculus grade–achievement test score data, Table 11.2, may be obtained by using the formula for r and the quantities

$$SS_{xy} = 1894$$
$$SS_x = 2474$$

and

$$SS_y = 2056,$$

which were computed previously.
Then,

$$r = \frac{SS_{xy}}{\sqrt{SS_x SS_y}} = \frac{1894}{\sqrt{2474(2056)}} = .84.$$

A study of the coefficient of correlation, r, yields rather interesting results and explains the reason for its selection as a measure of linear correlation. We note that the denominators used in calculating r and $\hat{\beta}_1$ will always be positive, since they both involve sums of squares of numbers. Since the numerator used in calculating r is identical to the numerator of the formula for the slope, $\hat{\beta}_1$, the coefficient of correlation, r, will assume exactly the same sign as $\hat{\beta}_1$ and will equal zero when $\hat{\beta}_1 = 0$. Thus $r = 0$ implies no linear correlation between y and x. A positive value for r will imply that the line slopes upward to the right; a negative value indicates that it slopes downward to the right (see Figure 11.9).

Figure 11.9 Some typical scatter diagrams with approximate values of r

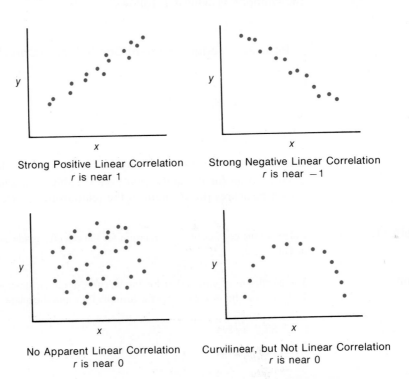

Strong Positive Linear Correlation
r is near 1

Strong Negative Linear Correlation
r is near -1

No Apparent Linear Correlation
r is near 0

Curvilinear, but Not Linear Correlation
r is near 0

The interpretation of nonzero values of r may be obtained by comparing the errors of prediction for the prediction equation

$$\hat{y} = \hat{\beta}_0 + \hat{\beta}_1 x$$

with the predictor of y, $\bar{y}$, that would be employed if x were ignored. Figure 11.10(a) and (b) shows the lines $\hat{y} = \hat{\beta}_0 + \hat{\beta}_1 x$ and $\hat{y} = \bar{y}$ fit to the same set of data. Certainly, if x is of any value in predicting y, then SSE, the sum of squares of deviations of y about the least squares line, should be less than the sum of squares about the predictor, $\bar{y}$, which would be

$$SS_y = \sum_{i=1}^{n} (y_i - \bar{y})^2.$$

Figure 11.10 Two models fit to the same data

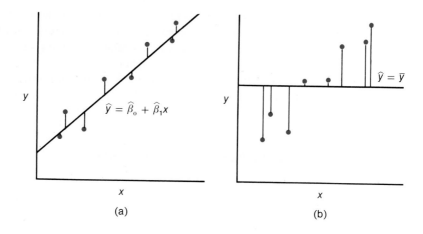

(a)   (b)

Indeed, we see that SSE can *never* be larger than

$$SS_y = \sum_{i=1}^{n} (y_i - \bar{y})^2$$

because

$$SSE = SS_y - \hat{\beta}_1 SS_{xy} = SS_y - \left(\frac{SS_{xy}}{SS_x}\right)SS_{xy}$$

$$= SS_y - \frac{(SS_{xy})^2}{SS_x}.$$

Therefore, SSE is equal to SS_y minus a positive quantity.

Furthermore, with the aid of a bit of algebraic manipulation, you can show that

$$r^2 = 1 - \frac{SSE}{SS_y}$$

$$= \frac{SS_y - SSE}{SS_y}.$$

In other words, r^2 will lie in the interval

$$0 \le r^2 \le 1$$

and r will equal $+1$ or -1 only when all the points fall exactly on the fitted line, that is, when SSE equals zero. Actually, we see that r^2 is equal to the ratio of the **reduction in the sum of squares of deviations obtained by using the linear model to the total sum of squares of deviations about the sample mean** $\bar{y}$**, which would be the predictor of** y **if** x **were ignored.** Thus r^2, called the coefficient of determination, would seem to give a more meaningful interpretation of the strength of the relation between y and x than would the correlation coefficient, r.

The Coefficient of Determination

$$r^2 = \frac{SS_y - SSE}{SS_y} = \frac{\sum_{i=1}^{n} (y_i - \bar{y})^2 - SSE}{\sum_{i=1}^{n} (y_i - \bar{y})^2}$$

You will observe that the sample correlation coefficient, r, is an estimator of a population correlation coefficient, ρ, which would be obtained if the coefficient of correlation were calculated using all the points in the population. A discussion of a test of an hypothesis concerning the value of ρ is omitted as well as a bound on the error of estimation. Ordinarily, we would be interested in testing the null hypothesis that $\rho = 0$ and, since this is algebraically equivalent to testing the hypothesis that $\beta_1 = 0$, we have already considered this problem. If the evidence in the sample suggests that y and x are related, it would seem that we would redirect our attention to the ultimate objective of our data analysis, using the prediction equation to obtain interval estimates for $E(y|x)$ and prediction intervals for y.

While r gives a rather nice measure of the goodness of fit of the least-squares line to the fitted data, its use in making inferences concerning ρ may be of dubious practical value in many situations. It would seem unlikely that a phenomenon y, observed in the physical sciences, and especially the social sciences, would be a function of a single variable. Thus the correlation coefficient between a student's predicted grade-point average and any variable likely would be quite small and of questionable value. A larger reduction in SSE could possibly be obtained by constructing a predictor of y based upon a set of variables $x_1, x_2, \ldots$.

One further reminder concerning the interpretation of r is worthwhile. It is not uncommon for researchers in some fields to speak proudly of sample correlation coefficients, r, in the neighborhood of .5 (and, in some cases, as low as .1) as indicative of a "relation" between y and x. Certainly, even if these values were *accurate* estimates of ρ, only a very weak relation would be indicated. A value $r = .5$ would imply that the use of x in predicting y reduced the sum of squares of deviations about the prediction line by only $r^2 = .25$, or 25 percent.

A correlation coefficient, $r = .1$, would imply only an $r^2 = .01$, or 1 percent, reduction in the total sum of squares of deviations that could be explained by x.

If the linear coefficients of correlation between y and each of two variables, x_1 and x_2, were calculated to be .4 and .5, respectively, it does not follow that a predictor using both variables would account for a $[(.4)^2 + (.5)^2] = .41$, or a 41 percent reduction in the sum of squares of deviations. Actually, x_1 and x_2 might be highly correlated and therefore contribute the same information for the prediction of y.

Finally, we remind you that r is a measure of *linear* correlation and that x and y could be perfectly related by some curvilinear function when the observed value of r is equal to zero.

Exercises

11.54. How does the coefficient of correlation measure the strength of the linear relationship between two variables, $\hat{y}$ and x?

11.55. Describe the significance of the algebraic sign and the magnitude of r.

11.56. What value does r assume if all the sample points fall on the same straight line and if

a. the line has positive slope?

b. the line has negative slope?

11.57. Given the data:

x	-2	-1	0	1	2
y	2	2	3	4	4

a. Find the least-squares line for the data.

b. Plot the data points and graph the least-squares line. Based on your graph, what will be the sign of the sample correlation coefficient?

c. Calculate r and r^2 and interpret their values.

11.58. For the following data:

x	1	2	3	4	5	6
y	7	5	5	3	2	0

a. Find the least-squares line.

b. Sketch the least-squares line on graph paper and plot the six points.

c. Calculate the sample coefficient of correlation, r, and interpret.

d. By what percentage was the sum of squares of deviations reduced by using the least-squares predictor, $\hat{y} = \hat{\beta}_0 + \hat{\beta}_1 x$, rather than $\bar{y}$ as a predictor of y?

11.59. Reverse the slope of the line, Exercise 11.58, by reordering the y observations:

x	1	2	3	4	5	6
y	0	2	3	5	5	7

Repeat the steps of Exercise 11.58. Notice the change in sign of r and the relation between the values of r^2 of Exercise 11.58 and this exercise.

11.60.   An experiment was conducted in a supermarket to observe the relation between the amount of display space allotted to a brand of coffee (brand A) and its weekly sales. The amount of space allotted to brand A was varied over 3-, 6-, and 9-square-feet displays in a random manner over 12 weeks while the space allotted to competing brands was maintained at a constant 3 square feet for each. The following data were observed:

Weekly Sales, y (dollars)	526	421	581	630	412	560	434	443	590	570	346	672
Space Allotted, x (ft^2)	6	3	6	9	3	9	6	3	9	6	3	9

a. Find the least-squares line appropriate for the data.

b. Calculate r and r^2. Interpret.

c. Find a 90 percent confidence interval for the mean weakly sales given that 6 square feet is allotted for display.

d. Use a 90 percent prediction interval to predict the weekly sales at some time in the future if 6 square feet is allotted for display.

e. By what percentage was the sum of squares of deviations reduced by using the least-squares predictor, $\hat{y} = \hat{\beta}_0 + \hat{\beta}_1 x$, rather that $\bar{y}$ as a predictor of y for these data?

f. Would you expect the relation between y and x to be linear if x were varied over a wider range (say $x = 1$ to $x = 30$)?

11.61.   Geothermal power could be an important source of energy in future years. Since the amount of energy contained in a pound of water is a function of its temperature, you might wonder whether water obtained from deeper wells contains more energy per pound. The data in the table are reproduced from an article on geothermal systems by A. J. Ellis.*

Location of Well	Average (Max.) Drill Hole Depth (m)	Average (Max.) Temperature (°C)
El Tateo, Chile	650	230
Ahuachapan, El Salvador	1000	230
Namafjall, Iceland	1000	250
Larderello (region), Italy	600	200
Matsukawa, Japan	1000	220
Cerro Prieto, Mexico	800	300
Wairakei, New Zeaiand	800	230
Kizildere, Turkey	700	190
The Geysers, U.S.	1500	250

a. Find the least-squares line.

* Ellis, A.J., "Geothermal Systems," *Amer. Scientist*, September–October 1975.

b. Sketch the least-squares line on graph paper and plot the nine points.

c. Calculate the sample coefficient of correlation, r, and interpret it.

d. By what percentage was the sum of squares of deviations reduced by using the least-squares predictor, $\hat{y} = \hat{\beta}_0 + \hat{\beta}_1 x$, rather than $\bar{y}$?

11.62.   Refer to the Peruvian anchoveta fishery data, Exercise 11.53.

a. Calculate the sample coefficient of correlation between the catch y and the fishing effort x. Interpret its value.

b. Calculate the coefficient of determination and interpret its value.

c. Noting the values of r and r^2 calculated in parts (a) and (b), reexamine your plot of the data points and your graph of the least-squares line, Exercise 11.53, part (c). Do you think that r and r^2 adequately describe how well the least-squares line fits the data?

11.63.   Refer to the Peruvian anchoveta fishery data, Exercise 11.53.

a. Calculate the sample coefficient of correlation r between the catch y and the number of fishing boats x. Interpret its value. (Delete the years in which data are missing.)

b. Find the coefficient of determination and interpret its value.

c. Plot the data points. Do they suggest a linear relationship between y and x? Does the plot agree with your interpretation for r and r^2?

11.64.   Will acid rain eventually eliminate the Eastern tiger salamander? Tests indicate that the jelly-like substance surrounding the salamander eggs absorbs acid water, leading to the eventual deaths of the embryos. Laboratory tests show a mortality rate of embryos of .6 percent when the water surrounding the eggs is neutral (a pH of 7.0). It increases to .9 percent at a pH of 6, to 44 percent at a pH of 5.5, and to approximately 65 percent when the pH is 5.0.*

a. Calculate the sample coefficient of correlation, r, between mortality rate y of salamander embryos and the pH x of the water surrounding the salamander eggs. Interpret its value.

b. Although it seems likely that the mortality rate y and the pH x are closely related, the coefficient of determination r^2 is not very large. Can you explain this apparent inconsistency?

11.9
A Multivariable Predictor

A prediction equation based upon a number of variables, $x_1, x_2, \ldots, x_k$, could be obtained by the method of least squares in exactly the same manner as that employed for the simple linear model. For example, we might wish to fit the model

$$y = \beta_0 + \beta_1 x_1 + \beta_2 x_2 + \beta_3 x_3 + \epsilon,$$

where $y =$ student grade-point average at the end of the freshman year, $x_1 =$ rank in high school class divided by the number in class, $x_2 =$ score on a mathematics achievement test, $x_3 =$ score on a verbal and written achievement test, to data on the achievement of freshmen college students. (Note that we could add other variables as well as the squares, cubes, and cross products of $x_1, x_2,$ and x_3.)

* "Acid Rain," EPA Publication #600/9-79-036, July 1980.

We would require a random sample of n freshmen selected from the population of interest and would record the values of y, x_1, x_2, and x_3, which, for each student, could be regarded as coordinates of a point in four-dimensional space. Then, ideally, we would like to possess a multidimensional "ruler" (in our case, a plane) that we could visually move about among the n points until the deviations of the observed values of y from the predicted values of y would in some sense be a minimum. Although we cannot graph points in four dimensions, the student readily recognizes that this device is provided by the method of least squares, which, mathematically, performs the task for us.

The sum of squares of deviations of the observed values of y from the fitted model would be

$$\text{SSE} = \sum_{i=1}^{n} (y_1 - \hat{y}_i)^2$$

$$= \sum_{i=1}^{n} [y_i - (\hat{\beta}_0 + \hat{\beta}_1 x_{1i} + \hat{\beta}_2 x_{2i} + \hat{\beta}_3 x_{3i})]^2,$$

where $\hat{y} = \hat{\beta}_0 + \hat{\beta}_1 x_1 + \hat{\beta}_2 x_2 + \hat{\beta}_3 x_3$ is the fitted model and $\hat{\beta}_0$, $\hat{\beta}_1$, $\hat{\beta}_2$, and $\hat{\beta}_3$ are estimates of the model parameters. We would then use the calculus to find the estimates, $\hat{\beta}_0$, $\hat{\beta}_1$, $\hat{\beta}_2$, and $\hat{\beta}_3$, that make SSE a minimum. The estimates, as for the simple linear model, would be obtained as the solution of a set of four simultaneous linear equations known as the least-squares equations.

The reasoning employed in obtaining the least-squares multivariable predictor is identical to the procedure developed for the simple linear model, but the presentation of the least-squares equations and their solutions is beyond the scope of this text. Since most multiple-regression analyses are performed by computers, we will introduce you to the computer printout for a simple linear regression analysis in Section 11.12 and will discuss the computer output for a multiple-regression analysis in Chapter 12.

11.10
Case Study: A Computer Analysis of Whooping Crane Sightings

The calculations associated with a multiple (or even a simple) regression analysis are so complex and time consuming, they are often performed on a computer. Since you may wish to use a computer to solve some of the problems in this chapter, we shall show you a computer printout of a regression analysis for an interesting set of data. This will help to familiarize you with the format of a computer printout and will give you a real-life application of regression analysis.

Our regression analysis will pertain to the sightings of whooping cranes, an endangered and protected species that is near extinction. The whooping crane, a native of North America, is the tallest living bird on the continent. According to wildlife officials, in 1941 there were only 21 known whooping cranes in the wilds and this number had dwindled to 15 by 1948. It is believed that the number of whooping cranes living, both in the wilds and in captivity, in 1977 was 100.

The birds, which winter in known wildlife refuges in the southwestern and south central United States, are monitored individually each year. The data we shall analyze are contained in an article by R.S. Miller and D.B. Botkin, which gives the number of whooping cranes sighted in the Arkansas National Wildlife Refuge from 1938 to 1972.* Miller and Botkin use the data to develop a prediction equation to forecast the number of whooping crane sightings in the Arkansas National Wildlife Refuge for 1975, 1980, etc. We shall use only a portion of their data, the number of whooping crane sightings at two-year intervals, for the period 1952 to 1972, and shall perform a computer regression analysis of the data. We shall then use our prediction equation to forecast the number of sightings in 1980 and shall compare our forecast with the forecast of Miller and Botkin.

The data for the 1952-to-1972 period are shown in Table 11.4. To simplify the numbers involved, we let x = year $-$ 1950. The annual numbers of sightings appear in the "Count" column of Table 11.4.

Table 11.4 Whooping crane sightings, Arkansas National Wildlife Refuge, 1952 to 1972

Year	x = Year $-$ 1950	Count y
1952	2	21
1954	4	21
1956	6	24
1958	8	32
1960	10	36
1962	12	32
1964	14	42
1966	16	43
1968	18	50
1970	20	57
1972	22	51

The printouts for computer multiple regression analyses of data vary from one computer program to another but they are similar. Once you become familiar with the printout for one computer regression program, you will be able to quickly adapt this knowledge to others. The whooping crane data were analyzed using an SAS (Statistical Analysis System)† multiple-regression computer program package. The printout is shown in Table 11.5. Many of the numbers appearing in the printout appear in the format used for a multiple-regression analysis (the analysis for a multivariable linear model) and hence are not relevant to our discussion. The pertinent items in the printout are shaded and numbered individually or in groups. We will discuss each of these groups and show you how to extract the pertinent numbers from the printout.

* Miller, R.S., and D.B. Botkin, "Endangered Species: Models and Predictions," *Amer. Scientist,* 62, no. 2 (1974).

† See references.

Table 11.5 SAS computer printout for the regression analysis of whooping crane sightings

STATISTICAL ANALYSIS SYSTEM
GENERAL LINEAR MODELS PROCEDURE

DEPENDENT VARIABLE: COUNT

SOURCE	DF	SUM OF SQUARES	MEAN SQUARE	F VALUE	PR > F	⑤ R-SQUARE	C.V.
MODEL	1	1454.54545455	1454.54545455	126.98	0.0001	0.933816	9.1025
ERROR	9	103.09090909	11.45454545		STD DEV ②		COUNT MEAN
CORRECTED TOTAL	10	1557.63636364			3.38445645		37.18181818

SOURCE	DF	TYPE I SS	F VALUE	PR > F	DF	TYPE IV SS	F VALUE	PR > F
YEAR	1	1454.54545455	126.98	0.0001	1	1454.54545455	126.98	0.0001

PARAMETER	① ESTIMATE	③ T FOR HO: PARAMETER = 0	PR > |T|	④ STD ERROR OF ESTIMATE
INTERCEPT	15.36363636	7.02	0.0001	2.18862574
YEAR	1.81818182	11.27	0.0001	0.16134763

1. Least-squares estimates of β_0 and β_1. The estimates are shown in the bottom left corner of the printout. Thus, to the nearest hundredth,

$$\hat{\beta}_0 = 15.36, \qquad \hat{\beta}_1 = 1.82,$$

and the prediction equation is

$$\hat{y} = 15.36 + 1.82x \qquad \text{where } x = \text{year} - 1950.$$

2. The degrees of freedom for s^2, the values of SSE, s^2, and s. At the top left of the printout, you will notice that the first four columns are marked "SOURCE," "DF," "SUM OF SQUARES," and "MEAN SQUARE." Go to the row of the table marked "ERROR" and proceed to the right. You will find d.f. = 9 in the DF column, SSE = 103.0909 . . . in the SUM OF SQUARES column, and $s^2 = 11.4545 . . .$ in the MEAN SQUARE column. The value of s is shown to be 3.384. . . .

3. Computed value of t for the test, $H_0 : \beta_1 = 0$. The computed value of t for the t test, Section 11.5,

$$t = \frac{\hat{\beta}_1}{s_{\hat{\beta}_1}},$$

is given at the bottom of the third column under "T FOR HO: PARAMETER = 0." The first number under this heading, 7.02, gives the computed value of t to test the null hypothesis that the intercept, β_0, equals 0. Since this test is of little use to us, we ignore this value of t. The second number in the column (in the YEAR row) gives the value of t to test $H_0 : \beta_1 = 0$. The computed value of t for this test is $t = 11.27$. Clearly, this value of t is so large, we reject the null hypothesis, $\beta_1 = 0$, and conclude $\beta_1 \neq 0$.

If you want the observed significance level for this test, it is given to the right of the t value under the column headed "PR $> |T|$." This column gives the probability of observing a value of t greater than or equal to the observed value, 11.27 (or less than $t = -11.27$). This value, .0001, is the significance level for the test.

4. Confidence interval for β_1. A confidence interval for β_1 is not given directly in the printout but it can be easily calculated. From Section 11.5, the confidence interval for β_1 is

$$\hat{\beta}_1 \pm t_{\alpha/2} \text{ (estimated standard deviation of } \hat{\beta}_1)$$

or

$$\hat{\beta}_1 \pm t_{\alpha/2}\, s_{\hat{\beta}_1}.$$

The quantity $t_{\alpha/2}$ is given in Table 4, the Appendix. For d.f. $= 9$ and $\alpha = .05$, $t_{.025} = 2.262$. The value of $s_{\hat{\beta}_1}$ is given in the last line of the printout under STD ERROR OF ESTIMATE. Thus, $s_{\hat{\beta}_1} = .1613 \ldots$ Substituting these values and the value of $\hat{\beta}_1$ into the formula for the confidence interval,

$$\hat{\beta}_1 \pm t_{\alpha/2}\, s_{\hat{\beta}_1},$$
$$1.82 \pm (2.262)(.161),$$

or

$$1.82 \pm .36.$$

Therefore, the 95 percent confidence interval for β_1 is 1.46 to 2.18. Since β_1 is the mean increase (or decrease) in the crane population over a two-year period, we estimate this increase to fall in the interval 1.46 to 2.18 cranes per two-year period.

5. The coefficient of determination, r^2. The value of r^2 is given at the top right of the printout under R-SQUARE. Thus $r^2 = .9338 \ldots$ The coefficient of correlation is not given in the printout but it will take the same sign (positive) as β_1 and can easily be calculated. Thus,

$$r = \sqrt{r^2} = \sqrt{.9338}$$

or

$$r = +.97.$$

Miller and Botkin fit a model (not a simple linear probabilistic model) to their data. Using their prediction equation, they obtain forecasts for the number of cranes that will be observed in 1975, 1980, etc. and give the standard deviations for their predictions. Particularly, they forecast 70 cranes for 1980. Let us see how a forecast obtained from our least-squares prediction equation compares with Miller and Botkin's forecast.

In 1980,

$$x = \text{year} - 1950 = 30.$$

Then the predicted value of y for 1980 is

$$\hat{y} = 15.36 + 1.82x$$
$$= 15.36 + 1.82(30)$$

or

$$\hat{y} = 69.96.$$

You will notice that this value is very close to the forecast of 70 cranes given by Miller and Botkin.

How good are these forecasts? Miller and Botkin give a value of 15.6 which they indicate by the symbol s.d. We will use the method of Section 11.7 to find a 95 percent prediction interval for our forecast. From Section 11.7, a 95 percent prediction interval for y is

$$\hat{y} \pm t_{.025}s \sqrt{1 + \frac{1}{n} + \frac{(x_p - \bar{x})^2}{\text{SS}_x}}.$$

Since $\bar{x}$ and SS_x are not given explicitly in the printout, we can calculate them from Table 11.4. You can verify that

$$\sum_{i=1}^{11} x_i = 132 \quad \text{and} \quad \sum_{i=1}^{11} x_i^2 = 2024.$$

Then

$$\bar{x} = \frac{\sum_{i=1}^{11} x_i}{n} = \frac{132}{11} = 12$$

and

$$\text{SS}_x = \sum_{i=1}^{11} x_i^2 - \frac{\left(\sum_{i=1}^{11} x_i\right)^2}{n} = 2024 - \frac{(132)^2}{11} = 440.$$

Since we wish to predict the whooping crane population for $x = 30$ (the year 1980), we set $x_p = 30$. Substituting this value along with s, $t_{.025}$, n, $\bar{x}$, and

SS_x into the formula for the prediction interval, we obtain

$$\bar{y} \pm t_{.025}s \sqrt{1 + \frac{1}{n} + \frac{(x_p - \bar{x})^2}{SS_x}}$$

$$69.96 \pm (2.262)(3.38) \sqrt{1 + \frac{1}{11} + \frac{(30 - 12)^2}{440}}$$

or

$$69.96 \pm 10.33.$$

As noted before, our least-squares prediction equation forecasts the same number of whooping crane sightings as forecast by Miller and Botkin. We give an error of prediction of 10.33. Clearly, this forecast must be viewed with caution because, as noted earlier, it is dangerous to predict y for values of x that fall outside the range of the x values used in fitting the model. Some very basic change in the system could have occurred between the end of the study and 1980 that would have changed the form of the relationship between x and y. Some new predator might have appeared on the scene, some pesticide might have reduced their population, or efforts to clean up our environment might have caused the whooping crane population to rapidly increase. Nevertheless, we have done our best. And in any case, all forecasting models, both deterministic and probabilistic, are subject to this risk.

11.11

Assumptions

The assumptions for a regression analysis are as follows:

Assumptions for a Regression Analysis

1. The response y can be represented by the probabilistic model,

$$y = \beta_0 + \beta_1 x + \epsilon.$$

2. x is measured without error.
3. ϵ is a random variable such that, for a given value of x,

$$E(\epsilon) = 0$$
$$\sigma_\epsilon^2 = \sigma^2,$$

and all pairs, ϵ_i, ϵ_j, are independent in a probabilistic sense.

4. ϵ possesses a normal probability distribution.

At first glance, you might forget assumption 1 or fail to catch its significance. Models, deterministic or otherwise, are, as the name implies, only models for real relationships that occur in nature. Consequently, there will always be some error (slight if we are careful) due to the fact that there is curvature in the response curve. If you have obtained a good fit to the data, then the only difficulty that inherent lack of fit can cause is some problem if you attempt to use the model to predict y for some value of x outside the range of values used to fit the least-squares equation. Of course, this will always occur if x is time and you attempt to forecast y at some point in the future. But then this problem occurs with any kind of model for future time predictions. Consequently, you make the forecast but keep the limitations in mind.

The assumption that the variance of the ϵ's (and consequently y) is constant and equal to σ^2 will not be true for some types of data. Similarly, if x is time, say year, it is possible that y values measured over adjacent years will tend to be dependent (an overly large value of y in 1982 might signal a large value of y in 1983). Substantial departure from either of these assumptions will affect the confidence coefficients and significance levels associated with interval estimates and tests described in this chapter.

Like unequal variances and correlation of the random errors, if the normality assumption (4) is not satisfied, the confidence coefficients and significance levels for interval estimates and tests will not be what we expect them to be. But modest departures from normality will not seriously disturb these values.

11.12
Does the Personal Savings Rate Increase as Investment Income Tax Liability Increases?

Now that we have learned about the straight-line regression model and the concept of simple linear correlation, we are ready to return to our case study and see whether Dr. Blotnick's data agree with his theory; that is, that the personal savings rate (on a national basis) decreases as the tax on investment income increases.

You will recall that the data, given in the case study, consist of eight observations, each observation giving the personal savings rate y (%) and the investment tax rate x (%) for a particular country. A plot of the data points is shown in Figure 11.11 and an SAS regression analysis computer printout for the data is shown in Table 11.6.

Figure 11.11 Plot of the data points

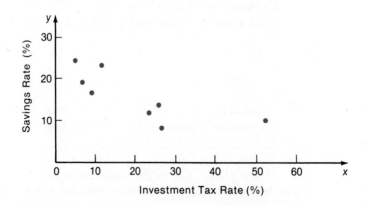

Table 11.6 Regression analysis SAS computer printout for the savings rate–
investment income data

Y = PERSONAL SAVINGS RATE   X = TAXATION RATE
GENERAL LINEAR MODELS PROCEDURE

DEPENDENT VARIABLE: Y

SOURCE	DF	SUM OF SQUARES	MEAN SQUARE	F VALUE	PR > F	R-SQUARE	C.V.
MODEL	1	158.78381752	158.78381752	10.54	0.0175	0.637238	27.1901
ERROR	6	90.39118248	15.06519708		STD DEV		Y MEAN
CORRECTED TOTAL	7	249.17500000			3.88139113		14.27500000

SOURCE	DF	TYPE I SS	F VALUE	PR > F	DF	TYPE IV SS	F VALUE	PR > F
X	1	158.78381752	10.54	0.0175	1	158.78381752	10.54	0.0175

PARAMETER	ESTIMATE	T FOR H0: PARAMETER = 0	PR > |T|	STD ERROR OF ESTIMATE
INTERCEPT	21.18093677	8.37	0.0002	2.53142243
X	−0.29293475	−3.25	0.0175	0.09023088

OBSERVATION	OBSERVED VALUE	PREDICTED VALUE	RESIDUAL	LOWER 95% CL INDIVIDUAL	UPPER 95% CL INDIVIDUAL
1	23.10000000	19.30615436	3.79384564	8.54253007	30.06977865
2	21.50000000	16.96267635	4.53732365	6.68747537	27.23787733
3	17.20000000	19.04251309	−1.84251309	8.34727669	29.73774948
4	14.50000000	17.72430670	−3.22430670	7.32070154	28.12791187
5	12.20000000	11.66055734	0.53944266	1.39609558	21.92501911
6	10.30000000	12.39289422	−2.09289422	2.21996387	22.56582457
7	9.10000000	5.74327535	3.35672465	−6.20773099	17.69428169
8	6.30000000	11.36762259	−5.06762259	1.05849775	21.67674743
9*		15.32224174		5.21782565	25.42665782

* OBSERVATION WAS NOT USED IN THIS ANALYSIS

From the printout, you see that $\hat{\beta}_0 = 21.093677$, $\hat{\beta}_1 = -0.29293475$, and, therefore, that the prediction equation is,

$$\hat{y} = 21.18 - 0.29x.$$

The coefficient of determination, given in the printout, is $r^2 = .637238$. Therefore, since r takes the same sign as the slope, $\hat{\beta}_1$, $r = -\sqrt{r^2} = -.80$.

Do the data (and the negative simple correlation coefficient $r = -.80$) support the theory that the population coefficient of correlation, ρ, is less than 0; that is, that y and x are negatively correlated? We noted in Section 11.8 that a test of the null hypothesis, $H_0: \rho = 0$ is equivalent to a test of the slope, $H_0: \beta_1 = 0$. Consequently, if we wish to detect a negative correlation between y and x, if it exists, we will test

$$H_0: \beta_1 = 0$$

against the alternative hypothesis,

$$H_a: \beta_1 < 0.$$

This alternative hypothesis implies a one-tailed t test with the rejection region in the lower tail of the t distribution.

The calculated value of the t statistic is shown on the SAS printout, Table 11.6, as $t = -3.25$, and the observed significance level for a *two-tailed* test (that is, $P\{t > 3.25 \text{ or } t < -3.25\}$) is shown as .0175. But, we want the observed significance level for a *one-tailed* test; that is, we want $P\{t < -3.25\}$. This will equal half of the value of the observed significance level for a two-tailed test. Thus, the p value for a one-tailed test is,

$$p \text{ value} = \frac{.0175}{2} = .00875.$$

If we were to choose α equal to .05, or even a value as small as .01, we would reject H_0 and accept H_a. There is ample evidence to indicate that $\rho < 0$; that is, that personal savings rate y and investment tax rate x are negatively correlated.

If we were to set the investment tax rate x for a particular country at, say, $x = 20$ percent, could we predict the personal savings rate y? The answer is "yes" but the prediction error would be quite large. You can see that $r^2 = .637238$ and, therefore, that the straight-line model accounts for only (approximately) 64 percent of the variability of y about its mean. Thus, as we would expect, there are many other variables, in addition to the investment income tax rate, that contribute information for the prediction of y.

You can see from the printout, Table 11.6, that $s = 3.88139113$ or $s = \sqrt{s^2} \approx 3.88$. A 95 percent prediction interval for y when $x = x_p$ is,

$$\hat{y} \pm t_{.025} s \sqrt{1 + \frac{1}{n} + \frac{(x_p - \bar{x})^2}{\sum\limits_{i=1}^{n} (x_i - \bar{x})^2}}$$

where $t_{.025} = 2.447$, $n = 8$, $\bar{x} = 23.575$,* and

$$SS_x = \sum_{i=1}^{8} (x_i - \bar{x})^2 = 1850.395.*$$

The prediction intervals for several values of x_p are:

x_p	Lower Prediction Limit	Upper Prediction Limit
6.4	8.54	30.07
20.0	5.22	25.43
30.0	2.22	22.57

* Note that $\bar{x}$ and SS_x are calculated from the original data and are not shown on the printout.

These wide prediction intervals confirm our earlier expectation. The personal savings rate *does* decrease as the investment income tax increases, but the investment tax rate alone does not contribute sufficient information to obtain accurate predictions of the personal savings rate.

Our analysis of the relationship between savings rate y and investment tax rate x aptly illustrates the mechanics of a regression analysis, but you may wonder whether the data satisfy the assumptions upon which the methodology is based. The savings rate y and tax rate x are percentages, and it is possible (as in the case of a binomial sample proportion $\hat{p}$) that the variance of y depends upon the value of x. If it is true that the variance of y depends upon the value of x, our data would violate the assumptions of the regression model (Sections 11.2 and 11.11).

If you are concerned about the validity of our conclusion, that savings rate y and investment tax rate x are negatively correlated, we can employ a nonparametric statistical test for correlation, one which does not require the assumptions of Section 11.2. We will explain the mechanics of this test in Chapter 16 and will apply it to the savings rate–tax rate data in Section 16.8. Other methods for coping with departures from the model assumptions are discussed in the references.

11.13 Summary

Although it was not stressed, you will observe that the prediction of a particular value of a random variable, y, was considered for the most elementary situation in Chapters 8 and 10. Thus, if we possessed no information concerning variables related to y, the sole information available for predicting y would be provided by its probability distribution. As we noted in Chapter 5, the probability that y would fall between two specific values, say y_1 and y_2, would equal the area under the probability distribution curve over the interval $y_1 \leq y \leq y_2$. And, if we were to select randomly one member of the population, we would most likely choose μ, or some other measure of central tendency, as the most likely value of y to be observed. Thus, we would wish to estimate μ, and this of course was considered in Chapters 8 and 10.

Chapter 10 is concerned with the problem of predicting y when auxiliary information is available on other variables, say $x_1, x_2, x_3, \ldots$, which are related to y and hence assist in its prediction. We have concentrated on the problem of predicting y as a linear function of a single variable, x, which provides the simplest extension of the prediction problem beyond that considered in Chapters 8 and 10. We will consider the multi-variable prediction problem in Chapter 12.

References

Dixon, W.J., and M.B. Brown, eds. *BMDP Biomedical Computer Programs, P Series.* Berkeley, Calif.: University of California, 1979.

Draper, N.R., and H. Smith, *Applied Regression Analysis.* New York: John Wiley & Sons, Inc., 1966.

Freund, J.E., *Mathematical Statistics*, 2nd ed. Englewood Cliffs, N.J.: Prentice-Hall, Inc., 1971.

Helwig, J.P., and K.A. Council, eds. *SAS User's Guide*. Cary, N.C.: SAS Institute, Inc., P.O. Box 8000, 1979.

Kleinbaum, D. and L. Kupper, *Applied Regression Analysis and Other Multivariable Methods*. Boston: Duxbury Press, 1978.

Mendenhall, W. and J.T. McClave, *A Second Course in Business Statistics: Regression Analysis*. San Francisco: Dellen Publishing Company, 1981.

Mendenhall, W., *An Introduction to Linear Models and the Design and Analysis of Experiments*. Belmont, Calif.: Wadsworth Publishing Company, Inc., 1968.

Mendenhall, W., R.L. Scheaffer, and D. Wackerly, *Mathematical Statistics with Applications*, 2nd ed. Boston: Duxbury Press, 1981.

Neter, J.W. and W. Wasserman, *Applied Linear Statistical Models*. Homewood, Ill.: Irwin, 1974.

Nie, N., C.H. Hull, J.G. Jenkins, K. Steinbrenner, and D.H. Bent, *Statistical Package for the Social Sciences*, 2nd ed. New York: McGraw-Hill Book Company, 1979.

Younger, M.S., *A Handbook for Linear Regression*. Boston: Duxbury Press, 1979.

Supplementary Exercises

11.65.   Graph the line corresponding to the equation $y = 2x + 1$ by locating points corresponding to $x = 0, 1,$ and 2.

11.66.   Given the linear equation $2x - 3y - 5 = 0$:

a. Give the y intercept and slope for the line.

b. Graph the line corresponding to the equation.

11.67.   Given the following data for corresponding values of two variables, y and x:

y	2	1.5	1	2.5	2.5	4	5
x	-3	-2	-1	0	1	2	3

a. Find the least-squares line for the data.

b. As a check on the calculations in (a), plot the seven points and graph the line.

11.68.   Calculate s^2 for the data in Exercise 11.67.

11.69.   Do the data in Exercise 11.67 present sufficient evidence to indicate that y and x are linearly related? (Test the hypothesis that $\beta_1 = 0$, using $\alpha = .05$.)

11.70.   For what configurations of sample points will s^2 be zero?

11.71.   For what parameter is s^2 an unbiased estimator? Explain how this parameter enters into the description of the probabilistic model $y = \beta_0 + \beta_1 x + \epsilon$.

11.72.   Find a 95 percent confidence interval for the slope of the line in Exercise 11.67.

11.73.   Refer to Exercise 11.67. Obtain a 95 percent confidence interval for the expected value of y when $x = -1$.

11.74.   Refer to Exercise 11.67. Given that $x = 2$, find an interval estimate for a particular value of y. Use a confidence coefficient equal to .90.

11.75.   Calculate the coefficient of correlation for the data in Exercise 11.67.

11.76.   By what percentage was the sum of squares of deviations reduced by using the least-squares predictor, $\hat{y} = \hat{\beta}_0 + \hat{\beta}_1 x$, rather than $\bar{y}$ as a predictor of y for the data in Exercise 11.67?

11.77.   An experiment was conducted to observe the effect of an increase in temperature on the potency of an antibiotic. Three 1-ounce portions of the antibiotic were stored for equal lengths of time at each of the following temperatures: 30°, 50°, 70°, and 90°. The potency readings observed at the temperature of the experimental period were:

Potency Readings, y	38, 43, 29	32, 26, 33	19, 27, 23	14, 19, 21
Temperature, x	30°	50°	70°	90°

a. Find the least-squares line appropriate for this data.
b. Plot the points and graph the line as a check on your calculations.
c. Calculate s^2.

11.78.   Refer to Exercise 11.77. Estimate the change in potency for a one-unit change in temperature. Use a 90 percent confidence interval.

11.79.   Refer to Exercise 11.77. Estimate the mean potency corresponding to a temperature of 50°. Use a 90 percent confidence interval.

11.80.   Refer to Exercise 11.77. Suppose that a batch of the antibiotic were stored at 50° for the same length of time as the experimental period. Predict potency of the batch at the end of the storage period. Use a 90 percent confidence interval.

11.81.   Calculate the coefficient of correlation for the data in Exercise 11.77. Interpret the results.

11.82.   A psychological experiment was conducted to study the relationship between the length of time necessary for a human being to reach a decision and the number of alternatives presented. The questions presented to the participants required a classification of an object into two or more classes, similar to the situation that one might encounter in grading potatoes. Five individuals classified one item each for a two-class, two-decision situation. Five each were also allotted to three-class and four-class categories. The length of time necessary to reach a decision is recorded for the 15 participants.

Length of Reaction Time, y (sec)	1, 3, 3, 2, 4	2, 4, 3, 4, 5	5, 6, 5, 7, 4
Number of Alternatives, x	2	3	4

a. Find the least-squares line appropriate for this data.
b. Plot the points and graph the line as a check on your claculations.
c. Calculate s^2.

11.83.   Do the data in Exercise 11.82 present sufficient evidence to indicate that the length of reaction time is linearly related to the number of alternatives? (Test at the $\alpha = .05$ level of significance.)

11.84.   A comparison of 12 student grade-point averages at the end of the college freshman year with corresponding scores on an IQ test produced the following results:

G.P.A., y	2.1	2.2	3.1	2.3	3.4	2.9	2.9	2.7	2.1	1.7	3.3	3.5
IQ Score, x	116	129	123	121	131	134	126	122	114	118	132	129

a. Find the least-squares prediction equation appropriate for the data.

b. Graph the points and the least-squares line as a check on your calculations.

c. Calculate s^2.

11.85. Do the data in Exercise 11.84 present sufficient evidence to indicate that x is useful in predicting y? (Test by using $\alpha = .05$.)

11.86. Calculate the coefficient of correlation for the data in Exercise 11.84.

11.87. Refer to Exercise 11.84. Obtain a 90 percent confidence interval for the expected grade-point average given an IQ score equal to 120.

11.88. Use the least-squares equation in Exercise 11.84 to predict the grade-point average of a particular student whose IQ score is equal to 120. Use a 90 percent prediction interval.

11.89. An experiment was conducted to investigate the effect of a training program on the length of time for a typical male college student to complete the 100-yard dash. Nine students were placed in the program. The reduction y in time to complete the 100-yard dash was measured for three students at the end of two weeks, for three at the end of four weeks, and for three at the end of six weeks of training. The data are shown below:

Reduction in Time, y (seconds)	1.6, .8, 1.0	2.1, 1.6, 2.5	3.8, 2.7, 3.1
Length of Training, x (weeks)	2	4	6

a. Find the least-squares line for these data.

b. Estimate the mean reduction in running time after four weeks of training. Use a 90 percent confidence interval.

11.90. Refer to Exercise 11.89. Suppose that only three students had been employed in the experiment and that the reduction in running time was measured for each student at the end of two, four, and six weeks. Would the assumptions required for the confidence interval, Exercise 11.89(b), be satisfied? Explain.

11.91. Suppose that the following data were collected on emphysema patients: the number of years the patient smoked and inhaled (x) and a physician's subjective evaluation of the extent of lung damage (y). The latter variable is measured on a scale of 0 to 100. Measurements taken on ten patients are as follows:

Patient	Years Smoking, x	Lung Damage, y
1	25	55
2	36	60
3	22	50
4	15	30
5	48	75
6	39	70
7	42	70
8	31	55
9	28	30
10	33	35

a. Calculate the coefficient of correlation, r, between years smoking (x) and lung damage (y).

b. Calculate the coefficient of determination, r^2. Interpret r^2.

c. Fit a least-squares line to the data. Graph the line and plot the data points. Compare with your computed least-squares line and your computed values of r and r^2.

11.92. Some varieties of nematodes, round worms that live in the soil and frequently are so small as to be invisible to the naked eye, feed upon the roots of lawn grasses and other plants. This pest, which is particularly troublesome in warm climates, can be treated by the application of nematicides. Data collected on the percent kill of nematodes for various rates of application (dosages given in pounds per acre of active ingredient) are as follows:

Rate of Application, x	2	3	4	5
Percent Kill, y	50, 56, 48	63, 69, 71	86, 82, 76	94, 99, 97

a. Calculate the coefficient of correlation, r, between rates of application (x) and percent kill (y).

b. Calculate the coefficient of determination, r^2, and interpret.

c. Fit a least-squares line for the data.

d. Suppose that you wish to estimate the mean percent kill for an application of 4 pounds of the nematicide per acre. Do the data satisfy the assumptions that are required for the confidence intervals of Section 11.6?

11.93. If you play tennis, you know that tennis rackets vary in their physical characteristics. The data shown in the table give measures of bending stiffness and twisting stiffness, as measured by engineering tests, for twelve tennis rackets.*

Racket	Bending Stiffness x	Twisting Stiffness y
Dunlop Maxply Fort	419	227
Garcia 240	407	231
Bancroft Bjorn Borg	363	200
Wilson Jack Kramer	360	211
Davis Classic	257	182
Spalding Smasher III	622	304
Yonex T-7500	424	384
Prince	359	194
Wilson T-4000	346	158
Yamaha YFG-30	556	225
Head Competition II	474	305
Adidas Adistar	441	235

* Gillen, R, ed., "Equipment Preview 1976: How to Pick the Right Racquet for Your Game," *Tennis USA*, February 1976, pp. 87–91, 99.

a. If a racket possesses bending stiffness, is it also likely to possess twisting stiffness? Do the data provide evidence that x and y are correlated? (*Hint:* Test $H_0: \beta_1 = 0$. As indicated in the text, this test is equivalent to testing the null hypothesis that $\rho = 0$.) Find the p value for the test and interpret its value.

b. Calculate the coefficient of determination, r^2, and interpret its value.

11.94.   Soybean meal production, a major source of protein, varies with the weather, rainfall, and the production of competing products. The data in the table show the annual U.S. production (in 100,000 tons) for the years 1960 to 1977.*

Year	Year–1960 x	Soybean Production y
1960	0	9.45
1961	1	10.3
1962	2	11.1
1963	3	10.6
1964	4	11.3
1965	5	12.9
1966	6	13.5
1967	7	13.7
1968	8	14.6
1969	9	17.6
1970	10	18.0
1971	11	17.0
1972	12	16.7
1973	13	19.7
1974	14	16.7
1975	15	20.8
1976	16	18.5
1977	17	20.1

a. Fit a least-squares line to the data.

b. Forecast the U.S. soybean meal production for 1978 using a 90 percent prediction interval.

c. Notice that you have forecast a value of y outside the range of the x values used to develop the prediction equation. How might this affect the interpretation of your prediction interval?

11.95.   Drivers suspected of drinking may face a new type of breath analyzer, a Vacu-Sampler. A.H. Principe reports on this new device and gives data linking Vacau-Sampler readings to those of the device currently in use, the Breathanalyzer.† The data shown in the table give corresponding Breathanalyzer and Vacu-Sampler readings for 15 drinkers.

* *Fats and Oils Situation*, Economic Research Service, U.S. Department of Agriculture, October 1977.

† Principe, A.H., "The Vacu-Sampler: A New Device for the Encapsulation of Breath and Other Gaseous Samples," *J. Police Sci. and Adm.*, 2, no. 4 (1974).

Breathanalyzer y	Vacu-Sampler x
.150	.154
.100	.085
.090	.079
.140	.144
.080	.078
.110	.078
.120	.097
.100	.088
.090	.082
.090	.072
.090	.080
.090	.092
.080	.067
.080	.077
.060	.053

a. Find the least-squares line relating Breathanalyzer readings, y, to corresponding readings on the Vacu-Sampler, x.

b. Plot the data and graph the least-squares line.

c. Do the data provide sufficient evidence to indicate that Breathanalyzer readings are linearly related to Vacu-Sampler readings? Find the p value for the test and interpret its value.

d. Suppose that a person's breath was analyzed using a Vacu-Sampler and gave a reading of .010. Predict the corresponding reading that would be recorded for a Breathanalyzer using a 90 percent prediction interval.

12. Multiple Regression Analysis

Chapter Objectives

General Objective

In Chapter 11 we introduced you to the concept of simple linear regression and correlation, the ultimate goal being to estimate the mean value of y or to predict a value of y using information contained in a single independent (predictor) variable x. In this chapter, we will extend this idea and relate the mean value of y to one or more independent variables, $x_1, x_2, \ldots x_k$, in models that are more flexible than the straight-line model of Chapter 11. The process of finding the least-squares prediction equation, testing the adequacy of the model, and conducting tests about and estimating the values of the model parameters is called a multiple regression analysis.

Specific Objectives

1. To identify the objectives of a multiple regression analysis. *Section 12.1*
2. To describe a multiple regression linear statistical model and to state the assumptions upon which a multiple regression analysis is based. *Section 12.2*
3. To assist you in writing an appropriate regression model for a practical situation. *Section 12.2, (Optional) Section 12.5*
4. To identify and interpret the output of a popular multiple regression computer program package. *Section 12.3*
5. To help you interpret the outputs from other multiple regression computer program packages. *Section 12.4*
6. To develop an understanding of how multiple regression analysis can solve applied problems. *Sections 12.3, 12.6*

Case Study

Predicting Worker Absenteeism

If you have ever been responsible for managing a high school or college function, or a recreational facility, or have worked in business or industry, you have probably experienced one of the major problems facing business managers, namely, worker absenteeism. Can you imagine being responsible for the planning and production of college homecoming activities when the participants, seemingly at random, miss the organizational and functional meetings? Or can you imagine coaching a professional football team in which the players, seemingly at random, are absent from games or practice sessions? Or can you imagine managing an automobile assembly line from which various members of the assembly team are absent on any given day? As worker absenteeism increases, the productivity of an operating group, an office force, a manufacturing plant, etc., decreases and the quality of the product usually decreases also. Consequently, business managers seek to identify the causes of worker absenteeism so that they can take action to control or reduce it.

Regression analysis is one method of determining variables that are related to worker absenteeism. The first step would be to define a measure y of worker absenteeism—say the number of worker absences per month, the number of absences for a given worker per month, or some other suitable criterion. The next step would be to define a set of

independent variables that you think might be related to y. Finally, we would relate y to the independent variables using a multiple regression model and would fit the model to a set of data. If the resulting least squares prediction equation provided a good fit to the data; that is, if it enabled you to predict the measure y of absenteeism with a small error of prediction, you would conclude that at least one of the independent variables contributes information for the prediction of y. This would not tell you *which* of the independent variables contributed the most information for the prediction of y, and it would not imply that any of the independent variables *caused* y to increase or decrease. It would only identify a set of variables important for the prediction of y. To conclude your study, you would need to don your detective's hat and attempt to determine which, if any, of the variables were causally related to the absenteeism measure y.

K. Constas and R.P. Vichas* employ a multiple regression analysis in an attempt to identify variables related to worker absenteeism at two different types of overseas industrial plants. They confine their attention to independent variables that characterize a worker's "life space," that is, variables that personally characterize the worker, as opposed to the work space or working conditions to which the worker is exposed. We will learn how to interpret the computer printout for a multiple

---

* Constas, K., and R.P. Vichas, "An Interpretative Policy-Making Model of Absenteeism with Reference to the Marginal Worker in Overseas Plants," *Journal of Management Studies*, 17, no. 2 (1980).

regression analysis in this chapter, and we will specifically examine the Constas and Vichas regression analysis in Section 12.6.

12.1
The Objective of a Multiple Regression Analysis

The objective of a multiple regression analysis is to relate a response variable y to a set of predictor variables using a multiple regression model. Presumably, we want to be able to estimate the mean value of y and/or predict particular values of y to be observed in the future when the predictor variables assume specific values.

To illustrate, we might wish to relate a company's regional sales y of a product to the amount x_1 of the company's television advertising expenditures, to the amount x_2 of newspaper advertising expenditures, and to the number x_3 of sales representatives assigned to the region. Thus, we would want to use data collected on y, x_1, x_2, and x_3 to obtain a mathematical prediction equation relating y to x_1, x_2, and x_3. We would use this prediction equation to predict product sales for the region for specific advertising expenditures (values of x_1 and x_2) and a specific number of sales representatives. If we are successful in developing a good prediction equation—one that makes accurate sales predictions—we may, by examining the equation, obtain a better understanding of the manner in which these controllable predictor variables, x_1, x_2, and x_3, affect the product's sales.

This chapter is intended to be a brief introduction to multiple regression analysis, to help you understand what it is, and to aid you in interpreting the results of a multiple regression computer printout. We do not intend to cover all of the topics usually covered in a discussion of multiple regression analysis, nor do we intend to spend much time discussing the difficult task of formulating the regression model. For additional information on multiple regression, we recommend the references listed at the end of the chapter.

12.2
The Multiple Regression Model and Associated Assumptions

The general linear model for a multiple regression analysis will take the form:

The General Linear Model and Assumptions

$$y = \beta_0 + \beta_1 x_1 + \beta_2 x_2 + \cdots + \beta_k x_k + \epsilon$$

where

1. y = the response variable that you wish to predict.
2. $\beta_0, \beta_1, \beta_2, \ldots, \beta_k$ are constants.
3. $x_1, x_2, \ldots, x_k$ are independent variables that are measured without error.

4. ϵ is a random error that, for any given set of values for $x_1, x_2, \ldots, x_k$, is normally distributed with mean 0 and variance equal to σ^2.

5. The random errors, say ϵ_i and ϵ_j, associated with any pair of y values, are independent.

With these assumptions, it follows that the mean value of y for a given set of values for $x_1, x_2, \ldots, x_k$, is equal to,

$$E(y) = \beta_0 + \beta_1 x_1 + \beta_2 x_2 + \cdots \beta_k x_k.$$

The variables, $x_1, x_2, \ldots, x_k$ that appear in the general linear model need not represent *different* predictor variables. Our assumption requires only that when we observe a value of y, the values of $x_1, x_2, \ldots, x_k$ can be recorded without error. For example, if we wish to relate the mean measure $E(y)$ of worker absenteeism to two predictor variables, say

$$x_1 = \text{worker age}$$

$$x_2 = \text{worker hourly wage rate,}$$

we could construct any of a number of models. One possibility is

$$E(y) = \beta_0 + \beta_1 x_1 + \beta_2 x_2.$$

This model graphs as a *plane*, a surface in the three-dimensional space defined by y, x_1, and x_2. Or, if we suspect that the response surface relating $E(y)$ to x_1 and x_2 possesses curvature, we might use the model

$$E(y) = \beta_0 + \beta_1 x_1 + \beta_2 x_2 + \beta_3 x_1 x_2$$

or

$$E(y) = \beta_0 + \beta_1 x_1 + \beta_2 x_2 + \beta_3 x_1 x_2 + \beta_4 x_1^2 + \beta_5 x_2^2.$$

Notice that all three of these models relate $E(y)$ to only two predictor variables, x_1 and x_2, but the numbers of terms and interpretations of the models differ.

Choosing a good model relating y to a set of predictor variables is a difficult and important task—much more difficult than fitting the model to a set of data (this is usually done automatically by a computer). Even if your model includes all of the important predictor variables, it still may provide a poor fit to your data if the form of the model is not properly specified.

For example, suppose that $E(y)$ is *perfectly* related to a single predictor variable x_1 by the relation,

$$E(y) = \beta_0 + \beta_1 x_1 + \beta_2 x_1^2.$$

If you fit the straight-line model,

$$E(y) = \beta_0 + \beta_1 x_1,$$

to the data, you will obtain the *best fitting least-squares line*, but it may still provide a poor fit to your data and may be of little value for estimation or prediction (see Figure 12.1(a)). In contrast, if you fit the model,

$$E(y) = \beta_0 + \beta_1 x_1 + \beta_2 x_1^2,$$

you obtain a perfect fit to the data (see Figure 12.1(b)).

Figure 12.1 Two models, each employing one predictor variable

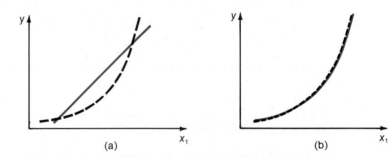

(a)   (b)

The lesson is quite clear. Including all of the important predictor variables in the model as first-order terms,* that is, $x_1, x_2, \ldots, x_k$, may not (and, most probably, will not) produce a model that provides a good fit to your data. You may have to include second-order terms, such as x_1^2, x_2^2, x_3^2, $x_1 x_2$, and $x_1 x_3$.

This discussion of linear models is purposely brief. We want you to understand the importance of model selection as a step that precedes a multiple regression analysis. Secondly, we want to help you understand the logic employed in selecting the model for the multiple regression analysis that follows in Section 12.3. Although we offer additional comments on model formulation in (Optional) Section 12.5, we have no intention of covering this complex topic in a short introductory chapter on regression analysis. Our aim is to help you understand the output of a multiple regression analysis and to understand some of the problems that can be solved using this statistical methodology. For a more thorough discussion of model building, we refer you to Mendenhall and McClave (1981).

Exercises

12.1. Graph the following equations:

a. $E(y) = 3 + 2x$.   b. $E(y) = -1 + x$.   c. $E(y) = 4 - \dfrac{x}{2}$.

* The order of a term is determined by the sum of the exponents of variables making up that term. Terms involving x_1 or x_2 are first-order. Terms involving x_1^2, x_2^2, or $x_1 x_2$ are second-order.

12.2.   Graph the following equations (they graph as parabolas):

a. $E(y) = 2x^2$.

b. $E(y) = 1 + 2x^2$.

c. $E(y) = -1 + 2x^2$.

d. $E(y) = -2x^2$.

e. How does the sign of the coefficient of x^2 affect the graph of a parabola?

12.3   Graph the following equation:

a. $E(y) = 2x^2 - 4x + 2$.

b. Compare the graph, part (a), with the graph of Exercise 12.2, part (a). What effect does the inclusion of the first-order term, $(-4x)$, have on the graph?

c. Suppose that $E(y) = 2x^2 + 4x + 2$. Can you deduce the effect on the parabola of replacing $-4x$ by $+4x$?

12.4.   Suppose that $E(y)$ is related to two predictor variables, x_1 and x_2, by the equation,

$$E(y) = 3 + x_1 - 2x_2.$$

a. Graph the relationship between $E(y)$ and x_1 when $x_2 = 2$. Repeat for $x_2 = 1$ and for $x_2 = 0$.

b. What relationship do the lines, part (a), have to each other?

12.5.   Refer to Exercise 12.4.

a. Graph the relationship $E(y)$ and x_2 when $x_1 = 0$. Repeat for $x_1 = 1$ and for $x_1 = 2$.

b. What relationship do the lines, part (a), have to each other?

c. Suppose that in a practical situation, you wanted to model the relationship between $E(y)$ and two predictor variables, x_1 and x_2. What would be the implication of using the first-order model, $E(y) = \beta_0 + \beta_1 x_1 + \beta_2 x_2$?

12.6.   Suppose that $E(y)$ is related to two predictor variables, x_1 and x_2, by the equation,

$$E(y) = 3 + x_1 - 2x_2 + x_1 x_2.$$

a. Graph the relationship between $E(y)$ and x_1 when $x_2 = 0$. Repeat for $x_2 = 2$ and for $x_2 = -2$.

b. Note that the equation for $E(y)$ is exactly the same as the equation, Exercise 12.4, except that we have added the term, $x_1 x_2$. How does the addition of the $x_1 x_2$ term affect the graphs of the three lines?

c. What flexbility is added to the first-order model, $E(y) = \beta_0 + \beta_1 x_1 + \beta_2 x_2$, by the addition of the term, $\beta_3 x_1 x_2$, using the model, $E(y) = \beta_0 + \beta_1 x_1 + \beta_2 x_2 + \beta_3 x_1 x_2$?

12.3
A Multiple Regression Analysis

A multiple regression analysis is performed in much the same manner as a simple linear regression analysis. That is, a multiple regression model, say, $E(y) = \beta_0 + \beta_1 x_1 + \beta_2 x_2 + \cdots + \beta_k x_k$, is fitted to a set of data using the method of least squares, a procedure that finds the prediction equation,

$$\hat{y} = \hat{\beta}_0 + \hat{\beta}_1 x_1 + \hat{\beta}_2 x_2 + \cdots + \hat{\beta}_k x_k,$$

which minimizes SSE, the sum of squares of deviations of the observed values of y from their predicted values. The major differences between a simple and a multiple regression analysis are that the multiple regression model contains more parameters, and the computation required for a multiple regression analysis is so complicated and time consuming that it is usually performed on an electronic computer. In this section we will present a set of data, formulate a model, show you the Statistical Analysis System (SAS) multiple regression computer printout for the data, and interpret it. You will see that the interpretation will be closely linked to the interpretations given to corresponding quantities that appeared in the computer output for a simple linear regression analysis.

Example 12.1

A study was conducted to examine the relationship between university salary y, the number of years of experience of the faculty member, and the sex of the faculty member. If we expect a straight-line relationship between mean salary and years of experience for both males and females, we write the model relating mean salary to the two predictor variables:

1. years of experience (quantitative)
2. sex of the professor (qualitative).

Solution

Since we might suspect the mean salary lines for females and males to be different, we want to construct a model for mean salary $E(y)$ that will appear as shown in Figure 12.2.

Figure 12.2
Hypothetical relationship between mean salary $E(y)$, years of experience (x_1), and sex (x_2)

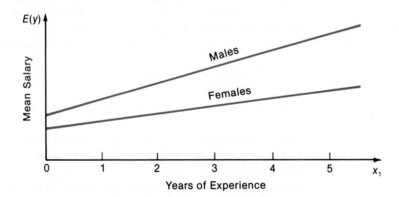

A straight-line relationship between $E(y)$ and years of experience, x_1, would imply the model,

$$E(y) = \beta_0 + \beta_1 x_1 \qquad \text{(graphed as straight line)}.$$

The qualitative variable,* sex, can only assume two "values," male and female. Therefore, we can enter the predictor variable "sex" into the model using one dummy (or indicator)

---

* Qualitative data, observations on a qualitative variable, were discussed in Section 2.1. We will have more to say about qualitative variables in Optional Section 12.5.

variable, x_2, as,

$$E(y) = \beta_0 + \beta_1 x_1 + \beta_2 x_2 \qquad \text{(graphs as two parallel lines)}$$

$$x_2 = \begin{cases} 1 & \text{if male} \\ 0 & \text{if female.} \end{cases}$$

The fact that we wish to allow the slopes of the two lines to differ means that we think that the two predictor variables **interact**; that is, the change in $E(y)$ corresponding to a change in x_1 depends upon whether the professor is male or female. To allow for this interaction (difference in slopes), we introduce the interaction term, $x_1 x_2$, into the model. The complete model that would characterize the graph in Figure 12.2 is,

$$
\begin{array}{c}
\text{dummy} \\
\text{variable} \\
\text{for sex} \\
\downarrow
\end{array}
$$

$$E(y) = \beta_0 + \underset{\underset{\substack{\text{years of} \\ \text{experience}}}{\uparrow}}{\beta_1 x_1} + \beta_2 x_2 + \underset{\underset{\text{interaction}}{\uparrow}}{\beta_3 x_1 x_2}$$

where

$$x_1 = \text{years of experience}$$

$$x_2 = \begin{cases} 1 & \text{if a male} \\ 0 & \text{if a female.} \end{cases}$$

The interpretation of the model parameters can be seen by assigning values to the dummy variable x_2. Thus, when you wish to acquire the line for females, the dummy variable $x_2 = 0$ (according to our coding) and

$$E(y) = \beta_0 + \beta_1 x_1 + \beta_2(0) + \beta_3 x_1(0)$$
$$= \beta_0 + \beta_1 x_1.$$

Therefore, β_0 is the y intercept for the female line and β_1 is the slope of the line relating expected salary to years of experience for *females only*.

Similarly, the line for males is obtained by letting $x_2 = 1$. Then,

$$E(y) = \beta_0 + \beta_1 x_1 + \beta_2(1) + \beta_3 x_1(1)$$
$$= \underset{y \text{ intercept}}{\underbrace{(\beta_0 + \beta_2)}} + \underset{\text{slope}}{\underbrace{(\beta_1 + \beta_3)}} x_1.$$

The y intercept of the male line is $(\beta_0 + \beta_2)$, and the slope, the coefficient of x_1, is equal to $(\beta_1 + \beta_3)$.

Since the slope of the male line is $(\beta_1 + \beta_3)$, and the slope of the female line is β_1, it follows that $(\beta_1 + \beta_3) - \beta_1 = \beta_3$ must be the difference in the slopes of the two lines. Similarly, β_2 is equal to the difference in the y intercepts for the two lines.

Example 12.2

Random samples of six female and six male assistant professors were selected from among the assistant professors in a College of Arts and Sciences. The data on salary and years of experience are shown below. Note that both samples contained two professors with three years of experience but no male professor had two years of experience.

	Years of Experience x_1				
	1	2	3	4	5
Salary y/Males	20,710		23,160	24,140	25,760
			23,210		25,590
Salary y/Females	19,510	20,440	21,340	22,750	23,200
			21,760		

a. Explain the output of the SAS multiple regression computer printout.

b. Graph the predicted salary lines.

Solution

a. The SAS computer printout is shown in Table 12.1. You will see that this printout is similar to the SAS printout contained in Section 11.10, but that it contains more entries. The relevant portions of the printout are shaded and numbered.

1. The value of

$$R^2 = 1 - \frac{\text{SSE}}{\sum\limits_{i=1}^{n}(y_i - \bar{y})^2} = 1 - \frac{\text{SSE}}{\text{SS}_y},$$

the multiple coefficient of determination, provides a measure of how well the model fits the data. As you can see, 99.2 percent of the sum of squares of deviations of the y values about $\bar{y}$ is explained by all of the terms in the model.

2. If the model contributes information for the prediction of y, at least one of the model parameters, β_1, β_2, or β_3, will differ from 0. Consequently, we wish to test the null hypothesis, $H_0: \beta_1 = \beta_2 = \beta_3 = 0$, against the alternative hypothesis, H_a: at least one of the parameters, β_1, β_2, or β_3, differs from 0. The test statistic for this test, an F statistic with $v_1 = $ (the number of parameters in H_0) = 3 and $v_2 = n - $ (number of parameters in the model) = $12 - 4 = 8$, is shown in shaded area 2 to be 346.24. Since this value exceeds the tabulated value for F, with $v_1 = 3$, $v_2 = 8$, and $\alpha = .05$, $F = 4.07$ (Table 8, the Appendix), we reject H_0 and conclude that at least one of the parameters, β_1, β_2, β_3, differs from 0. There is evidence to indicate

Table 12.1 The SAS multiple regression analysis computer printout for Example 12.2

STATISTICAL ANALYSIS SYSTEM
GENERAL LINEAR MODELS PROCEDURE

DEPENDENT VARIABLE: Y

SQUARE	DF	SUM OF SQUARES	MEAN SQUARE	F VALUE	PR > F ②	R-SQUARE ①	C.V.
MODEL	3	42108777.02898556	14036259.00966185	346.24	0.0001	0.992357	0.8897
ERROR	8	324314.63768142	40539.32971018 ⑦		STD DEV ⑧		Y MEAN
CORRECTED TOTAL	11	42433091.66666698 ⑨			201.34380971		22630.83333333

SOURCE	DF	TYPE I SS	F VALUE	PR > F	DF	TYPE IV SS	F VALUE	PR > F
X1	1	33294036.23595509	821.28	0.0001	1	9389610.00000008	231.62	0.0001
X2	1	8452796.51598297	208.51	0.0001	1	326808.74399183	8.06	0.0218
X1*X2	1	361944.27704750	8.93	0.0174	1	361944.27704750	8.93	0.0174

PARAMETER	ESTIMATE ③	T FOR H0: PARAMETER = 0 ④	PR > |T| ⑤	STD ERROR OF ESTIMATE ⑥
INTERCEPT	18593.00000000	89.41	0.0001	207.94699250
X1	969.00000000	15.22	0.0001	63.67050315
X2	866.71014493	2.84	0.0218	305.25678646
X1*X2	260.13043478	2.99	0.0174	87.05798112

Note: The computer uses a star to indicate multiplication. Thus, $x_1 x_2$ is printed as $x_1 * x_2$. Similarly, x_1^2 is printed as $x_1 * x_1$.

that the model contributes information for the prediction of y. The observed significance level (p value) for the test, $PR > F$, shown to the right of the value of the F statistic, is equal to .0001.

3. The estimates of the four model parameters are shown under the heading, ESTIMATE. Thus, $\hat{\beta}_0 = 18593.00000000$, $\hat{\beta}_1 = 969.000\ldots$, $\hat{\beta}_2 = 866.71014493$, and $\hat{\beta}_3 = 260.13043478$ and, rounding these estimates, we obtain the prediction equation $\hat{y} = 18593.0 + 969.0 x_1 + 866.7 x_2 + 260.1 x_1 x_2$.

4. Tests of hypotheses concerning each of the individual model parameters can be conducted using Student's t tests. The test statistic is similar* to the t statistic used for the test, $H_0: \beta_1 = 0$, for the slope β_1 in a simple linear model,

$$t = \frac{\text{parameter estimate}}{\text{estimated standard deviation of the estimator}}.$$

The computed value of the t statistic for each of the model parameters is shown in shaded area 4 under the column, T FOR H_0: PARAMETER = 0.

For example, the value of the t statistic for the test, $H_0: \beta_1 = 0$, is 15.22. The number of degrees of freedom for SSE, s^2, and therefore, t, will always be based on $v = n - $ (number of parameters in the model) degrees of freedom. For our example, this will be $v = 12 - 4 = 8$ degrees of freedom. For $\alpha = .05$, the tabulated value

---

* The test statistic looks the same, but the standard deviation in the denominator is calculated in a different and more complicated manner.

of t, $t_{.025}$, for a two-tailed test is given in Table 5, the Appendix, as 2.306. Since the computed t value corresponding to β_1 exceeds this value, there is evidence to indicate that β_1 differs from 0.

5. The observed significance levels (p values) for the t tests described in (4) are shown in shaded area 5 under the column headed PR > |T|. These values are calculated for two-tailed tests. If you wish to conduct a one-tailed test, the p value is half of the value shown in the printout. For example, the observed significance level for a test of the hypothesis $H_0: \beta_2 = 0$, $H_a: \beta_2 \neq 0$, is .0218 for this two-tailed test. If the alternative hypothesis is $H_a: \beta_2 > 0$, that is, you wish to conduct a one-tailed test of H_0, then the p value is equal to $(.0218)/2 = .0109$. If you were to test H_0 using $\alpha = .05$, these small p values would imply rejection of H_0.

6. The estimated standard deviations of the estimators are shown in shaded area 6 under the heading, STD ERROR OF ESTIMATE. For example,

$$s_{\hat{\beta}_1} = 63.67050315.$$

This quantity is useful in calculating $100(1 - \alpha)$ percent confidence intervals for the model parameters. For example, a $100(1 - \alpha)$ percent confidence interval for β_1 is equal to,

$$\hat{\beta}_1 \pm t_{\alpha/2} s_{\hat{\beta}_1}.$$

Therefore, a 95 percent confidence interval for β_1 is

$$969.0 \pm (2.306)(63.7)$$

or

$$969.0 \pm 146.9.$$

Confidence intervals for the other model parameters are constructed in the same way. Thus, a $100(1 - \alpha)$ percent confidence interval for β_2 is,

$$\hat{\beta}_2 \pm t_{\alpha/2} s_{\hat{\beta}_2}$$

where $t_{\alpha/2}$ is based on $v = n -$ (number of parameters in the model) $= 12 - 4 = 8$ degrees of freedom.

7. The values of SSE = 324314.63768142 and $s^2 = 40539.32971018$ are shown in shaded area 7 in the row corresponding to ERROR and under the columns headed, respectively, SUM OF SQUARES and MEAN SQUARE.

8. The standard deviation $s = 201.34380971$ is shown in shaded area 8, labeled STD DEV.

9. The total sum of squares of deviations, SS_y, of the y-values about the mean is shown in shaded area 9 to be 42433091.66666698. You can verify that the tabulated value of R^2 is equal to $1 - SSE/SS_y$.

b. A graph of the two salary lines is shown in Figure 12.3. Note that the salary line corresponding to the male faculty members appears to be rising at a more rapid rate than the

Figure 12.3 A graph of the faculty salary prediction lines

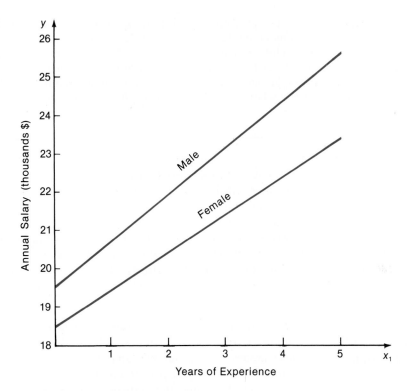

Years of Experience

salary line for females. Is this real or could it be due to chance? The following example will answer this question.

Example 12.3

Refer to Example 12.2. Do the data provide sufficient evidence to indicate that the annual rate of increase in male junior faculty salaries exceeds the annual rate of increase for women faculty salaries? Thus, we want to know whether the data provide sufficient evidence to indicate that the slope of the men's faculty salary line exceeds the slope of the women's faculty salary line.

Solution

Since β_3 measures the difference in slopes, the slopes of the two lines will be identical if $\beta_3 = 0$. Therefore, we wish to test,

$$H_0: \beta_3 = 0, \quad \text{that is, the slopes of the two lines are identical}$$

against the alternative hypothesis,

$$H_a: \beta_3 > 0, \quad \text{that is, the slope of the male faculty salary line is greater than the slope of the female faculty salary line.}$$

The calculated value of t corresponding to β_3, shown in shaded area 4 of the computer printout (Table 12.1) is 2.99. Since we wish to detect values, $\beta_3 > 0$, we will conduct a one-tailed test and reject H_0 if $t > t_\alpha$. The tabulated t value, Table 5, the Appendix, for $\alpha = .05$ and $v = 8$ degrees of freedom is $t_{.05} = 1.860$. The calculated value of t exceeds this value, and thus there is evidence to indicate that the annual rate of increase in men's faculty salaries exceeds the corresponding annual rate of increase in faculty salaries for women.*

Some computer program multiple regression packages (SAS is one of them) can be instructed to print confidence intervals for the mean value of y and/or prediction intervals for y for specific values of the independent variables. You can also instruct some programs to print a test statistic for testing an hypothesis that each parameter in a subset of the model parameters, say β_2 and β_3, simultaneously equal 0, that is, $H_0 : \beta_2 = \beta_3 = 0$. This latter topic is addressed in Optional Exercise 12.14.

12.4
A Comparison of Computer Printouts

Several popular statistical program packages contain multiple regression programs. Three of these, Minitab, SAS, and SPSS, are referenced at the end of this chapter. Some programs can only be used on specific computers; others are more versatile. If you plan to conduct a multiple regression analysis, you will want to determine the statistical program packages available at your computer center so that you may become familiar with their output.

We presented and explained an SAS multiple regression computer output in Section 12.3. Most other computer outputs are very similar. They vary somewhat in the format in which the output is presented and, in some cases, the type of output that can be requested. In this section we will present the output of three multiple regression computer packages, SAS, Minitab, and SPSS, for the same set of data, the data of Example 12.2. The three computer printouts for this set of data are shown in Table 12.2.

Corresponding output is shaded and marked with the same number in each printout. For example, parameter estimates on the Minitab printout appear under the column heading COEFFICIENT, and on the SPSS printout they appear under the column heading B.

The Minitab printout does not give the value of F for testing the complete model. You can calculate this directly using the formula,

$$F = \frac{\text{Mean Square for Regression}}{s^2}.$$

---

* If we wish to determine whether the data provide sufficient evidence to indicate that the male faculty members start at higher salaries, we would test $H_0 : \beta_2 = 0$ against the alternative hypothesis, $H_a : \beta_2 > 0$.

Table 12.2

(a) SAS Printout

DEPENDENT VARIABLE: Y

SQUARE	DF	SUM OF SQUARES	MEAN SQUARE	F VALUE ②	PR > F	R-SQUARE ①	C.V.
MODEL	3	42108777.02898556	14036259.00966185	346.24	0.0001	0.992357	0.8897
ERROR	8	324314.63768142	40539.32971018 ⑦		STD DEV ⑧		Y MEAN
CORRECTED TOTAL	11	42433091.66666698 ⑨			201.34380971		22630.83333333

SOURCE	DF	TYPE I SS	F VALUE	PR > F	DF	TYPE IV SS	F VALUE	PR > F
X1	1	33294036.23595509	821.28	0.0001	1	9389610.00000008	231.62	0.0001
X2	1	8452796.51598297	208.51	0.0001	1	326808.74399183	8.06	0.0218
X1*X2	1	361944.27704750	8.93	0.0174	1	361944.27704750	8.93	0.0174

PARAMETER	ESTIMATE ③	T FOR HO: PARAMETER = 0 ④	PR > |T| ⑤	STD ERROR OF ESTIMATE ⑥
INTERCEPT	18593.00000000	89.41	0.0001	207.94699250
X1	969.00000000	15.22	0.0001	63.67050315
X2	866.71014493	2.84	0.0218	305.25678646
X1*X2	260.13043478	2.99	0.0174	87.05798112

(b) Minitab Printout

MTB >

— REGR SALARY C1, USING 3 PREDICTORS YEARS C2 SEX C3 YR*SEX C4

THE REGRESSION EQUATION IS
Y = 18593. + 969. X1 + 867. X2
   + 260. X3

	COLUMN	COEFFICIENT ③	ST. DEV. OF COEF. ⑥	T-RATIO = COEF/S.D. ④
	—	18593.0	207.9	89.41
X1	YEARS	968.99	63.67	15.21
X2	SEX	866.7	305.2	2.83
X3	YR*SEX	260.13	87.05	2.98

THE ST. DEV. OF Y ABOUT REGRESSION LINE IS
S = 201.3   ⑧
WITH (12 − 4) = 8 DEGREES OF FREEDOM
R-SQUARED = 99.2 PERCENT ①
R-SQUARED = 98.9 PERCENT, ADJUSTED FOR D.F.

ANALYSIS OF VARIANCE

DUE TO	DF	SS	MS = SS/DF
REGRESSION	3	42108770	14036259
RESIDUAL	8	324314	40539 ⑦
TOTAL	11	42433090 ⑨	

FURTHER ANALYSIS OF VARIANCE
SS EXPLAINED BY EACH VARIABLE WHEN ENTERED IN THE ORDER GIVEN

DUE TO	DF	SS
REGRESSION	3	42108770
YEARS	1	33294010
SEX	1	8452821
YR*SEX	1	361944

ROW	X1 YEARS	Y SALARY	PRED. Y VALUE	ST. DEV. PRED. Y	RESIDUAL	ST. RES
1	1.00	20710.0	20688.8	169.6	21.1	0.19 X

X DENOTES AN OBS. WHOSE X VALUE GIVES IT LARGE INFLUENCE.

DURBIN-WATSON STATISTIC = 2.24
MTB >

— OUTFILE
% Minitab-I Current outfile closed

Table 12.2 (*continued*)

(c) SPSS printout

```
* * * * * * * * * * * * * * * * * * * * * * * * * * * * * *  MULTIPLE REGRESSION  * * * * * * * * * * * * * * * * * * *      VARIABLE LIST   1
                                                                                                                           REGRESSION LIST  1

DEPARTMENT VARIABLE. .    Y        SALARY

VARIABLE(S) ENTERED ON STEP NUMBER 1 . .    X3
                                            X1
                                            X2
```

			ANALYSIS OF VARIANCE	DF	SUM OF SQUARES	MEAN SQUARE	F	
MULTIPLE R	0.99617							②
R SQUARE	0.99236	①	REGRESSION	3.	42108777.02899	14036259.00966	346.23806	
ADJUSTED R SQUARE	0.98949		RESIDUAL	3.	324314.63768	40539.32971 ⑦		
STANDARD ERROR	201.34381	⑧						

```
--------------VARIABLES IN THE EQUATION--------------          ----------VARIABLES NOT IN THE EQUATION----------
              ③                    ⑥             ⑨
```

VARIABLE	B	BETA	STD ERROR B	F	VARIABLE	BETA IN	PARTIAL	TOLERANCE	F
X3	260.1304	0.27739	87.05798	8.928					
X1	969.0000	0.70168	63.67050	231.617					
X2	866.7101	0.23045	305.25679	8.062					
(CONSTANT)	18593.00								

ALL VARIABLES ARE IN THE EQUATION

STATISTICS WHICH CANNOT BE COMPUTED ARE PRINTED AS ALL NINES.

The denominator, s^2 (the estimator of σ^2), is shown in shaded area 7. The value of the quantity called, "Mean Square for Regression," appears under the column heading, MS = SS/DF, and in the REGRESSION row, as 14036259. Therefore,

$$F = \frac{\text{Mean Square for Regression}}{s^2} = \frac{14036259. \cdots}{40539. \cdots}$$

$$= 346.24.$$

You can see that this value of F is identical to the values shown on the SAS and the SPSS printouts.

Both the Minitab and the SPSS printouts give a value of R^2 that is adjusted for the number of degrees of freedom associated with SSE. It appears as R-SQUARED ADJUSTED FOR DF on the Minitab printout and ADJUSTED R SQUARE on the SPSS printout. We have not utilized this quantity in our analysis of the data, Section 12.3, and will therefore omit it from our discussion.

One difference between the SPSS and the two other printouts is that the tests of hypotheses concerning the individual model parameters are conducted in SPSS using an F statistic with $v_1 = 1$ and $v_2 = n -$ (number of parameters in the model) $= 12 - 4 = 8$ degrees of freedom. This is because an F statistic with $v_1 = 1$ numerator degrees of freedom is related to a Student's t. Specifically,

$$F = t^2,$$

where the degrees of freedom for F are,

$$v_1 = 1 \quad \text{and} \quad v_2 = n - \text{(number of parameters in the model)},$$

and the degrees of freedom for t are

$$v = v_2 = n - \text{(number of parameters in the model)}.$$

The calculated F values are shown in shaded area 9 on the SPSS printout. You can see that these values are equal to the squares of the corresponding t's shown on the SAS and the Minitab printouts.

A note of caution should be made concerning the interpretation of the SPSS F tests for the individual model parameters. These F tests are always two-tailed tests; that is, a test of $H_0: \beta_2 = 0$ is designed to detect the alternative, $H_a: \beta_2 \neq 0$. If you wish to conduct a one-tailed test, say, to detect $\beta_2 > 0$, you would need to calculate the value of the t statistic,

$$t = \sqrt{F},$$

and then proceed with your test (Example 12.3 required a one-tailed t test). The sign of the t statistic will be the same as the sign of the estimate, $\hat{\beta}_2$.

The preceding discussion should enable you to read the multiple regression computer output for most statistical program packages. They are very similar. If you have difficulty understanding the output of the package available at your computer center, consult the package instruction manual.

Exercises

12.7. A publisher of college textbooks conducted a study to relate profit per text to cost of sales over a six-year period in which its sales force (and sales costs) were growing rapidly. The following inflation-adjusted data (in thousands of dollars) were collected:

Profit per text y	16.5	22.4	24.9	28.8	31.5	35.8
Sales cost per text x	5.0	5.6	6.1	6.8	7.4	8.6

Expecting profit per book to rise and then plateau, the publisher fitted the model, $E(y) = \beta_0 + \beta_1 x + \beta_2 x^2$, to the data.

a. What sign would you expect the actual value of β_2 to assume? The SAS computer printout is shown in Table 12.3. Find the value of $\hat{\beta}_2$ on the printout and see if the sign agrees with your answer.

b. Find SSE and s^2 on the printout.

c. How many degrees of freedom do SSE and s^2 possess? Show that $s^2 = \text{SSE}/(\text{degrees of freedom})$.

d. Do the data provide sufficient evidence to indicate that the model contributes information for the prediction of y? Test using $\alpha = .05$.

Table 12.3 SAS output for Exercise 12.7

DEPENDENT VARIABLE: Y

SOURCE	DF	SUM OF SQUARES	MEAN SQUARE	F VALUE	PR > F	R-SQUARE	C.V.
MODEL	2	234.95514252	117.47757126	332.53	0.0003	0.995509	2.2303
ERROR	3	1.05985748	0.35328583		STD DEV		Y MEAN
CORRECTED TOTAL	5	236.01500000			0.59437852		26.65000000

SOURCE	DF	TYPE I SS	F VALUE	PR > F	DF	TYPE IV SS	F VALUE	PR > F
X	1	227.81864814	644.86	0.0001	1	15.20325554	43.03	0.0072
X*X	1	7.13649437	20.20	0.0206	1	7.13649437	20.20	0.0206

PARAMETER	ESTIMATE	T FOR H0: PARAMETER = 0	PR > |T|	STD ERROR OF ESTIMATE
INTERCEPT	-44.19249551	-5.33	0.0129	8.28688218
X	16.33386317	6.56	0.0072	2.48991042
X*X	-0.81976920	-4.49	0.0206	0.18239471

OBSERVATION	OBSERVED VALUE	PREDICTED VALUE	RESIDUAL	LOWER 95% CL FOR MEAN	UPPER 95% CL FOR MEAN
1	16.50000000	16.98259044	-0.48259044	15.33066719	18.63451369
2	22.40000000	21.56917626	0.83082374	20.56714686	22.57120566
3	24.90000000	24.94045805	-0.04045805	23.93483233	25.94608377
4	28.80000000	28.97164643	-0.17164643	27.81528749	30.12800537
5	31.50000000	31.78753079	-0.28753079	30.65188830	32.92317327
6	35.80000000	35.64859804	0.15140196	33.81460946	37.48258661
7*		27.34236657		26.23203462	28.45269852

* OBSERVATION WAS NOT USED IN THIS ANALYSIS

SUM OF RESIDUALS	-0.00000000
SUM OF SQUARED RESIDUALS	1.05985748
SUM OF SQUARED RESIDUALS — ERROR SS	-0.00000000
PRESS STATISTIC	12.11621807
FIRST ORDER AUTOCORRELATION	-0.39797399
DURBIN-WATSON D	2.55457960

e. Find the observed significance level for the test, part (d), and interpret its value.

f. Do the data provide sufficient evidence to indicate curvature in the relationship between $E(y)$ and x (i.e., evidence to indicate that β_2 differs from 0)? Test using $\alpha = .05$.

g. Find the observed significance level for the test, part (f), and interpret its value.

h. Find the prediction equation and graph the relationship between $\hat{y}$ and x.

i. Find R^2 on the printout and interpret its value.

j. Use the prediction equation to estimate the mean profit per text when the sales cost per text is $6500. (Express the sales cost in thousands of dollars before substituting into the prediction equation). We instructed the SAS program to print this confidence interval. The confidence interval when $x = 6.5$, 26.23203462 to 28.45269852, is shown at observation 7 in the printout, Table 12.3.

12.8.　Refer to Example 12.2. The t value for testing the hypothesis, $H_0: \beta_1 = 0$ is,

$$t = \frac{\hat{\beta}_1 - 0}{s_{\hat{\beta}_1}} = \frac{\hat{\beta}_1}{s_{\hat{\beta}_1}}$$

where $\hat{\beta}_1$ is the estimate of β_1 and $s_{\hat{\beta}_1}$ is the estimated standard deviation (or standard error) of $\hat{\beta}_1$. Both of these quantities are shown on the SAS, Minitab, and SPSS printouts, Table 12.2. Examine the printouts and verify that the corresponding printed values for $\hat{\beta}_1$ and $s_{\hat{\beta}_1}$ are identical except for rounding errors. Calculate $t = \hat{\beta}_1/s_{\hat{\beta}_1}$. Verify that your calculated value of t is equal to the value printed on the SAS and the Minitab printouts, that is, $t = 15.22$.

12.9. Refer to Exercise 12.8. We note in Section 12.4 that t^2 with v degrees of freedom is equal to an F statistic with $v_1 = 1$ and $v_2 = v$ degrees of freedom. To illustrate, show that the computed value of the F statistic, shown in the SPSS printout, for the test, $H_0: \beta_1 = 0$, is equal to the square of the t value computed in Exercise 12.8. Find the tabulated t value, $t_{.025}$, Table 5, the Appendix, that would be employed for a two-tailed test for $H_0: \beta_1 = 0$. Find the corresponding tabulated value for $F_{.05}$. Show that $F_{.05} = t_{.025}^2$.

12.10. When will an F statistic equal the square of a t statistic?

12.11. The waiting time y that elapses between the time a computing job is submitted to a large computer and the time at which the job is initiated (computing commences) is a function of many variables, including the priority assigned to the job, the number and sizes of the jobs already on the computer, the size of the job being submitted, etc. A study was initiated to investigate the relationship between waiting time y (in hours) for a job and x_1, the estimated CPU time (in seconds) for the job, and x_2, the CPU utilization factor. The estimated CPU time x_1 is an estimate of the amount of time that a job will occupy a portion of the computer's central processing unit's memory. The CPU utilization factor x_2 is the percentage of the memory bank of the central processing unit that is occupied at the time that the job is submitted. We would expect the waiting time y to increase as the size of the job x_1 increases and as the CPU utilization factor x_2 increases. To conduct the study, 15 jobs of varying sizes were submitted to the computer at randomly assigned times throughout the day. The job waiting time y, estimated CPU time x_1, and CPU utilization factor x_2 were recorded for each job. The data* are shown below:

							Job								
	1	2	3	4	5	6	7	8	9	10	11	12	13	14	15
x_1	2.0	9.3	5.6	3.7	12.4	18.1	13.5	26.6	34.2	38.8	56.1	60.3	4.4	2.6	20.9
x_2	45	80	23	25	67	30	55	21	79	40	22	37	50	66	42
y	0.001	1.140	0.030	0.001	0.780	0.300	0.600	0.200	2.240	0.440	0.001	0.320	0.160	0.290	0.490

A second-order model, $E(y) = \beta_0 + \beta_1 x_1 + \beta_2 x_2 + \beta_3 x_1 x_2 + \beta_4 x_1^2 + \beta_5 x_2^2$, was selected to model mean waiting time $E(y)$. The SAS multiple regression analysis for the data is shown in Table 12.4.

a. Find the values of SSE and s^2.

* Waiting time data frequently violate the assumptions required for significance tests and confidence intervals in a regression analysis. The probability distribution for waiting times is often skewed and its variance increases as the mean waiting time increases. Methods are available for coping with this problem (see Mendenhall and McClave (1981)), but we will ignore it for the purposes of this introductory discussion.

Table 12.4 SAS output for Exercise 12.11

DEPENDENT VARIABLE: Y

SOURCE	DF	SUM OF SQUARES	MEAN SQUARE	F VALUE	PR > F	R-SQUARE	C.V.
MODEL	5	4.74848837	0.94969767	159.23	0.0001	0.988822	16.5655
ERROR	9	0.05367803	0.00596423		STD DEV		Y MEAN
CORRECTED TOTAL	14	4.80216640			0.07722840		0.46620000

SOURCE	DF	TYPE I SS	F VALUE	PR > F	DF	TYPE IV SS	F VALUE	PR > F
X1	1	0.08032055	13.47	0.0052	1	0.00448637	0.75	0.4083
X2	1	3.21553051	539.14	0.0001	1	0.13905895	23.32	0.0009
X1*X2	1	0.96272106	161.42	0.0001	1	0.48988534	82.14	0.0001
X1*X1	1	0.22455767	37.65	0.0002	1	0.16181969	27.13	0.0006
X2*X2	1	0.26535859	44.49	0.0001	1	0.26535859	44.49	0.0001

PARAMETER	ESTIMATE	T FOR HO: PARAMETER = 0	PR > |T|	STD ERROR OF ESTIMATE
INTERCEPT	0.43806816	2.71	0.0239	0.16147813
X1	0.00526474	0.87	0.4083	0.00607025
X2	−0.03017242	−4.83	0.0009	0.00624867
X1*X2	0.00068770	9.06	0.0001	0.00007588
X1*X1	−0.00037952	−5.21	0.0006	0.00007286
X2*X2	0.00040726	6.67	0.0001	0.00006106

b. Find the prediction equation.

c. Find R^2 and interpret its value.

d. Do the data provide sufficient evidence to indicate that the model contributes information for the prediction of y? Test using $\alpha = .10$.

e. Find the observed significance level for the test, part (d), and interpret its value.

f. Note that the observed significance level for the test, $H_0: \beta_1 = 0$, is large (p value = .4083). Does this mean that there is little evidence to indicate that x_1 contributes information for the prediction of y?

g. Predict the waiting time when the estimated CPU time for a job is 30 seconds and the CPU utilization factor is 55 percent.

h. Graph the estimated mean waiting time as a function of estimated CPU time x_1 for CPU utilization, $x_2 = 30$ percent. Also graph the curves for $x_2 = 50$ percent and $x_2 = 70$ percent. Observe the behavior of the estimated mean waiting time as x_1 and x_2 increase in value.

12.5
Some Comments on Model Formulation (Optional)

The variables encountered in a regression analysis can be one of two types, **quantitative** or **qualitative**. For example, the age of a person is a quantitative variable because its values express the quantity or amount of something, in this case, age. In contrast, the nationality of a person is a variable that varies from person to person, but the values of the variable cannot be quantified. They can only be classified. To utilize the methods we present, the response variable y must always be (according to the assumptions of Section 12.2) a quantitative variable. In contrast, predictor variables may be either quantitative or qualitative. For example, suppose that we wish to predict the annual income of a person. Both the age and the nationality of the person could be important predictor variables, and there are probably many others.

Definition

A **quantitative** variable is one whose values correspond to the quantity or the amount of something.

Definition

A **qualitative** variable is one that assumes values that cannot be quantified. They can only be categorized.

As noted in Section 12.2, quantitative predictor variables, say x_1, x_2, etc., are usually entered into a model using first-order terms, those involving x_1, x_2, x_3, etc., and second-order terms, those involving x_1^2, x_2^2, x_3^2, $x_1 x_2$, $x_1 x_3$, $x_2 x_3$, etc. In general, the second-order terms allow for curvature in the response surface, but the cross-product second-order terms possess a special significance. They are often used to model the **interaction** between two predictor variables in their effect on the response variable.

To understand the concept of interaction, consider the first-order (planar) model,

$$E(y) = \beta_0 + \beta_1 x_1 + \beta_2 x_2.$$

If you were to graph $E(y)$ as a function of x_1 *with x_2 held constant*, you would obtain a straight line. Repeating this process for other values of x_2, you would obtain a set of parallel lines (with slopes β_1 but intercepts depending on the particular value of x_2) that might appear as shown in Figure 12.4. (Graphs of

Figure 12.4 No interaction between x_1 and x_2

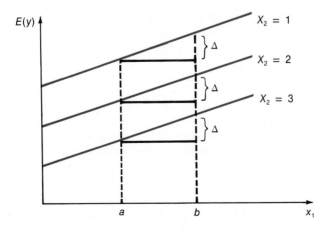

$E(y)$ versus x_2 for various values of x_1 would also produce a set of parallel lines but with common slope β_2.)

Figure 12.4 shows that, regardless of the value of x_2, a change in x_1, say from $x_1 = a$ to $x_1 = b$, will always produce the same change, Δ, in $E(y)$. When this situation occurs, that is, when the change in $E(y)$ for a change in one variable does not depend on the value of the second variable, we say that the variables *do not interact*.

In contrast, consider the second-order (interaction) model,

$$E(y) = \beta_0 + \beta_1 x_1 + \beta_2 x_2 + \beta_3 x_1 x_2.$$

If you were to again graph $E(y)$ versus x_1 for various values of x_2, you would obtain a set of lines, but they would not be parallel (see Figure 12.5).

Figure 12.5 Interaction between x_1 and x_2

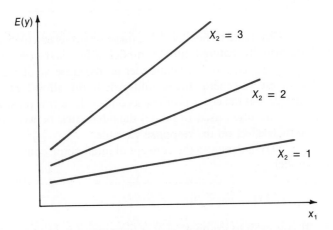

In Figure 12.5, you can see that $E(y)$ rises very slowly as x_1 increases when $x_2 = 1$, more rapidly when $x_2 = 2$, and even more rapidly when $x_2 = 3$. Therefore, the effect on $E(y)$ of a change in x_1 **depends** upon the value of x_2. When this situation occurs, we say that the predictor variables **interact**. If you attempt to fit the non-interactive first-order model to data that are graphed as shown in Figure 12.5, you obtain a very poor fit to your data. The warning is clear. You may need interaction terms in your model to obtain a good fit to a set of data.

> **Definition**
> Two predictor variables are said to **interact** if the change in $E(y)$ corresponding to a change in one predictor variable **depends** upon the value of the other variable.

In contrast to quantitative predictor variables, qualitative predictor variables are entered into a model using dummy (or indicator) variables. For example, suppose that you were attempting to relate the mean salary of a group of employees to a set of predictor variables, and that one of the variables that you wish to include is the employee's race. If each employee included in your study belongs to one of three races, say A, B, or C, you will wish to enter the qualitative predictor variable, race, into your model as follows:

$$E(y) = \beta_0 + \beta_1 x_1 + \beta_2 x_2$$

where

$$x_1 = \begin{cases} 1 & \text{if Race } B \\ 0 & \text{if not} \end{cases}$$

$$x_2 = \begin{cases} 1 & \text{if Race } C \\ 0 & \text{if not} \end{cases}$$

If you wish to find $E(y)$ for race A, we examine our coding for the dummy variables x_1 and x_2 and note that $x_1 = 0$ and $x_2 = 0$. Therefore, for race A,

$$E(y) = \beta_0 + \beta_1 x_1 + \beta_2 x_2 = \beta_0 + \beta_1(0) + \beta_2(0)$$
$$= \beta_0.$$

The value of $E(y)$ for race B is obtained by letting $x_1 = 1$ and $x_2 = 0$, that is, for race B,

$$E(y) = \beta_0 + \beta_1 x_1 + \beta_2 x_2 + \beta_0 = \beta_1(1) + \beta_2(0)$$
$$= \beta_0 + \beta_1.$$

Similarly, the value of $E(y)$ for race C is obtained by letting $x_1 = 0$ and $x_2 = 1$. Therefore, for race C,

$$E(y) = \beta_0 + \beta_1 x_1 + \beta_2 x_2 = \beta_0 + \beta_1(0) + \beta_2(1)$$
$$= \beta_0 + \beta_2.$$

Models with qualitative predictor variables that may assume any number, say k, of values may be constructed in a similar manner using $(k - 1)$ terms involving dummy variables. These terms may be added to models containing other predictor variables, quantitative or qualitative, and you can also include terms involving the cross products (interaction terms) of the dummy variables with other variables that appear in the model.

This optional section introduced two important aspects of model formulation—the concept of predictor variable interaction (and how to cope with it) and the method for introducing qualitative predictor variables into a model. Clearly, this is not enough to make you proficient in model formulation, but it will help you to understand why and how some terms are included in regression models, and it may help you to avoid some pitfalls encountered in model construction. An elementary but fairly complete discussion of this topic can be found in Mendenhall and McClave (1981).

12.6
A Multiple Regression Analysis for the Worker Absenteeism Case Study

As noted in the case study, Constas and Vichas (1980) investigated the relationship between worker absenteeism and four life-space predictor variables—those that characterized the worker rather than the work space. The measure of worker absenteeism was the number of days the worker was absent from his or her job during a one-year period of time. The predictor variables and their type—qualitative or quantitative—and the symbols which represent them were:

1. Sex (qualitative)
   Coding: $x_1 = 0$ if male, $x_1 = 1$ if female
   This variable was selected because (according to Constas and Vichas) "women tend to manifest higher absence rates—both in frequency and duration—than men in comparable working situations."

2. Marital Status (qualitative)
   Coding: $x_2 = 0$ if married or divorced
   $x_2 = 1$ if single
   Marital status was viewed as a measure of a worker's responsibilities. The suggestion is that workers with strong family ties and responsibilities might be less prone to absenteeism than workers with fewer responsibilities. You can see that marital status can actually assume any one of three values corresponding to the three categories, single (unmarried), married, and divorced. By using only one dummy variable and combining the "married" and "divorced" categories of workers (categories which may represent "greater" and "lesser" family responsibilities, respectively), the usefulness of marital status as a predictor variable could be nullified.

3. Age of the Employee (quantitative)
   Coding: x_3 = age of the employee
   Age is probably related to absenteeism, but the nature of the relationship is in debate. Most researchers agree that young employees, those younger than age 25, might be expected to have higher rates of absenteeism and that this rate might decrease as the worker approaches age 40. Some researchers theorize that the relationship between absenteeism and worker age can be characterized by an "envelope-shaped curve, decreasing to age mid-forties, then gradually rising." If this latter theory were true, we would expect to enter both first- and second-order terms, those involving x_3 and x_3^2, into the model to allow for curvature in the relationship between $E(y)$ and x_3.

4. Number of Dependents (quantitative)
   Coding: x_4 = number of dependents
   The expectation, in introducing this predictor variable, is that the number of dependents is a measure of the worker's responsibilities. If correct, we would expect absenteeism to decrease as the number of dependents increased.

5. Automobile Ownership (qualitative)
   Coding: $x_5 = 0$ if employee owns a vehicle, $x_5 = 1$ if not.
   The logic of including this variable is that car ownership may be associated with the difficulty the worker may have in traveling to the workplace.

The data for the Constas–Vichas study consist of observations on 50 workers randomly selected from among the workforce of an electronics assembly plant and observations on 120 workers at a petrochemical plant, both located in southern Puerto Rico. The authors fitted separate first-order regression models,

$$E(y) = \beta_0 + \beta_1 x_1 + \beta_2 x_2 + \cdots + \beta_5 x_5,$$

to each of the two sets of data. The sample sizes, their computed values of R^2 and F values for tests of the complete models, are shown in Table 12.5. The parameter estimates, their estimated standard deviations (standard errors), and the computed values of the t statistics are presented in Table 12.6.

Table 12.5 Results of two multiple regression analyses

	Plant A	Plant B
Sample Size n	50	120
Coefficient of Determination R^2	0.452	0.282
F	2.266	1.969

Table 12.6

Plant A			Variable			
		Sex	Marital Status	Age of Employee	Number of Dependents	Automobile Ownership
Parameter	β_0	β_1	β_2	β_3	β_4	β_5
Parameter Estimate	12.388	5.113	4.756	-0.315	0.675	1.906
Estimated Standard Deviation of Estimate		3.139	2.856	0.152	0.729	2.209
Computed Value of t		1.63	1.67	$-2.07*$	0.93	0.86

Plant B			Variable			
		Sex	Marital Status	Age of Employee	Number of Dependents	Automobile Ownership
Parameter	β_0	β_1	β_2	β_3	β_4	β_5
Parameter Estimate	13.180	-13.101	3.043	-0.064	-1.218	10.714
Estimated Standard Deviation of Estimate		8.914	5.831	0.178	0.933	4.150
Computed Value of t		-1.47	0.52	-0.36	-1.31	2.58*

* Stars attached to t values indicate that they are statistically significant for $\alpha = .05$.

Does the model contribute information for the prediction of worker absenteeism y for plants A and B? The F statistics for testing the complete model, 2.266 for plant A and 1.969 for plant B, are based on $v_1 = 5$ and $v_2 = n - 6$ ($v_2 = 44$ for plant A and 114 for plant B). The corresponding critical (upper-tail) F values for $\alpha = .10$ are approximately 1.99 and 1.90, respectively. Since the computed values of the F statistics exceed the corresponding critical values, there is evidence, at the $\alpha = .10$ level of significance, to indicate that at least one of the five predictor variables contributes information for the prediction of worker absenteeism *when the variables are employed in a first-order model*. Examining the t tests for each of the five individual parameters, $\beta_1, \beta_2, \ldots, \beta_5$, you will observe (Table 12.6) that two of the parameters—age of the worker in the plant A model, and car ownership in the plant B model—are statistically significant for $\alpha = .05$.

One interesting aspect of the analysis concerns the relationship between absenteeism and worker age, x_3. If, as Constas and Vichas note, the relationship between absenteeism and worker age is "an envelope shaped curve" as shown in Figure 12.6, then a straight line (equivalent to entering x_3 into the model as a first-order term) should possess a slope near 0 and should provide a very poor fit to data. Why, then, did the t test for the age variable indicate statistical significance for plant A? The answer is that the workers included in the plant A regression analysis were relatively young (the mean age for plant A was 32.54 years, and the standard deviation was 7.65 years). Consequently, a first-order (straight-line) model might be expected to provide an adequate fit to these data (see Figure 12.7(a)). In contrast, the ages of the workers included in the regression analysis for plant B varied from very young to near retirement (the mean age was 34.80 years, and the standard deviation was 19.01 years). A test of the significance of the slope (test of $H_0: \beta_3 = 3$) for plant B produced a negligible t value (equal to -0.36), a value that may be explained by "an envelope shaped" curvilinear relation between absenteeism and age of worker (see Figure 12.7(b)).

So, what can we conclude from the Constas–Vichas study? There is evidence to indicate that the first-order model provides some information for

Figure 12.6 A theoretical relation between worker absenteeism and age of worker

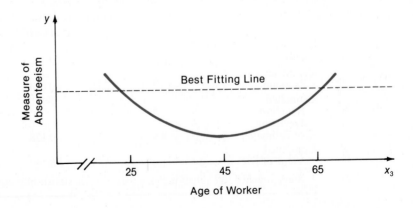

Figure 12.7

Hypothetical relations between absenteeism and age of worker

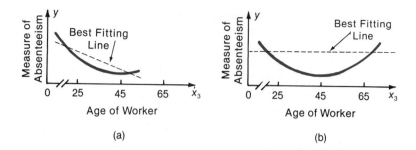

(a)

(b)

the prediction of the rate of worker absenteeism, but it is also clear that there is ample room for improvement in the model. The computed values of R^2 indicate that the prediction equations for plants A and B provide very poor fits to their repsective sets of data. The prediction equation for plant A accounts for only 45.2 percent of the variability of the y values about the mean, and the percentage corresponding to plant B is only 28.2 percent.

Prediction equations that provide poor fits to data sets are not uncommon and they are prevalent when the response is some aspect of human behavior. This is because the number of variables that affect or are related to the response is often large, and some of the variables may be unmeasurable. In general, the cause of poor-fitting prediction equations can always be traced to the model. The model may not contain the appropriate information-contributing predictor variables, and/or the manner in which they have been entered into the model (i.e., the model formulation) may not permit the predictor variables to contribute the information that they contain.

12.7
Summary

Chapter 12 provides an introduction to an extremely useful area of statistical methodology, one that enables us to relate the mean value of a response variable y to a set of predictor variables. We commence by postulating a model that expresses $E(y)$ as the sum of a number of terms, each term involving the product of a single unknown parameter and a function of one or more of the predictor variables. Values of y are observed when the predictor variables assume specified values, and these data are used to estimate the unknown parameters in the model using the method of least squares. The procedure for estimating the unknown parameters of the model, testing its utility, or testing hypotheses about sets of the model parameters is known as a multiple regression analysis.

Perhaps the most important result of a multiple regression analysis is the use of the fitted model—the prediction equation—to estimate the mean value of y or to predict a particular value of y for a given set of values of the predictor variables. Scientists, sociologists, engineers, and business managers all face the same problem—forecasting some quantitative response based on the values of a set of predictor variables that describe current or future conditions. When the appropriate assumptions are satisfied, a multiple regression analysis can be used to solve this important problem.

References

Dixon, W.J., and M.B. Brown, ed. *BMDP Biomedical Computer Programs, P Series.* Berkeley, Calif.: University of California, 1979.

Constas, K., and R.P. Vichas. "An Interpretative Policy-Making Model of Absenteeism with Reference to the Marginal Worker in Overseas Plants." *Journal of Management Studies,* 17, no. 2 (1980).

Draper, N., and H. Smith. *Applied Regression Analysis.* New York: John Wiley & Sons, Inc., 1966.

Hamburg, M. *Statistical Analysis for Decision Making,* 2nd ed. New York: Harcourt Brace Jovanovich, 1977.

Helwig, J.P., and K.A. Council, eds. *SAS User's Guide.* Cary, N.C.: SAS Institute, Inc., P.O. Box 8000, 1979.

Kleinbaum, D., and L. Kupper. *Applied Regression Analysis and Other Multivariable Methods.* Boston: Duxbury Press, 1978.

Mendenhall, W. and J.T. McClave. *A Second Course in Business Statistics: Regression Analysis.* San Francisco: Dellen Publishing Co., 1981.

Miller, R.B., and D.W. Wichern. *Intermediate Business Statistics: Analysis of Variance, Regression, and Time Series.* New York: Holt, Rinehart, and Winston, 1977.

Neter, J., and W. Wasserman. *Applied Linear Statistical Models.* Homewood, Ill.: Richard D. Irwin, 1974.

Nie, N.H., C.H. Hull, J.G. Jenkins, K. Steinbrenner, and D.H. Bent. *SPSS Statistical Package for Social Sciences,* 2nd ed. New York: McGraw-Hill, 1979.

Ryan, T.A., B.L. Joiner, and B.F. Ryan. *Minitab Student Handbook.* Boston: Duxbury Press, 1979.

Wonnacott, R.J., and T.H. Wonnacott. *Econometrics,* 2nd ed. New York: John Wiley & Sons, Inc., 1979.

Ya-lun Chou. *Statistical Analysis with Business and Economic Applications,* 2nd ed. New York: Holt, Rinehart, and Winston, 1975.

Younger, M.S. *A Handbook for Linear Regression.* Boston: Duxbury Press, 1979.

Supplementary Exercises

[Starred (*) exercises are optional.]

12.12. Utility companies, which must plan the operation and expansion of electricity generation, are vitally interested in predicting customer demand over both short and long periods of time. A short-term study was conducted to investigate the effect of mean monthly daily temperature x_1 and cost per kilowatt x_2 on the mean daily consumption (KWH) per household. The company expected the demand for electricity to rise in cold weather (due to heating), fall when the weather was moderate, and rise again when the temperature rose and there was a need for air conditioning. They expected demand to decrease as the cost per kilowatt hour increased, reflecting greater attention to conservation. Data were available for two years, a period in which the cost per kilowatt hour, x_2, increased due to the increasing costs of fuel. The company fitted the model,

$$E(y) = \beta_0 + \beta_1 x_1 + \beta_2 x_1^2 + \beta_3 x_2 + \beta_4 x_1 x_2 + \beta_5 x_1^2 x_2$$

to the data shown below:

Price Per KWH x_2		Mean Daily Consumption (KWH) Per Household											
8¢	Mean Daily Temperature (°F)x_1	31	34	39	42	47	56	62	66	68	71	75	78
	Mean Daily Consumption y	55	49	46	47	40	43	41	46	44	51	62	73
10¢	Mean Daily Temperature x_1	32	36	39	42	48	56	62	66	68	72	75	79
	Mean Daily Consumption y	50	44	42	42	38	40	39	44	40	44	50	55

Before we analyze the data, let us examine the logic employed in formulating the model.

a. Examine the relationship between $E(y)$ and temperature x_1 for a fixed price per kilowatt hour (KWH), x_2. Substitute a value for x_2, say 10¢, into the equation for $E(y)$. What type of curve will the model relating $E(y)$ to x_1 represent?

b. If the company's theory regarding the relationship between mean daily consumption $E(y)$ and temperature x_1 is correct, what should be the sign (positive or negative) of the coefficient of x_1^2?

c. Examine the relationship between $E(y)$ and price per KWH, x_2, when temperature x_1 is held constant. Substitute a value for x_1 into the equation for $E(y)$, say $x_1 = 50°F$. What type of curve will the model relating $E(y)$ to x_2 represent?

d. Refer to (c). If the company's theory regarding the relationship between mean daily consumption $E(y)$ and price per KWH x_2 is correct, what should be the sign of the coefficient of x_2?

e. What effect do the terms $\beta_4 x_1 x_2$ and $\beta_5 x_1^2 x_2$ have on the curves relating $E(y)$ to x_1 for various values of price per KWH?

12.13. The SAS multiple regression computer printout for the data, Exercise 12.12, is shown in Table 12.7.

a. Find SSE and s^2 on the printout.

b. How many degrees of freedom do SSE and s^2 possess? Show that $s^2 = $ SSE/(degrees of freedom).

c. Do the data provide sufficient evidence to indicate that the model contributes information for the prediction of mean daily KWH consumption per household? Test using $\alpha = .10$.

d. Use Tables 6, 7, 8, 9, and 10, the Appendix, to find the approximate observed significance level for the test, part (c).

e. Find the total SS, SS_y, on the printout. Find the value of R^2 and interpret its value. Show that $R^2 = 1 - $ SSE/SS_y.

Table 12.7 SAS output for Exercise 12.13

DEPENDENT VARIABLE: Y

SOURCE	DF	SUM OF SQUARES	MEAN SQUARE	F VALUE	PR > F	R-SQUARE	C.V.
MODEL	5	1346.44751546	269.28950309	31.85	0.0001	0.898455	6.2029
ERROR	18	152.17748454	8.45430470		STD DEV		Y MEAN
CORRECTED TOTAL	23	1498.62500000			2.90762871		46.87500000

SOURCE	DF	TYPE I SS	F VALUE	PR > F	DF	TYPE IV SS	F VALUE	PR > F
X1	1	140.71101071	16.64	0.0007	1	104.43456772	12.35	0.0025
X1*X1	1	892.77914943	105.60	0.0001	1	125.58263370	14.85	0.0012
X2	1	192.44266992	22.76	0.0002	1	46.78392523	5.53	0.0302
X1*X2	1	57.83521130	6.84	0.0175	1	50.03623552	5.92	0.0256
X1*X1*X2	1	62.67947420	7.41	0.0140	1	62.67947420	7.41	0.0140

PARAMETER	ESTIMATE	T FOR H0: PARAMETER = 0	PR > |T|	STD ERROR OF ESTIMATE
INTERCEPT	325.60644505	3.92	0.0010	83.06412820
X1	−11.38255606	−3.51	0.0025	3.23859476
X1*X1	0.11319735	3.85	0.0012	0.02944828
X2	−21.69920900	−2.35	0.0302	9.22432344
X1*X2	0.87302921	2.43	0.0256	0.35886030
X1*X1*X2	−0.00886945	−2.72	0.0140	0.00325742

    f. Use the prediction equation to predict mean daily consumption when the mean daily temperature is $60\,°F$ and the price per KWH is 8¢.

    g. If you have access to a computer, use it to perform a multiple regression analysis for the data, Exercise 12.12. Compare your computer output with the SAS output, Table 12.7.

* 12.14.   Refer to Exercises 12.12 and 12.13.

    a. Graph the curve depicting $\hat{y}$ as a function of temperature x_1 when the cost per KWH is $x_2 = 8$¢. Construct a similar graph for the case when $x_2 = 10$¢ per KWH. Does it appear that the consumption curves differ?

    b. If cost per KWH is unimportant in predicting usage, then we do not need the terms involving x_2 in the model. Therefore, the null hypothesis, H_0: "x_2 does not contribute information for the prediction of y" is equivalent to the hypothesis, $H_0: \beta_3 = \beta_4 = \beta_5 = 0$ (if $\beta_3 = \beta_4 = \beta_5 = 0$, the terms involving x_2 disappear from the model). This hypothesis is tested by computing SSE_1, the SSE that you would obtain if $\beta_3 = \beta_4 = \beta_5 = 0$, that is, using the Reduced Model,

        Model 1 (Reduced Model):
        $E(y) = \beta_0 + \beta_1 x_1 + \beta_2 x_1^2$

and then computing SSE_2, the SSE for the Complete Model,

        Model 2 (Complete Model):
        $E(y) = \beta_0 + \beta_1 x_1 + \beta_2 x_1^2 + \beta_3 x_2 + \beta_4 x_1 x_2 + \beta_5 x_1^2 x_2.$

If the terms involving β_3, β_4, and β_5 contribute information for the prediction of y, then SSE_2 should be much less than SSE_1. Therefore, the test of $H_0 : \beta_3 = \beta_4 = \beta_5$ is conducted as follows:

Null Hypothesis: $H_0 : \beta_3 = \beta_4 = \beta_5 = 0$.

Alternative Hypothesis: H_a: at least one of the parameters, β_3, β_4, or β_5, differs from 0.

Test Statistic:
$$F = \frac{\dfrac{SSE_1 - SSE_2}{v_1}}{\dfrac{SSE_2}{v_2}} = \frac{\dfrac{SSE_1 - SSE_2}{v_1}}{s^2}$$

where v_1 = number of parameters involved in H_0 (in this exercise, $v_1 = 3$)

v_2 = number of degrees of freedom for SSE_2 and s^2 (in this exercise, $v_2 = 18$)

s^2 = estimate of σ^2 obtained for the complete model.

Rejection Region: Reject H_0 if $F > F_\alpha$ where F_α is the tabulated value of F (Tables 6, 7, 8, 9, and 10, the Appendix) based on v_1 and v_2 degrees of freedom.

The SAS multiple regression computer printout, obtained by fitting the Reduced Model to the data, is shown in Table 12.8. The following steps enable you to conduct a test to determine whether the data provide sufficient evidence to indicate that price per KWH, x_2, contributes information for the prediction of y. Test using $\alpha = .05$.

Table 12.8 SAS output for Exercise 12.14(b)

DEPENDENT VARIABLE: Y

SOURCE	DF	SUM OF SQUARES	MEAN SQUARE	F VALUE	PR > F	R-SQUARE	C.V.
MODEL	2	1033.49016004	516.74908002	23.33	0.0001	0.689626	10.0401
ERROR	21	465.13432996	22.14927809		STD DEV		Y MEAN
CORRECTED TOTAL	23	1498.62500000			4.70630196		46.87500000

SOURCE	DF	TYPE I SS	F VALUE	PR > F	DF	TYPE IV SS	F VALUE	PR > F
X1	1	140.71101071	6.35	0.0199	1	810.33211666	36.59	0.0001
X1*X1	1	892.77914933	40.31	0.0001	1	892.77914933	40.31	0.0001

PARAMETER	ESTIMATE	T FOR H0: PARAMETER = 0	PR > |T|	STD ERROR OF ESTIMATE
INTERCEPT	130.00929147	8.74	0.0001	14.87580615
X1	−3.30171859	−6.05	0.0001	0.57893460
X1*X1	0.03337133	6.35	0.0001	0.00525631

i. Find SSE_1 and SSE_2.

ii. Give the values for v_1 and v_2.

iii. Calculate the value of F.

iv. Find the tabulated value of $F_{.05}$.

v. State your conclusions.

*✎ 12.15. A department store conducted an experiment to investigate the effects of advertising expenditures on the weekly sales for its men's wear, children's wear, and women's wear

departments. Five weeks were randomly selected for each department and an advertising budget x_1 (hundreds of dollars) was assigned for each. The weekly sales (thousands of dollars) are shown in the table for each of the fifteen 1-week sales periods.

	Advertising Expenditure (Hundreds of Dollars)				
Department	1	2	3	4	5
Men's Wear A	5.2	5.9	7.7	7.9	9.4
Children's Wear B	8.2	9.0	9.1	10.5	10.5
Women's Wear C	10.0	10.3	12.1	12.7	13.6

If we expect weekly sales $E(y)$ to be linearly related to advertising expenditure x_1, and if we expect the slopes of the lines corresponding to the three departments to differ, then an appropriate model for $E(y)$ is,

$$E(y) = \underbrace{\beta_0 + \beta_1 x_1}_{\substack{\text{Advertising} \\ \text{expenditure}}} + \underbrace{\beta_2 x_2 + \beta_3 x_3}_{\substack{\text{Dummy variables} \\ \text{used to introduce} \\ \text{the qualitative} \\ \text{variable, "Department,"} \\ \text{into the model}}} + \underbrace{\beta_4 x_1 x_2 + \beta_5 x_1 x_3}_{\substack{\text{Interaction terms} \\ \text{that introduce} \\ \text{differences in} \\ \text{slopes}}}$$

where

$x_1 = $ advertising expenditure

$$x_2 = \begin{cases} 1 & \text{if children's wear department } B \\ 0 & \text{if not} \end{cases}$$

$$x_3 = \begin{cases} 1 & \text{if women's wear department } C \\ 0 & \text{if not.} \end{cases}$$

a. Find the equation of the line relating $E(y)$ to advertising expenditure x_1 for the men's wear department A. (*Hint:* According to the coding employed for the dummy variables, the model represents mean sales $E(y)$ for the men's wear department A when $x_2 = x_3 = 0$. Substitute $x_2 = x_3 = 0$ into the equation for $E(y)$ to find the equation of this line.)

b. Find the equation of the line relating $E(y)$ to x_1 for the children's wear department B. (*Hint:* According to the coding, the model represents $E(y)$ for the children's wear department when $x_2 = 1$ and $x_3 = 0$.)

c. Find the equation of the line relating $E(y)$ to x_1 for the women's wear department C.

d. Find the difference between the intercepts of the $E(y)$ lines corresponding to the children's wear B and men's wear A departments.

e. Find the difference in slopes between $E(y)$ lines corresponding to the women's wear C and men's wear A departments.

f. Refer to part (e). Suppose that you wish to test the null hypothesis that the slopes of the lines corresponding to the three departments are equal. Express this as a test of an hypothesis about one or more of the model parameters.

* 12.16. If you have access to a computer and a multiple regression computer program package, perform a multiple regression analysis for the data, Exercise 12.15.

   a. Verify that SSE $= 1.2190$, that $s^2 = .1354$, and that $s = .368$.

   b. How many degrees of freedom will SSE and s^2 possess?

   c. Verify that the total SS $= SS_y = 76.0493$.

   d. Calculate R^2 and interpret its value.

   e. Verify that the parameter estimates (values rounded) are approximately equal to

   $$\hat{\beta}_0 \approx 4.10 \qquad \hat{\beta}_1 \approx 1.04 \qquad \hat{\beta}_2 \approx 3.53$$
   $$\hat{\beta}_3 \approx 4.76 \qquad \hat{\beta}_4 \approx -.43 \qquad \hat{\beta}_5 \approx -.08.$$

   f. Find the prediction equation and graph the three department sales lines.

   g. Examine the graphs, part (f). Do the slopes of the lines corresponding to the children's wear (B) and the men's wear (A) departments appear to differ? Test the null hypothesis that the slopes do not differ (i.e., $H_0: \beta_4 = 0$) against the alternative hypothesis that the slopes do differ (i.e., $H_a: \beta_4 \neq 0$). Test using $\alpha = .05$. (*Note:* Verify on your computer printout that the estimated standard deviation of $\hat{\beta}_4$ is $s_{\hat{\beta}_4} \approx .165$.)

   h. Find a 95 percent confidence interval for the difference between the slopes of lines B and A.

   i. Do the data provide sufficient evidence to indicate a difference in the slopes between lines C and A? Test using $\alpha = .05$. (*Hint*: $s_{\hat{\beta}_5} \approx .165$.)

12.17. According to Buescher, *et al.**, "Cucumber pickles usually are preserved by fermentation followed by storage in brine until it is convenient for processors to manufacture various finished products. The primary function of high salt content concentration levels in brine is to retard softening of the pickles and to prevent freezing." Recent studies by Buescher, *et al.* suggest that pickle firmness can be maintained for longer periods by adding small amounts of calcium chloride to the brine solution. To compare this brine with the standard salt (sodium chloride) solution, pickles were stored in the two types of brine for 4, 14, 32, and 52 weeks. Pickle firmness y was measured at the start of the experiment and at each of four time periods. The data shown below:

	Weeks in Storage at 72°F				
Brine	**0**	**4**	**14**	**32**	**52**
			Firmness in Pounds		
Standard	19.8	16.5	12.8	8.1	7.5
Low Salt with Calcium Chloride	19.8	22.0	19.6	20.0	20.0

* Buescher, R.W., J.M. Hudson, J.R. Adams, and D.H. Wallace, "Calcium Makes It Possible to Store Cucumber Pickles in Low Salt Brine," *Arkansas Farm Research*, 30, no. 4 (July/August 1981).

a. Plot the ten data points and sketch the curve that relates hardness to time for each of the two types of brine.

b. Explain why the following model might be a good choice for characterizing the relationship between mean firmness $E(y)$ and the two predictor variables, time (weeks in storage) and "brine type."

$$E(y) = \beta_0 + \beta_1 x_1 + \beta_2 x_1^2 + \beta_3 x_2 + \beta_4 x_1 x_2 + \beta_5 x_1^2 x_2$$

x_1 = time in weeks

x_2 = 1 if standard brine

x_2 = 0 if not.

The qualitative variable brine type is entered into the model using the dummy (indicator) variable, x_2.

c. The SAS computer printout for the regression analysis is shown in Table 12.9. Does the model contribute information for the prediction of y? Test using $\alpha = .05$.

d. Find the observed significance level for the test, part (c), and interpret its value.

e. On the same piece of graph paper, graph the predictor value of y as a function of time of storage x_1 for: 1) the standard brine,   2) the brine containing calcium chloride.

Table 12.9 SAS output for Exercise 12.17

DEPENDENT VARIABLE: Y

SOURCE	DF	SUM OF SQUARES	MEAN SQUARE	F VALUE	PR > F	R-SQUARE	C.V.
MODEL	5	247.22102336	49.44420467	48.86	0.0011	0.983890	6.0565
ERROR	4	4.04797664	1.01199416		STD DEV		Y MEAN
CORRECTED TOTAL	9	251.26900000			1.00597920		16.61000000

SOURCE	DF	TYPE I SS	F VALUE	PR > F	DF	TYPE IV SS	F VALUE	PR > F
X1	1	54.85657530	54.21	0.0018	1	0.22880167	0.23	0.6592
X1*X1	1	8.66901488	8.57	0.0429	1	0.11034618	0.11	0.7578
X2	1	134.68900000	133.09	0.0003	1	1.62363682	1.60	0.2740
X1*X2	1	42.88308477	42.37	0.0029	1	16.72909295	16.53	0.0153
X1*X1*X2	1	6.12334841	6.05	0.0697	1	6.12334841	6.05	0.0697

PARAMETER	ESTIMATE	T FOR HO: PARAMETER = 0	PR > |T|	STD ERROR OF ESTIMATE
INTERCEPT	20.71339998	26.19	0.0001	0.79080905
X1	−0.04272103	−0.48	0.6592	0.08984648
X1*X1	0.00055598	0.33	0.7578	0.00168371
X2	−1.41658315	−1.27	0.2740	1.11837288
X1*X2	−0.51661055	−4.07	0.0153	0.12706211
X1*X1*X2	0.00585716	2.46	0.0697	0.00238112

12.18. Refer to Exercise 12.14. If there is no difference in mean response for the two types of brine, then the variable x_2 need not be included in the model. A test of this null hypothesis is equivalent to a test of the hypothesis, $H_0: \beta_3 = \beta_4 = \beta_5 = 0$. The SAS computer printout appropriate for fitting the model

$$E(y) = \beta_0 + \beta_1 x_1 + \beta_2 x_1^2$$

to the data is shown in Table 12.10. Use the method of Exercise 12.14 to test the null hypothesis that there is no difference in mean firmness $E(y)$ between the two brine types. Test using $\alpha = .05$. State your conclusions.

Table 12.10 SAS regression analysis: reduced model

DEPENDENT VARIABLE: Y

SOURCE	DF	SUM OF SQUARES	MEAN SQUARE	F VALUE	PR > F	R-SQUARE	C.V.
MODEL	2	63.52559018	31.76279509	1.18	0.3606	0.252819	31.1791
ERROR	7	187.74340982	26.82048712		STD DEV		Y MEAN
CORRECTED TOTAL	9	251.26900000			5.17884998		16.61000000

SOURCE	DF	TYPE I SS	F VALUE	PR > F	DF	TYPE IV SS	F VALUE	PR > F
X1	1	54.85657530	2.05	0.1958	1	22.72033479	0.85	0.3880
X1*X1	1	8.66901488	0.32	0.5874	1	8.66901488	0.32	0.5874

PARAMETER	ESTIMATE	T FOR H0: PARAMETER = 0	PR > |T|	STD ERROR OF ESTIMATE
INTERCEPT	20.00510841	6.95	0.0002	2.87873018
X1	−0.30102630	−0.92	0.3880	0.32706224
X1*X1	0.00348455	0.57	0.5874	0.00612908

13. Analysis of Enumerative Data

Chapter Objectives

General Objective

Many types of surveys and experiments result in the observation of qualitative rather than quantitative response variables. That is, the responses can be classified but not quantified. Consequently, data from these experiments will be the number of response observations falling in each of the data classes included in the experiment. The objective of this chapter is to present methods for analyzing count (or enumerative) data.

Specific Objectives

1. To present the properties of the multinomial experiment. Except for minor exceptions, which are noted, all the techniques presented in this chapter will assume that the sampling satisfies the properties of a multinomial experiment. *Section 13.1*

2. To present the chi-square test, which will be used to test hypotheses concerning the class (or cell) probabilities. *Section 13.2*

3. To present a test of an hypothesis concerning specific values of the cell probabilities. *Section 13.3*

4. To present a method for testing for dependence between two methods of classification. *Sections 13.4, 13.5*

5. To familiarize you with typical computer printouts for contingency table analyses. *Section 13.6*

6. To suggest other applications of the general statistical test of Section 13.2. *Sections 13.7, 13.8*

7. To emphasize the assumptions you must make in order for the test procedures described in this chapter to be valid. *Section 13.9*

Case Study

No Wine Before Its Time

In an article titled, "They Will Can No Wine before Its Time," *Washington Post* staff writer, Nicholas D. Kristof, surveys industry response to a novel method for packaging wine—canning!*

Both the Geyser Creek Winery (of California) and Wine Spectrum, a subsidiary of Coca Cola, envision great sales prospects for this new method of marketing wine. Packaged in six-packs, similar to beer, the cans are more compact, lighter in weight, and less susceptible to damage than bottles. For example, a case of wine in bottles weighs 48 pounds, while an equivalent amount packed in aluminum cans weighs 22 pounds and occupies 30 percent less space. This feature is particularly appealing to the airlines, where weight and space play an important role in a flight's profit equation.

But what about the concept and the taste? Will it be appealing to consumers? "Trés, trés tacky," is the reported response of Jean Jacques Moreau, a wine connoisseur and producer of fine wines in the Chablis region of France. Moreau is further quoted as saying, "C'est ridicule, ça ne va pas. Ça n'a pas de sense d'acheter du bon vin en autre chose qu'uane bouteille de verre. Ça ne marcherait pas en Europe," which Kristof translates to mean, in short, "Canned wine is gauche."

But what does the consumer say? Reynolds Metals Company, which makes the aluminum cans, claims that the cans are coated on the inside to prevent the wine from acquiring a metallic taste. Reynolds claims to have conducted a blind taste test, one in which the participants were given two glasses of the same wine, one obtained from a can and the other from a bottle. Fifty percent of the participants preferred the wine from the can; 39 percent from the bottle, and 11 percent could detect no difference between the two.

How can we use the Reynolds Metals Company taste test data to determine whether consumers can detect a difference in taste between canned and bottled wine? Noting that we wish to test, $H_0: p_1 = p_2$, where p_1 and p_2 are the proportions of consumers who prefer canned and bottled wine, respectively, we might think of employing the large-sample z test of Chapter 9. However, the z test is inappropriate because it is based on independent random samples, selected from two different binomial populations. Further examination will reveal that the Reynolds taste test was analogous to a binomial experiment, the difference being that the response of each person fell in one of three, rather than two, categories, namely (1) favor the canned wine, (2) favor the bottled wine, and (3) have no preference. We call this a *multinomial* experiment.

This chapter will introduce you to the analysis of data obtained from a multinomial experiment. We will reexamine the data from this taste test in Section 13.7.

* *Washington Post*, August 23, 1981.

13.1
A Description of the Experiment

Many experiments, particularly in the social sciences, result in enumerative (or count) data. For instance, the classification of people into five income brackets would result in an enumeration or count corresponding to each of the five income classes. Or, we might be interested in studying the reaction of a mouse to a particular stimulus in a psychological experiment. If a mouse will react in one of three ways when the stimulus is applied and if a large number of mice were subjected to the stimulus, the experiment would yield three counts, indicating the number of mice falling in each of the reaction classes. Similarly, a traffic study might require a count and classification of the type of motor vehicles using a section of highway. An industrial process manufactures items that fall into one of three quality classes: acceptable, seconds, and rejects. A student of the arts might classify paintings in one of k categories according to style and period in order to study trends in style over time. We might wish to classify ideas in a philosophical study or style in the field of literature. The results of an advertising campaign would yield count data indicating a classification of consumer reaction. Indeed, many observations in the physical sciences are not amenable to measurement on a continuous scale and hence result in enumerative or classificatory data.

The illustrations in the preceding paragraph exhibit, to a reasonable degree of approximation, the following characteristics that define a **multinomial experiment**:

The Characteristics of a Multinomial Experiment

1. The experiment consists of n identical trials.
2. The outcome of each trial falls into one of k classes or cells.
3. The probability that the outcome of a single trial will fall in a particular cell, say cell i, is p_i $(i = 1, 2, \ldots, k)$, and remains the same from trial to trial. Note that

$$p_1 + p_2 + p_3 + \cdots + p_k = 1.$$

4. The trials are independent.
5. We are interested in $n_1, n_2, n_3, \ldots, n_k$, where n_i $(i = 1, 2, \ldots, k)$ is equal to the number of trials in which the outcome falls in cell i. Note that $n_1 + n_2 + n_3 + \cdots + n_k = n$.

The above experiment is analogous to tossing n balls at k boxes, where each ball must fall in one of the boxes. The boxes are arranged such that the probability that a ball will fall in a box varies from box to box but remains the same for a particular box in repeated tosses. Finally, the balls are tossed in such a way that the trials are independent. At the conclusion of the experiment, we observe n_1 balls in the first box, n_2 in the second, $\ldots$, and n_k in the kth. The total number of balls is equal to

$$\sum_{i=1}^{k} n_i = n.$$

Note the similarity between the binomial and multinomial experiments and, in particular, that the binomial experiment represents the special case for the multinomial experiment when $k = 2$. The two cell probabilities, p and q, of the binomial experiment are replaced by the k cell probabilities, $p_1, p_2, \ldots, p_k$, of the multinomial experiment. The objective of this chapter is to make inferences about the cell probabilities, $p_1, p_2, \ldots, p_k$. The inferences will be expressed in terms of a statistical test of an hypothesis concerning their specific numerical values or their relationship, one to another.

If we were to proceed as in Chapter 6, we would derive the probability of the observed sample $(n_1, n_2, \ldots, n_k)$ for use in calculating the probability of the type I and type II errors associated with a statistical test. Fortunately, we have been relieved of this chore by the British statistician Karl Pearson, who proposed a very useful test statistic for testing hypotheses concerning $p_1, p_2, \ldots, p_k$, and gave its approximate sampling distribution.

13.2
The Chi-Square Test

Suppose that $n = 100$ balls were tossed at the cells and that we knew that p_1 was equal to .1. How many balls would be expected to fall in the first cell? Referring to Chapter 6 and utilizing knowledge of the binomial experiment, we would calculate

$$E(n_1) = np_1 = 100(.1) = 10.$$

In like manner, the expected number falling in the remaining cells may be calculated using the formula

$$E(n_i) = np_i, \quad i = 1, 2, \ldots, k.$$

Now suppose that we hypothesize values for $p_1, p_2, \ldots, p_k$ and calculate the expected value for each cell. Certainly, if our hypothesis is true, the cell counts, n_i, should not deviate greatly from their expected values, np_i ($i = 1, 2, \ldots, k$). Hence it would seem intuitively reasonable to use a test statistic involving the k deviations,

$$(n_i - np_i), \quad i = 1, 2, \ldots, k.$$

In 1900, Karl Pearson proposed the following test statistic, which is a function of the squares of the deviations of the observed counts from their expected values, weighted by the reciprocals of their expected values:

$$X^2 = \sum_{i=1}^{k} \frac{[n_i - E(n_i)]^2}{E(n_i)}$$

$$= \sum_{i=1}^{k} \frac{[n_i - np_i]^2}{np_i}.$$

Although the mathematical proof is beyond the scope of this text, it can be shown that when n is large, X^2 will possess, approximately, a chi-square probability distribution in repeated sampling. Experience has shown that the cell counts, n_i, should not be too small in order that the chi-square distribution provide an adequate approximation to the distribution of X^2. As a rule of thumb, we will require that all expected cell counts equal or exceed five, although Cochran (see the references) has noted that this value can be as low as one for some situations.

You will recall the use of the chi-square probability distribution for testing an hypothesis concerning a population variance, σ^2, in Section 10.7. Particularly, we stated that the shape of the chi-square distribution would vary depending upon the number of degrees of freedom associated with s^2, and we discussed the use of Table 5, the Appendix, which presents the critical values of χ^2 corresponding to various right-hand tail areas of the distribution. Therefore, we must know which χ^2 distribution to use—that is, the number of degrees of freedom—in approximating the distribution of X^2, and we must know whether to use a one-tailed or two-tailed test in locating the rejection region for the test. The latter problem may be solved directly. Since large deviations of the observed cell counts from those expected would tend to contradict the null hypothesis concerning the cell probabilities $p_1, p_2, \ldots, p_k$, we would reject the null hypothesis when X^2 is large and employ a one-tailed statistical test using the upper tail values of χ^2 to locate the rejection region.

The determination of the appropriate number of degrees of freedom to be employed for the test can be rather difficult and therefore will be specified for the physical applications described in the following sections. In addition, we will state the principle involved (which is fundamental to the mathematical proof of the approximation) so that the reader may understand why the number of degrees of freedom changes with various applications. The principle states that **the appropriate number of degrees of freedom will equal the number of cells, k, less one degree of freedom for each independent linear restriction placed upon the observed cell counts.** For example, one linear restriction is always present because the sum of the cell counts must equal n; that is,

$$n_1 + n_2 + n_3 + \cdots + n_k = n.$$

Other restrictions will be introduced for some applications because of the necessity for estimating unknown parameters required in the calculation of the expected cell frequencies or because of the method in which the sample is collected. These will become apparent as we consider various practical examples.

13.3
A Test of an Hypothesis Concerning Specified Cell Probabilities

The simplest hypothesis concerning the cell probabilities would be one that specifies numerical values for each. For example, suppose that we were to consider a simple extension of the rat experiment discussed in Exercise 6.49. One or more rats proceed down a ramp to one of three doors. We wish to test the hypothesis that the rats have no preference concerning the choice of a door and therefore that

$$H_0 : p_1 = p_2 = p_3 = \frac{1}{3},$$

where p_i is the probability that a rat will choose door i, $i = 1, 2,$ or 3.

Suppose that the rat was sent down the ramp $n = 90$ times and that the three observed cell frequencies were $n_1 = 23$, $n_2 = 36$, and $n_3 = 31$. The expected cell frequency would be the same for each cell: $E(n_i) = np_i = 90(1/3) = 30$. The observed and expected cell frequencies are presented in Table 13.1. Noting the discrepancy between the observed and expected cell frequency, we would wonder whether the data present sufficient evidence to warrant rejection of the hypothesis of no preference.

Table 13.1 Observed and expected cell counts for the rat experiment

	Door		
	1	2	3
Observed cell frequency	$n_1 = 23$	$n_2 = 36$	$n_3 = 31$
Expected cell frequency	30	30	30

The chi-square test statistic for our example will possess $(k - 1) = 2$ degrees of freedom since the only linear restriction on the cell frequencies is that

$$n_1 + n_2 + \cdots + n_k = n,$$

or, for our example,

$$n_1 + n_2 + n_3 = 90.$$

Therefore, if we choose $\alpha = .05$, we would reject the null hypothesis when $X^2 > 5.99147$ (see Table 5, the Appendix).

Substituting into the formula for X^2, we obtain

$$X^2 = \sum_{i=1}^{k} \frac{[n_i - E(n_i)]^2}{E(n_i)} = \sum_{i=1}^{k} \frac{[n_i - np_i]^2}{np_i}$$

$$= \frac{(23 - 30)^2}{30} + \frac{(36 - 30)^2}{30} + \frac{(31 - 30)^2}{30}$$

$$= 2.87.$$

Since X^2 is less than the tabulated critical value of χ^2, 5.991, the null hypothesis is not rejected and we conclude that the data do not present sufficient evidence to indicate that the rat has a preference for a particular door.

Exercises

13.1.   List the characteristics of a multinomial experiment.

13.2.   Suppose that a response can fall in one of $k = 5$ categories with probabilities, $p_1, p_2, \ldots, p_5$, respectively, and that $n = 300$ responses produced the following category counts:

Category	1	2	3	4	5
Observed Count	47	63	74	51	65

a. If you were to test $H_0: p_1 = p_2 = \cdots = p_5$, using the chi-square test, how many degrees of freedom would the test statistic possess?

b. If $\alpha = .05$, find the rejection region for the test.

c. What is your alternative hypothesis?

d. Conduct the test, part (a), using $\alpha = .05$. State your conclusions.

e. Find the approximate observed significance level for the test and interpret its value.

13.3.   A city expressway utilizing four lanes in each direction was studied to see whether drivers preferred to drive on the inside lanes. A total of 1000 automobiles were observed during the heavy early morning traffic and their respective lanes recorded. The results were as follows:

Lane	1	2	3	4
Observed Count	294	276	238	192

Do the data present sufficient evidence to indicate that some lanes are preferred over others? (Test the hypothesis that $p_1 = p_2 = p_3 = p_4 = 1/4$, using $\alpha = .05$.)

13.4.   The Mendelian theory states that the number of peas of a certain type falling into the classifications round and yellow, wrinkled and yellow, round and green, and wrinkled and green should be in the ratio $9:3:3:1$. Suppose that 100 such peas revealed 56, 19, 17, and 8 in the respective classes. Do these data disagree with the Mendelian theory? Use $\alpha = .05$.

13.5.   Medical statistics show that deaths due to four major diseases—call them A, B, C, and D—account for 15, 21, 18, and 14 percent, respectively, of all nonaccidental deaths. A study of the causes of 308 nonaccidental deaths at a hospital serving primarily indigent black patients gave the following counts of patients dying of diseases A, B, C, and D.

Disease	Number of Deaths
A	43
B	76
C	85
D	21
Other	83
	308

Do these data provide sufficient evidence to indicate that the proportions of people dying of diseases *A*, *B*, *C*, and *D* at this hospital differ from the proportions accumulated for the population at large?

13.6. You may recall the exercise in Chapter 9 (Exercise 9.55) that reported on research that linked the prevalence of schizophrenia with birth during particular months of the year. Suppose you are working on a similar problem and you suspect a linkage between a disease observed in later life with month of birth. You have records of 400 cases of the disease and you classify them according to month of birth. The data appear in the table. Do the data present sufficient evidence to indicate that the proportion of cases of the disease per month varies from month to month? Test with $\alpha = .10$.

Month	Jan	Feb	Mar	Apr	May	June	July	Aug	Sept	Oct	Nov	Dec
No. of Births	38	31	42	46	28	31	24	29	33	36	27	35

13.7. Officials in a particular community are seeking a federal program that they hope will boost local income levels. As justification, the city claims that its local income distribution differs substantially from the national distribution and that incomes tend to be lower than expected. A random sample of 2000 family incomes was classified and compared with the corresponding national percentages. They are shown in the table. Do the data provide sufficient evidence to indicate that the distribution of family incomes within the city differs from the national distribution? Test with $\alpha = .05$.

Income	National Percentages	City Salary Class Frequency
more than $50,000	2	27
$25,000 to $50,000	16	193
$20,000 to $25,000	13	234
$15,000 to $20,000	19	322
$10,000 to $15,000	20	568
$5,000 to $10,000	19	482
below $5,000	11	174
Total	100	2000

13.4
Contingency
Tables

A problem frequently encountered in the analysis of count data concerns the independence of two methods of classification of observed events. For example, we might wish to classify defects found on furniture produced in a manufacturing plant, first, according to the type of defect and, second, according to the production shift. Ostensibly, we wish to investigate a contingency—a dependence between the two classifications. Do the proportions of various types of defects vary from shift to shift?

A total of $n = 309$ furniture defects was recorded and the defects were classified according to one of four types: A, B, C, or D. At the same time, each piece of furniture was identified according to the production shift in which it was manufactured. These counts are presented in Table 13.2, which is known as a **contingency table**. (*Note:* Numbers in parentheses are the expected cell frequencies.)

Table 13.2 Contingency table

| Shift | Type of Defect | | | | Total |
	A	B	C	D	
1	15 (22.51)	21 (20.99)	45 (38.94)	13 (11.56)	94
2	26 (22.99)	31 (21.44)	34 (39.77)	5 (11.81)	96
3	33 (28.50)	17 (26.57)	49 (49.29)	20 (14.63)	119
Total	74	69	128	38	309

Let p_A equal the unconditional probability that a defect will be of type A. Similarly, define p_B, p_C, and p_D as the probabilities of observing the three other types of defects. Then these probabilities, which we will call the column probabilities of Table 13.2, will satisfy the requirement

$$p_A + p_B + p_C + p_D = 1.$$

In like manner, let p_i ($i = 1, 2,$ or 3) equal the row probability that a defect will have occurred on shift i, where

$$p_1 + p_2 + p_3 = 1.$$

Then, if the two classifications are independent of each other, a cell probability will equal the product of its respective row and column probabilities in accordance with the multiplicative law of probability. For example, the probability that a particular defect will occur on shift 1 and be of type A is $(p_1)(p_A)$. Thus, we observe that the numerical values of the cell probabilities are unspecified in the problem under consideration. The null hypothesis specifies only that each cell probability will equal the product of its respective row and column probabilities and therefore imply independence of the two classifications.

The analysis of the data obtained from a contingency table differs from the problem discussed in Section 13.3 because we must *estimate* the row and column probabilities to estimate the expected cell frequencies.

If proper estimates of the cell probabilities are obtained, the estimated expected cell frequencies may be substituted for the $E(n_i)$ in X^2, and X^2 will continue to possess a distribution in repeated sampling that is approximated by the chi-square probability distribution. The proof of this statement as well as a discussion of the methods for obtaining the estimates are beyond the scope of this text. Fortunately, the procedures for obtaining the estimates, known as the **method of maximum likelihood and the method of minimum chi-square,** yield estimates that are intuitively obvious for our relatively simple applications.

It can be shown that the maximum likelihood estimator of a column probability will equal the column total divided by $n = 309$. If we denote the total of column j as c_j, then

$$\hat{p}_A = \frac{c_1}{n} = \frac{74}{309},$$

$$\hat{p}_B = \frac{c_2}{n} = \frac{69}{309},$$

$$\hat{p}_C = \frac{c_3}{n} = \frac{128}{309},$$

$$\hat{p}_D = \frac{c_4}{n} = \frac{38}{309}.$$

Likewise, the row probabilities p_1, p_2, and p_3 may be estimated using the row totals r_1, r_2, and r_3:

$$\hat{p}_1 = \frac{r_1}{n} = \frac{94}{309},$$

$$\hat{p}_2 = \frac{r_2}{n} = \frac{96}{309},$$

$$\hat{p}_3 = \frac{r_3}{n} = \frac{119}{309}.$$

Denote the observed frequency of the cell in row i and column j of the contingency table as n_{ij}. Then the estimated expected value of n_{11} will be

$$\hat{E}(n_{11}) = n[\hat{p}_1\hat{p}_A] = n\left(\frac{r_1}{n}\right)\left(\frac{c_1}{n}\right)$$

$$= \frac{r_1 c_1}{n},$$

where $(\hat{p}_1 \hat{p}_A)$ is the estimated cell probability. Likewise, we may find the estimated expected value for any other cell, say $\hat{E}(n_{23})$,

$$\hat{E}(n_{23}) = n[\hat{p}_2 \hat{p}_3] = n\left(\frac{r_2}{n}\right)\left(\frac{c_3}{n}\right)$$

$$= \frac{r_2 c_3}{n}.$$

In other words, we observe that the estimated expected value of the observed cell frequency, n_{ij}, for a contingency table is equal to the product of its respective row and column totals divided by the total frequency; that is,

---

Estimated Expected Cell Frequency

$$\hat{E}(n_{ij}) = \frac{r_i c_j}{n},$$

where

$$r_i = \text{total for row } i$$

$$c_j = \text{total for column } j$$

The estimated expected cell frequencies for our example are shown in parentheses in Table 13.2.

We may now use the expected and observed cell frequencies shown in Table 13.2 to calculate the value of the test statistic:

$$X^2 = \sum_{j=1}^{4} \sum_{i=1}^{3} \frac{[n_{ij} - \hat{E}(n_{ij})]^2}{\hat{E}(n_{ij})}$$

$$= \frac{(15 - 22.51)^2}{22.51} + \frac{(26 - 22.99)^2}{22.99} + \cdots + \frac{(20 - 14.63)^2}{14.63}$$

$$= 19.18.$$

The only remaining obstacle involves the determination of the appropriate number of degrees of freedom associated with the test statistic. We will give this as a rule that we will attempt to justify. **The degrees of freedom associated with a contingency table possessing r rows and c columns will always equal $(r - 1)(c - 1)$.** Thus, for our example, we will compare X^2 with the critical value of χ^2 with $(r - 1)(c - 1) = (3 - 1)(4 - 1) = 6$ degrees of freedom.

You will recall that the number of degrees of freedom associated with the X^2 statistic will equal the number of cells (in this case, $k = rc$) less one degree of freedom for each independent linear restriction placed upon the observed cell frequencies. The total number of cells for the data of Table 13.2 is $k = 12$. From this we subtract one degree of freedom because the sum of the observed cell frequencies must equal n; that is,

$$n_{11} + n_{12} + \cdots + n_{34} = 309.$$

In addition, we used the cell frequencies to estimate three of the four column probabilities. Note that the estimate of the fourth column probability will be determined once we have estimated p_A, p_B, and p_C because

$$p_A + p_B + p_C + p_D = 1.$$

Thus, we lose $(c - 1) = 3$ degrees of freedom for estimating the column probabilities.

Finally, we used the cell frequencies to estimate $(r - 1) = 2$ row probabilities and, therefore, we lose $(r - 1) = 2$ additional degrees of freedom. The total number of degrees of freedom remaining will be

$$\text{d.f.} = 12 - 1 - 3 - 2 = 6.$$

And, in general, we see that the total number of degrees of freedom associated with an $r \times c$ contingency table will be

$$\text{d.f.} = rc - 1 - (c - 1) - (r - 1) = rc - c - r + 1$$

or

$$\text{d.f.} = (r - 1)(c - 1).$$

Therefore, if we use $\alpha = .05$, we will reject the null hypothesis that the two classifications are independent if $X^2 > 12.5916$. Since the value of the test statistic, $X^2 = 19.18$, exceeds the critical value of χ^2, we will reject the null hypothesis. The data present sufficient evidence to indicate that the proportion of the various types of defects varies from shift to shift. A study of the production operations for the three shifts would likely reveal the cause.

Example 13.1   A survey was conducted to evaluate the effectiveness of a new flu vaccine that had been administered in a small community. The vaccine was provided free of charge in a two-shot sequence over a period of two weeks to those wishing to avail themselves of it. Some people received the two-shot sequence, some appeared only for the first shot, and others received neither.

A survey of 1000 local inhabitants the following spring provided the information shown in Table 13.3. Do the data present sufficient evidence to indicate that the vaccine was successful in reducing the number of flu cases in the community?

Table 13.3 Data tabu-
lation for Example 13.1

	No Vaccine	One Shot	Two Shots	Total
Flu	24 (14.4)	9 (5.0)	13 (26.6)	46
No flu	289 (298.6)	100 (104.0)	565 (551.4)	954
Total	313	109	578	1000

Solution

The question stated above asks whether the data provide sufficient evidence to indicate a dependence between the vaccine classification and the occurrence or nonoccurrence of flu. We therefore analyze the data as a contingency table.

The estimated expected cell frequencies may be calculated using the appropriate row and column totals,

$$\hat{E}(n_{ij}) = \frac{r_i c_j}{n}.$$

Thus,

$$\hat{E}(n_{11}) = \frac{r_1 c_1}{n} = \frac{46(313)}{1000} = 14.4,$$

$$\hat{E}(n_{12}) = \frac{r_1 c_2}{n} = \frac{46(109)}{1000} = 5.0,$$

$$\vdots$$

$$\hat{E}(n_{23}) = \frac{r_2 c_3}{n} = \frac{954(578)}{1000} = 551.4.$$

These values are shown in parentheses in Table 13.3.

The value of the test statistic, X^2, will now be computed and compared with the critical value of χ^2 possessing $(r - 1)(c - 1) = (1)(2) = 2$ degrees of freedom. Then, for $\alpha = .05$, we will reject the null hypothesis when $X^2 > 5.99147$. Substituting into the formula for X^2, we obtain

$$X^2 = \frac{(24 - 14.4)^2}{14.4} + \frac{(289 - 298.6)^2}{298.6} + \cdots + \frac{(565 - 551.4)^2}{551.4}$$

$$= 17.35.$$

Observing that X^2 falls in the rejection region, we reject the null hypothesis of independence of the two classifications. A comparison of the percentage incidence of flu for each of the three categories would suggest that those receiving the two-shot sequence were less susceptible to the disease. Further analysis of the data could be obtained by deleting one of the three categories, the second column, for example, to compare the effect of the vaccine with that of no vaccine. This could be done by using a 2×2 contingency table or treating the two categories as two binomial populations and using the methods of Section 9.7. Or, we might wish to analyze the data by comparing the results of the two-shot vaccine sequence with those of the combined no vaccine–one shot group. That is, we would combine the first two columns of the 2 × 3 table into one.

Exercises

13.8.   Suppose that a consumer survey summarizes the responses of $n = 307$ people in a contingency table that contains three rows and five columns. How many degrees of freedom will be associated with the chi-square test statistic?

13.9.   A survey of 400 respondents produced the following cell counts in a 2 × 3 contingency table.

	Columns			
Rows	**1**	**2**	**3**	**Total**
1	37	34	93	164
2	66	57	113	236
Total	103	91	206	400

a. If you wish to test the null hypothesis of "independence"—that the probability that a response falls in any one row is independent of the column it will fall in—and you plan to use a chi-square test, how many degrees of freedom will be associated with the χ^2 statistic?

b. Find the value of the test statistic.

c. Find the rejection region for $\alpha = .10$.

d. Conduct the test and state your conclusions.

e. Find the approximate observed significance level for the test and interpret its value.

13.10.   Wilson and Mandelbrote (1978)* report on factors related to conviction for a criminal offense after persons have received treatment for drug abuse. One part of their study of 60 graduates of a drug rehabilitation program categorizes them according to whether or not they were convicted after treatment and according to years of education. The data are shown below.

Education	Convicted	Not Convicted	Total
16 years or more	6	18	24
15 years or less	16	20	36
Total	22	38	60

* Wilson, Stephen and Bertram Mandelbrote, "Drug Rehabilitation and Criminality," *British Journal of Criminality*, 18, no. 4 (1978).

a. The data appear to suggest that persons with less education are more prone to conviction. Do the data support a theory that the probability of conviction after treatment is dependent upon a person's educational level? Use the chi-square test with $\alpha = .05$.

b. Conduct the same test using the large-sample z test, Section 9.7. Compare your results with part (a). Show that $\chi^2 = z^2$.

13.11.　An experiment was conducted to investigate the effect of general hospital experience on the attitudes of physicians toward lower-class people. A random sample of 50 physicians who had just completed 4 weeks of service in a general hospital were categorized according to their concern for lower-class people before and after their general hospital experience. The data are shown below. Do the data provide sufficient evidence to indicate a change in "concern" due to the general hospital experience?

Concern Before Experience in General Hospital	After Experience in a General Hospital		
	High	**Low**	**Total**
Low	27	5	32
High	9	9	18

13.12.　A survey of 477 hospitals by the American College of Surgeons' Commission on Cancer produced the data in the table, which classifies women with liver tumors into 6 classes.* The women are classified according to whether they used oral contraceptives and according to type of liver tumor. Do the data provide sufficient evidence to indicate a dependence between type of tumor and whether or not they used oral contraceptives?

	Type of Tumor		
	Benign	**Malignant**	**Total**
Contraceptive users	138	49	187
Nonusers	39	41	80
Use not known	35	76	111
Total	212	166	378

13.13.　Reporting on a Gallup Poll, the *San Francisco Chronicle* (August 31, 1981) notes that cigarette smoking is on the decline in the United States, "apparently because growing numbers of Americans have become convinced there is a causal relationship between smoking and such diseases as lung cancer, throat cancer, heart disease, and birth defects." Do smokers see less harm in smoking than do nonsmokers? In response to the question,

* J. Vana, G. Murphy, B. Aronoff, and H. Baker, "Primary Liver Tumors and Oral Contraceptives," *J. Amer. Med. Assoc., 238,* no. 20 (1977).

"Is smoking harmful to your health?" the percentage of smokers responding "no" (17 percent) appears to be much higher than the corresponding percentage of nonsmokers (2 percent). The percentages are shown below.

	Yes	No	No Opinion
Smokers	80	17	3
Nonsmokers	96	2	2

The survey involved 1535 adults, approximately 35 percent of whom were smokers. Therefore, the approximate numbers in the response categories were:

	Yes	No	No Opinion	Total
Smokers	430	91	16	537
Nonsmokers	958	20	20	998
Total	1388	111	36	1535

a. Do the cell counts shown above indicate that the response to the question, "Is smoking harmful to your health?" depends upon whether the respondent is a smoker or a nonsmoker? Test using $\alpha = .05$.

b. Find the observed significance level for the test, part (a), and interpret its value.

13.14. The following data represent a small portion of a telephone survey conducted by Jerry R. Lynn* to determine the characteristics of the newspaper readership in small towns, rural areas, and farms in Tennessee, as well as the impact of advertising on that market. A sample of 1486 persons was selected from among the small towns, rural areas, and farms of Tennessee. One survey question asked whether the respondent did or did not read a newspaper. The numbers of respondents in the six community-readership categories are shown below:

Community	Readers	Nonreaders	Total
Urban	529	121	650
Rural	373	137	510
Farm	237	89	326
Total	1139	347	1486

* Lynn, Jerry R., "Newspaper Ad Impact in Nonmetropolitan Markets," *Journal of Advertising Research*, 21, no. 4 (1981).

a. Do the data provide sufficient evidence to indicate that the proportions of readers differ among the three community groups? Test using $\alpha = .05$.

b. Find the approximate p value for the test and interpret its value.

13.15. In a study of changing evaluations of flood plain hazards, Payne and Pigram* describe the attitudes of people at risk of flood hazards in the Hunter River Valley, Australia. In part of the study, each respondent was asked to classify the damage (major or minor) that the respondent expected to incur if a flood occurred. The respondent was also asked whether his/her preparation for a flood would result in a low, moderate, or high cost. The numbers of responses in the six cost–damage categories are shown below:

	Damage		
Cost	**Major**	**Minor**	**Total**
Low	43	16	59
Moderate	10	28	38
High	4	9	13
Total	57	53	110

a. Do the proportions of people taking the three levels of preparation prior to a flood differ depending upon whether the respondents perceive the prospective flood damage as major or minor? Test using $\alpha = .05$.

b. Find the observed significance level for the test and interpret its value.

13.5
r × *c* Tables with Fixed Row or Column Totals

In the previous section we have described the analysis of an $r \times c$ contingency table using examples that, for all practical purposes, fit the multinomial experiment described in Section 13.1. Although the methods of collecting data in many surveys may adhere to the requirements of a multinomial experiment, other methods do not. For example, we might not wish to sample randomly the population described in Example 13.1 because we might find that, owing to chance, one category is completely missing. People who have received no flu shots might fail to appear in the sample. Thus we might decide beforehand to interview a specified number of people in each column category, thereby fixing the column totals in advance. Although these restrictions tend to disturb somewhat our visualization of the experiment in the multinomial context, they have no effect on the analysis of the data. As long as we wish to test the hypothesis of independence of two classifications and none of the row or column prob-

* Payne, Robert J. and John J. Pigram, "Changing Evaluations of Flood Plain Hazard: The Hunter Valley, Australia," *Environment and Behavior*, 13, no. 4 (1981), pp. 461–480, © 1981 Sage Publications, Inc., with permission.

abilities are specified in advance, we may analyze the data as an $r \times c$ contingency table. It can be shown that the resulting X^2 will possess a probability distribution in repeated sampling that is approximated by a chi-square distribution with $(r - 1)(c - 1)$ degrees of freedom.

To illustrate, suppose that we wish to test an hypothesis concerning the equivalence of four binomial populations as indicated in the following example.

Example 13.2

A survey of voter sentiment was conducted in four midcity political wards to compare the fraction of voters favoring candidate A. Random samples of 200 voters were polled in each of the four wards with results as shown in Table 13.4. Do the data present sufficient evidence to indicate that the fractions of voters favoring candidate A differ in the four wards?

Table 13.4 Data tabulation for Example 13.2

	Ward				
	1	2	3	4	Total
Favor A	76 (59)	53 (59)	59 (59)	48 (59)	236
Do not favor A	124 (141)	147 (141)	141 (141)	152 (141)	564
Total	200	200	200	200	800

Solution

You will observe that the test of an hypothesis concerning the equivalence of the parameters of the four binomial populations corresponding to the four wards is identical to an hypothesis implying independence of the row and column classifications. Thus, if we denote the fraction of voters favoring A as p and hypothesize that p is the same for all four wards, we imply that the first- and second-row probabilities are equal to p and $(1 - p)$, respectively. The probability that a member of the sample of $n = 800$ voters falls in a particular ward will equal one-fourth, since this was fixed in advance. Then the cell probabilities for the table would be obtained by multiplication of the appropriate row and column probabilities under the null hypothesis and be equivalent to a test of independence of the two classifications.

The estimated expected cell frequencies, calculated using the row and column totals, appear in parentheses in Table 13.4. We see that

$$X^2 = \sum_{j=1}^{4} \sum_{i=1}^{2} \frac{[n_{ij} - \hat{E}(n_{ij})]^2}{\hat{E}(n_{ij})}$$

$$= \frac{(76 - 59)^2}{59} + \frac{(124 - 141)^2}{141} + \cdots + \frac{(152 - 141)^2}{141}$$

$$= 10.72.$$

The critical value of χ^2 for $\alpha = .05$ and $(r - 1)(c - 1) = (1)(3) = 3$ degrees of freedom is 7.81473. Since X^2 exceeds this critical value, we reject the null hypothesis and conclude that the fraction of voters favoring candidate A is not the same for all four wards.

Exercises

13.16. Suppose that you wish to test the null hypothesis that three binomial parameters, p_A, p_B, and p_C, are equal, against the alternative that at least two of the parameters differ. Independent random samples of 100 observations were selected from each of the populations. The data are shown below:

	Population			
	A	B	C	Total
Number of Successes	24	19	33	76
Number of Failures	76	81	67	224
Total	100	100	100	300

a. Find the value of X^2, the test statistic.

b. Give the rejection region for the test for $\alpha = .05$.

c. State your test conclusions.

d. Find the observed significance level for the test and interpret its value.

13.17. A study to determine the effectiveness of a drug (serum) for arthritis resulted in the comparison of two groups each consisting of 200 arthritic patients. One group was inoculated with the serum, the other received a placebo (an inoculation that appears to contain serum but actually is nonactive). After a period of time, each person in the study was asked to state whether his arthritic condition was improved. The following results were observed:

	Treated	Untreated
Improved	117	74
Not improved	83	126

Do these data present sufficient evidence to indicate that the serum was effective in improving the condition of arthritic patients?

a. Test by means of the X^2 statistic. Use $\alpha = .05$.

b. Test by use of the z test in Section 9.6.

13.18. A particular poultry disease is thought to be noncommunicable. To test this theory, 30,000 chickens were randomly partitioned into three groups of 10,000. One group had no contact with diseased chickens, one had moderate contact, and the third had heavy contact. After a six-month period, the following data were collected on the number of diseased chickens in each group of 10,000. Do the data provide sufficient evidence to indicate a dependence between the amount of contact between diseased and nondiseased fowl and the incidence of the disease?

	No Contact	Moderate Contact	Heavy Contact
Number of diseased chickens	87	89	124
Number of nondiseased chickens	9,913	9,911	9,876
Total	10,000	10,000	10,000

13.19. A study of the purchase decisions for three stock portfolio managers, A, B, and C, was conducted to compare the rates of stock purchases that resulted in profits over a time period that was less than or equal to one year. One hundred randomly selected purchases obtained for each of the managers gave the following results:

	Manager		
	A	B	C
Purchases that resulted in a profit	63	71	55
Purchases that resulted in no profit	37	29	45
Total	100	100	100

Do the data provide evidence of differences among the rates of successful purchases for the three managers?

13.20. In a 1976 article in *American Scientist* H.W. Menard writes about manganese nodules, a mineral-rich concoction found abundantly on the deep-sea floor.* In one portion of his report, Menard provides data relating the magnetic age of the earth's crust to "the probability of finding manganese nodules." The data shown in the table give the number of samples of the earth's core and the percentage of those that contain manganese nodules

* Menard, H.W., "Time, Chance and the Origin of Manganese Nodules," *Amer. Scientist*, September–October 1976.

for each of a set of magnetic-crust ages. Do the data provide sufficient evidence to indicate that the probability of finding manganese nodules in the deep-sea earth's crust is dependent upon the magnetic-age classification? Test with $\alpha = .05$.

Age	Number of Samples	Percentage with Nodules
Miocene–recent	389	5.9
Oligocene	140	17.9
Eocene	214	16.4
Paleocene	84	21.4
Late Cretaceous	247	21.1
Early and Middle Cretaceous	1,120	14.2
Jurassic	99	11.0

13.21. According to the *Wall Street Journal* (October 23, 1981), "most bosses shun symbols of status, help take care of household tasks." The *Wall Street Journal*–Gallup survey finds that 49 percent of the 307 heads of the nation's largest firms, who were included in the survey, sometimes do grocery shopping, in comparison to 47 percent of the executives of 309 medium-sized firms and 45 percent of the executives of 208 small companies. We want to know whether the percentages suggest a difference in the proportions of the chief executives of small, medium, and large firms who sometimes do grocery shopping.

a. Do the data provide sufficient evidence to indicate that the proportions of chief executives who sometimes do grocery shopping differ for the three categories of company size? Test using $\alpha = .05$. (*Hint:* Construct a 2×3 contingency table showing the numbers of executives in each of the six cell categories.)

b. Suppose that, prior to observing the data, you had planned to compare the proportions of executives of large corporations and small corporations. Do the data provide sufficient evidence to indicate a difference in proportions for these two groups? Test using $\alpha = .05$.

13.6
A Computer Analysis for an $r \times c$ Contingency Table

Most statistical program packages include a program for analyzing data contained in an $r \times c$ contingency table. To illustrate, the SAS, SPSS, and Minitab computer printouts for the analysis of the data, Example 13.1, are shown in Table 13.5. The computed value of X^2 and its observed significance level are boxed on each of the three computer printouts.

At the top of the SAS printout, Table 13.5(a), is the chi-square table showing the observed and expected values in their respective cells. The computed value of the test statistic, $X^2 = 17.313$, is shown directly below the table. The number of degrees of freedom for X^2, DF = 2, and the observed significance level, PROB = 0.0002, are shown to the right of X^2. The last four lines of the printout, PHI, CONTINGENCY COEFFICIENT, CRAMER'S V, and LIKELIHOOD RATIO CHISQUARE, are not pertinent to our analysis.

Table 13.5 SAS, SPSS, and Minitab computer printouts for the chi-square analysis of the data, Example 13.1

(a) SAS

TABLE OF FLU BY SHOTS

FLU		SHOTS			
FREQUENCY EXPECTED		NONE	ONE	TWO	TOTAL
NO		289 298.6	100 104.0	565 551.4	954
YES		24 14.4	9 5.0	13 26.6	46
TOTAL		313	109	578	1000

STATISTICS FOR 2-WAY TABLES

CHI-SQUARE	17.313	DF =	2	PROB = 0.0002
PHI	0.132			
CONTINGENCY COEFFICIENT	0.130			
CRAMER'S V	0.132			
LIKELIHOOD RATIO CHISQUARE	17.252	DF =	2	PROB = 0.0002

(b) SPSS

*   CROSSTABULATION OF   *
FLU BY SHOTS

* PAGE 1 OF 1

	SHOTS			
COUNT ROW PCT COL PCT TOT PCT	NONE	ONE	TWO	ROW TOTAL
FLU				
NO	289 30.3 92.3 28.9	100 10.5 91.7 10.0	565 59.2 97.8 56.5	954 95.4
YES	24 52.2 7.7 2.4	9 19.6 8.3 0.9	13 28.3 2.2 1.3	46 4.6
COLUMN TOTAL	313 31.3	109 10.9	578 57.8	1000 100.0

Table 13.5 *(continued)*

```
CHI SQUARE =      17.31296 WITH  2 DEGREES OF FREEDOM   SIGNIFICANCE = 0.0002
CRAMER'S V =     0.13158
CONTINGENCY COEFFICIENT =      0.13045
LAMBDA (ASYMMETRIC) =   0.0 WITH FLU DEPENDENT.   =   0.02607 WITH SHOTS DEPENDENT.
LAMBDA (SYMMETRIC) =    0.02350
UNCERTAINTY COEFFICIENT (ASYMMETRIC) = 0.04624 WITH FLU DEPENDENT.   =   0.00936 WITH SHOTS DEPENDENT.
UNCERTAINTY COEFFICIENT (SYMMETRIC) = 0.01556
KENDALL'S TAU B =    −0.12119  SIGNIFICANCE =   0.0000
KENDALL'S TAU C =    −0.05355  SIGNIFICANCE =   0.0000
GAMMA =    −0.46640
SOMERS'S D (ASYMMETRIC) = −0.04815 WITH FLU DEPENDENT.   = −0.30505 WITH SHOTS DEPENDENT.
SOMERS'S D (SYMMETRIC) = −0.08317
```

(c) Minitab

EXPECTED FREQUENCIES ARE PRINTED BELOW OBSERVED FREQUENCIES

	ZERO	ONE	TWO	TOTALS
1	24	9	13	46
	14.4	5.0	26.6	
2	289	100	565	954
	298.6	104.0	551.4	
TOTALS	313	109	578	1000

TOTAL CHI SQUARE =

$$6.40 + 3.17 + 6.94 +$$
$$0.31 + 0.15 + 0.33 +$$

$$= 17.31$$

DEGREES OF FREEDOM $= (2-1) \times (3-1) = 2$

Four numbers are shown in each cell of the 2×3 table for the SPSS printout, Table 13.5(b). These are, top to bottom, the cell frequency, and the percentages of the row total, the column total, and the grand total (n) that this cell frequency represents. For example, the cell in the first row, first column, had a cell frequency of 289. This frequency represented 30.3 percent of the row total, 954, 92.3 percent of the column total, 313, and 28.9 percent of the total, $n = 1000$ persons included in the experiment. The value, $X^2 = 17.31296$, its degrees of freedom, d.f. $= 2$, and observed significance level, $p = .0002$, are shown directly below the table. The quantities printed beneath the value of X^2 are irrelevant to our discussion.

The Minitab printout, Figure 13.5(c), shows the same 2×3 contingency table except that row 1 corresponds to "Yes" and row 2 to "No." The numbers shown in the cells are the observed and expected cell frequencies, respectively. The computed value, $X^2 = 17.31$, is shown below the table along with its degrees of freedom, d.f. $= 2$.

The information shown on the printouts differs slightly from one printout to another, as do some of the computed numbers (due to rounding), but the basic information is the same. All three show the calculated value of X^2 and its degrees of freedom. These two quantities, along with the table of the critical values of chi-square, Table 5, the Appendix, are all that we need to test in order to detect dependence between the two qualitative variables represented in the contingency table. The value of the observed significance level, given in the SAS and the SPSS printouts, eliminates the need for the chi-square table and gives us a measure of the weight of evidence favoring rejection of the null hypothesis.

13.7
Wine Preferences: Canned or Bottled?

Now that we understand the basic concepts involved in a chi-square test about the probabilities associated with a multinomial experiment, let us reexamine the Reynolds Metals Company taste test data for comparing consumer preferences for canned versus bottled wine. We will compare the observed with expected cell counts but, as you will see, this application does not fit the stereotypes of the preceding sections.

For this particular experiment, each taste tester's response fell in one of $k = 3$ categories, as shown in Figure 13.1. The theoretical cell frequencies, p_1, p_2, and p_3 ($p_1 + p_2 + p_3 = 1$), are shown at the tops of the category boxes, and the observed frequencies, corresponding to the preference percentages, are shown at the bottom.

Figure 13.1 The $k = 3$ categories for the wine taste test

p_1	p_2	p_3
prefer canned wine	prefer bottled wine	no preference
.50	.39	.11

We will want to test the null hypothesis, $H_0:p_1 = p_2$, namely, that the proportion of consumers preferring the canned wine is equal to the proportion preferring the bottled wine. The alternative hypothesis is that one of the two types of wines, canned or bottled, is preferred; that is, $H_a:p_1 \neq p_2$. Do the data provide sufficient evidence to indicate that $p_1 \neq p_2$?

To answer this question, we need to know the numbers of responses, n_1, n_2, and n_3, that fell in the three categories of Figure 13.1. In his article

Kristof reports the observed cell relative frequencies, $n_1/n = .50$, $n_2/n = .39$, and $n_3/n = .11$, but fails to give us the total number n of taste testers. Consequently, we can only conduct the chi-square test if we supply a value for n. For example, if $n = 100$ taste testers were involved in the experiment, then $n_1 = (.50)n = (.50)(100) = 50$. Similarly, $n_2 = (.39)n = 39$ and $n_3 = (.11)n = 11$.

The next step is to estimate the values of p_1, p_2, and p_3 and calculate the estimated expected cell frequencies. It can be shown that if $p_1 = p_2$ (proof omitted) then the approximate estimates* are,

$$\hat{p}_1 = \hat{p}_2 = \frac{1}{2}\left(\frac{n_1 + n_2}{n}\right) \quad \text{and} \quad \hat{p}_3 = 1 - \hat{p}_1 - \hat{p}_2$$

or,

$$\hat{p}_1 = \hat{p}_2 = \frac{1}{2}\left(\frac{50 + 39}{100}\right) = .445 \quad \text{and} \quad \hat{p}_3 = 1 - \hat{p}_1 - \hat{p}_2 = .11.$$

The resulting expected cell frequencies are:

$$\hat{E}(n_1) = \hat{E}(n_2) = n\hat{p}_1 = (100)(.445) = 44.5$$

and

$$\hat{E}(n_3) = n\hat{p}_3 = (100)(.11) = 11.$$

These values, along with the observed cell frequencies, are shown in Table 13.6.

Table 13.6 The observed and estimated expected cell frequencies for the wine tasting experiment

	Prefer Canned	Prefer Bottled	No Preference
Observed	50	39	11
Estimated Expected	44.5	44.5	11

We have estimated one parameter p_1 (and, consequently, p_2, since we are hypothesizing that $p_1 = p_2$) and $n_1 + n_2 + n_3 = n$. Therefore, we are placing two linear restrictions on the cell frequencies, and the chi-square test statistic

* These are known in statistics as maximum likelihood estimates. They represent the values of p_1, p_2, and p_3 that give the highest probability of observing this particular set of sample data. Maximum likelihood estimates often agree with our intuition. Thus, if we assume that $p_1 = p_2$, it seems reasonable to find the estimate of p_1 by letting it equal 1/2 of the estimated probability for the *combined* two cells; that is, 1/2 of $(n_1 + n_2)/n$.

will be based on

$$\begin{aligned} \text{d.f.} &= k - (\text{number of linearly independent restrictions on } n_1, n_2, \text{ and } n_3) \\ &= 3 - 2 = 1. \end{aligned}$$

For $\alpha = .05$, we will reject the null hypothesis, $H_0 : p_1 = p_2$, if the computed value of X^2 exceeds the tabulated value of $\chi^2_{.05}$, Table 5, the Appendix; namely, $\chi^2_{.05} = 3.84146$.

Using the observed and estimated expected values of the cell frequencies, the computed value of chi-square is,

$$\begin{aligned} X^2 &= \frac{\sum\limits_{i=1}^{k} [n_i - \hat{E}(n_i)]^2}{\hat{E}(n_i)} = \frac{\sum\limits_{i=1}^{3} (n_i - n\hat{p}_i)^2}{n\hat{p}_i} \\ &= \frac{(50 - 44.5)^2}{44.5} + \frac{(39 - 44.5)^2}{44.5} + \frac{(11 - 11)^2}{11} \\ &= 1.58. \end{aligned}$$

Since this computed value is less than the critical value, $\chi^2_{.05} = 3.84146$, there is insufficient evidence to indicate a difference in the proportions, p_1 and p_2, that favor the canned and bottled wines.

The conclusion "insufficient evidence to show a difference between wine preferences" may disagree with your intuition, because the observed percentage of taste testers favoring canned wine, 50 percent, seems to be much larger (and different) from the percentage, 39 percent, favoring bottled wine. Failure to detect a difference in preference for the two types of wine may be because there really is no difference in preference, or it may be due to the fact that the sample size,* $n = 100$ taste testers, is not large enough to detect a difference.

13.8
Other Applications

The applications of the chi-square test in analyzing enumerative data described in Sections 13.3, 13.4, 13.5 and 13.7 represent only a few of the interesting classificatory problems that may be approximated by one or more multinomial experiments and for which our method of analysis is appropriate. By and large, these applications are complicated to a greater or lesser degree because the numerical values of the cell probabilities are unspecified and hence require the estimation of one or more population parameters. Then, as in Sections 13.4,

---

* Recall that we did not know how many taste testers were involved in the Reynolds Metals Company taste test experiment. For purposes of explanation, we assumed that $n = 100$.

13.5, and 13.7, we can estimate the cell probabilities. Although we omit the mechanics of the statistical tests, several additional applications of the chi-square test are worth mentioning as a matter of interest.

For example, suppose that we wish to test an hypothesis stating that a population possesses a normal probability distribution. The cells of a sample frequency histogram (for example, Figure 3.2) would correspond to the k cells of the multinomial experiment, and the observed cell frequencies would be the number of measurements falling in each cell of the histogram. Given the hypothesized normal probability distribution for the population, we could use the areas under the normal curve to calculate the theoretical cell probabilities and hence the expected cell frequencies. The difficulty arises when μ and σ are unspecified for the normal population and these parameters must be estimated to obtain the estimated cell probabilities. This difficulty can, of course, be surmounted.

The construction of a two-way table to investigate dependency between two classifications can be extended to three or more classifications. For example, if we wish to test the mutual independence of three classifications, we would employ a three-dimensional "table" or rectangular parallelepiped. The reasoning and methodology associated with the analysis of both the two- and three-way tables are identical, although the analysis of the three-way table is a bit more complex.

A third and interesting application of our methodology would be its use in the investigation of the rate of change of a multinomial (or binomial) population as a function of time. For example, we might study the decision-making ability of a human (or any animal) as he or she is subjected to an educational program and tested over time. If, for instance, he or she is tested at prescribed intervals of time and the test is of the yes or no type yielding a number of correct answers, y, that would follow a binomial probability distribution, we would be interested in the behavior of the probability of a correct response, p, as a function of time. If the number of correct responses was recorded for c time periods, the data would fall in a $2 \times c$ table similar to that in Example 13.2 (Section 13.5). We would then be interested in testing the hypothesis that p is equal to a constant (that is, that no learning has occurred) and we would then proceed to more interesting hypotheses to determine whether the data present sufficient evidence to indicate a gradual (say, linear) change over time as opposed to an abrupt change at some point in time. The procedures we have described could be extended to decisions involving more than two alternatives.

You will observe that our learning example is common to business, to industry, and to many other fields, including the social sciences. For example, we might wish to study the rate of consumer acceptance of a new product for various types of advertising campaigns as a function of the length of time that the campaign has been in effect. Or, we might wish to study the trend in the lot fraction defective in a manufacturing process as a function of time. Both of these examples, as well as many others, require a study of the behavior of a binomial (or multinomial) process as a function of time.

The examples that we have just described are intended to suggest the relatively broad application of the chi-square analysis of enumerative data, a fact that should be borne in mind by the experimenter concerned with this type of data. The statistical test employing X^2 as a test statistic is often called a "goodness-of-fit" test. Its application for some of these examples requires care in the determination of the appropriate estimates and the number of degrees of freedom for X^2, which, for some of these problems, may be rather complex.

13.9
Assumptions

The following assumptions must be satisfied if X^2 is to possess approximately a chi-square distribution and, consequently, if the tests described in this chapter are to be valid:

Assumptions

1. The cell counts, $n_1, n_2, \ldots, n_k$, satisfy the conditions of a multinomial experiment (or a set of multinomial experiments created by restrictions on row or column totals).
2. The expected values of all cell counts should equal or exceed 5.

It is absolutely essential that assumption 1 be satisfied. The chi-square goodness-of-fit tests, of which these tests are special cases, compare observed frequencies with expected frequencies and apply only to data generated by a multinomial experiment.

The larger the sample size n, the closer the chi-square distribution will approximate the distribution of X^2. We have stated in assumption 2 that n must be large enough so that all the expected cell frequencies will be equal to 5 or more. This is a safe figure. Actually, the expected cell frequencies can be smaller for some tests than for others. For information on the minimum expected cell frequencies for specific goodness-of-fit tests, see the paper by Cochran listed in the references.

13.10
Summary

The preceding material has been concerned with a test of an hypothesis regarding the cell probabilities associated with one or more multinomial experiments. When the number of observations, n, is large, the test statistic, X^2, can be shown to possess, approximately, a chi-square probability distribution in repeated sampling, the number of degrees of freedom being dependent upon the particular application. In general we assume that n is large and that the minimum expected cell frequency is equal to or greater than five.

Several words of caution concerning the use of the X^2 statistic as a method of analyzing enumerative-type data are appropriate. The determination of the correct number of degrees of freedom associated with the X^2 statistic is very important in locating the rejection region. If the number is incorrectly specified, erroneous conclusions might result. Also, note that nonrejection of the null hypothesis does not imply that it should be accepted. We would have difficulty in stating a meaningful alternative hypothesis for many practical applications and, therefore, would lack knowledge of the probability of making a type II error. For example, we hypothesize that the two classifications of a contingency table are independent. A specific alternative would have to specify some measure of dependence, which may or may not possess practical significance to the experimenter. Finally, if parameters are missing and the expected cell frequencies must be estimated, the estimators of missing parameters should be of a particular type in order that the test be valid. In other words, the application of the chi-square test for other than the applications outlined in Sections 13.3, 13.4, 13.5, and 13.7 will require experience beyond the scope of this introductory presentation of the subject.

References

Anderson, R.L., and T.A. Bancroft, *Statistical Theory in Research*. New York: McGraw-Hill Book Company, 1952.

Chapman, D.G., and R.A. Schaufele, *Elementary Probability Models and Statistical Inference*. New York: John Wiley & Sons, Inc., 1970.

Cochran. W.G., "The χ^2 Test of Goodness of Fit." *Ann. Math. Stat.*, *23* (1952), pp. 315–345.

Dixon, W.J., and M.B. Brown, eds. *BMDP Biomedical Computer Programs, P Series*. Berkeley, Calif.: University of California, 1979.

Dixon, W.J., and F.J. Massey, Jr., *Introduction to Statistical Analysis*, 3rd ed. New York: McGraw-Hill Book Company, 1969.

Helwig, J.P., and K.A. Council, eds. *SAS User's Guide*. Cary, N.C.: SAS Institute, Inc., P.O. Box 8000, 1979.

Kendall, M.G., and A. Stuart, *The Advanced Theory of Statistics*, Vol. 2, rev. 3rd ed. New York: Hafner Press, 1974.

Nie, N., C.H. Hull, J.G. Jenkins, K. Steinbrenner, and D.H. Bent *Statistical Package for the Social Sciences*, 2nd ed. New York: McGraw-Hill Book Company, 1979.

Ryan, T.A., B.L. Joiner, and B.F. Ryan, *Minitab Student Handbook*. Boston: Duxbury Press, 1979.

Supplementary Exercises

13.22.   A manufacturer of floor polish conducted a consumer-preference experiment to see whether a new floor polish, *A*, was superior to those produced by four of his competitors. A sample of 100 housewives viewed five patches of flooring that had received the five polishes, and each indicated the patch that she considered superior in appearance. The lighting, background, etc., were approximately the same for all five patches. The results of the survey are as follows:

Polish	*A*	*B*	*C*	*D*	*E*
Frequency	27	17	15	22	19

Do these data present sufficient evidence to indicate a preference for one or more of the polished patches of floor over the others? If one were to reject the hypothesis of "no preference" for this experiment, would this imply that polish A is superior to the others? Can you suggest a better method of conducting the experiment?

13.23. A survey was conducted to investigate interest of middle-aged adults in physical-fitness programs in Rhode Island, Colorado, California, and Florida. The objective of the investigation was to determine whether adult participation in physical-fitness programs varies from one region of the United States to another. A random sample of people was interviewed in each state and the following data were recorded.

	Rhode Island	Colorado	California	Florida
Participate	46	63	108	121
Do not participate	149	178	192	179

Do the data indicate a difference in adult participation in physical-fitness programs from one state to another?

13.24. An analysis of accident data was made to determine the distribution of numbers of fatal accidents for automobiles of three sizes. The data for 346 accidents are as follows.

	Size of Auto		
	Small	Medium	Large
Fatal	67	26	16
Not fatal	128	63	46

Do the data indicate that the frequency of fatal accidents is dependent on the size of automobiles?

13.25. A group of 306 people were interviewed to determine their opinion concerning a particular current American foreign-policy issue. At the same time their political affiliation was recorded. The data are as follows:

	Approve of Policy	Do Not Approve of Policy	No Opinion
Republicans	114	53	17
Democrats	87	27	8

Do the data present sufficient evidence to indicate a dependence between party affiliation and the opinion expressed for the sampled population?

13.26. A survey was conducted to determine student, faculty, and administration attitudes on a new university parking policy. The distribution of those favoring or opposed to the policy is shown below.

	Student	Faculty	Administration
Favor	252	107	43
Opposed	139	81	40

Do the data provide sufficient evidence to indicate that attitudes regarding the parking policy are independent of student, faculty, or administration status?

13.27. The chi-square test used in Exercise 13.17 is equivalent to the two-tailed z test of Section 9.3 provided α is the same for the two tests. Show algebraically that the chi-square test statistic X^2 is the square of the test statistic z for the equivalent test.

13.28. It is often not clear whether all properties of a binomial experiment are actually met in a given application. A "goodness-of-fit" test is desirable for such cases. Suppose that an experiment consisting of four trials was repeated 100 times. The number of repetitions on which a given number of successes was obtained is recorded in the following table:

Possible Results (Number of Successes)	0	1	2	3	4
Number of Times Obtained	11	17	42	21	9

Estimate p (assuming that the experiment was binomial), obtain estimates of the expected cell frequencies, and test for goodness of fit. To determine the appropriate number of degrees of freedom for X^2, note that p was estimated by a linear combination of the frequencies obtained.

13.29. A problem that sometimes occurs during surgical operations is the occurrence of infections during blood transfusions. An experiment was conducted to determine whether the injection of antibodies reduced the probability of infection. An examination of the records of 138 patients produced the data shown in the accompanying table. Do the data provide sufficient evidence to indicate that injections of antibodies affect the likelihood of transfusional infections? Test by using $\alpha = .10$.

	Infection	No Infection
Antibody	4	78
No antibody	11	45

13.30. By tradition U.S. labor unions have been content to leave the management of the company to the managers and corporate executives. But in Europe worker participation in management decision making is an accepted idea and one that is continually spreading. To study the effect of worker satisfaction with worker participation in managerial decision making, 100 workers were interviewed in each of two separate West German manufacturing plants. One plant had active worker participation in managerial decision making; the other did not. Each selected worker was asked whether he or she generally approved of the managerial decisions made within the firm. The results of the interviews are shown in the table.

	Participative Decision Making	No Participative Decision Making
Generally approve of the firm's decisions	73	51
Do not approve of the firm's decisions	27	49

a. Do the data provide sufficient evidence to indicate that approval or disapproval of management's decisions depends on whether workers participate in decision making? Test by using the X^2 test statistic. Use $\alpha = .05$.

b. Do these data support the hypothesis that workers in a firm with participative decision making more generally approve of the firm's managerial decisions than those employed by firms without participative decision making? Test by using the z test presented in Section 9.7. This problem requires a one-tailed test. Why?

13.31. A study reported in the *New England Journal of Medicine* (December 8, 1977) suggested that aspirin can protect male surgery patients from forming postsurgical blood clots in their veins. Of 23 men who received 4 aspirin tablets per day, only 4 developed blood clots in comparison with 14 of 25 men who took placebos intended to look like aspirin but which in fact were inert. Do these data provide sufficient evidence to indicate a link between the use of aspirin and the frequency of formation of postsurgical blood clots?

13.32. An occupant-traffic study was conducted to aid in the remodeling of an office building that contains three entrances. The choice of entrance was recorded for a sample of 200 persons entering the building. Do the data shown in the table indicate that there is a difference in preference for the three entrances? Find a 90 percent confidence interval for the proportion of persons favoring entrance 1.

	Entrance		
	1	2	3
Number entering	83	61	56

13.33. The computer, which at its advent was expected to play its primary role in scientific computation, has become essential in banking, in the recording of personal data, in merchandising, and in many areas of our daily lives. Along with this growth in computer applications have come numerous cases of computer abuse, financial fraud, theft of information, etc. The data in the accompanying table give four different types of computer abuse that were reported and verified for the years 1970 to 1973.* The frequency of computer abuses would be expected to increase as the years go by unless safeguards are found to prevent their occurrence. But are the proportions of the four types of abuses changing over time? Test by using $\alpha = .10$.

Year	Financial Fraud	Theft of Information or Property	Unauthorized Use	Vandalism	Total
1970	7	5	9	8	29
1971	22	18	6	6	52
1972	12	15	6	12	45
1973	21	15	16	9	61
Total	62	53	37	35	187

13.34.   Exercise 9.53 gave the national statistics shown in the accompanying table on admission to medical colleges, classified according to major, for the 1979-1980 academic year. Do the data provide sufficient evidence to indicate that the proportion of applicants admitted to medical schools is dependent upon the major field of study? Explain.

Major	Number of Admissions	Percentage of Total of Major Applications Admitted
biology	6,139	44.2
chemistry	2,273	53.5
sociology	886	43.7
premed	658	50.3
biochemistry	859	59.2
interdisciplinary	86	62.8

13.35.   As part of a study of small-town police, Galliher and colleagues questioned police in towns of varying sizes to determine what types of people they watched most carefully because of their potential for criminal activity.[†] Particularly, they were interested in determining

* Parker, D.B., S. Nycum, and S.S. Oura, *Computer Abuse*. Menlo Park, Calif.: Stanford Research Institute, 1973.

† Galliher, J.F., L.P. Donavan, and D.L. Adams, "Small Town Police: Trouble, Tasks and Publics," *J. Police Sci. and Adm.*, 3, no. 1 (1975).

whether the "most watched" type of person depended upon the size of the town. They interviewed police in 224 small towns and they obtained the data shown in the table. Do the data provide sufficient information to indicate a dependence between the type of persons most watched and town size? Test by using $\alpha = .05$.

Type of Persons Most Watched	Town Size		
	Under 5,000	5,000–10,000	10,000–25,000
known criminals	21	26	25
young people	39	15	20
strangers, and other suspicious persons	13	5	11
bar crowds	11	2	2
others	10	11	13
Total	94	59	71

13.36. A recent complaint of blacks and minorities has resulted in at least one lawsuit. According to the complaint, culturally biased IQ tests result in many black students being assigned to EMR (educable mentally retarded) classes. Typical of data in support of this contention would be the following. A school district has an enrollment of 6000 elementary and junior high school students of which 1640 are black and 4360 are white. A total of 280 students are enrolled in EMR classes and of these 163 are black. The data are given in the table. Does it appear that student assignment to EMR classes is dependent on race? Test with $\alpha = .01$.

School Class Assignment	Race		Total
	Black	White	
EMR	163	117	280
non-EMR	1477	4243	5720
Total	1640	4360	6000

13.37. Edgar L. Jackson, in a study of human response to earthquake hazard, presents data indicating how each of 302 respondents view the prospect of future earthquake damage as a function of the amount of damage each respondent has experienced in the past.*

* Jackson, Edgar L., "Response to Earthquake Hazard: The West Coast of North America." *Environment and Behavior*, 13, no. 4 (1981), pp. 387–416, © 1981 Sage Publications, Inc., with permission.

The data are shown below:

| | Past Experience | | | |
Expectations	Major Losses	Minor Losses	No Losses	Total
Earthquake and Damage Expected	11	8	8	27
Uncertainty About Occurrence	38	69	98	205
No Earthquake Expected	26	6	38	70
Total	75	83	144	302

a. Is there a linkage between a respondent's expectations and past experiences, that is, do the proportions of respondents in the expectations categories differ depending upon the category of past experience? Test using $\alpha = .05$.

b. Find the approximate observed significance level for the test and interpret its value.

14. Considerations in Designing Experiments

Chapter Objectives

General Objective

In the preceding chapters, we studied the mechanisms and the concepts involved in making inferences about a population based on information contained in a sample. In this chapter we shall demonstrate to you how certain factors affect the quantity of information contained in a sample and how you can use this information to minimize the cost of sampling.

Specific Objectives

1. To explain how to measure the quantity of information in a sample that pertains to a specific parameter.   *Section 14.1*

2. To identify the factors that affect the quantity of information in an experiment. *Section 14.1*

3. To familiarize you with the terminology employed in the design of experiments. *Section 14.2*

4. To explain how to use random-number tables to select random samples.   *Section 14.3*

5. To explain how the selection of treatments (volume-increasing experimental designs) increases the amount of information in an experiment.   *Sections 14.4, 14.6*

6. To explain how blocking can be used to remove sources of random variation (noise-reducing experimental designs) and thereby increase the amount of information in an experiment.   *Section 14.5*

Case Study

Does Depression Cause You to Eat More?

Are depressed persons more inclined to seek relief through excessive eating? Baucom and Aiken* conducted an experiment to investigate the effect of three qualitative predictor variables on a person's tendency to eat:

1. Mood: whether or not the person is in a state of depression
2. Obesity: whether or not the person is obese
3. Dieting: whether or not the person is dieting.

In Chapter 12, we considered methods for analyzing data obtained from multivariable experiments of this type. Now we take a step backward and consider the problem that faces an experimenter at the onset of an experimental investigation of some phenomenon, namely, the *design* of the experiment.

A statistical design of an experiment is a plan for selecting sample data. We learned in Chapter 10 that a paired-difference experiment will often provide more information on the difference between a pair of population means than data obtained from independent random samples. The design of experiments, although unmentioned, also was apparent in Chapters 11 and 12, when the standard errors of the regression model parameters depended upon the selected settings of the predictor variables. Thus, without stressing the point, we have seen

that *how* the sample is selected affects the *amount* of information contained in sample data.

The Baucom–Aiken experiment is interesting because it demonstrates an aspect of experimental design that is present in some types of experimentation—namely, the problem of choosing effective measures of the response and predictor variables. For example, how do you measure "tendency to eat"; that is, what should you choose for the variable y that measures this response? How do you induce depression in a subject—that is, in a person employed in the experiment—and be fairly certain that the subject is in fact depressed? Questions of this type are psychological, rather than statistical, but they address an important aspect of experimental design, namely, making reasonably certain that the variables employed in an experiment do in fact measure the phenomena involved in a study.

The statistical design of an experiment comes *after* the experimenter has chosen the variables to be included in a study and knows how to measure them. The design concerned with choosing the combinations of values of the predictor variables for which y will be measured, deciding how many observations will be acquired for each combination, and, finally, deciding whether the observations will be made on randomly selected subjects or upon matched groups of subjects.

* Baucom, Donald H., and Pamela A. Aiken, "Effect of Depressed Mood on Eating Among Obese and Nonobese Dieting and Nondieting Persons," *Journal of Personality and Social Psychology*, 41, no. 3 (1981). Copyright 1981 by the American Psychological Association. Reprinted by permission of the author.

This chapter presents an introduction to the basic concepts of the statistical design of experiments. We will identify the factors that affect the quantity of information in an experiment and will demonstrate, with some elementary statistical designs, how you can control them. Finally, we will describe the Baucom–Aiken experiment in greater detail in Section 12.6. It will introduce you to a new and important type of experimental design and show you why it is useful.

14.1
The Factors Affecting the Information in a Sample

The information in a sample that is available to make an inference about a population parameter can be measured by the width (or half-width) of the confidence interval that could be constructed from the sample data. Recall that a large-sample confidence interval for a population mean is

$$\bar{y} \pm 1.96 \frac{\sigma}{\sqrt{n}}.$$

The widths of almost all the commonly employed confidence intervals are, like the confidence interval for a population mean, dependent on the population variance, σ^2, and the sample size, n. The less variation in the population, measured by σ^2, the shorter will be the confidence interval. Similarly, the width of the confidence interval will decrease as n increases. This interesting phenomenon would lead us to believe that two factors affect the quantity of information in a sample pertinent to a parameter, namely, the variation of the data and the sample size, n. We shall find this deduction to be slightly oversimplified but essentially true.

A strong similarity exists between the audio theory of communication and the theory of statistics. Both are concerned with the transmission of a message (signal) from one point to another and, consequently, both are theories of information. For example, the telephone engineer is responsible for transmitting a verbal message that might originate in New York City and be received in New Orleans. Or equivalently, a speaker may wish to communicate with a large and noisy audience. If static or background noise is sizable for either example, the receiver may acquire only a sample of the complete signal, and from this partial information, must infer the nature of the complete message. Similarly, scientific experimentation is conducted to verify certain theories about natural phenomena, or simple to explore some aspect of nature, and hopefully to deduce—either exactly or with a good approximation—the relationship of certain natural variables. Thus one might think of experimentation as the communication between nature and a scientist. The message about the natural phenomenon is contained, in garbled form, in the experimenter's sample data. Imperfections in his measuring instruments, nonhomogeneity of experimental material, and many other factors contribute background noise (or static) that tends to obscure nature's signal and cause the observed response to vary in a random manner. **For both the communications engineer and the statistician, two factors affect the quantity of information in an experiment: the magnitude of the background noise (or variation) and the volume of the signal. The greater the noise or, equivalently, the variation, the less information will be contained in the sample. Likewise, the louder the signal, the greater the amplification will be and hence it is more likely that the message will penetrate the noise and be received.**

The design of experiments is a very broad subject concerned with methods of sampling to reduce the variation in an experiment, to amplify nature's signal, and thereby to acquire a specified quantity of information at minimum cost. Despite the complexity of the subject, some of the important considerations in the design of good experiments can be easily understood and should be presented to the beginner. We take these considerations as our objective in the succeeding discussion.

14.2
The Physical Process of Designing an Experiment

We commence our discussion of experimental design by clarifying terminology and then, through examples, by identifying the steps that one must take in designing an experiment.

Definition

The objects upon which measurements are taken are called **experimental units.**

If an experimenter subjects a set of $n = 10$ rats to a stimulus and measures the response of each, a rat is the experimental unit. The collection of $n = 10$ measurements is a sample. Similarly, if a set of $n = 10$ items is selected from a list of hospital supplies in an inventory audit, each item is an experimental unit. The observation made on each experimental unit is the dollar value of the item actually in stock, and the set of 10 measurements constitutes a sample.

What one does to the experimental units that makes them differ from one population to another is called a treatment. Thus, one might wish to study the density of a specific kind of cake when baked at $x = 350\,°F$, $x = 400\,°F$, and $x = 450\,°F$ in a given oven. An experimental unit would be a single mix of batter in the oven at a given point in time. The three temperatures, $x = 350°$, $400°$, and $450°$, would represent three treatments. The millions and millions of cakes that conceptually *could* be baked at $350\,°F$ would generate a population of densities, and one could similarly generate populations corresponding to $400\,°F$ and $450\,°F$. The objective of the experiment would be to compare the cake density, y, for the three populations. Or, we might wish to study the effect of temperature of baking, x, on cake density by fitting a linear or curvilinear model to the data points as described in Chapter 11.

In another experiment we might wish to compare tire wear for two manufacturers, A and B. Each tire–wheel combination tested at a particular time would represent an experimental unit, and each of the two manufacturers would represent a treatment. Note that one does not physically treat the tires to make them different. They receive two different treatments by the very fact that they are manufactured by two different companies in different locations and in different factories.

As a third example, consider an experiment conducted to investigate the effect of various amounts of nitrogen and phosphate on the yield of a variety of corn. An experimental unit would be a specified acreage, say 1 acre, of corn. A treatment would be a fixed number of pounds of nitrogen, x_1, and phosphate, x_2, applied to a given acre of corn. For example, one treatment might be to use $x_1 = 100$ pounds of nitrogen per acre and $x_2 = 200$ pounds of phosphate. A second treatment could be $x_1 = 150$ and $x_2 = 100$. Note that the experimenter could experiment with different amounts (x_1, x_2) of nitrogen and phosphate and that each combination would represent a treatment.

Most experiments involve a study of the effect of one or more independent variables on a response.

Definition

Independent experimental variables are called **factors**.

Factors can be quantitative or qualitative.

Definition

A **quantitative factor** is one that can take values corresponding to points on a real line. Factors that are not quantitative are said to be **qualitative**.

Oven temperature, pounds of nitrogen, and pounds of phosphate are examples of quantitative factors. In contrast, manufacturers, types of drugs, or physical locations are factors that cannot be quantified and are called qualitative.

Definition

The intensity setting of a factor is called a **level**.

Thus the three temperatures, 350°, 400°, and 450°, represent three levels of the quantitative factor "oven temperature." Similarly, the two treatments, manufacturer A and manufacturer B, represent two levels of the qualitative factor, "manufacturers." Note that a third or fourth tire manufacturer could have been included in the tire-wear experiment and resulted in either three or four levels of the factor "manufacturers."

We noted previously that what one does to experimental units that makes them differ from one population to another is called a treatment. Since every treatment implies a combination of one or more factor levels, we have a more precise definition for the term.

Definition

A **treatment** is a specific combination of factor levels.

The experiment may involve only a single factor such as temperature in the baking experiment. Or, it could be composed of combinations of levels of two (or more) factors as for the corn-fertilizing experiment. Each combination would represent a treatment. One of the early steps in the design of an experiment is the selection of factors to be studied and a decision regarding the combinations of levels (treatments) to be employed in the experiment.

The design of an experiment implies one final problem after selecting the factor combinations (treatments) to be employed in an experiment. One must decide how the treatments should be assigned to the experimental units. Should the treatments be randomly assigned to the experimental units or should some semirandom pattern be employed? For example, should the tires corresponding to manufacturers A and B be randomly assigned to all the automobile wheels, or should one each of tire types A and B be assigned to the rear wheels of each car?

The foregoing discussion suggests that the design of an experiment involves four steps:

Steps Employed in Designing an Experiment

1. The selection of factors to be included in the experiment and a specification of the population parameter(s) of interest.

2. Deciding how much information is desired pertinent to the parameter(s) of interest. (For example, how accurately do you wish to estimate the parameters?)

3. The selection of the treatments to be employed in the experiment (combination of factor levels) and deciding on the number of experimental units to be assigned to each.

4. Deciding on the manner in which the treatments should be applied to the experimental units.

Steps 3 and 4 correspond to the two factors that affect the quantity of information in an experiment. First, how one selects the treatments (combinations of factor levels) and the number of experimental units assigned to each affects the intensity of nature's signal pertinent to the population parameter(s) of interest to the experimenter. Second, the method of assigning the treatments to the experimental units affects the background noise or, equivalently, the variation of the experimental units. We will examine each of these assertions in detail in Sections 14.4 and 14.5 after digressing briefly in Section 14.3 to consider the implications of random sampling and how to draw a random sample.

14.3
Random Sampling and the Completely Randomized Design

Random sampling—that is, giving every possible sample in a population an equal probability of selection—has two purposes. First, it avoids the possibility of bias introduced by a nonrandom selection of sample elements. For example, a sample selection of voters from telephone directories in 1936 indicated a clear win for Landon. However, the sample did not represent a random selection from the whole population of eligible voters, because the majority of telephone users in 1936 were Republicans. As another example, suppose that we sample to determine the percent of homeowners favoring the construction of a new city park and modify our original random sample by ignoring owners who are not at home. The result may yield a biased response because those at home will likely have children and may be more inclined to favor the new park. **The second purpose of random sampling is to provide a probabilistic basis for the selection of a sample.** That is, it treats the selection of a sample as an experiment (Chapter 4), enabling the statistician to calculate the probability of an observed sample and to use this probability in making inferences. Thus, we learned that under fairly general conditions, the mean, $\bar{y}$, of a random sample of n elements will possess a probability distribution that is approximately normal when n is large (the Central Limit Theorem). Fundamental to the proof of the Central Limit Theorem is the assumption of random sampling. Similarly, the confidence intervals and tests of hypotheses based on Student's t, Chapter 10, required the assumption of random sampling.

The random selection of independent samples to compare two or more populations is the simplest type of experimental design. The populations differ because we have applied different treatments (factor combinations) and we now wish to consider step 4 in the design of an experiment—deciding how to apply the treatments to the experimental units, or equivalently, how to select samples.

Definition

The selection of independent random samples from p populations is called a **completely randomized design.**

The comparison of five brands of aspirin, A, B, C, D, and E, by randomly selecting 100 pills from the production of each manufacturer would be a completely randomized design with $n_A = n_B = \cdots = n_E = 100$. Similarly, we might wish to compare five teaching techniques, A, B, C, D, and E, using 25 students in each class ($n_A = n_B = \cdots = n_E = 25$). The populations corresponding to A, B, C, D, and E are nonexistent (that is, they are conceptual) because students taught by the five techniques either do not exist or they are unavailable. Consequently, random samples are obtained for the five conceptual populations by randomly selecting 125 students of the type envisioned for the study and then *randomly* assigning 25 students to each of the five teaching techniques. The scheme will yield independent random samples of 25 students subjected to each technique and will result in a completely randomized design.

As you might surmise, not all designs are completely randomized, but all good experiments rely on randomization to some extent. For example, the paired-difference experiment of Chapter 10 implied the application of the treatments A and B to pairs of experimental units that were nearly homogeneous. Then A and B were *randomly* assigned, one each, to the experimental units in the pair. This design, which uses sampling that is partially, but not completely, random, was employed to reduce the experimental error (reduce the background noise) and, thereby, to increase the information in the experiment. We will comment further on noise-reducing designs in Section 14.5.

Before concluding this section on randomization, let us consider a method for selecting a random sample or randomly assigning experimental units to a set of treatments.

The simplest and most reliable way to select a random sample of n elements from a large population is to employ a table of random numbers such as that shown in Table 15, the Appendix. Random-number tables are constructed so that integers occur randomly and with equal frequency. For example, suppose that the population contains $N = 1000$ elements. Number the elements in sequence, 1 to 1000. Then turn to a table of random numbers such as the excerpt shown in Table 14.1.

Table 14.1 Portion of a table of random numbers

15574	35026	98924
45045	36933	28630
03225	78812	50856
88292	26053	21121

Select n of the random numbers in order. The population elements to be included in the random sample will be given by the first three digits of the random numbers (unless the first four digits are 1000). Thus, if $n = 5$, we would include elements numbered 155, 450, 32, 882, and 350. So as not to use the same sequence of random numbers over and over again, the experimenter should select different starting points in Table 15 to begin the selection of random numbers for different samples.

The random assignment of 40 experimental units, 10 to each of four treatments A, B, C, and D, is equally easy using the random-number table. Number the experimental units 1 to 40. Select a sequence of random numbers and refer only to the first two digits since $n = 40$. Discard random numbers greater than 40. Then assign the experimental units associated with the first 10 numbers that appear to A, those associated with the second 10 to B, and so on. To illustrate, refer to Table 14.1. Starting in the first column and moving top to bottom, we acquire the numbers 15 and 3 (discard 45 and 88 because no experimental units possess those numbers). This procedure would be continued until all integers, 1 to 40, were selected. The resulting numbers would occur in random order.

The analysis of data for a completely randomized design is treated in Section 15.3.

Exercises

14.1.   Suppose that a telephone company wishes to select a random sample of $n = 20$ (we select this small number to simplify this exercise) out of 7000 customers for a survey of customer attitudes concerning service. If the customers are numbered for identification purposes, indicate the customers whom you will include in your sample. Use the random-number table and explain how you selected your sample.

14.2.   A small city contains 20,000 voters. Use the random-number table to identify the voters to be included in a random sample of $n = 15$.

14.3.   Twelve people are each to receive a vaccination in random order. Use the random-number table to select the ordering.

14.4.   If a survey of a newspaper's readership is obtained by requesting readers to respond to a questionnaire published in the newspaper, are the resulting responses likely to give a random sample of readership opinion? Explain.

14.5.   A random sample of the public opinion for a small town was obtained by selecting every tenth person to pass by the busiest corner in the downtown area. Will this sample have the characteristics of a random sample selected from the town's citizenry? Explain.

14.6.   Suppose that you decide to conduct a telephone public opinion survey, randomly sampling members from the telephone directory. The survey is conducted from 9 A.M. to 5 P.M. Will the resulting responses represent a random sample of adult public opinion in the community? Explain.

14.4
Volume-Increasing Experimental Designs

You will recall that designing to increase the volume of a signal, that is, the information pertinent to one or more population parameters, depends on the selection of treatments and the number of experimental units assigned to each (step 3 in designing an experiment). The treatments, or combinations of factor levels, identify points at which one or more response measurements will be made and indicate the general location in which the experimenter is focusing his attention. As we shall see, some designs contain more information concerning specific population parameters than others for the same number of observations. Other very costly experiments contain no information concerning certain population parameters. And no single design is best in acquiring information concerning all types of population parameters. Indeed, the problem of finding the best design for focusing information on a specific population parameter has been solved in only a few specific cases.

The purpose of this section is not to present a general theory or find the best selection of factor-level combinations for a given experiment but rather to present a few examples to illustrate the principles involved. The optimal design providing the maximum amount of information pertinent to the parameter(s) of interest (for a fixed sample size, n) will be given for the following two examples.

The simplest example of an information-focusing experiment is the problem of estimating the difference between a pair of population means, $(\mu_1 - \mu_2)$, based on independent random samples. In this instance, the two treatments have already been selected and the question concerns the allocation of experimental units to the two samples. If the experimenter plans to invest money sufficient to sample a total of n experimental units, how many units should he select from populations 1 and 2, say n_1 and n_2 respectively $(n_1 + n_2 = n)$, so as to maximize the information in the data pertinent to $(\mu_1 - \mu_2)$? Thus if $n = 10$, should he select $n_1 = n_2 = 5$ observations from each population or would an allocation of $n_1 = 4$ and $n_2 = 6$ be better?

Recall that the estimator of $(\mu_1 - \mu_2)$, $(\bar{y}_1 - \bar{y}_2)$, has a standard deviation

$$\sigma_{(\bar{y}_1 - \bar{y}_2)} = \sqrt{\frac{\sigma_1^2}{n_1} + \frac{\sigma_2^2}{n_2}}.$$

The smaller $\sigma_{(\bar{y}_1 - \bar{y}_2)}$, the smaller will be the corresponding error of estimation and the greater will be the quantity of information in the sample pertinent to $(\mu_1 - \mu_2)$. If, as we frequently assume, $\sigma_1^2 = \sigma_2^2 = \sigma^2$, then

$$\sigma_{(\bar{y}_1 - \bar{y}_2)} = \sigma \sqrt{\frac{1}{n_1} + \frac{1}{n_2}}.$$

You can verify that for a fixed total number $(n = n_1 + n_2)$ of observations, this quantity is a minimum for $n_1 = n_2$. Consequently, the sample contains a maximum of information on $(\mu_1 - \mu_2)$ when the n experimental units are equally divided between the two treatments. We illustrate (but do not prove) our point with an example.

Example 14.1   Calculate the standard deviation for $(\bar{y}_1 - \bar{y}_2)$ when $\sigma_1^2 = \sigma_2^2 = \sigma^2$ and

a. $n = 10, n_1 = n_2 = 5$.

b. $n = 10, n_1 = 4, n_2 = 6$.

Solution $$\sigma_{(\bar{y}_1 - \bar{y}_2)} = \sigma \sqrt{\frac{1}{n_1} + \frac{1}{n_2}}.$$

a. When $n_1 = n_2 = 5$,

$$\sigma_{(\bar{y}_1 - \bar{y}_2)} = \sigma \sqrt{\frac{1}{5} + \frac{1}{5}} = \sigma \sqrt{.4}.$$

b. When $n_1 = 4$, $n_2 = 6$,

$$\sigma_{(\bar{y}_1 - \bar{y}_2)} = \sigma\sqrt{\frac{1}{4} + \frac{1}{6}} = \sigma\sqrt{.42}.$$

Note that $\sigma_{(\bar{y}_1 - \bar{y}_2)}$ is smaller for an equal allocation, $n_1 = n_2 = 5$.

Equal allocation of experimental units to the two treatments is *not* best when $\sigma_1^2 \neq \sigma_2^2$. Then, the allocation of the n experimental units to give the maximum information concerning $(\mu_1 - \mu_2)$ assigns values to n_1 and n_2 that are proportional to σ_1 and σ_2, respectively. Proof of this statement is omitted.

As a second example, consider the problem of fitting a straight line through a set of n points using the least-squares method of Chapter 11 (see Figure 14.1).

Figure 14.1 Fitting a straight line by the method of least squares

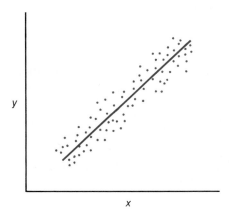

Further, suppose that we are primarily interested in the slope of the line, β_1, in the linear model

$$y = \beta_0 + \beta_1 x + \epsilon.$$

If we have the option of selecting the n values of x for which y will be observed, which values of x will maximize the quantity of information on β_1? Thus we have one single quantitative factor, x, and have the problem of deciding on the levels to employ $(x_1, x_2, \ldots, x_n)$ as well as the number of observations to be taken at each.

A strong lead to the best design for fitting a straight line can be achieved by viewing Figure 14.2(a) and (b). Suppose that y was linearly related to x and generated data similar to that shown in Figure 14.2(a) for the interval $x_1 < x < x_2$. Note the approximate range of variation for a given value of x.

Figure 14.2 Two different level selections for fitting a straight line

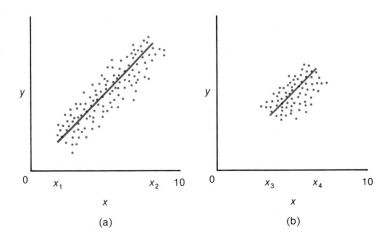

(a)   (b)

Now suppose that instead of the wide range for x employed in Figure 14.2(a), the experimenter selected data from the same population but over the very narrow range $x_3 < x < x_4$, as shown in Figure 14.2(b). The variation in y, given x, is the same as for Figure 14.2(a). Which distribution of data points would provide the greater amount of information concerning the slope of the line, β_1? You might guess (correctly) that the best estimate of slope will occur when the levels of x are selected farther apart as shown in Figure 14.2(a). The data for 14.2(b) could yield a very inaccurate estimate of the slope and, as a matter of fact, might leave a question as to whether the slope is positive or negative.

The best design for estimating the slope, β_1, can be determined by considering the standard deviation of $\hat{\beta}_1$,

$$\sigma_{\hat{\beta}_1} = \frac{\sigma}{\sqrt{\displaystyle\sum_{i=1}^{n} (x_i - \bar{x})^2}}.$$

The larger the sum of squares of deviations of $x_1, x_2, \ldots, x_n$ about their mean, the smaller will be the standard deviation of $\hat{\beta}_1$. The experimenter will usually have some experimental region, say $x_1 < x < x_2$, over which he wishes to observe y, and this range will frequently be selected prior to experimentation. Then, the smallest value for $\sigma_{\hat{\beta}_1}$ will occur when the n data points are equally divided, with half located at the lower boundary of the region, x_1, and half at the upper boundary, x_2. (Proof is omitted.) Thus, if an experimenter wishes to fit a line using $n = 10$ data points in the interval $2 < x < 6$, he should select five data points at $x = 2$ and five at $x = 6$. Before concluding discussion of this example, you should note that observing all values of y at only two values of x will not provide information on curvature of the response curve in case the assumption of linearity in the relation of y and x is incorrect. It is frequently safer to select a few points (as few as one or two) somewhere near the middle of the experimental region to detect curvature if it should be present (see Figure 14.3).

Figure 14.3 Good design for fitting a straight line ($n = 10$)

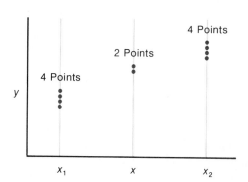

To summarize, we have given optimal designs (factor-level combinations and the allocation of experimental units per combination) for comparing a pair of means and fitting a straight line. These two simple designs illustrate the manner in which information in an experiment can be increased or decreased by the selection of the factor-level combinations that represent treatments and by changing the allocation of observations to treatments. Thus, we have demonstrated that factor-level selection and the allocation of experimental units to treatments can greatly affect the information in an experiment pertinent to a particular population parameter and, thereby, can amplify nature's signal. Thus step 3, Section 14.2, is an important consideration in the design of good experiments.

Exercises

14.7.   Suppose that one wishes to compare the means for two populations. If $\sigma_1^2 = \sigma_2^2$, what is the optimal allocation of n_1 and n_2 ($n_1 + n_2 = n$) for a fixed sample size n? Does this operation employ the principle of noise reduction or signal amplification?

14.8.   Refer to Exercise 14.7. Suppose that $\sigma_1^2 = 9$, $\sigma_2^2 = 25$, and $n = 90$. What allocation of $n = 90$ to the two samples will result in the maximum amount of information on $(\mu_1 - \mu_2)$?

14.9.   Refer to Exercise 14.8. Suppose that we allocate $n_1 = n_2$ observations to each sample. How large must n_1 and n_2 be in order to obtain the same amount of information as that implied by the solution to Exercise 14.8?

14.10.   Suppose that you wish to estimate the slope of a line using the method of least squares, that you can afford only 12 experimental points, and that the assumptions of Section 11.2 are satisfied. How can you select the locations of the points (the settings of the independent variable, x) so as to minimize the variance of $\hat{\beta}_1$? Note that this selection of settings accomplishes an increase in "volume" without increasing the sample size. Essentially, it shifts information within the design so that it focuses on the slope, β_1.

14.5
Noise-Reducing
Experimental
Designs

Noise-reducing experimental designs increase the information in an experiment by decreasing the background noise (variation) caused by uncontrolled nuisance variables. Thus, by serving as filters to screen out undesirable noise, they permit nature's signal to be received more clearly. Reduction of noise can be accomplished in step 4 of the design of an experiment, namely, in the method for assigning treatments to the experimental units.

Designing for noise reduction is based on the single principle of making all comparisons of treatments within relatively homogeneous groups of experimental units. The more homogeneous the experimental units, the easier it is to detect a difference in treatments. Because noise-reducing, or filtering, designs work with blocks of relatively homogeneous experimental units, they are called block designs.

The paired-difference experiment, Section 10.5, employed the blocking principle to achieve a very sizable reduction in noise and a large increase in information pertinent to the difference between treatment means. The objective of the experiment was to make an inference about the difference in mean wear for two types of automobile tires, A and B. Five tires of each type were to be mounted on the rear wheels of five automobiles subjected to a large and specified number of miles of driving and the wear recorded for each. The experimenter realized that tire wear might vary substantially from automobile to automobile, owing primarily to the difference in drivers and the completely different sequence of starts and stops to which each car would be subjected over the test distance. In contrast, the abrasions applied to the rear wheels of the same car would be relatively equivalent. Thus, he decided to block his design on automobiles, assigning an A and a B tire type in a random manner to the rear wheels of each automobile. This eliminated auto-to-auto variability in comparing treatments and produced approximately 400 times as much information concerning $(\mu_A - \mu_B)$ in comparison with a completely randomized design employing the same number of tires. This fact was made evident by comparing the width of the confidence intervals based on the data generated by the paired and unpaired analyses.

The paired-difference design is a special case of the most elementary noise-reducing design, the randomized block design. The randomized block design is employed to compare the means of any number of treatments, say p, within relatively homogeneous blocks of p experimental units. Any number of blocks may be employed, but each treatment must appear exactly once in each block. Thus the paired-difference experiment is a randomized block design containing $p = 2$ treatments.

Definition

A **randomized block design** containg b blocks and p treatments consists of b blocks of p experimental units each. The treatments are randomly assigned to the units in each block, with each treatment appearing exactly once in every block.

The difference between a randomized block design and the completely randomized design can be demonstrated by considering an experiment designed to compare subject reaction to a set of four stimuli (treatments) in a stimulus–response psychological experiment. We shall denote the treatments as T_1, T_2, T_3, and T_4.

Suppose that eight subjects are to be randomly assigned to each of the four treatments. Random assignment of subjects to treatments (or vice versa) randomly distributes errors due to person-to-person variability to the four treatments and yields four samples that are, for all practical purposes, random and independent. This would be a completely randomized experimental design.

The experimental error associated with a completely randomized design is composed of a number of components. Some of these are due to the difference between subjects, to the failure of repeated measurements within a subject to be identical (owing to variations in physical and psychological conditions), to the failure of the experimenter to administer a given stimulus with exactly the same intensity in repeated measurements, and, finally, to errors of measurement. Reduction of any of these causes of error will increase the information in the experiment.

The subject-to-subject variation in the above experiment can be eliminated by using subjects as blocks. Thus each subject would receive each of the four treatments assigned in a random sequence. The resulting randomized block design would appear as in Figure 14.4. Now only eight subjects are required to obtain eight response measurements per treatment. Note that each treatment occurs exactly once in each block.

Figure 14.4 Randomized block design

Subjects

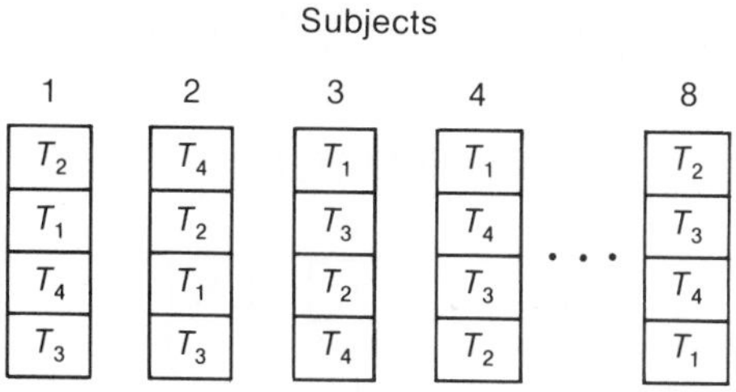

The word "randomization" in the name of the design implies that the treatments are randomly assigned within a block. For our experiment, position in the block would pertain to the position in the sequence when assigning the stimuli to a given subject over time. The purpose of the randomization (that is, position in the block) is to eliminate bias caused by fatigue or learning.

Blocks may represent time, location, or experimental material. Thus, if three treatments are to be compared and there is a suspected trend in the mean response over time, a substantial part of the time–trend variation may be removed by blocking. All three treatments would be randomly applied to experimental units in one small block of time. This procedure would be repeated in succeeding blocks of time until the required amount of data is collected. A comparison of the sale of competitive products in supermarkets should be made within supermarkets, thus using the supermarkets as blocks and removing market-to-market variability. Animal experiments in agriculture and medicine often utilize animal litters as blocks, applying all the treatments, one each, to animals within a litter. Because of heredity, animals within a litter are more homogeneous than those between litters. This type of blocking removes the litter-to-litter variation. The analysis of data generated by a randomized block design is discussed in Section 15.8.

The randomized block design is only one of many types of block designs. Blocking in two directions can be accomplished using a Latin square design. Suppose that the subjects of the preceding example became fatigued as the stimuli were applied so that the last stimulus always produced a lower response than the first. If this trend (and consequent lack of homogeneity of the experimental units in a block) were true for all subjects, a Latin square design would be appropriate. The design would be constructed as shown in Figure 14.5. Each stimulus is applied once to each subject and occurs exactly once in each position of the order of presentation. Thus all four stimuli occur in each row and in each column of the 4×4 configuration. The resulting design is called a 4×4 Latin square. A Latin square design for three treatments will require a 3×3 configuration and, in general, p treatments will require a $p \times p$ array of experimental units. If more observations are desired per treatment, the experimenter would utilize several Latin square configurations in one experiment. In the example above, it would require running two Latin squares to obtain eight observations per treatment. The experiment would then contain the same number of observations per treatment as the randomized block design, Figure 14.4.

Figure 14.5 Latin square design

Subjects (Columns)

	1	2	3	4
Order of Presentation of Stimuli (Rows) 1	T_1	T_2	T_3	T_4
2	T_2	T_3	T_4	T_1
3	T_3	T_4	T_1	T_2
4	T_4	T_1	T_2	T_3

A comparison of means for any pair of stimuli would eliminate the effect of subject-to-subject variation but would, in addition, eliminate the effect of the fatigue trend within each stimulus because each treatment was applied in each position of the stimuli time administering sequence. Consequently, the effect of trend would be canceled in comparing the means.

A more extensive discussion of block designs and their analyses is contained in the references. The objective of this section is to make the reader aware of the existence of block designs, how they work, and how they can produce substantial increases in the quantity of information in an experiment by reducing nuisance variation.

Exercises

14.11. Consider the tire-wear experiment of Section 14.5. Why do we regard "automobiles" as blocks rather than levels of a factor?

14.12. Could an independent variable be a factor in one experiment and a nuisance variable (a noise contributor) in another?

14.13. Suppose that you intended to compare the mean responses of humans to five drugs that were not expected to produce a long-lasting effect that would carry over from one part of the experiment to another. Each drug is to be administered to each of seven people. The intent is to make a comparison of drug effects within each person.

   a. What type of experimental design is to be used and how is it supposed to increase the quantity of information in the experiment?

   b. Use the random-number table to obtain an ordering of drug applications within the seven people.

14.14. An experiment was conducted to compare the mean yield (in soybeans) of soybean plants subjected to six different fertilizer combinations. Six plots 50 feet by 50 feet square were selected at each of eight different locations (farms), and the plots at each farm were located as near as possible to each other. The six fertilizer combinations were randomly assigned to the plots at each of the eight farms, the plots were seeded with soybeans using the same rate of planting for all plots, and the crop was harvested for each plot at the end of the growing season.

   a. Identify the experimental units.

   b. Identify the treatments.

   c. What type of experimental design was employed?

   d. What principle was employed to increase the quantity of information in the experiment?

14.15. Describe an experimental situation for which a randomized block design would be appropriate.

14.16. What is a Latin square design? What is the objective of the design? Describe an experimental situation for which a Latin square design would be appropriate.

14.6
An Example of a Factorial Experiment

In this section we will return to the case study and reexamine the Baucom–Aiken experiment from the standpoint of statistical design. We will briefly describe how Baucom and Aiken conducted their experiment, omitting some details such as the procedure they employed to verify that a subject was depressed, the justification for their choice of response variable, etc. Whether the variables in the experiment do, in fact, measure the phenomena that they are intended to measure is a question that we will leave for psychologists. Instead, we will concentrate our attention on the statistical design employed in the experiment.

You will recall that Baucom and Aiken investigated the effect of three qualitative variables—mood (depressed or nondepressed), obesity (obese or nonobese), and dieting (whether or not the subject was trying to lose weight)—on a response variable y that was intended to measure "tendency to eat." Subjects—college students, who were selected for the experiment—were not told of the nature of the experiment or its intended purpose. Depression was induced in a subject by requiring the subject to perform a number of frustrating, unperformable tasks. A student who was not to receive the depression "treatment" was assigned a similar number of easily performable tasks. Both types of students were then led to believe that they were participating in a cracker taste testing experiment. They were asked to compare and rate the tastes of three types of crackers, while being viewed through a one-way mirror, and the number y of crackers consumed by each subject was recorded. Thus y was the response variable chosen to measure "tendency to eat."

Each of the three qualitative predictor variables was investigated at each of two levels. For example, the qualitative variable "mood" was set, for each subject, at one of two levels, namely, "depressed" or "nondepressed." Similarly, the qualitative variables, obesity and diet, could assume only one of two levels, obese or nonobese and dieting or nondieting, respectively. The first aspect of the experimental design required the experimenters to choose the combinations of levels (i.e., experimental conditions) that they planned to employ in their experiment. For example, they could choose to measure y on subjects who were not dieting, not obese, and not depressed. They could choose to measure others who were not dieting, not obese, and depressed. These are two of the $2 \times 2 \times 2 = 8$ combinations of levels of the three qualitative variables that could be investigated in the experiment. These eight experimental conditions are listed in Table 14.2.

Table 14.2 A complete list of possible settings for qualitative variables, each of which can assume one of two levels

Experimental Condition	Variable Setting		
	Depression	Obesity	Diet
1	nondepressed	nonobese	nondieting
2	nondepressed	nonobese	dieting
3	nondepressed	obese	nondieting
4	nondepressed	obese	dieting
5	depressed	nonobese	nondieting
6	depressed	nonobese	dieting
7	depressed	obese	nondieting
8	depressed	obese	dieting

One way to design the experiment would be to observe subjects only for conditions 1, 2, 3, and 5. All students in the experimental condition groups 1 and 2 would be nondepressed and nonobese. Consequently, the only difference in conditions to which the students in groups 1 and 2 were exposed is whether they had been dieting prior to the experiment. A comparison of the mean number of crackers consumed by these two groups should provide an estimate of the effect of the qualitative variable "diet" on "tendency to eat." Using a similar argument, we could reason that a comparison of the mean number of crackers consumed by group 1 and group 3 should enable us to estimate the effect of the qualitative variable "obesity" on "tendency to eat." Comparing the means for groups 1 and 5 would enable us to observe the effect of "depression" on "tendency to eat." This choice of experimental conditions is called a "one-at-a-time" experimental design because all variables are fixed at their low levels, condition 1, and the other three conditions are chosen by raising the level of each variable, one at a time.

In contrast to the one-at-a-time experimental design, Baucom and Aiken included all eight of the experimental conditions, Table 14.2, in their experimental design. They classified students into each of four groups of 14 students, each according to whether the students were or were not obese and whether they had or had not been dieting prior to the experiment. Then depression was induced in half (7) of the students in each group, and the measure y for each of the 28 depressed and the 28 nondepressed students was recorded. This design, employing all eight experimental conditions of Table 14.2, is known as a $2 \times 2 \times 2$ *factorial experiment*.

The mean values of y corresponding to the right experimental conditions of Table 14.2 are shown in Table 14.3. The means corresponding to the four one-at-a-time experimental conditions are shaded.

Table 14.3 Mean number of crackers eaten as a function of depressed/nondepressed, obesity/nonobesity, and dieting/nondieting

Condition	Dieting	Nondieting
Nondepressed		
Obese	6.57 ③	12.57
Nonobese	9.43 ②	13.14 ①
Depressed		
Obese	15.86	8.71
Nonobese	15.43	8.29 ⑤

You can see the flaw in the one-at-a-time design by examining the four means that would have been generated if the experimenters had used that approach. Comparing sample means of conditions # 1, ($\bar{y}_1 = 13.14$), and # 5, ($\bar{y}_5 = 8.29$), you would be led to believe that depression reduces one's "tendency to eat." You obtain the same impression by comparing the means for students who were obese and nondieting (although these means would not have been required using the one-at-a-time design). Those who were nondepressed had a mean equal to 12.57, in comparison to only 8.71 for those who were depressed.

Now compare corresponding means for conditions in the Dieting column of Table 14.3. You will see that depression had a completely opposite effect on these students—one which would never have been detected using the data obtained from the one-at-a-time design. The means for students in the depressed groups are much higher—15.86 versus 6.57, and 15.43 versus 9.43—than those in the nondepressed groups. In other words, the effect of depression on "tendency to eat" *depends* upon whether the student was dieting or not dieting prior to the experiment. This condition (encountered previously in Section 12.5), when the effect of one predictor variable on the response depends upon the setting of one (or more) other predictor variables, is known as variable *interaction*.

The conclusions that we draw from an analysis of the Baucom–Aiken experimental design and resulting data is that one-at-a-time experimentation can lead to fallacious conclusions. Their experiment shows how a factorial experiment can be used to detect factor interactions. Perhaps most importantly, it demonstrates the need to detect factor interactions when they exist. We must compare means for the various variable combinations when interaction is present rather than lumping means together to assess the overall effects of individual variables. Otherwise, as Table 14.3 demonstrates, we can be led to some very faulty conclusions.

How does one classify a factorial experiment? Is it a volume-increasing or a noise-reducing experimental design? The answer is "volume increasing." Factorial experimental designs specify a particular type of treatment selection and consequently are volume-increasing experimental designs. They focus information on factor interactions. In contrast, the amount of information on factor interactions contained in a one-at-a-time design is zero.

14.7 Summary

The objective of this chapter is to identify the factors that affect the quantity of information in an experiment and to use this knowledge to design better experiments. The subject, design of experiments, is very broad and is certainly not susceptible to condensation into a single chapter in an introductory text. In contrast, the philosophy underlying design, the methods for varying information in an experiment, and desirable strategies for design are easily explained and comprise the objective of Chapter 14.

Two factors affect the quantity of information in an experiment—the volume of nature's signal and the magnitude of variation caused by uncontrolled variables. The volume of information pertinent to a parameter of interest depends on the selection of factor-level combinations (treatments) to be included in the experiment and on the allocation of the total number of experimental units to the treatments. This choice determines the focus of attention of the experimenter.

The second method for increasing the information in an experiment concerns the method for assigning treatments to the experimental units. Blocking, comparing treatments within relatively homogeneous blocks of experimental material, can be used to eliminate block-to-block variation when comparing treatments. As such, it serves as a filter to reduce the unwanted variation that tends to obscure nature's signal.

Thus the selection of factors and the selection of factor levels are important considerations in shifting information in an experiment to amplify the information on a population parameter. The use of blocking in assigning treatments to experimental units reduces the noise created by uncontrolled variables and, consequently, increases the information in an experiment.

The analysis of some elementary experimental designs is given in Chapter 15. A more extensive treatment of the design and analysis of experiments is a course in itself. The reader interested in exploring this subject is directed to the references.

References

Cochran, W.G., and G.M. Cox, *Experimental Designs*, 2nd ed. New York: John Wiley & Sons, Inc., 1957.

Hicks, C.R., *Fundamental Concepts in the Design of Experiments*, 2nd ed. New York: Holt, Rinehart and Winston, Inc., 1973.

Mendenhall, W., *An Introduction to Linear Models and the Design and Analysis of Experiments*. Belmont, Calif.: Wadsworth Publishing Company, Inc., 1968.

Scheaffer, R., W. Mendenhall, and L. Ott, *Elementary Survey Sampling*, 2nd ed. Boston: Duxbury Press, 1979.

Winer, B.J., *Statistical Principles in Experimental Design*, 2nd ed. New York: McGraw-Hill Book Company, 1971.

Supplementary Exercises

14.17. How can one measure the quantity of information in a sample pertinent to a specific population parameter?

14.18. Give the two factors that affect the quantity of information in an experiment.

14.19. What is a random sample?

14.20. Give two reasons for the use of random samples.

14.21. A political analyst wishes to select a sample of $n = 20$ people from a population of 2000. Use the random-number table to identify the people to be included in the sample.

14.22. Two drugs, A and B, are to be applied to five rats each. Suppose that the rats are numbered from 1 to 10. Use the random-number table to randomly assign the rats to the two treatments.

14.23. Refer to Exercise 14.22, and suppose that the experiment involved three drugs, A, B, and C, with five rats assigned to each. Use the random-number table to randomly assign the 15 rats to the three treatments.

14.24. A population contains 50,000 voters. Use the random-number table to identify the voters to be included in a random sample of $n = 15$.

14.25. What is a factor?

14.26. State the steps involved in designing an experiment.

14.27. If one were to design an experiment, what part of the design procedure would result in signal amplification?

14.28. Refer to Exercise 14.27. What part of the design procedure would result in noise reduction?

14.29.   Complete the assignment of treatments for the following 3×3 Latin square design:

	A	
C		
		B

14.30.   Suppose that one wishes to study the effect of the stimulant digitalis on the blood pressure of rats over a dosage range of $x = 2$ to $x = 5$ units. The response is expected to be linear over the region. Six rats are available for the experiment, and each rat can receive only one dose. What dosages of digitalis should be employed in the experiment and how many rats should be run at each dosage to maximize the quantity of information on β_1, the slope of the regression line? Which aspect of design—noise reduction or signal amplification—is implied in this experiment?

14.31.   Refer to Exercise 14.30. Consider two methods for selecting the dosages. Method 1 assigns three rats to the dosage $x = 2$ and three rats to $x = 5$. Method 2 equally spaces the dosages between $x = 2$ and $x = 5$ ($x = 2, 2.6, 3.2, 3.8, 4.4, 5.0$). Suppose that σ is known and that the relationship between $E(y)$ and x is truly linear (see Chapter 10). How much larger will be the confidence interval for the slope (β_1) for method 2 in comparison with method 1? Approximately how many observations would be required to obtain the same confidence interval as obtained by the optimal assignment of method 1?

14.32.   Refer to Exercise 14.30. Why might it be advisable to assign one or two points at $x = 3.5$?

14.33.   An experiment is to be conducted to compare the effect of digitalis on the contraction of the heart muscle of a rat. The experiment is conducted by removing the heart from a live rat, slicing the heart into thin layers, and treating the layers with a dosage of digitalis. The muscle contraction is then measured. If four dosages ($A, B, C,$ and D) are to be employed, what advantage might be derived by applying $A, B, C,$ and D to a slice of tissue from each rat heart? What principle of design is illustrated by this example?

14.34.   Describe the factors that affect the quantity of information in an experiment and the design procedures that control these factors.

14.35.   When one statistic tends to support a theory and a second statistic, based on a completely different survey, tends to refute it, the reason may not be the occurrence of a highly improbable event. The apparent disagreement of statistics may be due to the fact that the researchers are sampling different populations. For example, researchers at the Florida Technological University conducted a survey to determine the effect of Florida's new mandatory three-year sentence for gun-related crimes.* To determine the effectiveness of the law, the researchers conducted a survey of 277 prisoners convicted of the use of handguns and found that 73 percent said that they would continue to use handguns upon release from prison. They therefore concluded that the law was ineffective. In response, the sponsor of the law pointed out that Florida Department of Criminal Law Enforcement statistics show a 30 percent drop in gun-related crimes since the law was enacted. Which of the two populations is the more pertinent in deciding whether the law was effective in reducing the use of handguns in crimes? Discuss the two statistics and explain their relevance to the researchers' objective.

* "Senator Says Gun Research Is Wrong." *Gainesville Sun,* December 23, 1977.

15. The Analysis of Variance

Chapter Objectives

General Objective

Methods for comparing two population means, based on independent random samples and on a paired-difference experiment, were presented in Chapter 10. Chapter 15 extends these analyses to the comparison of any number of population means using a technique called an analysis of variance. In this chapter we shall explain the logic of an analysis of variance and give the analyses for three experimental designs.

Specific Objectives

1. To explain the logic of an analysis of variance. *Section 15.2*

2. To present the analysis of variance for a comparison of two treatment means and to relate the resulting F test to the t test of Chapter 9. *Section 15.2*

3. To give the analysis of variance for comparing two or more population means by using a completely randomized design. *Sections 15.3, 15.4, 15.5, 15.6, 15.7*

4. To give the analysis of variance for a randomized block design and to explain how to interpret the results of the analysis. *Sections 15.8, 15.9, 15.10, 15.11*

5. To give the analysis of variance for a Latin square design and to explain how to interpret the results of the analysis. *Sections 15.12, 15.13, 15.14*

6. To familiarize you with typical computer printouts for analyses of variance. *Sections 15.7, 15.10, 15.11, 15.13*

7. To explain how to select sample sizes to obtain a specific amount of information concerning the difference between a pair of treatment means. *Section 15.15*

8. To explain how blocking increases the information in an experiment. *Sections 15.8, 15.9, 15.12, 15.13, 15.16*

9. To summarize the assumptions upon which the tests and confidence intervals of an analysis of variance are based and to explain what happens and what to do when they are not satisfied. *Section 15.17*

Case Study

A Comparison of Car-Insurance Costs for Different Locales

In an article* titled "Sure-Fire Ways to Save on Car-Insurance Costs," Barbara Gilder Quint notes that some of us are paying more for car insurance than our vehicles are worth. In addition to providing helpful hints on how to save, she provides some data on the costs of $50,000/$100,000/$10,000 liability and $50 deductible comprehensive insurance on a 1981 standard Chevrolet Citation. The costs of the same type of policy (excluding claim-handling reliability, etc.) for six different insurance companies were acquired for four different locales in the United States. The data are shown in Table 15.1.

You can see that the differences in mean car-insurance costs are substantial—at least they appear to be. But how can we tell for certain and how can we place a confidence interval on the difference between any pair of locale means?

You will notice that this sampling situation is similar to the paired-difference experiment of Section 10.5. The major difference is that we are comparing four population means, in contrast to two in the paired-difference experiment. The pairing, in this case "matching," occurs because the insurance costs for the four locales are compared within individual insurance companies. Since procedures for establishing rates may vary substantially from one company to another, comparing locale rates *within* a single company, and repeating the procedure for the six companies, enables us to remove *between*-company variation from the comparison. This extension of the paired-difference experiment to the comparison of more than two

Table 15.1 A comparison of car-insurance costs for a standard policy

	BIG CITY Chicago Business District	MIDDLE-SIZED CITY Topeka, Kan.	RURAL Dillsboro, Ind.	SUBURBAN Seattle (Suburban)
ALLSTATE	$597	$280	$245	$339
CONTINENTAL	768	260	240	275
HOME (GOLD KEY PACKAGE)	776	284	257	304
NATIONWIDE	739	334	262	317
STATE FARM	562	338	250	335
TRAVELERS	698	315	330	350

* Quint, Barbara Gilder, "Sure-Fire Ways to Save on Car-Insurance Costs," *Family Circle*, February 20, 1979. © 1979 THE FAMILY CIRCLE, INC.

population means, is a randomized block design (discussed in Section 14.5). Each insurance company represents a block. The four locales are the treatments.

In this chapter, we will extend the methods of Chapter 10 to the comparison of more than two population means. We will return to an analysis of the car-insurance data in Section 15.11.

15.1
Introduction

Most experiments involve a study of the effect of one or more independent variables on a response. In Chapters 12 and 14 we learned that a response, y, can be affected by two types of independent variables, quantitative and qualitative. Those independent variables that can be controlled in an experiment are called factors.

The analysis of data generated by a multivariable experiment requires identification of the independent variables in the experiment. These not only will be factors (controlled independent variables) but also will be directions of blocking. If one studies the wear for three types of tires, A, B, and C, on each of four automobiles, "tire types" is a factor representing a single qualitative variable at three levels. Automobiles are blocks and represent a single qualitative variable at four levels. The response for a Latin square design depends upon the factors that represent treatments, but it also is affected by two qualitative independent block variables, "rows" and "columns."

It is not possible to present a comprehensive treatment of the analysis of multivariable experiments in a single chapter of an introductory text. However, it is possible to introduce the reasoning upon which one method of analysis, the analysis of variance, is based and to show how the technique is applied to a few common experimental designs.

The application of noise reduction and signal amplification to the design of experiments was illustrated in Chapter 14. In particular, the completely randomized and the randomized block designs were shown to be generalizations of simple designs for the unpaired and paired comparisons of means discussed in Chapter 10. Treatments correspond to combinations of factor levels and identify the different populations of interest to the experimenter. Chapter 15 presents an introduction to the analysis of variance and gives methods for the analysis of the completely randomized, the randomized block, and the Latin square designs.

15.2
The Analysis of Variance

The methodology for the analysis of experiments involving several independent variables can best be explained in terms of the linear probabilistic model of Chapter 11. Although elementary and unified, this approach is not susceptible to the condensation necessary for inclusion in an elementary text. Instead, we shall attempt an intuitive discussion using a procedure known as the analysis of variance. Actually, the two approaches are connected and the analysis of variance can easily be explained in a general way in terms of the linear model. The interested reader can consult an introductory text on this subject by Mendenhall (1968), which is listed in the references.

As the name implies, the analysis-of-variance procedure attempts to analyze the variation of a response and to assign portions of this variation to each of a set of independent variables. The reasoning is that response variables vary only because of variation in a set of unknown independent variables. Since the experimenter will rarely, if ever, include all the variables affecting the response in his experiment, random variation in the response is observed even though all independent variables considered are held constant. The objective of the analysis of variance is to locate important independent variables in a study and to determine how they interact and affect the response.

The rationale underlying the analysis of variance can be indicated best with a symbolical discussion. The actual analysis of variance—that is, "how to do it"—can be illustrated with an example.

You will recall that the variability of a set of n measurements is proportional to the sum of squares of deviations, $SS_y = \sum_{i=1}^{n} (y_i - \bar{y})^2$, and that this quantity is used to calculate the sample variance. **The analysis of variance partitions SS_y, called the total sum of squares of deviations, into parts, each of which is attributed to one of the independent variables in the experiment, plus a remainder that is associated with random error.** This may be shown diagrammatically as indicated in Figure 15.1 for three independent variables.

Figure 15.1 Partitioning of the total sum of squares of deviations

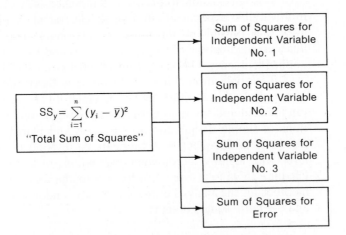

If a multivariable linear regression model were written for the response y, as suggested in Chapter 12, the portion of the total sum of squares of deviations assigned to error would be the sum of squares of deviations of the y values about their respective predicted values obtained from the prediction equation, $\hat{y}$. You will recall that this quantity, represented by the sum of squares of deviations of the y values from their predicted values, was denoted as SSE in Chapters 11 and 12.

For the cases that we consider, and when the response is unrelated to the independent variables, it can be shown that each of the pieces of the total sum of squares of deviations, divided by an appropriate constant, provides an independent and unbiased estimator of σ^2, the variance of the experimental error. When a variable is highly related to the response, its portion (called the "sum of squares" for the variable) of variability will be inflated. This condition can be detected by comparing the estimate of σ^2 for a particular independent variable with that obtained from SSE using an F test (similar to the one encountered in Section 10.8). If the estimate for the independent variable is significantly larger, the F test will reject an hypothesis of "no effect for the independent variable" and thereby produce evidence to indicate a relation to the response.

The logic behind an analysis of variance can be illustrated by considering a familiar example, the comparison of two population means for an unpaired experiment (completely randomized design), which was analyzed in Chapter 10 using a Student's t statistic. We shall commence by giving a graphic and intuitive explanation of the procedure.

Suppose that we have selected random samples of five observations each from two populations, I and II, and that the y values are plotted as shown in Figure 15.2. Note that the $n_1 = 5$ observations from population I lie to the left; the $n_2 = 5$ observations from population II lie to the right. The sample means, $\bar{y}_1$ and $\bar{y}_2$, are shown as horizontal lines in the figure and the deviations of the y values about their respective means are the vertical line segments. Now examine Figure 15.2. Do you think the data provide sufficient evidence to indicate a difference between the two population means? Before we explain, let us look at another figure.

The same two sets of five points are plotted in Figure 15.3 except that the relative distance between the two sets is greater in (b) than in (a) and even greater in (c). Therefore, the distance between $\bar{y}_1$ and $\bar{y}_2$ increases as you move

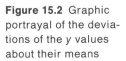

Figure 15.2 Graphic portrayal of the deviations of the y values about their means

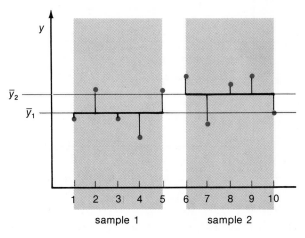

sample 1   sample 2

Observation Number

Figure 15.3 Three fictitious sets of measurements, $n_1 = n_2 = 5$ (the relative positions of points within each set is held constant)

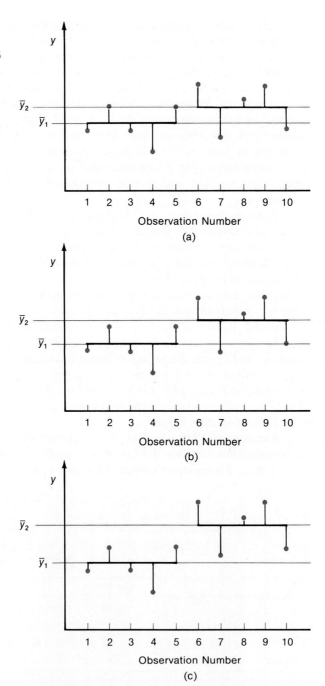

from Figure 15.3(a) to (b) and then to (c), but the relative variation within each set is held constant.

Now view the three plots, (a), (b), and (c), and decide which situation, (a), (b), or (c), provides the greatest evidence to indicate a difference between μ_1 and μ_2. We think you will choose Figure 15.3(c) because that plot shows the greatest difference between sample means in comparison with the variation of the points about their respective sample means. This latter variation was held constant for all three plots.

Note that the population means actually may differ for Figure 15.3(a), but this fact would not be apparent, because the variation of the points about their respective sample means is too large in comparison with the difference between $\bar{y}_1$ and $\bar{y}_2$. Figure 15.4 shows the same difference between sample means as for Figure 15.3(a), but the variation within samples has been reduced. Now it appears that a difference does exist between μ_1 and μ_2.

Figure 15.4 Small amount of within-sample variation in comparison with the difference between sample means

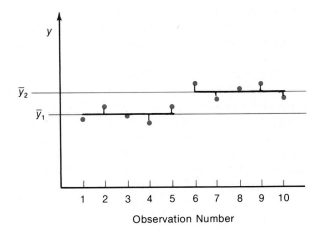

Now let us leave our intuitive discussion and consider the two-sample comparison of means for sample sizes n_1 and n_2. Particularly, we shall want to see how the total sum of squares of deviations can be partitioned into portions corresponding to the difference between the means and another to the variation within the two samples. The total sum of squares of deviations of all $(n_1 + n_2)$ y values about the general mean is

$$\text{Total SS} = \sum_{i=1}^{2} \sum_{j=1}^{n_i} (y_{ij} - \bar{y})^2,$$

where $\bar{y}$ is the average of all $(n_1 + n_2)$ observations contained in the two samples. Then with a bit of algebra you can show that

$$\text{Total SS} = \sum_{i=1}^{2} \sum_{j=1}^{n_i} (y_{ij} - \bar{y})^2 = \underbrace{\sum_{i=1}^{2} n_i(\bar{y}_i - \bar{y})^2}_{\text{SST}} + \underbrace{\sum_{i=1}^{2} \sum_{j=1}^{n_i} (y_{ij} - \bar{y}_i)^2}_{\text{SSE}}$$

where $\bar{y}_i$ is the average of the observations in the ith sample, $i = 1, 2$. The first quantity to the right of the equal sign, called the **sum of squares for treatments, and denoted by the symbol SST,** can be shown (with a bit of algebra) to equal

$$SST = \frac{n_1 n_2}{n_1 + n_2} (\bar{y}_1 - \bar{y}_2)^2.$$

Thus SST, which increases as the difference between $\bar{y}_1$ and $\bar{y}_2$ increases, measures the variation between the sample means. Consequently, SST would be larger for the data of Figure 15.3(c) than for Figure 15.3(a).

The second quantity to the right of the equal sign is the familiar pooled sum of squares of deviations computed in the t test of Section 10.4; namely,

$$SSE = \sum_{j=1}^{n_1} (y_{1j} - \bar{y}_1)^2 + \sum_{j=1}^{n_2} (y_{2j} - \bar{y}_2)^2.$$

It is the sum of the sum of squares of deviations of the y values about their respective sample means. This pooled sum of squares measures within sample variation, a variation that is usually attributed to experimental error, and is consequently called "sum of squares for error" (denoted by the symbol SSE).

The quantities SST and SSE measure the two kinds of variation that we viewed in the graphic representation of Figure 15.3, the **variation between the means of the samples** and the **variation of the observations within samples.** The greater the variation between means (the larger SST) in comparison with the variation within samples (SSE), the greater the weight of evidence to indicate a difference between μ_1 and μ_2. How large is large? When will SST be large enough (relative to SSE) to indicate a real difference between μ_1 and μ_2?

As indicated in Chapter 10,

$$s^2 = MSE = \frac{SSE}{n_1 + n_2 - 2}$$

provides an unbiased estimator of σ^2. (In the language of analysis of variance, s^2 is usually denoted as **MSE,** meaning "mean square for error.") Also, when the null hypothesis is true (that is, $\mu_1 = \mu_2$), SST divided by an appropriate number of degrees of freedom yields a second unbiased estimator of σ^2, which we shall denote as **MST.** For this example, the number of degrees of freedom for MST is equal to 1.

When the null hypothesis is true (that is, $\mu_1 = \mu_2$), **MSE (the mean square for error)** and **MST (the mean square for treatments)** both estimate the same quantity, σ^2, and should be "roughly" of the same magnitude. When the null hypothesis is false and $\mu_1 \neq \mu_2$, MST will probably be larger than MSE.

The preceding discussion, along with a review of the variance ratio, Section 10.8, suggests the use of

$$\frac{MST}{MSE}$$

as a test statistic to test the hypothesis $\mu_1 = \mu_2$ against the alternative, $\mu_1 \neq \mu_2$. Indeed, when both populations are normally distributed, it can be shown that MST and MSE are independent in a probabilistic sense and that

$$F = \frac{\text{MST}}{\text{MSE}}$$

possesses the F probability distribution first discussed in Section 10.8. Disagreement with the null hypothesis is indicated by a large value of F, and hence the rejection region for a given α will be

$$F \geq F_\alpha.$$

Thus **the analysis-of-variance test results in a one-tailed F test.** The degrees of freedom for the F will be those associated with MST and MSE, which we will denote as v_1 and v_2, respectively. Although we have not indicated how one determines v_1 and v_2, in general, $v_1 = 1$ and $v_2 = (n_1 + n_2 - 2)$ for the two-sample experiment described.

Example 15.1

The coded values for the measure of elasticity in plastic, prepared by two different processes, are given below for samples of six drawn randomly from each of the two processes.

Process A	Process B
6.1	9.1
7.1	8.2
7.8	8.6
6.9	6.9
7.6	7.5
8.2	7.9
Total 43.7	48.2

Do the data present sufficient evidence to indicate a difference in mean elasticity for the two processes?

Solution

Although the Student's t could be used as the test statistic for this example, we shall use the analysis-of-variance F test.

The two desired sums of squares of deviations* are

$$\text{SST} = n_1 \sum_{i=1}^{2} (\bar{y}_i - \bar{y})^2$$

$$= \frac{n_1}{2} (\bar{y}_1 - \bar{y}_2)^2 = \frac{6}{2} \left(\frac{43.7}{6} - \frac{48.2}{6} \right)^2$$

$$= 1.6875$$

* *Note:* This formula for SST applies to the special case where $n_1 = n_2$.

and

$$SSE = \sum_{i=1}^{2} \sum_{j=1}^{6} (y_{ij} - \bar{y}_i)^2 = \sum_{j=1}^{6} (y_{1j} - \bar{y}_1)^2 + \sum_{j=1}^{6} (y_{2j} - \bar{y}_2)^2$$

$$= 5.8617.$$

(The reader may verify that **SSE is the pooled sum of squares of the deviations for the two samples discussed in Section 10.4.**) The mean squares for treatment and error are, respectively,

$$MST = \frac{SST}{1} = 1.6875,$$

$$s^2 = MSE = \frac{SSE}{2n_1 - 2} = \frac{5.8617}{10} = .5862.$$

To test the null hypothesis $\mu_1 = \mu_2$, we compute the test statistic

$$F = \frac{MST}{MSE} = \frac{1.6875}{.5862} = 2.88.$$

The critical value of the F statistic for $\alpha = .05$ is 4.96. Although the mean square for treatments is almost three times as large as the mean square for error, it is not large enough to reject the null hypothesis. Consequently, there is not sufficient evidence to indicate a difference between μ_1 and μ_2.

As noted, the purpose of the preceding example was to illustrate the computations involved in a simple analysis of variance. **The F test for comparing two means is equivalent to a Student's t test, because an F statistic with one degree of freedom in the numerator is equal to t^2** (proof omitted). Had the t test been used for Example 15.1, we would have found $t = -1.6967$, which we see satisfies the relationship, $t^2 = (-1.6967)^2 = 2.88 = F$. Similarly, you can see that the relationship, $t^2 = F$, holds for the critical values. The square of $t_{.025} = 2.228$ (used for the two-tailed test with $\alpha = .05$ and $\nu = 10$ degrees of freedom) is equal to $F_{.05} = 4.96$. Since each value of F corresponds to two values of t, one positive and one negative, the F test with one degree of freedom in the numerator always corresponds to a two-tailed t test.

Of what value is the Total SS? The answer is that it provides an easy way to compute SSE. Since the Total SS partitions into SST and SSE, that is,

$$\text{Total SS} = \text{SST} + \text{SSE},$$

then

$$\text{SSE} = \text{Total SS} - \text{SST}.$$

Both the Total SS and SST are easy to compute. Hence one can easily find SSE by substituting into the expression above. For this example

$$\text{Total SS} = \sum_{i=1}^{2} \sum_{j=1}^{6} (y_{ij} - \bar{y})^2 = \sum_{i=1}^{2} \sum_{j=1}^{6} y_{ij}^2 - \frac{\left(\sum_{i=1}^{2} \sum_{j=1}^{6} y_{ij} \right)^2}{12}$$

$$= (\text{sum of squares of all } y \text{ values}) - \text{CM}$$

$$= 711.35 - \frac{(91.9)^2}{12} = 7.5492$$

(the term CM denotes "correction for the mean"). Then

$$\text{SSE} = \text{Total SS} - \text{SST}$$
$$= 7.5492 - 1.6875$$
$$= 5.8617.$$

This is exactly the same value obtained by the tedious computation and pooling of the sums of squares of deviations from the individual samples.

Exercise

15.1. Analyze the data of Exercise 10.56 by the procedure outlined in Section 15.2. Determine whether there is evidence of a difference in mean reaction times for the two stimuli. Test at the $\alpha = .05$ level of significance. If you have not worked Exercise 10.56, analyze the same data using Student's t test (as described in Exercise 10.56). Note that both methods lead to the same conclusion and that the computed values of F and t, for the two methods, are related. That is, $F = t^2$ (this will hold true only for the comparison of *two* population means).

15.3
A Comparison of More Than Two Means

An analysis of variance to detect a difference in a set of more than two population means is a simple generalization of the analysis of variance of Section 15.2. You will recall (Chapter 14) that the random selection of independent samples from p populations is known as a **completely randomized experimental design.**

Assume that independent random samples have been drawn from p normal populations with means $\mu_1, \mu_2, \ldots, \mu_p$, respectively, and variance σ^2. Thus, all populations are assumed to possess equal variances. And, to be completely general, we will allow the sample sizes to be unequal and let n_i, $i = 1, 2, \ldots, p$, be the number in the sample drawn from the ith population. The total number of observations in the experiment will be $n = n_1 + n_2 + \cdots + n_p$.

Let y_{ij} denote the measured response on the jth experimental unit in the ith sample and let T_i and $\bar{T}_i$ represent the total and mean, respectively, for the observations in the ith sample. (The modification in the symbols for sample totals and averages will simplify the computing formulas for the sums of squares.) Then, as in the analysis of variance involving two means,

Total SS = SST + SSE

(proof given in Section 15.4), where

$$\text{Total SS} = \sum_{i=1}^{p} \sum_{j=1}^{n_i} (y_{ij} - \bar{y})^2 = \sum_{i=1}^{p} \sum_{j=1}^{n_i} y_{ij}^2 - \text{CM}$$

$$= \text{(sum of squares of all } y \text{ values)} - \text{CM}$$

$$\text{CM} = \frac{\text{(total of all observations)}^2}{n} = \frac{\left(\sum_{i=1}^{p} \sum_{j=1}^{n_i} y_{ij}\right)^2}{n} = n\bar{y}^2$$

$$\text{SST} = \sum_{i=1}^{p} n_i(\bar{T}_i - \bar{y})^2 = \sum_{i=1}^{p} \frac{T_i^2}{n_i} - \text{CM}$$

$$= \text{(sum of squares of the treatment totals with each square}$$
$$\text{divided by the number of observations in that particular}$$
$$\text{total)} - \text{CM}$$

$$\text{SSE} = \text{Total SS} - \text{SST}.$$

Although the easy way to compute SSE is by subtraction, as just shown, it is interesting to note that SSE is the pooled sum of squares for all p samples and is equal to

$$\text{SSE} = \sum_{i=1}^{p} \sum_{j=1}^{n_i} (y_{ij} - \bar{T}_i)^2.$$

The unbiased estimator of σ^2 based on $(n_1 + n_2 + \cdots + n_p - p)$ degrees of freedom is

$$s^2 = \text{MSE} = \frac{\text{SSE}}{n_1 + n_2 + \cdots + n_p - p}.$$

The mean square for treatments will possess $(p - 1)$ degrees of freedom, that is, one less than the number of means, and

$$\text{MST} = \frac{\text{SST}}{p - 1}.$$

To test the null hypothesis

$$H_0 : \mu_1 = \mu_2 = \cdots = \mu_p$$

against the alternative that at least one of the equalities does not hold, MST is compared with MSE using the F statistic based upon $v_1 = p - 1$ and $v_2 = \left(\sum_{i=1}^{p} n_i - p \right) = n - p$ degrees of freedom. The null hypothesis will be rejected if

$$F = \frac{\text{MST}}{\text{MSE}} > F_\alpha,$$

where F_α is the critical value of F, based on $(p - 1)$ and $(n - p)$ degrees of freedom, for probability of a type I error, α.

Intuitively, the greater the difference between the observed treatment means, $\bar{T}_1, \bar{T}_2, \ldots, \bar{T}_p$, the greater will be the evidence to indicate a difference between their corresponding population means. It can be seen from the formula for SST that $\text{SST} = 0$ when all the observed treatment means are identical because then $\bar{T}_1 = \bar{T}_2 = \cdots = \bar{T}_p = \bar{y}$, and the deviations appearing in SST, $(\bar{T}_i - \bar{y})$, $i = 1, 2, \ldots, p$, will equal zero. As the treatment means get farther apart, the deviations $(\bar{T}_i - \bar{y})$ will increase in absolute value and SST will increase in magnitude. Consequently, the larger the value of SST, the greater will be the weight of evidence favoring a rejection of the null hypothesis. This same line of reasoning will apply to the F tests employed in the analysis of variance for all designed experiments.

The test is summarized as follows:

F Test for Comparing *p* Population Means

1. *Null Hypothesis:* $H_0 : \mu_1 = \mu_2 = \cdots = \mu_p$.
2. *Alternative Hypothesis:* H_a: One or more pairs of population means differ.
3. *Test Statistic:* $F = \text{MST}/\text{MSE}$, where F is based on $v_1 = (p - 1)$ and $v_2 = (n - p)$ degrees of freedom.
4. *Rejection Region:* Reject if $F > F_\alpha$, where F_α lies in the upper tail of the F distribution (with $v_1 = p - 1$ and $v_2 = n - p$) and satisfies the expression $P(F > F_\alpha) = \alpha$.

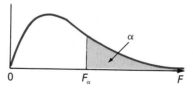

Assumptions:

1. The samples have been randomly and independently selected from their respective populations.
2. The populations are normally distributed with means, $\mu_1, \mu_2, \ldots, \mu_p$ and equal variances, $\sigma_1^2 = \sigma_2^2 = \cdots = \sigma_p^2 = \sigma^2$.

The assumptions underlying the analysis-of-variance F tests should receive particular attention. The samples are assumed to have been randomly selected from the p populations in an independent manner. The populations are assumed to be normally distributed with equal variances, σ^2, and means $\mu_1, \mu_2, \ldots, \mu_p$. Moderate departures from these assumptions will not seriously affect the properties of the test. This is particularly true of the normality assumption.

Example 15.2   Four groups of students were subjected to different teaching techniques and tested at the end of a specified period of time. Because of dropouts in the experimental groups (sickness, transfers, etc.), the number of students varied from group to group. Do the following data present sufficient evidence to indicate a difference in the mean achievement for the four teaching techniques?

	Techniques			
	1	2	3	4
	65	75	59	94
	87	69	78	89
	73	83	67	80
	79	81	62	88
	81	72	83	
	69	79	76	
		90		
T_i	454	549	425	351
$\bar{T}_i$	75.67	78.43	70.83	87.75

Solution

$$CM = \frac{\left(\sum_{i=1}^{4} \sum_{j=1}^{n_i} y_{ij} \right)^2}{n} = \frac{(\text{total of all observations})^2}{n}$$

$$= \frac{(1779)^2}{23} = 137{,}601.8,$$

$$\text{Total SS} = \sum_{i=1}^{4} \sum_{j=1}^{n_i} y_{ij}^2 - CM$$

$$= (\text{sum of squares of all } y \text{ values}) - CM$$

$$= (65)^2 + (87)^2 + (73)^2 + \cdots + (88)^2 - CM$$

$$= 139{,}511 - 137{,}601.8 = 1909.2,$$

$$SST = \sum_{i=1}^{4} \frac{T_i^2}{n_i} - CM$$

= (sum of squares of the treatment totals with each square divided by the number of observations in that particular total) − CM

$$= \frac{(454)^2}{6} + \frac{(549)^2}{7} + \frac{(425)^2}{6} + \frac{(351)^2}{4} - CM$$

$$= 138{,}314.4 - 137{,}601.8$$

$$= 712.6,$$

$$SSE = \text{Total SS} - SST = 1196.6.$$

The mean squares for treatment and error are

$$MST = \frac{SST}{p-1} = \frac{712.6}{3} = 237.5,$$

$$MSE = \frac{SSE}{n_1 + n_2 + \cdots + n_p - p} = \frac{SSE}{n - p} = \frac{1196.6}{19} = 63.0.$$

The test statistic for testing the hypothesis, $\mu_1 = \mu_2 = \mu_3 = \mu_4$, is

$$F = \frac{MST}{MSE} = \frac{237.5}{63.0} = 3.77,$$

where

$$v_1 = (p-1) = 3, \qquad v_2 = \sum_{i=1}^{p} n_i - p = 19.$$

The critical value of F for $\alpha = .05$ is $F_{.05} = 3.13$. Since the computed value of F exceeds $F_{.05}$, we reject the null hypothesis and conclude that the evidence is sufficient to indicate a difference in mean achievement for the four teaching techniques.

You may feel that the above conclusion could have been made on the basis of visual observation of the treatment means. It is not difficult to construct a set of data that will lead the "visual" decision maker to erroneous results.

15.4
Proof of Additivity of the Sums of Squares (Optional)

The proof that

$$\text{Total SS} = SST + SSE$$

for the completely randomized design is presented in this section for the benefit of the interested reader. It may be omitted without loss of continuity.

The proof utilizes the three summation theorems of Chapter 2 and the device of adding and subtracting $\bar{T}_i$ within the expression for the Total SS.

Thus

$$\text{Total } SS = \sum_{i=1}^{p} \sum_{j=1}^{n_i} (y_{ij} - \bar{y})^2$$

$$= \sum_{i=1}^{p} \sum_{j=1}^{n_i} (y_{ij} - \bar{T}_i + \bar{T}_i - \bar{y})^2$$

$$= \sum_{i=1}^{p} \sum_{j=1}^{n_i} [(y_{ij} - \bar{T}_i) + (\bar{T}_i - \bar{y})]^2$$

$$= \sum_{i=1}^{p} \sum_{j=1}^{n_i} [(y_{ij} - \bar{T}_i)^2 + 2(y_{ij} - \bar{T}_i)(\bar{T}_i - \bar{y}) + (\bar{T}_i - \bar{y})^2].$$

Summing first over j, we obtain

$$\text{Total SS} = \sum_{i=1}^{p} \left[\sum_{j=1}^{n_i} (y_{ij} - \bar{T}_i)^2 + 2(\bar{T}_i - \bar{y}) \sum_{j=1}^{n_i} (y_{ij} - \bar{T}_i) + n_i(\bar{T}_i - \bar{y})^2 \right],$$

where

$$\sum_{j=1}^{n_i} (y_{ij} - \bar{T}_i) = T_i - n_i\bar{T}_i = T_i - T_i = 0.$$

Consequently, the middle term in the expression for the Total SS is equal to zero. Then, summing over i, we obtain

$$\text{Total SS} = \sum_{i=1}^{p} \sum_{j=1}^{n_i} (y_{ij} - \bar{T}_i)^2 + \sum_{i=1}^{p} n_i(\bar{T}_i - \bar{y})^2.$$

The first expression on the right is SSE, the pooled sum of squares of deviations of the sample measurements about their respective means. The second is the formula for SST. Therefore,

$$\text{Total SS} = \text{SSE} + \text{SST}.$$

Proof of the additivity of the analysis-of-variance sums of squares for other experimental designs can be obtained in a similar manner. The proofs are omitted.

15.5
An Analysis-of-Variance Table for a Completely Randomized Design

The calculations of the analysis of variance are usually displayed in an analysis-of-variance (ANOVA or AOV) table. The table for the design of Section 15.3 involving p treatment means is shown in Table 15.2. Column 1 shows the source of each sum of squares of deviations; column 2 gives the respective degrees of freedom; columns 3 and 4 give the corresponding sums of squares and mean squares, respectively. A calculated value of F, comparing MST and

Table 15.2 ANOVA table for a comparison of means

Source	d.f.	SS	MS	F
Treatments	$p - 1$	SST	$MST = SST/(p - 1)$	MST/MSE
Error	$n - p$	SSE	$MSE = SSE/(n - p)$	
Total	$n - 1$	Total SS		

MSE, is shown in column 5. Note that the degrees of freedom and sums of squares add to their respective totals.

The ANOVA table for Example 15.2, shown in Table 15.3, gives a compact presentation of the appropriate computed quantities for the analysis of variance.

Table 15.3 ANOVA table for Example 15.2

Source	d.f.	SS	MS	F
Treatments	3	712.6	237.5	3.77
Error	19	1196.6	63.0	
Total	22	1909.2		

15.6
Estimation for the Completely Randomized Design

Confidence intervals for a single treatment mean and the difference between a pair of treatment means, Section 15.3, are very similar to those given in Chapter 10. The confidence interval for the mean of treatment i or the difference between treatments i and j are, respectively,

---

Completely Randomized Design: 100 $(1 - \alpha)$ Percent Confidence Intervals for a Single Treatment Mean and the Difference between Two Treatment Means

A Single Treatment Mean: $\bar{T}_i \pm t_{\alpha/2} s / \sqrt{n_i}$.

The Difference between Two Treatment Means:

$$(\bar{T}_i - \bar{T}_j) \pm t_{\alpha/2} s \sqrt{\frac{1}{n_i} + \frac{1}{n_j}},$$

where

$$s = \sqrt{s^2} = \sqrt{MSE} = \sqrt{\frac{SSE}{n_1 + n_2 + \cdots + n_p - p}}$$

$$n = n_1 + n_2 + \cdots + n_p$$

and $t_{\alpha/2}$ is based upon $(n - p)$ degrees of freedom.

Note that the confidence intervals stated above are appropriate for single treatment means or a comparison of a pair of means selected prior to observation of the data. The stated confidence coefficients are based on random sampling. If one were to look at the data and always compare the largest and smallest sample means, the assumption of randomness would be disturbed. This is because the difference between the largest and smallest sample means you would expect to be larger than for a pair selected at random.

Example 15.3   Find a 95 percent confidence interval for the mean score for teaching technique 1, Example 15.2.

Solution   The 95 percent confidence interval for the mean score is

$$\bar{T}_1 \pm \frac{t_{.025}s}{\sqrt{n_1}}$$

or

$$75.67 \pm \frac{(2.093)(7.94)}{\sqrt{6}}$$

or

$$75.67 \pm 6.78.$$

Thus we infer that the interval 75.67 ± 6.78 or 68.89 to 82.45 encloses the mean score for students subjected to teaching technique 1.

Example 15.4   Find a 95 percent confidence interval for the difference in mean score for teaching techniques 1 and 4, Example 15.2.

Solution   The 95 percent confidence interval for $(\mu_1 - \mu_4)$ is

$$(\bar{T}_1 - \bar{T}_4) \pm t_{\alpha/2}s\sqrt{\frac{1}{n_1} + \frac{1}{n_4}}$$

or

$$(75.67 - (87.75) \pm (2.093)(7.94)\sqrt{\frac{1}{6} + \frac{1}{4}}$$

or

$$-12.08 \pm 10.73.$$

Thus we estimate that the interval -12.08 ± 10.73 or -22.81 to -1.35 encloses the difference in mean scores for teaching techniques 1 and 4. Because all points in the interval are negative, we infer that μ_4 is larger than μ_1. Also, note that the variability in scores within students is rather large. Consequently, the sample sizes should be increased if the experimenter wishes to reduce the width of the confidence interval.

Tips on Problem Solving

The following suggestions apply to all the analyses of variance in this chapter.

1. When calculating sums of squares, be certain to carry at least six significant figures before performing subtractions.
2. Remember, sums of squares can never be negative. If you obtain a negative sum of squares, you have made a mistake in arithmetic.
3. Always check your analysis-of-variance table to make certain that the degrees of freedom sum to the total degrees of freedom, $n - 1$, and that the sum of the sums of squares equals the Total SS.

15.7
A Computer Printout for a Completely Randomized Design

Computer packages for analyses of variance are readily available and are very similar. The purpose of this section is to familiarize you with a typical printout for the analysis of variance for a completely randomized design in case you would like to perform your computations on a computer. As for the regression analysis, Section 11.10, we will use the SAS (Statistical Analysis System) computer package (listed in the references).

So that you will more readily identify the elements of an SAS printout, we have given the analysis variance for Example 15.2. The SAS printout is shown in Table 15.4. You will note that it is broken into two parts, both shaded. The upper table breaks the total sum of squares (called CORRECTED TOTAL)

Table 15.4 SAS computer printout for Example 15.2

STATISTICAL ANALYSIS SYSTEM

ANALYSIS OF VARIANCE PROCEDURE

DEPENDENT VARIABLE: Y

SOURCE	DF	SUM OF SQUARES	MEAN SQUARE	F VALUE	PR > F	R-SQUARE	C.V.
MODEL	3	712.58643892	237.52881297	3.77	0.0280	0.373235	10.2602
ERROR	19	1196.63095238	62.98057644		STD DEV		Y MEAN
CORRECTED TOTAL	22	1909.21739130			7.93603027		77.34782609

SOURCE	DF	ANOVA SS	F VALUE	PR > F
TRTMENTS	3	712.58643892	3.77	0.0280

into two sources, MODEL and ERROR. You can see that the numbers appearing in the ERROR row (d.f., SS, MS) of Table 15.4 are identical (except for rounding) to the numbers appearing in the "Error" row of the ANOVA table, Table 15.3. The MODEL row in the printout, Table 15.4, corresponds to all effects other than error. In this case, there is only one other source of variation, namely treatments. Consequently, the row identified as MODEL gives the values of d.f., SST, MST and F. You can see that these numbers are identical to the numbers that appear in the "Treatments" row of the ANOVA table, Table 15.3.

The lower shaded table gives a breakdown of the MODEL source of variation into its component sources. In this particular case, there is only one source, TREATMENTS. Consequently, this line repeats the value of SST (712.58 . . .), gives the computed F value (3.77) for a test of the null hypothesis, "no difference between treatment means," and gives the probability of observing a value of F as large or larger than 3.77, given the null hypothesis is true. This probability, 0.0280, is the observed significance level or p value for the test.

The value of s, $s = 7.936 . . .$, is given at the right of the printout. This is useful in calculating confidence intervals. We omit discussion of the last two columns of the printout because they are not relevant to the analysis of variance that we have given in Section 15.5.

Exercises

ψ 15.2. A clinical psychologist wished to compare three methods for reducing hostility levels in university students. A certain psychological test (HLT) was used to measure the degree of hostility. High scores on this test were taken to indicate great hostility. Eleven students obtaining high and nearly equal scores were used in the experiment. Five were selected at random from among the 11 problem cases and treated by method A. Three were taken at random from the remaining six students and treated by method B. The other three students were treated by method C. All treatments continued throughout a semester. Each student was given the HLT test again at the end of the semester, with the following results:

Method	Scores on the HLT Tests				
A	73	83	76	68	80
B	54	74	71		
C	79	95	87		

a. Perform an analysis of variance for this experiment.

b. Do the data provide sufficient evidence to indicate a difference in mean student response for the three methods after treatment?

15.3. Refer to Exercise 15.2. Let μ_A and μ_B, respectively, denote the mean scores at the end of the semester for the populations of extremely hostile students who are treated throughout that semester by method A and method B.

a. Find a 95 percent confidence interval for μ_A.

b. Find a 95 percent confidence interval for μ_B.

c. Find a 95 percent confidence interval for $\mu_A - \mu_B$.

d. Is it correct to claim that the confidence intervals found in parts (a), (b), and (c) are jointly valid?

15.4. Three groups of fourth graders were randomly selected and assigned, one group each, to three different physical exercise programs to determine whether the programs were effective in increasing the childrens' abilities to throw an object. Twenty-eight students were involved in the experiment, ten in a control group (no exercise) and nine each assigned to two different exercise regimes, each of which lasted four weeks. The velocity at which a child was able to throw a test ball was measured before and after the four weeks of exercise and the gain (or loss) in velocity y (in feet per second) was recorded. A table of mean gains for the three groups and a partially completed analysis-of-variance table for the data are as shown.

	Sample Means				ANOVA Table		
Control	Exercise Regime *A*	Exercise Regime *B*		Source	d.f.	SS	MS
−1.34	.32	3.69		Exercise regimes	—	64.31	—
				Error	—	—	—
				Total	—	402.33	

a. Fill in the missing numbers in the analysis-of-variance table.

b. Do the data provide sufficient evidence to indicate a difference in population means for the three groups? Explain the implications of the test results.

c. Find a 95 percent confidence interval for the difference in mean gain between children in the control group versus those on exercise regime *B*.

d. Find a 95 percent confidence interval for the mean gain for children on exercise regime *B*.

15.5. An experiment was conducted to compare the price of a loaf of bread (a particular brand) at four city locations. Four stores were randomly sampled in locations 1, 2, and 3 but only two were selected from location 4 (only two carried the brand). Note that a completely randomized design was employed. Conduct an analysis of variance for the data.

Location	Prices (cents)			
1	79	83	85	81
2	78	81	84	83
3	74	79	75	78
4	89	90		

a. Do the data provide sufficient evidence to indicate a difference in mean price of the bread in stores located in the four areas of the city?

b. Suppose that prior to seeing the data, we wished to compare the mean prices between locations 1 and 4. Estimate the difference in means using a 95 percent confidence interval.

15.6. An experiment was conducted to compare the effectiveness of three training programs, A, B, and C, in training assemblers of a piece of electronic equipment. Fifteen employees were randomly assigned, 5 each, to the three programs. After completion of the courses, each person was required to assemble four pieces of the equipment and the average length of time required to complete the assembly was recorded. Due to resignation from the company, only 4 employees completed program A and only 3 completed B. The data are shown in the accompanying table. An SAS computer printout of the analysis of variance for the data is also shown below. Use the information in the printout to answer questions (a)–(d).

Training Program	Average Assembly Time (min)				
A	59	64	57	62	
B	52	58	54		
C	58	65	71	63	64

STATISTICAL ANALYSIS SYSTEM

ANALYSIS OF VARIANCE PROCEDURE

DEPENDENT VARIABLE: Y

SOURCE	DF	SUM OF SQUARES	MEAN SQUARE	F VALUE	PR > F	R-SQUARE	C.V.
MODEL	2	170.45000000	85.22500000	5.70	0.0251	0.559005	6.3802
ERROR	9	134.46666667	14.94074074		STD DEV		Y MEAN
CORRECTED TOTAL	11	304.91666667			3.86532544		60.58333333

SOURCE	DF	ANOVA SS	F VALUE	PR > F
TRTMENTS	2	170.45000000	5.70	0.0251

a. Do the data provide sufficient evidence to indicate a difference in mean assembly time for people trained by the three programs? Give the p value for the test and interpret its value.

b. Find a 90 percent confidence interval for the difference in mean assembly time between persons trained by programs A and B.

c. Find a 90 percent confidence interval for the mean assembly time for persons trained in program A.

d. Do you think the data will satisfy (approximately) the assumption that they have been selected from normal populations? Why?

15.7. An ecological study was conducted to compare rate of growth of vegetation at four swampy undeveloped sites and to determine the cause of any differences that might be observed. Part of the study involved the measurement of leaf lengths of a particular plant species at a preselected date in May. Six plants were randomly selected at each of the four sites to be used in the comparison. The following data represent the mean leaf length per plant, in centimeters, for a random sample of ten leaves per plant.

Location	Mean Leaf Length (cm)					
1	5.7	6.3	6.1	6.0	5.8	6.2
2	6.2	5.3	5.7	6.0	5.2	5.5
3	5.4	5.0	6.0	5.6	4.9	5.2
4	3.7	3.2	3.9	4.0	3.5	3.6

a. You will recall that the test and estimation procedures for an analysis of variance require that the observations be selected from normally distributed (at least, roughly so) populations. Why might you feel reasonably confident that your data satisfy this assumption?

b. Do the data provide sufficient evidence to indicate a difference in mean leaf length among the four locations?

c. Suppose, prior to seeing our data, that we decided to compare the mean leaf length between locations 1 and 4. Test the null hypothesis that $\mu_1 = \mu_4$ against the alternative that $\mu_1 \neq \mu_4$.

d. Refer to part (c). Find a 90 percent confidence interval for $(\mu_1 - \mu_4)$.

e. Rather than use an analysis-of-variance F test, it would seem simpler to examine one's data, select the two locations that have the smallest and largest sample mean lengths, and then compare these two means using a Student's t test. If there is evidence to indicate a difference in these means, there is clearly evidence of a difference among the four. (If you were to use this logic, there would be no need for the analysis-of-variance F test.) Explain why this procedure is invalid.

15.8. Water samples were taken at four different locations in a river to determine whether the quantity of dissolved oxygen, a measure of water pollution, varied from one location to another. Locations 1 and 2 were selected above an industrial plant, one near the shore and the other in midstream; location 3 was adjacent to the industrial water discharge for the plant, and location 4 was slightly downriver in midstream. Five water specimens were randomly selected at each location, but one specimen, corresponding to location 4, was lost in the laboratory. The data are shown below (the greater the pollution, the lower will be the dissolved oxygen readings).

Location	Mean Dissolved Oxygen Content				
1	5.9	6.1	6.3	6.1	6.0
2	6.3	6.6	6.4	6.4	6.5
3	4.8	4.3	5.0	4.7	5.1
4	6.0	6.2	6.1	5.8	

a. Do the data provide sufficient evidence to indicate a difference in mean dissolved oxygen content for the four locations?

b. Compare the mean dissolved oxygen content in midstream above the plant with the mean content adjacent to the plant (location 2 versus location 3). Use a 95 percent confidence interval.

15.9. An experiment was conducted to examine the effect of age on heart rate when a person is subjected to a specific amount of exercise. Ten male subjects were randomly selected from four age groups, 10–19, 20–39, 40–59, and 60–69. Each subject walked a treadmill at a fixed grade for a period of 12 minutes and the increase in heart rate, the difference before and after exercise, was recorded (in beats per minute). These data are shown in the table.

	Age			
10–19	**20–39**	**40–59**	**60–69**
29	24	37	28
33	27	25	29
26	33	22	34
27	31	33	36
39	21	28	21
35	28	26	20
33	24	30	25
29	34	34	24
36	21	27	33
22	32	33	32
Total 309	275	295	282

a. Do the data provide sufficient evidence to indicate a difference in mean increase in heart rate among the four age groups? Test by using $\alpha = .05$.

b. Find a 90 percent confidence interval for the difference in mean increase in heart rate between the 10–19-years age group and the 60–69-years age group.

c. Find a 90 percent confidence interval for the mean increase in heart rate for the 20–39-years age group.

d. Approximately how many people would you need in each group if you wished to be able to estimate a group mean correct to within 2 beats per minute with probability equal .95?

15.8
The Analysis of Variance for a Randomized Block Design

The method for constructing a randomized block design was presented in Section 14.5.

The randomized block design implies the presence of two qualititive independent variables, "blocks" and "treatments." Consequently, the total sum of squares of deviations of the response measurements about their mean may be partitioned into three parts: the sums of squares for blocks, treatments, and error.

Denote the total and mean of all observations in block i as B_i and $\bar{B}_i$, respectively. Similarly, let T_j and $\bar{T}_j$ represent the total and the mean for all observations receiving treatment j. Then, for a randomized block design involving b blocks and p treatments,

$$\text{Total SS} = \text{SSB} + \text{SST} + \text{SSE},$$

$$\text{Total SS} = \sum_{i=1}^{b} \sum_{j=1}^{p} (y_{ij} - \bar{y})^2 = \sum_{i=1}^{b} \sum_{j=1}^{p} y_{ij}^2 - \text{CM},$$

$$= (\text{sum of squares of all } y \text{ values}) - \text{CM}$$

$$\text{SSB} = p \sum_{i=1}^{b} (\bar{B}_i - \bar{y})^2 = \frac{\sum_{i=1}^{b} B_i^2}{p} - \text{CM},$$

$$= \frac{\text{sum of squares of all block totals}}{\text{number of observations in a single total}} - \text{CM}$$

$$\text{SST} = b \sum_{j=1}^{p} (\bar{T}_j - \bar{y})^2 = \frac{\sum_{j=1}^{p} T_j^2}{b} - \text{CM},$$

$$= \frac{\text{sum of squares of all treatment totals}}{\text{number of observations in a single total}} - \text{CM}$$

where

$$\bar{y} = (\text{average of all } n = bp \text{ observations}) = \frac{\sum_{i=1}^{b} \sum_{j=1}^{p} y_{ij}}{n}$$

and

$$\text{CM} = \frac{(\text{total of all observations})^2}{n} = \frac{\left(\sum_{i=1}^{b} \sum_{j=1}^{p} y_{ij}\right)^2}{n}, \qquad \text{where } n = bp.$$

The analysis of variance for the randomized block design is presented in Table 15.5. The degrees of freedom associated with each sum of squares is shown in the second column. Mean squares are calculated by dividing the sums of squares by their respective degrees of freedom.

Table 15.5 ANOVA table for a randomized block design

Source	d.f.	SS	MS	F
Blocks	$b - 1$	SSB	$\text{SSB}/(b - 1)$	MSB/MSE
Treatments	$p - 1$	SST	$\text{SST}/(p - 1)$	MST/MSE
Error	$n - b - p + 1$	SSE	MSE	
Total	$n - 1$	Total SS		

To test the null hypothesis "there is no difference in treatment means," we use the F statistic,

$$F = \frac{\text{MST}}{\text{MSE}}$$

and reject if $F > F_\alpha$ based on $v_1 = (p - 1)$ and $v_2 = (n - b - p + 1)$ degrees of freedom.

Blocking not only reduces the experimental error, it also provides an opportunity to see whether evidence exists to indicate a difference in the mean response for blocks. Under the null hypothesis that there is no difference in mean response for blocks, MSB provides an unbiased estimator for σ^2 based on $(b - 1)$ degrees of freedom. Where a real difference exists in the block means, MSB will tend to be inflated in comparison with MSE. Therefore,

$$F = \frac{\text{MSB}}{\text{MSE}}$$

can be used as a test statistic to test for differences among block means. As in the test for treatments, the rejection region for the test will be

$$F > F_\alpha,$$

based on $v_1 = b - 1$ and $v_2 = n - b - p + 1$ degrees of freedom.

F Test for Comparing p Treatments, Using a Randomized Block Design

1. *Null Hypothesis:* H_0: The population treatment means are equal.
2. *Alternative Hypothesis:* H_a: One or more pairs of population treatment means differ.
3. *Test Statistic:* $F = \text{MST/MSE}$, where F is based on $v_1 = (p - 1)$ and $v_2 = (n - b - p + 1)$ degrees of freedom.
4. *Rejection Region:*   Reject if $F > F_\alpha$.

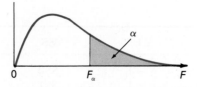

Assumptions:

1. The treatment populations are normally distributed with means, $\mu_1, \mu_2, \ldots, \mu_p$ with common variances equal to σ^2.

2. The p treatments are randomly assigned, one treatment to each of the p experimental units within a block.

Example 15.5

A stimulus–response experiment involving three treatments was laid out in a randomized block design using four subjects. The response was the length of time to reaction measured in seconds. The data (treatment identification numbers are circled) are as follows:

Subject

1	2	3	4
① 1.7	③ 2.1	① .1	② 2.2
③ 2.3	① 1.5	② 2.3	① .6
② 3.4	② 2.6	③ .8	③ 1.6

Do the data present sufficient evidence to indicate a difference in the mean response for stimuli (treatments)? Subjects?

Solution

The treatment and block totals are as follows: $T_1 = 3.9$, $T_2 = 10.5$, $T_3 = 6.8$, $B_1 = 7.4$, $B_2 = 6.2$, $B_3 = 3.2$, $B_4 = 4.4$.

The sums of squares for the analysis of variance are shown individually below, and jointly in the analysis-of-variance table. Thus

$$\text{CM} = \frac{(\text{Total})^2}{n} = \frac{(21.2)^2}{12} = 37.4533,$$

$$\text{Total SS} = \sum_{i=1}^{4}\sum_{j=1}^{3}(y_{ij} - \bar{y})^2 = \sum_{i=1}^{4}\sum_{j=1}^{3} y_{ij}^2 - \text{CM}$$

$$= (\text{sum of squares of all } y \text{ values}) - \text{CM}$$

$$= (1.7)^2 + (2.3)^2 + \cdots + (1.6)^2 - \text{CM}$$

$$= 46.8600 - 37.4533 = 9.4067,$$

$$\text{SSB} = \frac{\sum_{i=1}^{4} B_i^2}{3} - \text{CM}$$

$$= \frac{\text{sum of squares of all block totals}}{\text{number of observations in a single total}} - \text{CM}$$

$$= \frac{(7.4)^2 + (6.2)^2 + (3.2)^2 + (4.4)^2}{3} - \text{CM}$$

$$= 40.9333 - 37.4533 = 3.4800,$$

$$SST = \frac{\sum_{j=1}^{3} T_j^2}{4} - CM$$

$$= \frac{\text{sum of squares of all treatment totals}}{\text{number of observations in a single total}} - CM$$

$$= \frac{(3.9)^2 + (10.5)^2 + (6.8)^2}{4} - CM$$

$$= 42.9250 - 37.4533 = 5.4717,$$

$$SSE = \text{Total SS} - SSB - SST$$

$$= 9.4069 - 3.4800 - 5.4717 = .4550.$$

The analysis-of-variance table for Example 15.5 is shown in Table 15.6.

Table 15.6 ANOVA table for Example 15.5

Source	d.f.	SS	MS	F
Blocks	3	3.4767	1.16	15.26
Treatments	2	5.4767	2.74	36.05
Error	6	.4533	.076	
Total	11	9.4067		

We use the ratio of mean-square treatments to mean-square error to test an hypothesis of no difference in the expected response for treatments. Thus,

$$F = \frac{MST}{MSE} = \frac{2.74}{.076} = 36.05.$$

The critical value of the F statistic ($\alpha = .05$) for $v_1 = 2$ and $v_2 = 6$ degrees of freedom is $F_{.05} = 5.14$. Since the computed value of F exceeds the critical value, there is sufficient evidence to reject the null hypothesis and conclude that a real difference does exist in the expected response for the four stimuli.

A similar test may be conducted for the null hypothesis that no difference exists in the mean response for subjects. Rejection of this hypothesis would imply that subject-to-subject variability does exist, and that blocking is desirable. The computed value of F based on $v_1 = 3$ and $v_2 = 6$ degrees of freedom is

$$F = \frac{MSB}{MSE} = \frac{1.16}{.076} = 15.26.$$

Since this value of F exceeds the corresponding tabulated critical value, $F_{.05} = 4.76$, we reject the null hypothesis and conclude that a real difference exists in the expected response in the group of subjects.

15.9
Estimation for the Randomized Block Design

The confidence interval for the difference between a pair of means is exactly the same as for the completely randomized design, Section 15.5. It is:

Randomized Block Design: A $100(1 - \alpha)$ Percent Confidence Interval for the Difference between Two Treatment Means

$$(\bar{T}_i - \bar{T}_j) \pm t_{\alpha/2} \, s \sqrt{\frac{2}{b}},$$

where $n_i = n_j = b$, the number of observations contained in a treatment mean, and $t_{\alpha/2}$ is based upon $(n - b - p + 1)$ degrees of freedom.

The difference between the confidence intervals for the completely randomized and the randomized block designs is that s, appearing in the expression above, will tend to be smaller than for the completely randomized design.

Similarly, one may construct a $100(1 - \alpha)$ percent confidence interval for the difference between a pair of block means. Each block contains p observations corresponding to the p treatments. Therefore, the confidence interval is

---

Randomized Block Design: A $100(1 - \alpha)$ Percent Confidence Interval for the Difference between Two Block Means

$$(\bar{B}_i - \bar{B}_j) \pm t_{\alpha/2} \, s \sqrt{\frac{2}{p}},$$

where $t_{\alpha/2}$ is based upon $(n - b - p + 1)$ degrees of freedom.

Example 15.6

Construct a 95 percent confidence interval for the difference between treatments 1 and 2, Example 15.5.

Solution

The confidence interval for the difference in mean response for a pair of treatments is

$$(\bar{T}_i - \bar{T}_j) \pm t_{\alpha/2} \, s \sqrt{\frac{2}{b}},$$

where for our example, $t_{.025}$ is based upon 6 degrees of freedom. Then $t_{.025} = 2.447$ and $s = \sqrt{MSE} = \sqrt{.076} = .28$. Substituting these values along with the values of $\bar{T}_1$ and $\bar{T}_2$, we have

$$(.98 - 2.63) \pm (2.447)(.28)\sqrt{\frac{2}{4}}$$

or

$$-1.65 \pm .48.$$

Thus we infer that the interval $-1.65 \pm .48$ or -2.13 to -1.17, encloses the difference in mean reaction times for stimuli 1 and 2. Because all points in the interval are negative, we conclude that the mean time to react for stimulus 2 is larger than for stimulus 1.

Tips on Problem Solving

Be careful of this point: Unless the blocks have been randomly selected from a population of blocks, you cannot obtain a confidence interval for a single treatment mean. This is because the sample treatment mean is biased by the positive and negative effects that the blocks have on the response.

The same comment applies to the Latin square design of Section 15.12. Further discussion of this topic can be found in the references.

15.10
A Computer Printout for a Randomized Block Design

The SAS computer printout for the randomized block data, Example 15.5, is shown in Table 15.7. As for the printout of the analysis of variance for a completely randomized design, Table 15.4, we have shaded two portions of the printout.

Table 15.7 SAS computer printout for Example 15.5

STATISTICAL ANALYSIS SYSTEM

ANALYSIS OF VARIANCE PROCEDURE

DEPENDENT VARIABLE: Y

SOURCE	DF	SUM OF SQUARES	MEAN SQUARE	F VALUE	PR > F	R-SQUARE	C.V.
MODEL	5	8.95166667	1.79033333	23.61	0.0007	0.951630	15.5875
ERROR	6	0.45500000	0.07583333		STD DEV		Y MEAN
CORRECTED TOTAL	11	9.40666667			0.27537853		1.76666667

SOURCE	DF	ANOVA SS	F VALUE	PR > F
BLOCKS	3	3.48000000	15.30	0.0032
TRTMENTS	2	5.47166667	36.08	0.0005

The upper table gives a breakdown of the Total SS into two parts, one corresponding to MODEL and the second to ERROR. The numbers appearing in the row corresponding to ERROR are identical (except for rounding error) to the numbers in the "Error" row of the ANOVA table, Table 15.6. The row corresponding to MODEL gives the totals of d.f. and SS corresponding to BLOCKS and TREATMENTS. A breakdown of these totals is given in the lower table. There you see that the degrees of freedom and sums of squares for BLOCKS and TREATMENTS correspond to the quantities given in the ANOVA table, Table 15.6.

Notice that MSB and MST are not given in either table but that they can easily be calculated from the sums of squares if they are desired. The lower table gives the F values and significance levels for tests of the null hypotheses of "no difference between block means" ($F = 15.30$) and "no difference between treatment means" ($F = 36.08$). The value of s, $s = .275 \ldots$, given in the sixth column, can be used to calculate confidence intervals for the difference between any pair of treatment or block means. The slight discrepancies between the numbers in the ANOVA table, Table 15.6, and those in the computer printout, Table 15.7, are due to rounding errors.

Exercises

15.10. A study was conducted to compare automobile gasoline mileage for three brands of gasoline, A, B, and C. Four automobiles, all of the same make and model, were employed in the experiment and each gasoline brand was tested in each automobile. Using each brand within the same automobile has the effect of eliminating (blocking out) automobile-to-automobile variability. The data, in miles per gallon, are as follows:

Gasoline Brand	Automobile			
	1	2	3	4
A	15.7	17.0	17.3	16.1
B	17.2	18.1	17.9	17.7
C	16.1	17.5	16.8	17.8

a. Do the data provide sufficient evidence to indicate a difference in mean mileage per gallon for the three gasolines?

b. Is there evidence of a difference in mean mileage for the four automobiles?

c. Suppose that *prior to looking at the data*, we had decided to compare the mean mileage per gallon for gasoline brands A and B. Find a 90 percent confidence interval for this difference.

15.11. An experiment was conducted to compare the effect of four different chemicals, A, B, C, and D, in producing water resistance in textiles. A strip of material, randomly selected from a bolt, was cut into four pieces and the pieces were randomly assigned to receive one of the four chemicals, *A*, *B*, *C*, or *D*. This process was replicated three times, thus producing a randomized block design. The design, with moisture-resistance measurements, is as shown (low readings indicate low moisture penetration). An SAS computer printout of the analysis of variance for the data is also presented. Use the information in the printout to answer the following questions:

a. Do the data provide sufficient evidence to indicate a difference in the mean moisture penetration for fabric treated with the four chemicals?

b. Do the data provide evidence to indicate that blocking increased the amount of information in the experiment?

c. Find a 95 percent confidence interval for the difference in mean moisture penetration for fabrics treated by chemicals *A* and *D*. Interpret the interval.

Blocks (Bolt Samples)

1	2	3
C 9.9	D 13.4	B 12.7
A 10.1	B 12.9	D 12.9
B 1.1.4	A 12.2	C 11.4
D 12.1	C 12.3	A 11.9

STATISTICAL ANALYSIS SYSTEM

ANALYSIS OF VARIANCE PROCEDURE

DEPENDENT VARIABLE: Y

SOURCE	DF	SUM OF SQUARES	MEAN SQUARE	F VALUE	PR > F	R-SQUARE	C.V.
MODEL	5	12.37166667	2.47433333	27.75	0.0004	0.958549	2.5023
ERROR	6	0.53500000	0.08916667		STD DEV		Y MEAN
CORRECTED TOTAL	11	12.90666667			0.29860788		11.93333333

SOURCE	DF	ANOVA SS	F VALUE	PR > F
BLOCKS	2	7.17166667	40.21	0.0003
TRTMENTS	3	5.20000000	19.44	0.0017

15.12. An experiment was conducted to determine the effect of three methods of soil preparation on the first-year growth of slash pine seedlings. Four locations (state forest lands) were selected and each location was divided into three plots. Since it was felt that soil fertility within a location was more homogeneous than between locations, a randomized block design was employed using locations as blocks. The methods of soil preparation were *A* (no preparation), *B* (light fertilization), and *C* (burning). Each soil preparation was randomly applied to a plot within each location. On each plot the same number of seedlings were planted and the observation recorded was the average first-year growth of the seedlings on each plot.

Soil Preparation	Location			
	1	2	3	4
A	11	13	16	10
B	15	17	20	12
C	10	15	13	10

a. Conduct an analysis of variance. Do the data provide sufficient evidence to indicate a difference in the mean growth for the three soil preparations?

b. Is there evidence to indicate a difference in mean rates of growth for the four locations?

c. Use a 90 percent confidence interval to estimate the difference in mean growth for methods A and B.

15.13. A study was conducted to compare the effect of three levels of digitalis on the level of calcium in the heart muscle of dogs. A description of the actual experimental procedure is omitted, but it is sufficient to note that the general level of calcium uptake varies from one animal to another so that comparison of digitalis levels (treatments) had to be blocked on heart muscles. That is, the tissue for a heart muscle was regarded as a block and comparisons of the three treatments were made within a given muscle. The calcium uptakes for the three levels of digitalis, A, B, and C, were compared based on the heart muscle of four dogs. The results are as follows:

Dogs			
1	2	3	4
A 1342	C 1698	B 1296	A 1150
B 1608	B 1387	A 1029	C 1579
C 1881	A 1140	C 1549	B 1319

a. Calculate the sums of squares for this experiment and construct an analysis-of-variance table.

b. How many degrees of freedom are associated with SSE?

c. Do the data present sufficient evidence to indicate a difference in the mean uptake of calcium for the three levels of digitalis?

d. Do the data indicate a difference in the mean uptake in calcium for the four heart muscles?

e. Give the standard deviation of the difference between the mean calcium uptake for two levels of digitalis.

f. Find a 95 percent confidence interval for the difference in mean response between treatments A and B.

15.14. An experiment was conducted to investigate the toxic effect of three chemicals, A, B, and C, on the skin of rats. One-inch squares of skin were treated with the chemicals and then scored from 0 to 10 depending on the degree of irritation. Three adjacent 1-inch squares were marked on the backs of eight rats and each of the three chemicals was applied to each rat. Thus the experiment was blocked on rats to eliminate the variation in skin sensitivity from rat to rat. The data are as follows:

				Rats				
	1	**2**	**3**	**4**	**5**	**6**	**7**	**8**
	B 5	A 9	A 6	C 6	B 8	C 5	C 5	B 7
	A 6	C 4	B 9	B 8	C 8	A 5	B 7	A 6
	C 3	B 9	C 3	A 5	A 7	B 7	A 6	C 7

a. Do the data provide sufficient evidence to indicate a difference in the toxic effect of the three chemicals?

b. Estimate the difference in mean score for chemicals A and B using a 95 percent confidence interval.

15.15. A building contractor employs three construction engineers, A, B, and C, to estimate and bid on jobs. To determine whether one tends to be a more conservative (or liberal) estimator than the others, the contractor selects four projected construction jobs and has each estimator independently estimate the cost (dollars per square foot) of each job. The data are shown in the table.

Estimator (Treatments)	Construction Job (Blocks)				Total
	1	**2**	**3**	**4**	**Total**
A	35.10	34.50	29.25	31.60	130.45
B	37.45	34.60	33.10	34.40	139.55
C	36.30	35.10	32.45	32.90	136.75
Total	108.85	104.20	94.80	98.90	406.75

a. Do the data provide sufficient evidence to indicate a difference in the mean building costs estimated by the three estimators? Test by using $\alpha = .05$.

b. Find a 90 percent confidence interval for the difference in the mean of the estimates produced by estimators A and B. Interpret the interval.

c. Do the data support the contention that the mean estimate of the cost per square foot varies from job to job?

15.11
An Analysis of the Difference in Car-Insurance Costs for Four Locales

We mentioned in the case study that a randomized block design was employed for this experiment. An insurance quote for a standard policy was obtained for each of the four locales for the same insurance company. The reasoning was that a single company has an established rate-setting policy and that differences in quotes between locales will be more stable *within* companies than *among* companies. Whether or not this logic is reasonable can be examined by testing the block means.

The SAS printout for the analysis of variance is shown in Table 15.8. The key portions of the analysis of variance are shaded and numbered.

Table 15.8 SAS ANOVA printout for the car-insurance data

STATISTICAL ANALYSIS SYSTEM

ANALYSIS OF VARIANCE PROCEDURE

DEPENDENT VARIABLE: PRICE

SOURCE	DF	SUM OF SQUARES	MEAN SQUARE	F VALUE	PR > F	R-SQUARE	C.V.
MODEL	8	721941.33333333	90242.66666667	30.53	0.0001	0.942142	13.8000
ERROR	15	44355.62500000	2955.70833333		STD DEV		PRICE MEAN
CORRECTED TOTAL	27	766276.95833333			54.36642653		393.95833333

SOURCE	DF	ANOVA SS	F VALUE	PR > F
LOCALE	3	710920.12500000	80.17	0.0001
COMPANY	5	11021.20833333	0.75	0.6015

1. The ANOVA table, marked as shaded area 1, shows the Total SS as CORRECTED TOTAL, with 23 d.f., as 766276.95833333. SSE, with 15 d.f., is 44335.62500000. The sum of squares for MODEL is equal to the sum of the sums of squares for treatments and blocks as indicated in shaded area 2.

2. Shaded area 2 gives a breakdown of the sum of squares for MODEL into its two components: sum of squares for treatments (LOCALE), with 3 d.f.; and sum of squares for blocks (COMPANY), with 5 d.f. These sums of squares are, respectively,

SST = 710920.12500000,

SSB = 11021.20833333.

The F values for testing equality of treatment means or block means and their observed significance levels are shown under the columns marked F VALUE and $PR > F$ as,

	F	p value
Locale	80.17	0.0001
Company	0.75	0.6015

The analysis of variance, Table 15.8, shows that there is a substantial difference in the mean car-insurance rates from one locale to another. There is insufficient evidence to indicate a difference in the mean car-insurance rates among the six insurance companies.

3. Shaded area 3 gives the standard deviation, $s = 54.36642653$. This quantity would be needed to construct a confidence interval for the difference between a pair of treatment means.

Suppose that we want a 95 percent confidence interval for the difference in mean rates between Chicago and Topeka. The interval would be,

$$(\bar{T}_C - \bar{T}_T) \pm t_{.025}s \sqrt{\frac{1}{6} + \frac{1}{6}} = (\$690 - \$301.83) \pm (2.131)(54.37)\sqrt{\frac{1}{3}}$$

$$- \$388.17 \pm 66.89.$$

Thus, we estimate the difference in mean rates between Chicago and Topeka to be between $321.28 and $455.06.

15.12
The Analysis of Variance for a Latin Square Design

The method for constructing a Latin square design for comparing p treatments is presented in Section 14.5. The purpose of the design is to remove unwanted variation as might occur in the mechanized application of icing to cakes on a conveyor belt. Variation in the thickness of icing could occur across the belt due to the variation in pressure at the applicator nozzles. Similarly, the thickness of icing could vary somewhat along the length of the belt due to variations in the consistency of the icing supplied to the machine. Now suppose that we wish to compare three different types of cake mixes, A, B, and C, that result in different porosities, which affect absorption of the icing into the cakes. Then the thickness of the resulting icing, y, could be compared for the three treatments (mixes) by employing a 3×3 Latin square design. Each mix would appear in each column (across the conveyor belt) and in each row as one proceeds down the belt. The design configuration is shown in Figure 15.5.

The three independent variables in a Latin square design are "rows," "columns," and treatments. All are qualitative variables, although the treatments could be levels of a single quantitative factor or combinations of levels

Figure 15.5 A 3 × 3 Latin square design

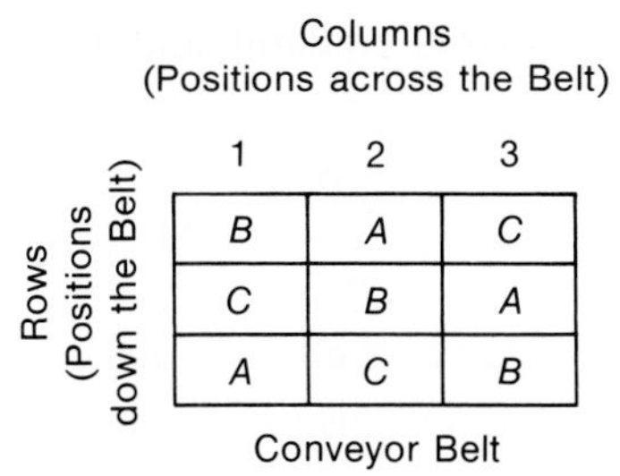

Columns
(Positions across the Belt)

Conveyor Belt

for two or more factors. Thus the total variation in an analysis of variance can be partitioned into four parts, one each corresponding to the variation in rows, columns, treatments, and experimental error. The analysis-of-variance table for a $p \times p$ Latin square design is shown in Table 15.9. As for previous designs, the four sums of squares add to the total sum of squares of deviations, Total SS.

Table 15.9 ANOVA table for a Latin square design

Source	d.f.	SS	MS	F
Rows	$p - 1$	SSR	$SSR/(p - 1)$	MSR/MSE
Columns	$p - 1$	SSC	$SSC/(p - 1)$	MSC/MSE
Treatments	$p - 1$	SST	$SST/(p - 1)$	MST/MSE
Error	$n - 3p + 2$	SSE	MSE	
Total	$n - 1$	Total SS		

The formulas for computing the total sum of squares, SSR, SSC, and SST, are identical to the corresponding formulas given for the randomized block design. Let y_{ij} denote an observation in row i and column j. Then

$$\text{Total SS} = \sum_{j=1}^{p} \sum_{i=1}^{p} (y_{ij} - \bar{y})^2 = \sum_{j=1}^{p} \sum_{i=1}^{p} y_{ij}^2 - \text{CM},$$

$$= (\text{sum of squares of all } y \text{ values}) - \text{CM},$$

where

$$\text{CM} = \frac{(\text{total of all observations})^2}{n} = \frac{\left(\sum_{j=1}^{p} \sum_{i=1}^{p} y_{ij}\right)^2}{n}, \quad \text{where } n = p^2.$$

Similarly, let R_i and $\bar{R}_i$, C_j and $\bar{C}_j$, and T_k and $\bar{T}_k$ represent the total and average for all observations in row i, column j, and treatment k, respectively.

Then all sums of squares for rows, columns, and treatments are

$$SSR = p \sum_{i=1}^{p} (\bar{R}_i - \bar{y})^2 = \frac{\sum_{i=1}^{p} R_i^2}{p} - CM$$

$$= \frac{\text{sum of squares of all row totals}}{\text{number of observations in a single total}} - CM,$$

$$SSC = p \sum_{j=1}^{p} (\bar{C}_j - \bar{y})^2 = \frac{\sum_{j=1}^{p} C_j^2}{p} - CM$$

$$= \frac{\text{sum of squares of all column totals}}{\text{number of observations in a single total}} - CM,$$

$$SST = p \sum_{k=1}^{p} (\bar{T}_k - \bar{y})^2 = \frac{\sum_{k=1}^{p} T_k^2}{p} - CM$$

$$= \frac{\text{sum of squares of all treatment totals}}{\text{number of observations in a single total}} - CM.$$

The sums of squares for error, SSE, can be obtained by subtraction. Thus

$$\text{Total SS} = SSR + SSC + SST + SSE.$$

Hence

$$SSE = \text{Total SS} - SSR - SSC - SST.$$

The mean squares corresponding to rows, columns, and treatments can be obtained by dividing the respective sum of squares by $(p - 1)$. Thus, for example,

$$MST = \frac{SST}{(p - 1)}.$$

The hypothesis, "no difference in mean response for treatments," is tested using the F statistic when

$$F = \frac{MST}{MSE}.$$

Similarly, the F statistic can be used to test an hypothesis of no difference between rows (or columns) by using the ratio of MSR (or MSC) to MSE.

F Test for Comparing p Treatments, Using a Latin Square Design

1. *Null Hypothesis:* H_0: The population treatment means are equal.
2. *Alternative Hypothesis:* H_a: One or more pairs of population means differ.
3. *Test Statistic:* $F = \text{MST}/\text{MSE}$, where F is based on $v_1 = (p - 1)$ and $v_2 = (n - 3p + 2)$ degrees of freedom.
4. *Rejection Region:*   Reject if $F > F_\alpha$.

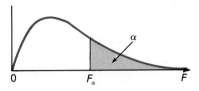

Example 15.7

An experiment was conducted to investigate the difference in mean time to assemble four different electronic devices, A_1, A_2, A_3, and A_4. Two sources of unwanted variation affect the response: the variation between people and the effect of fatigue if a person assembles a series of the devices over time. Consequently, four assemblers were selected and each assembled all four of the devices, A_1, A_2, A_3, and A_4, in the following Latin square design (the observed responses, in minutes, are shown in the cells):

Rows (Position in Assembly Sequence)	Columns (Assemblers)				Total
	1	2	3	4	
1	A_3 44	A_1 41	A_2 30	A_4 40	155
2	A_2 41	A_3 42	A_4 49	A_1 49	181
3	A_1 59	A_4 41	A_3 59	A_2 34	193
4	A_4 58	A_2 37	A_1 53	A_3 59	207
Total	202	161	191	182	736

Do the data provide sufficient evidence to indicate a difference in mean time to assemble the four devices? A difference in the mean time to assemble for people? Is there evidence of a fatigue factor (a difference in mean response for positions in the assembly sequence)?

Solution

The totals for rows, columns, and treatments are:

	i			
	1	**2**	**3**	**4**
R_i	155	181	193	207
C_i	202	161	191	182
T_i	202	142	204	188

$$CM = \frac{(736)^2}{16} = \frac{541{,}696}{16} = 33{,}856.0.$$

Then,

$$\text{Total SS} = \sum_{j=1}^{4} \sum_{i=1}^{4} y_{ij}^2 - CM = 35{,}186.0 - 33{,}856.0 = 1330.0,$$

$$SSR = \frac{\sum_{i=1}^{4} R_i^2}{4} - CM$$

$$= \frac{(155)^2 + (181)^2 + (193)^2 + (207)^2}{4} - CM$$

$$= 34{,}221.0 - 33{,}856.0 = 365.0,$$

$$SSC = \frac{\sum_{j=1}^{4} C_j^2}{4} - CM$$

$$= \frac{(202)^2 \, (161)^2 + (191)^2 + (182)^2}{4} - CM$$

$$= 34{,}082.5 - 33{,}856.0 = 226.5,$$

$$SST = \frac{\sum_{k=1}^{4} T_k^2}{4} - CM$$

$$= \frac{(202)^2 + (142)^2 + (204)^2 + (188)^2}{4} - CM$$

$$= 34{,}482.0 - 33{,}856.0 = 626.0.$$

Finally,

$$SSE = \text{Total SS} - SSR - SSC - SST$$
$$= 1330.0 - 365.0 - 226.5 - 626.0$$
$$= 112.5$$

The analysis-of-variance table for the example is shown in Table 15.10. Note that the mean squares were obtained by dividing the sums of squares by their respective degrees of freedom.

Table 15.10 ANOVA table for Example 15.7

Source	d.f.	SS	MS	F
Rows	3	365	121.67	6.49
Columns	3	226.5	75.50	4.03
Treatments	3	626	208.67	11.13
Error	6	112.5	18.75	
Total	15	1,330.0		

All the computed F statistics are based on $v_1 = 3$ and $v_2 = 6$ degrees of freedom. The corresponding tabulated critical value is $F_{.05} = 4.76$. A comparison of the computed F statistics with the tabulated value indicates that the computed F's for both rows and treatments exceed the critical value. Thus the data provide sufficient evidence to indicate a difference in the mean time to assemble the four devices. The data also show a difference in mean time to assemble for rows, or equivalently, positions in the sequence of assembly. Thus it appears that fatigue, boredom, or some other factor increases the mean time to assemble as the length of employment increases.

15.13
Estimation for the Latin Square Design

A confidence interval for the difference between a pair of treatment means, with confidence coefficient $(1 - \alpha)$, is obtained in the same manner as for the completely randomized and the randomized blocks designs. Since each treatment mean will contain p observations, $n_1 = n_2 = p$, the standard deviation of the difference in a pair of means is

$$\sigma_{\bar{T}_i - \bar{T}_j} = \sigma \sqrt{\frac{1}{p} + \frac{1}{p}}.$$

The $100(1 - \alpha)$ percent confidence interval is

Latin Square Design: A $100(1 - \alpha)$ Percent Confidence Interval for the Difference between Two Treatment Means

$$(\bar{T}_i - \bar{T}_j) \pm t_{\alpha/2}s \sqrt{\frac{2}{p}},$$

where $t_{\alpha/2}$ has $n - 3p + 2$ degrees of freedom.

Corresponding confidence intervals for the difference between a pair of row or column means are, respectively,

---

Latin Square Design: $100(1 - \alpha)$ Percent Confidence Intervals for the Difference between Two Row or Two Column Means

$$(\bar{R}_i - \bar{R}_j) \pm t_{\alpha/2}s \sqrt{\frac{2}{p}} \quad \text{and} \quad (\bar{C}_i - \bar{C}_j) \pm t_{\alpha/2}s \sqrt{\frac{2}{p}},$$

where $t_{\alpha/2}$ has $n - 3p + 2$ degrees of freedom.

Example 15.8

Refer to Example 15.7. Estimate the difference in mean response between treatments 1 and 2 using a 95 percent confidence interval.

Solution

From the analysis-of-variance table, Example 15.7, observe that $s^2 = \text{MSE} = 18.75$ is based on six degrees of freedom. Consequently, the tabulated value, $t_{\alpha/2}$, with six degrees of freedom is $t_{.025} = 2.447$. Then the 95 percent confidence interval for the difference between the treatment means, $(\mu_1 - \mu_2)$, is

$$(\bar{T}_1 - \bar{T}_2) \pm t_{\alpha/2}s \sqrt{\frac{2}{p}}$$

or

$$(50.5 - 35.5) \pm (2.447)(4.33) \sqrt{\frac{2}{4}}$$

or

$$15 \pm 7.49.$$

Thus we estimate the difference in mean time to assemble the two devices to be between 7.51 and 22.49 minutes.

15.14
A Computer Printout for a Latin Square Design

Now that you are familiar with the SAS computer printouts for the analyses of variance for the completely random and the randomized blocks designs (Sections 15.7 and 15.10), you will be able to identify the appropriate quantities in the SAS printout for the analysis of variance of the data for the Latin square design, Example 15.7. This printout is shown in Table 15.11.

Table 15.11 SAS computer printout for Example 15.7

STATISTICAL ANALYSIS SYSTEM

ANALYSIS OF VARIANCE PROCEDURE

DEPENDENT VARIABLE: Y

SOURCE	DF	SUM OF SQUARES	MEAN SQUARE	F VALUE	PR > F	R-SQUARE	C.V.
MODEL	9	1217.50000000	135.27777778	7.21	0.0129	0.915414	9.4133
ERROR	6	112.50000000	18.75000000		STD DEV		Y MEAN
CORRECTED TOTAL	15	1330.00000000			4.33012702		46.00000000

SOURCE	DF	ANOVA SS	F VALUE	PR > F
ROWS	3	365.00000000	6.49	0.0259
COLUMNS	3	226.50000000	4.03	0.0692
TRTMENTS	3	626.00000000	11.13	0.0073

The upper table (shaded) shown in the printout gives a breakdown of the Total SS into two parts, the first corresponding to MODEL and the second to ERROR. Note that the numbers appearing in the ERROR row are identical to those in the "Error" row of the ANOVA, Table 15.10. The lower table of the printout gives a breakdown of the MODEL degrees of freedom and sum of squares into portions corresponding to ROWS, COLUMNS, and TREAT-MENTS. Note that the degrees of freedom and sum of squares in this table correspond to those of the ANOVA, Table 15.10. As in the printout for the randomized block design, Section 15.10, the lower table does not give the mean squares, MSR, MSC, and MST, but they are not needed because the computed F values, necessary for the F tests, along with their observed significance levels, are shown. Note that they correspond with the values given in the ANOVA, Table 15.10. The value of s, $s = 4.33 \ldots$, given in the sixth column of the printout, can be used to calculate confidence intervals for the difference between any pair of treatment, row, or column means.

Exercises

15.16.   Give the analysis of variance for the following 3×3 Latin square design. Find a 95 percent confidence interval for the difference in mean response for treatments A and C. Assume that this is a preplanned comparison.

	Columns		
Rows	**1**	**2**	**3**
1	B 12	A 7	C 17
2	C 10	B 7	A 4
3	A 2	C 8	B 12

15.17. Paper machines distribute a thin mixture of wood fibers and water to a wide wire mesh belt that is traveling at a very high speed. Thus it is conceivable that the distribution of fibers, thickness, porosity, etc., will vary across the belt as well as along the belt and produce corresponding variations in the strength of the final paper product. A paper company designed an experiment to compare the strength of four coatings intended to improve the appearance of packaging paper. Because the uncoated paper strength could vary down as well as across a roll (because of the reasons described above), the coatings were applied in a Latin square design. The strength measurements for the four coatings, A, B, C, and D, were as follows:

Position Down the Roll	Position Across the Roll			
	1	**2**	**3**	**4**
1	B 12.4	A 10.4	C 13.1	D 11.8
2	C 13.4	B 12.4	D 11.8	A 10.9
3	A 10.5	D 11.4	B 12.3	C 12.9
4	D 11.4	C 13.3	A 10.7	B 12.0

a. Do the data present sufficient evidence to indicate a difference in mean strength for paper treated with the four paper coatings?

b. Suppose that prior to seeing our data, we wished to compare the mean strength between paper coated with coatings A and C. Estimate the difference using a 95 percent confidence interval.

c. Do the data provide sufficient evidence to indicate a difference in mean strength for locations across the roll?

d. Down the roll?

15.18. The voltage within an electronic system was to be investigated for four different conditions, A, B, C, and D, but unfortunately the voltage measurements were affected by the line voltage within the laboratory. Since the line voltage assumed the same pattern within each day due to the usage patterns of various industrial firms in the area and since the general level of line voltage varied from day to day, a Latin square design was employed. The data are shown below:

Time Period Within a Day	Days			
	1	2	3	4
1	A 116	B 108	C 126	D 112
2	C 111	D 124	A 122	B 121
3	B 120	C 115	D 126	A 109
4	D 118	A 125	B 116	C 127

a. Calculate the appropriate sums of squares and construct an analysis-of-variance table for this data.

b. Do the data present sufficient evidence to indicate a difference between treatments?

c. Estimate the difference in mean responses for voltage conditions A and B. Use a 90 percent confidence interval.

15.19.   A 4 × 4 Latin square design was employed to investigate the gasoline consumption at a fixed horsepower for four different brands of gasoline, A, B, C, and D. Four engines of the same design and specifications were employed in the experiment and each gasoline was tested in each of the engines. The tests, one gasoline per engine, were conducted on four different days. The data are shown in the table. An SAS computer printout for the analysis of variance is also shown. Use the information in the printout to answer the following questions:

Rows (Days)	Columns (Engines)			
	1	2	3	4
1	D 19.5	B 16.8	A 19.8	C 19.2
2	B 18.0	A 17.9	C 17.9	D 17.7
3	A 18.1	C 21.0	D 17.5	B 17.2
4	C 20.1	D 19.2	B 17.0	A 18.5

STATISTICAL ANALYSIS SYSTEM

ANALYSIS OF VARIANCE PROCEDURE

DEPENDENT VARIABLE: Y

SOURCE	DF	SUM OF SQUARES	MEAN SQUARE	F VALUE	PR > F	R-SQUARE	C.V.
MODEL	9	14.99750000	1.66638889	1.42	0.3465	0.679927	5.8754
ERROR	6	7.06000000	1.17666667		STD DEV		Y MEAN
CORRECTED TOTAL	15	22.05750000			1.08474267		18.46250000

SOURCE	DF	ANOVA SS	F VALUE	PR > F
ROWS	3	2.13250000	0.60	0.6360
COLUMNS	3	2.20250000	0.62	0.6252
TRTMENTS	3	10.66250000	3.02	0.1156

a. Do the data provide sufficient evidence to indicate a difference in mean gasoline consumption for the four brands of gasoline? Test by using $\alpha = .05$.

b. Do the data provide sufficient evidence to indicate that blocking on engines increased the amount of information in the experiment concerning mean gasoline consumption for the four brands of gasoline? Explain.

c. Do the data provide sufficient evidence to indicate that blocking on days increased the amount of information in the experiment concerning the mean gasoline consumption for the four brands of gasoline? Explain.

d. If you conducted a similar experiment in the future, should you block on engines and days? Explain.

e. Find a 90 percent confidence interval for the difference in mean gasoline consumption between brands A and C. Interpret the interval.

15.15
Selecting the Sample Size

Selecting the sample size for the completely randomized or the randomized block design is an extension of the procedure of Section 8.10. We confine our attention to the case of equal sample sizes, $n_1 = n_2 = \cdots = n_p$, for the p treatments of the completely randomized design. The number of observations per treatment is equal to b for the randomized block design. Thus the problem is to select n_1 or b for these two designs so as to purchase a specified quantity of information.

The selection of sample size follows a similar procedure for both designs; hence we will outline a general method. **First the experimenter must decide on the parameter (or parameters) of major interest. Usually, he will wish to compare a pair of treatment means. Second, he must specify a bound on the error of estimation that he is willing to tolerate. Once determined, he need only select n_i (the number of observations in a treatment mean) or, correspondingly, b (the number of blocks in the randomized block design) that will reduce the half-width of the confidence interval for the parameter so that it is less than or equal to the specified bound on the error of estimation.** It should be emphasized that the sample-size solution will *always* be an approximation, since σ is unknown and s

is unknown until the sample is acquired. The best available value will be used for s in order to produce an approximate solution. We will illustrate the procedure with an example.

Example 15.9

A completely randomized design is to be conducted to compare five teaching techniques on classes of equal size. Estimation of the difference in mean response to an achievement test is desired correct to within 30 test-score points, with probability equal to .95. It is expected that the test scores for a given teaching technique will possess a range approximately equal to 240. Find the approximate number of observations required for each sample in order to acquire the specified information.

Solution

The confidence interval for the difference between a pair of treatment means is

$$(\bar{T}_i - \bar{T}_j) \pm t_{\alpha/2} s \sqrt{\frac{1}{n_i} + \frac{1}{n_j}}.$$

Therefore, we will wish to select n_i and n_j so that

$$t_{\alpha/2} s \sqrt{\frac{1}{n_i} + \frac{1}{n_j}} \le 30.$$

The value of σ is unknown and s is a random variable. However, an approximate solution for $n_i = n_j$ can be obtained by guessing s to be roughly equal to one-fourth of the range. Thus $s \approx (240)/4 = 60$. The value of $t_{\alpha/2}$ will be based upon $(n_1 + n_2 + \cdots + n_5 - 5)$ degrees of freedom, and for even moderate values of n_i, $t_{.025}$ will be approximately equal to 2. Then

$$t_{.025} s \sqrt{\frac{1}{n_i} + \frac{1}{n_j}} \approx 2(60) \sqrt{\frac{2}{n_i}} \le 30$$

or, solving for n_i,

$$n_i \approx 32, \qquad i = 1, 2, \ldots, 5.$$

Example 15.10

An experiment is to be conducted to compare the toxic effect of three chemicals on the skin of rats. The resistance to the chemicals was expected to vary substantially from rat to rat. Therefore, all three chemicals were to be tested on each rat, thereby blocking out rat-to-rat variability.

The standard deviation of the experimental error was unknown, but prior experimentation involving several applications of a given chemical on the same rat suggested a range of response measurements equal to 5 units.

Find a value for b such that the error of estimating the difference between a pair of treatment means is less than one unit, with probability equal to .95.

Solution   A very approximate value for s would be one-fourth the range, or $s \approx 1.25$. Then we wish to select b so that

$$t_{.025} s \sqrt{\frac{1}{b} + \frac{1}{b}} = t_{.025} s \sqrt{\frac{2}{b}} \le 1.$$

Since $t_{.025}$ will depend upon the degrees of freedom associated with s^2, which will be $(n - b - p + 1)$, we will guess $t_{.025} \approx 2$. Then

$$2(1.25) \sqrt{\frac{2}{b}} \le 1$$

or

$$b \approx 13, \qquad i = 1, 2, 3.$$

Thus approximately 13 rats will be required to obtain the required information.

The degrees of freedom associated with s^2 will be 24 based on this solution. Therefore, the guessed value of t would seem to be adequate for this approximate solution.

The sample-size solutions for Examples 15.9 and 15.10 are very approximate, and are intended to provide only a rough estimate of approximate size and consequent cost of the experiment. The experimenter will obtain information on σ as the data are being collected and can recalculate a better approximation to n as he or she proceeds.

Selecting the sample size for Latin square designs follows essentially the same procedure as for the completely randomized and the randomized block designs. The only difference is that one must decide on the number of $p \times p$ Latin squares that will be needed to acquire the desired information. We omit discussion of this topic because we have not shown how to conduct an analysis of variance for more than one $p \times p$ Latin square. The reader interested in this topic should consult the references.

Exercises

15.20.   Refer to Exercise 15.7. How many plants would have to be selected and analyzed for each location if we wish to estimate the difference in mean leaf length for two locations with an error of less than .3 and a probability approximately equal to .95?

15.21.   Refer to Exercise 15.14. Approximately how many rats would be required to estimate the difference in mean scores for two chemicals correct to within 1 unit?

15.22.   Refer to Exercise 15.13. Approximately how many replications would be required for each treatment (observations per treatment) in order that the error of estimating the difference in mean response for a pair of treatments be less than 20 (with probability equal to .95)? Assume that additional observations would be made within a randomized block design.

15.23. A manufacturer wished to compare the mean daily production for four production lines, A, B, C, and D, which manufacture the same product. Since daily production is affected by the quality of raw materials and labor associated with supply to the lines, it was decided to make the comparison of line productivity by making comparisons *within* each day (thereby eliminating day to day variations associated with supply).

   a. What type of experimental design was the manufacturer using to collect his data?

   b. Suppose that the mean daily yield per line is in the neighborhood of 1000 units and that an earlier analysis of the differences in daily production, recorded each day over a long period of time, suggests that the variance of $(y_1 - y_2)$, the difference in daily yields for production lines A and B, is equal to 5000. For how many days must the daily production be recorded for the three production lines to estimate the difference in mean production between any pair of lines correct to within 25 units?

15.24. Refer to Exercise 15.11. Suppose you wish to estimate the difference in mean moisture absorption between fabrics treated by two different chemicals correct to within .2 units, with probability equal to .95. Approximately how many bolts of fabric (blocks) should be used for the experiment?

15.25. An experiment is to be conducted to compare the mean amount of effluents discharged by three chemical plants. Suppose that the variance of the effluent measurements at all three plants is approximately equal to .03. If you wish to estimate the mean amount of effluents (pounds/gallon) at a plant correct to within .05 lb/gal, with probability near .95, approximately how many effluent specimens would have to be randomly selected at a single point?

15.16
Some Cautionary
Comments on
Blocking

Two points need to be emphasized when using block designs. First, you should not use a randomized block design to investigate the effects of two factors on a response (letting treatments be the levels of one factor and blocks be the levels of the second factor). If the two factors do not affect the response independently of each other—that is, the effect of one factor on the response depends upon the level of the other factor (called "factor interaction")—a "randomized block" analysis of variance could lead to very erroneous conclusions about the relationship between the factors and the response. Similarly, you should not use a Latin square design to investigate the effects of three factors on a response. Remember, the object of block designs is to block out the variation of nuisance variables (blocks) and thereby reduce SSE. Information on designs for multivariable experiments can be found in any standard text on the design of experiments (see the references).

   The second point is that blocking is not always beneficial. Just as we noted for the paired-difference experiment (a randomized block design with two treatments), blocking produces a gain in information if the between-block variation is larger than the within-block variation. Then blocking removes this larger source of variation from SSE and s^2 assumes a smaller value (as does the population variance σ^2) because of the design. At the same time, you lose information because blocking reduces the number of degrees of freedom associated with SSE (and s^2). (To see this, compare the ANOVA tables for the completely randomized design, Table 15.2, and the randomized block design,

Table 15.4.) Consequently, if blocking is to be beneficial, the gain in information due to the elimination of block variation must outweigh the loss due to a reduction in the number of degrees of freedom associated with SSE. Unless blocking leaves you with only a small number of degrees of freedom for SSE, the loss in degrees of freedom causes only a small reduction in information in the experiment. Consequently, if you have a reason for suspecting that there is block-to-block variation, it will usually pay to block.

15.17
Assumptions

The assumptions about the probability distribution of the random response, y, are the same for the analysis of variance as for the regression models of Chapter 11. These are as follows:

Assumptions

1. For any treatment or block combination, y is normally distributed with variance equal to σ^2.
2. The observations have been selected independently so that any pair of y values, y_i and y_j, are independent in a probabilistic sense for all i and j ($i \neq j$).

In a practical situation, you can never be certain that the assumptions are satisfied, but you will often have a fairly good idea whether the assumptions are reasonable for your data. To illustrate, it has been mentioned that the inferential methods of Chapters 10, 11, 12, and 15 are not seriously affected by moderate departures from the assumptions of normality, but you would want the probability distribution of y at least to be mound-shaped. So if y is a discrete random variable that only assumes three values, say $y = 10, 11, 12$, then it is unreasonable to assume that the probability distribution of y is approximately normal!

The assumption of a constant variance for y for the various experimental conditions should be approximately satisfied, although violation of even this assumption is not too serious if the sample sizes for the various experimental conditions are equal. However, suppose the response is binomial, say the proportion $\hat{p}$ of people who favor a particular food product. We know that the variance of a proportion (Section 6.4) is

$$\sigma_{\hat{p}}^2 = \frac{pq}{n}, \qquad \text{where } q = 1 - p,$$

and therefore that the variance is dependent on p. The variance of the response is a function of the mean response and will change from one experimental setting to another, and the assumptions of the analysis of variance have been violated. A similar situation occurs when the response measurements are

Poisson data (mentioned in the case study and in the exercises, Chapter 5). If the response possesses a Poisson probability distribution, the variance of the response will equal the mean. Consequently, Poisson response data also violate the analysis-of-variance assumptions.

Many kinds of data are not measurable and hence are unsuitable for an analysis of variance. For example, many responses cannot be measured but can be ranked. Consumer preference studies yield data of this type: you know you prefer product A to B and you prefer B to C, but you have difficulty assigning an exact value to the strength of your preference.

So what do you do when the assumptions of an analysis of variance are not satisfied? For example, suppose the variances of the responses for various experimental conditions are not equivalent. This situation can be remedied sometimes by transforming the response measurements. That is, instead of using the original response measurements, we might use their square roots, logarithms, or some other function of the response, y. Transformations intended to stabilize the variance of the response have been found to make the probability distributions of the transformed responses more nearly normal. See the references for discussions of these topics.

When nothing can be done to satisfy (even approximately) the assumptions of the analysis of variance, or if the data are rankings, you should use a nonparametric testing and estimation procedure. When our assumptions are satisfied, these procedures, which rely on the comparative magnitudes of measurements (often ranks), are almost as powerful in detecting treatment differences as the tests presented in this chapter. When the assumptions are not satisfied, the nonparametric procedures may be more powerful.

15.18
Summary and Comments

The completely randomized, the randomized block, and the Latin square designs are illustrations of experiments involving one, two, and three qualitative independent variables, respectively. The analysis of variance partitions the total sum of squares of deviations of the response measurements about their mean into portions associated with each independent variable and the experimental error. The former may be compared with the sum of squares for error, using mean squares and the F statistics, to see whether the mean squares for the independent variables are unusually large and, thereby, indicative of an effect on the response.

This chapter has presented a very brief introduction to the analysis of variance and its associated subject, the design of experiments. Experiments can be designed to investigate the effect of many quantitative and qualitative variables on a response. These may be variables of primary interest to the experimenter as well as nuisance variables, such as blocks, which we attempt to separate from the experimental error. They are subject to an analysis of variance when properly designed. A more extensive coverage of the basic concepts of experimental design and the analysis of experiments will be found in the references.

References

Cochran, W.G., and G.M. Cox, *Experimental Designs*, 2nd ed. New York: John Wiley & Sons, Inc., 1957.

Dixon, W.J., and M.B. Brown, eds. *BMDP Biomedical Computer Programs, P Series*. Berkeley, Calif.: University of California, 1979.

Guenther, W.C., *Analysis of Variance*. Englewood Cliffs, N.J.: Prentice-Hall, Inc., 1964.

Helwig, J.P., and K.A. Council, eds. *SAS User's Guide*. Cary, N.C.: SAS Institute, Inc., P.O. Box 8000, 1979.

Hicks, C.R., *Fundamental Concepts in the Design of Experiments*, 2nd ed. New York: Holt, Rinehart and Winston, Inc., 1973.

Mendenhall, W., *An Introduction to Linear Models and the Design and Analysis of Experiments*. Belmont, Calif.: Wadsworth Publishing Company, Inc., 1968.

Nie, N., C.H. Hull, J.G. Jenkins, K. Steinbrenner, and D.H. Bent, *Statistical Package for the Social Sciences*, 2nd ed. New York: McGraw-Hill Book Company, 1979.

Ott, L., *An Introduction to Statistical Methods and Data Analysis*. Boston: Duxbury Press, 1977.

Ryan, T.A., B.L. Joiner, and B.F. Ryan. *Minitab Student Handbook*. Boston: Duxbury Press, 1979.

Snedecor, G.W., and W.G. Cochran, *Statistical Methods*, 7th ed. Ames, Iowa: The Iowa State University Press, 1980.

Steel, R.G.D., and J.H. Torrie, *Principles and Procedures of Statistics*. New York: McGraw-Hill Book Company, 1960.

Winer, B.J., *Statistical Principles in Experimental Design*, 2nd ed. New York: McGraw-Hill Book Company, 1971.

Supplementary Exercises

15.26. State the assumptions underlying the analysis of variance of a completely randomized design.

15.27. Refer to Example 15.2. Calculate SSE by pooling the sums of squares of deviations within each of the four samples, and compare with the value obtained by subtraction. Note that this is an extension of the pooling procedure used in the two-sample case discussed in Section 15.2.

15.28. A study was initiated to investigate the effect of two drugs, administered simultaneously, in reducing human blood pressure. It was decided to utilize three levels of each drug and to include all nine combinations in the experiment. Nine high-blood-pressure patients were selected for the experiment, and one was randomly assigned to each of the nine drug combinations. The response observed was the drop in blood pressure over a fixed interval of time.

a. Is this a randomized block design?

b. Suppose that two patients were assigned to each of the nine drug combinations. What type of experimental design is this?

15.29. Refer to Exercise 15.28. Suppose that prior experimentation suggests that $\sigma = 20$.

a. How many replications would be required to estimate any treatment (drug combination) mean correct to within ± 10?

b. How many degrees of freedom will be available for estimating σ^2 when using the number of replications determined in part (a)?

c. Give the approximate half-widths of a confidence interval for the difference in mean response for two treatments when using the number of replications determined in part (a).

15.30. A completely randomized design was conducted to compare the effect of five different stimuli on reaction time. Twenty-seven people were employed in the experiment that was conducted using a completely randomized design. Regardless of the results of the analysis of variance, it is desired to compare stimuli A and D. The results of the experiment were as follows:

Stimulus	Reaction Time (sec)							Total	Mean
A	.8	.6	.6	.5				2.5	.625
B	.7	.8	.5	.5	.6	.9	.7	4.7	.671
C	1.2	1.0	.9	1.2	1.3	.8		6.4	1.067
D	1.0	.9	.9	1.1	.7			4.6	.920
E	.6	.4	.4	.7	.3			2.4	.48

a. Conduct an analysis of variance and test for a difference in mean reaction time due to the five stimuli.

b. Compare stimuli A and D to see if there is a difference in mean reaction time.

15.31. The experiment in Exercise 15.30 might have been more effectively conducted using a randomized block design with people as blocks since we would expect mean reaction time to vary from one person to another. Hence four people were used in a new experiment and each person was subjected to each of the five stimuli in a random order. The reaction times (sec) were as follows:

	Stimulus				
Subject	A	B	C	D	E
1	.7	.8	1.0	1.0	.5
2	.6	.6	1.1	1.0	.6
3	.9	1.0	1.2	1.1	.6
4	.6	.8	.9	1.0	.4

Conduct an analysis of variance and test for differences in treatments (stimuli).

15.32. The following measurements are on the thickness of cake icing for the Latin square design discussed in Section 15.12. The only difference between this exercise and the text discussion is that five (not three) mixes, A, B, C, D, and E, were employed in a 5 × 5 Latin square design. Thickness measurements are given in hundredths of an inch. Do the data provide sufficient evidence to indicate a difference in mean thickness of icing for the five cake mixtures? Estimate the difference in mean thickness for mixtures A and B.

	Columns				
Rows	1	2	3	4	5
1	D 9	E 18	C 6	A 8	B 11
2	B 12	D 17	E 10	C 4	A 5
3	A 6	B 16	D 10	E 9	C 4
4	C 4	A 13	B 11	D 8	E 13
5	E 14	C 11	A 7	B 10	D 15

15.33.   Consider the following one-way classification consisting of three treatments, *A*, *B*, and *C*, where the number of observations per treatment varies from treatment to treatment.

A	B	C
24.2	24.5	26.0
27.5	22.7	
25.9		
24.7		

Do the data present sufficient evidence to indicate a difference among treatment means?

♀ 15.34.   Fifteen subjects with relatively equal reaction times and physical attributes were randomly grouped in five's. The first five were to react to a light stimulus, the second five to a sound stimulus, and the third five to a stimulus of both sound and light. The results are as follows:

Stimulus	Reaction Time (sec)				
Light	.36	.41	.45	.42	.31
Sound	.49	.42	.35	.48	.40
Both	.29	.37	.42	.41	.47

Is there sufficient evidence to indicate a difference in mean reaction times for the three stimuli? Use $\alpha = .05$.

15.35.   A company wished to study the differences among four sales-training programs on the sales abilities of their sales personnel. Thirty-two people were randomly divided into four groups of equal size and the groups were then subjected to the different sales-training programs. Because there were some dropouts (illness, etc.) during the training programs, the number of trainees completing the programs varied from group to group. At the end of the training programs, each salesperson was randomly assigned a sales area from a group of sales areas that were judged to have equivalent sales potentials. The numbers of sales made by each of the four groups of salespeople during the first week after completing the training program are listed in the table.

	Training Program			
	1	2	3	4
	78	99	74	81
	84	86	87	63
	86	90	80	71
	92	93	83	65
	69	94	78	86
	73	85		79
		97		73
		91		70
Total	482	735	402	588

a. Do the data present sufficient evidence to indicate a difference in the mean achievement for the four training programs?

b. Find a 90 percent confidence interval for the difference in the mean of sales that would be expected for persons subjected to training programs 1 and 4. Interpret the interval.

c. Find a 90 percent confidence interval for the mean number of sales by persons subjected to training program 2.

15.36. Four chemical plants, producing the same product and owned by the same company, discharge effluents into streams in the vicinity of their locations. To check on the extent of the pollution created by the effluents and to determine if this varies from plant to plant, the company collected random samples of liquid waste, five specimens for each of the four plants. The data are shown in the table. An SAS computer printout of the analysis of variance for the data is also shown. Use the information in the printout to answer questions a–c.

Plant	Polluting Effluents (lb/gal of waste)				
A	1.65	1.72	1.50	1.37	1.60
B	1.70	1.85	1.46	2.05	1.80
C	1.40	1.75	1.38	1.65	1.55
D	2.10	1.95	1.65	1.88	2.00

STATISTICAL ANALYSIS SYSTEM

ANALYSIS OF VARIANCE PROCEDURE

DEPENDENT VARIABLE: Y

SOURCE	DF	SUM OF SQUARES	MEAN SQUARE	F VALUE	PR > F	R-SQUARE	C.V.
MODEL	3	0.46489500	0.15496500	5.20	0.0107	0.493679	10.1515
ERROR	16	0.47680000	0.02980000		STD DEV		Y MEAN
CORRECTED TOTAL	19	0.94169500			0.17262677		1.70050000

SOURCE	DF	ANOVA SS	F VALUE	PR > F
TRTMENTS	3	0.46489500	5.20	0.0107

a. Do the data provide sufficient evidence to indicate a difference in the mean amount of effluents discharged by the four plants?

b. If the maximum mean discharge of effluents is 1.5 lb/gal, do the data provide sufficient evidence to indicate that the limit is exceeded at plant A?

c. Estimate the difference in the mean discharge of effluents between plants A and D, using a 95 percent confidence interval.

15.37.   The yields of wheat in bushels per acre were compared for five different varieties, A, B, C, D, and E, at six different locations. Each variety was randomly assigned to a plot at each location. The results of the experiment are shown in the table. An SAS computer printout for the analysis of variance for the data is also shown below. Use the information in the printout to answer questions a–c.

	Location					
Varieties	**1**	**2**	**3**	**4**	**5**	**6**
A	35.3	31.0	32.7	36.8	37.2	33.1
B	30.7	32.2	31.4	31.7	35.0	32.7
C	38.2	33.4	33.6	37.1	37.3	38.2
D	34.9	36.1	35.2	38.3	40.2	36.0
E	32.4	28.9	29.2	30.7	33.9	32.1

STATISTICAL ANALYSIS SYSTEM

ANALYSIS OF VARIANCE PROCEDURE

DEPENDENT VARIABLE: Y

SOURCE	DF	SUM OF SQUARES	MEAN SQUARE	F VALUE	PR > F	R-SQUARE	C.V.
MODEL	9	210.81166667	23.42351852	12.22	0.0001	0.846152	4.0499
ERROR	20	38.33000000	1.91650000		STD DEV		Y MEAN
CORRECTED TOTAL	29	249.14166667			1.38437712		34.18333333

SOURCE	DF	ANOVA SS	F VALUE	PR > F
BLOCKS	5	68.14166667	7.11	0.0006
TRTMENTS	4	142.67000000	18.61	0.0001

a. Do the data provide sufficient evidence to indicate a difference in the mean yield for the five varieties of wheat?

b. Do the data provide sufficient evidence to indicate that blocking on "location" increased the amount of information in the experiment?

c. Find a 95 percent confidence interval for the difference in mean yields for varieties C and E.

16. Nonparametric Statistics

Chapter Objectives

General Objective

Chapters 8, 9, and 10 presented statistical techniques for comparing two populations by comparing their respective population parameters. The techniques are applicable to data measured on a continuum and to data possessing normal population relative frequency distributions. The purpose of this chapter is to present a set of statistical test procedures to compare populations for the many types of data that do not satisfy these assumptions.

Specific Objectives

1. To explain the difference between parametric and nonparametric tests. *Section 16.1*
2. To explain why nonparametric tests are useful and, sometimes, essential. *Section 16.1*
3. To present a quick and easy method for comparing two populations, the sign test. *Section 16.2*
4. To show how statistical tests can be compared and to identify the advantages associated with nonparametric tests. *Sections 16.3, 16.9*
5. To present a nonparametric test for comparing two populations based on independent random samples, the Mann–Whitney U test. *Sections 16.4, 16.5*
6. To present a nonparametric test for comparing two populations for a paired-difference experiment, the Wilcoxon signed rank test. *Section 16.6*
7. To present a nonparametric test for correlation between two variables, based on Spearman's r_s. *Sections 16.7, 16.8*

Case Study

Does It Pay to Save? A Second Look

You may recall the case study of Chapter 11 that discussed Dr. Srully Blotnick's theory concerning the relationship between the percentage of total national income saved y and the approximate percentage of investment income x that must be paid in taxes. We calculated the simple coefficient of correlation r between y and x for $n = 8$ countries and found that there was sufficient evidence (p value $= .00875$) to indicate that y and x were negatively correlated; that is, that y tends to decrease as x increases. Then we issued a warning, that it was quite conceivable that the data did not satisfy the assumptions required for the regression analysis test for correlation (Chapter 11). So, let us ask the question again. Does the national rate of savings tend to decrease as the investment-income tax increases? Or, putting it another way, are y and x negatively correlated?

The case study, Chapter 11, emphasizes an important aspect of statistical inference. Most statistical methods are based on assumptions concerning the nature of the sampled population(s) and the manner in which the sample(s) was (or were) selected. Since the nature of the sampled population(s) is usually unknown, some doubt always exists concerning the validity of any statistical inferences based on unverified assumptions.

As discussed earlier in this text, studies have been conducted to determine the effect that moderate departures from the assumptions have on the measures of reliability associated with a method of inference. For example, we would want to know whether moderate departures from the assumption of normality would produce a substantial change in the confidence coefficient for a particular confidence interval or whether it would produce a substantial change in α and β, the probabilities of type I and type II errors for a particular test. Estimation and test procedures whose properties remain fairly stable when the assumptions are relaxed are said to be *robust*.

In addition to studying the robustness of statistical estimation and test procedures, efforts have been made to develop a methodology of statistical inference, estimation, and tests of hypotheses that requires very few inferences concerning the sampled populations. Some of these methods, called *nonparametric* statistical procedures, are presented in this chapter. Particularly, we will reexamine the relationship between the rate of savings y and the investment tax rate x in Section 16.8.

16.1 Introduction

Some experiments yield response measurements that defy quantification. That is, they generate response measurements that can be ordered (ranked), but the location of the response on a scale of measurement is arbitrary. For example, the data may only admit pair-wise directional comparisons; that is, whether one observation is larger than another or vice versa. Or, we may be able to rank all observations in a data set but may not know the exact values of the measurements. To illustrate, suppose that a judge is employed to evaluate and rank the sales abilities of four salespeople, the edibility and taste characteristics of five brands of cornflakes, or the relative appeal of five new automobile designs. Clearly it is impossible to give an exact measure of sales competence, palatability of food, or design appeal, but it is usually possible to rank the salespeople, brands, etc. according to which one we think is best, second best, etc. Thus the response measurements here differ markedly from those presented in preceding chapters. Experiments that produce this type of data occur in almost all fields of study, but they are particularly evident in social science research and in studies of consumer preference. The data that they produce can be analyzed using *nonparametric statistical methods*.

Nonparametric statistical procedures are also useful in making inferences in situations where serious doubt exists about the assumptions that underlie standard methodology. For example, the *t* test for comparing a pair of means, Section 10.4, is based on the assumption that both populations are normally distributed with equal variances. The experimenter will never know whether these assumptions hold in a practical situation but will often be reasonably certain that departures from the assumptions will be small enough so that the properties of the statistical procedure will be undisturbed. That is, α and β will be approximately what the experimenter thinks they are. On the other hand, it is not uncommon for the experimenter seriously to question the assumptions and wonder whether he or she is using a valid statistical procedure. This difficulty may be circumvented by using a nonparametric statistical test and thereby avoiding reliance on a very uncertain set of assumptions.

Research has shown that nonparametric statistical tests are almost as capable of detecting differences among populations as the parametric methods of preceding chapters when normality and other assumptions are satisfied. They may be, and often are, more powerful in detecting population differences when the assumptions are not satisfied. For this reason, many statisticians advocate the use of nonparametric statistical procedures in preference to their parametric counterparts.

In this chapter we will discuss only the most common situations in which nonparametric methods are used. For a more complete treatment of the subject, see the texts by Siegel (1956) or Conover (1971) listed in the references.

16.2
The Sign Test for Comparing Two Populations

Without emphasizing the point, we employed a nonparametric statistical test as an alternative procedure for determining whether evidence existed to indicate a difference in the mean wear for the two types of tires in the paired-difference experiment, Section 10.5. Each pair of responses was compared and y (the number of times A exceeded B) was used as the test statistic. This nonparametric test is known as the **sign test** because y is the number of positive (or negative) signs associated with the differences. The implied null hypothesis is that the two population distributions are identical and the resulting technique is completely independent of the form of the distribution of differences. Thus, regardless of the distribution of differences, the probability that A exceeds B for a given pair will be $p = .5$ when the null hypothesis is true (that is, when the distributions for A and B are identical). Then y will possess a binomial probability distribution and a rejection region for y can be obtained using the binomial probability distribution of Chapter 6.

The following example will help you to recall how the sign test was constructed and how it is used in a practical situation.

Example 16.1

The number of defective electrical fuses proceeding from each of two production lines, A and B, was recorded daily for a period of ten days, with the following results:

Day	Production Lines A	Production Lines B
1	170	201
2	164	179
3	140	159
4	184	195
5	174	177
6	142	170
7	191	183
8	169	179
9	161	170
10	200	212

Assume that both production lines produced the same daily output. Compare the number, y_A, of defectives produced each day by production line A with the number, y_B, produced by production line B and let y equal the number of days when y_A exceeded y_B. Do the data present sufficient evidence to indicate that one production line tends to produce more defectives than the other or, equivalently, that $P(y_A > y_B) \neq 1/2$? State the null hypothesis to be tested and use y as a test statistic.

Solution

Note that this is a paired-difference experiment. Let y be the number of times that the observation for production line A exceeds that for B in a given day. Under the null hypothesis that the two distributions of defectives are identical, the probability, p, that A exceeds B for a given pair is $p = .5$. Or, equivalently, we may wish to test the null hypothesis that the binomial parameter, p, equals .5. against the alternative hypothesis, $p \neq .5$.

Very large or very small values of y are most contradictory to the null hypothesis. Therefore, the rejection region for this two-tailed test will be located by including the most extreme values of y that at the same time provide an α that is feasible for the test.

Suppose that we would like α to be somewhere on the order of .05 or .10. We would commence the selection of the rejection region by including $y = 0$ and $y = 10$ and calculate the α associated with this region using $p(y)$ (the probability distribution for the binomial random variable, Chapter 6). Thus, with $n = 10$, $p = .5$,

$$\alpha = p(0) + p(10) = C_0^{10}(.5)^{10} + C_{10}^{10}(.5)^{10} = .002.$$

Since this value of α is too small, the rejection region is expanded by including the next pair of y values that are most contradictory to the null hypothesis, namely, $y = 1$ and $y = 9$. The value of α for this rejection region ($y = 0, 1, 9, 10$) is obtained from Table 1 in the Appendix:

$$\alpha = p(0) + p(1) + p(9) + p(10) = .022.$$

This α is also small, so we again expand the region. This time we include $y = 0, 1, 2, 8, 9, 10$. You may verify that the corresponding value of α is .11. At this point, we have a choice. We can either choose α equal to .022 or .11, depending on the risk we are willing to take of making a type I error. For this example, we will assume that you have chosen $\alpha = .11$ and, consequently, that you will employ $y = 0, 1, 2, 8, 9, 10$ as the rejection region for the test. (See Figure 16.1.)

Figure 16.1 Rejection region for Example 16.1

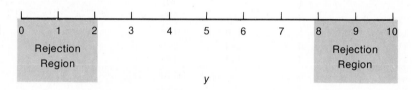

From the data, we observe that $y = 1$ and therefore we reject the null hypothesis. Thus we conclude that sufficient evidence exists to indicate that the population distributions of number of defects are not identical. A quick glance at the data suggests that the mean number of defects for production line B tends to exceed that for A. The probability of rejecting the null hypothesis when true is only $\alpha = .11$, and therefore we are reasonably confident of our conclusion.

The experimenter in this example is using the test procedure as a rough tool for detecting faulty production lines. The rather large value of α is not likely to be disturbing because additional data can easily be collected if the experimenter is concerned about making a type I error in reaching a conclusion.

One problem that may occur when conducting a sign test is that the observations associated with one or more pairs may be equal and therefore result in ties. When this situation occurs, delete the tied pairs and reduce n, the total number of pairs.

You will also encounter situations where n, the number of pairs, is large. Then the values of α associated with the sign test can be obtained by using the normal approximation to the binomial probability distribution discussed in

Section 7.6. You can verify (by comparison of exact probabilities with their approximations) that these approximations will be quite adequate for n as small as 10. This is due to the symmetry of the binomial probability distribution for $p = .5$. For $n \geq 25$, the z test of Chapters 7 and 9 will be quite adequate, where

$$z = \frac{y - np}{\sqrt{npq}} = \frac{y - .5n}{.5\sqrt{n}}.$$

This would be testing the null hypothesis, $p = .5$, against the alternative, $p \neq .5$, for a two-tailed test or against the alternative, $p > .5$ (or $p < .5$), for a one-tailed test. The tests would utilize the familiar rejection regions of Chapter 9.

Sign Test for Large Samples $n \geq 25$

1. *Null hypothesis:* $H_0 : p = .5$ (one treatment is not preferred to a second treatment, and vice versa).

2. *Alternative hypothesis:* $H_a : p \neq .5$, for a two-tailed test. (Note: We use the two-tailed test as an example. Many analyses might require a one-tailed test.)

3. *Test statistic:* $z = \dfrac{y - .5n}{.5\sqrt{n}}$.

4. *Rejection region:* Reject H_0 if $z \geq z_{\alpha/2}$ or if $z \leq -z_{\alpha/2}$, where $z_{\alpha/2}$ is the z-value from Table 3 of the Appendix corresponding to an area of $\alpha/2$ in the upper tail of the normal distribution.

Example 16.2

A production superintendent claims that there is no difference in the employee accident rates for the day and evening shifts in a large manufacturing plant. The number of accidents per day is recorded for both the day and evening shifts for $n = 100$ days. It is found that the number of accidents per day for the evening shift y_E exceeded the corresponding number of accidents on the day shift y_D on 63 of the 100 days. Do these results provide sufficient evidence to indicate that more accidents tend to occur on one shift than on the other, or equivalently, that $P(y_E > y_D) \neq 1/2$?

Solution

This is a paired-difference experiment, with $n = 100$ pairs of observations corresponding to the 100 days. To test the null hypothesis that the two distributions of accidents are identical, we use the test statistic

$$z = \frac{y - .5n}{.5\sqrt{n}},$$

where y represents the number of days in which the number of accidents on the evening shift exceeded the number of accidents on the day shift. Then for $\alpha = .05$ we will reject the null hypothesis if $z \geq 1.96$ or $z \leq -1.96$. Substituting into the formula for z, we have

$$z = \frac{y - .5n}{.5\sqrt{n}} = \frac{63 - (.5)(100)}{.5\sqrt{100}} = \frac{13}{5} = 2.60$$

Since the calculated value of z exceeds $z_{\alpha/2} = 1.96$, we reject the null hypothesis. The data provide sufficient evidence to indicate a difference in the accident rate distributions for the day and evening shifts.

You will note that the data, Example 16.1, are the result of a paired-difference experiment. Suppose that the paired differences are normally distributed with a common variance, σ^2. Will the sign test detect a shift in location of the two populations as effectively as the Student's t test? Intuitively, we would suspect that the answer is "no," and this is correct, because the Student's t test utilizes comparatively more information. In addition to giving the sign of the difference, the t test uses the magnitudes of the observations to obtain a more accurate value of the sample means and variances. Thus one might say that the sign test is not as "efficient" as the Student's t test, but this statement is meaningful only if the populations conform to the assumption stated above; that is, the differences in paired observations are normally distributed with a common variance, σ_d^2. The sign test might be more efficient when these assumptions are not satisfied.

Exercises

16.1. What significance levels between $\alpha = .01$ and $\alpha = .15$ are available for a two-tailed sign test utilizing 25 paired observations? (Make use of tabulated values in Table 1, the Appendix, for $n = 25$.) What are the corresponding rejection regions?

16.2. In Exercise 10.32, we compared property evaluations of two tax assessors, A and B. Their assessments for eight properties are shown in the table.

| Property | Assessor | |
	A	B
1	36.3	35.1
2	48.4	46.8
3	40.2	37.3
4	54.7	50.6
5	28.7	29.1
6	42.8	41.0
7	36.1	35.3
8	39.0	39.1

a. Use the sign test to determine whether the data present sufficient evidence to indicate that one of the assessors tends to be consistently more conservative than the other; i.e., $P(y_A > y_B) \neq 1/2$. Test by using a value of α near .05. Find the p value for the test and interpret its value.

b. Exercise 10.32 uses the t statistic to test the null hypothesis that there is no difference in the mean level of property assessments between assessors A and B. Check the answer (in the answer section) for Exercise 10.32 and compare it with your answer to part (a). Do the test results agree? Explain why the answers are (or are not) consistent.

16.3. Two gourmets rated 20 meals on a scale of 1 to 10. The data are shown in the table. Do the data provide sufficient evidence to indicate that one of the gourmets tends to give higher ratings than the other? Test by using the sign test with a value of α near .05.

Meal	A	B	Meal	A	B
1	6	8	11	6	9
2	4	5	12	8	5
3	7	4	13	4	2
4	8	7	14	3	3
5	2	3	15	6	8
6	7	4	16	9	10
7	9	9	17	9	8
7	7	8	18	4	6
9	2	5	19	4	3
10	4	3	20	5	5

16.4. Clinical data concerning the effectiveness of two drugs in treating a particular disease were collected from 10 hospitals. The numbers of patients treated with the drugs varied from one hospital to another as well as between drugs within a given hospital. The data, in percent recovery, are shown below:

| Hospital | Drug A | | | Drug B | | |
	Number in Group	Number Recovered	Percent Recovered	Number in Group	Number Recovered	Percent Recovered
1	84	63	75.0	96	82	85.4
2	63	44	69.8	83	69	83.1
3	56	48	85.7	91	73	80.2
4	77	57	74.0	47	35	74.5
5	29	20	69.0	60	42	70.0
6	48	40	83.3	27	22	81.5
7	61	42	68.9	69	52	75.4
8	45	35	77.8	72	57	79.2
9	79	57	72.2	89	76	85.4
10	62	48	77.4	46	37	80.4

Do the data present sufficient evidence to indicate a higher recovery rate for one of the two drugs?

a. Test using the sign test. Choose your rejection region so that α is near .10.

b. Why would it be inappropriate to use a Student's t test in analyzing the data?

16.5. An experiment was conducted to compare the tenderness of meat cuts subjected to two different meat tenderizers, A and B. To reduce the effect of extraneous variables, the data were paired by the specific meat cut, by applying the tenderizers to two cuts taken from the same steer, by cooking paired cuts together, and by using a single judge for each pair. After cooking, each cut was rated by a judge on a scale of 1 to 10, with 10 corresponding to the most tender condition of the cooked meat. The data, shown for a single judge, are given below. Do the data provide sufficient evidence to indicate that one of the two tenderizers tends to receive higher ratings than the other? Use a value of α near .05. [*Note:* $(1/2)^8 = .003906$.] Would a Student's t test be appropriate for analyzing these data? Explain.

| | Tenderizer | |
Cut	A	B
Shoulder roast	5	7
Chuck roast	6	5
Rib steak	8	9
Brisket	4	5
Club steak	9	9
Round steak	3	5
Rump roast	7	6
Sirloin steak	8	8
Sirloin tip steak	8	9
T-bone steak	9	10

16.3
A Comparison of Statistical Tests

We have introduced two statistical test procedures utilizing given data for the same hypothesis, and we are now faced with the problem of comparison of tests. Which, if either, is better? One method would be to hold the sample size and α constant for both procedures and compare β, the probability of a type II error. Actually, statisticians prefer a comparison of the **power of a test** where

power $= 1 - \beta.$

Since β is the probability of failing to reject the null hypothesis when it is false, the power of the test is the probability of rejecting the null hypothesis when it is false and some specified alternative is true. It is the probability that the test will do what it was designed to do, that is, detect a departure from the null hypothesis when a departure exists.

Probably the most common method of comparing two test procedures is in terms of the relative efficiency of a pair of tests. **Relative efficiency** is the ratio of the sample sizes for the two test procedures required to achieve the same α and β for a given alternative to the null hypothesis.

In some situations, you may not be too concerned whether you are using the most powerful test. For example, you might choose to use the sign test over a more powerful competitor because of its ease of application. Thus you might view tests as microscopes that are utilized to detect departures from an hypothesized theory. One need not know the exact power of a microscope to use it in a biological investigation, and the same applies to statistical tests. If the test procedure detects a departure from the null hypothesis, we are delighted. If not, we can reanalyze the data by using a more powerful test, or we can increase the power of the microscope (test) by increasing the sample size.

16.4
The Use of Ranks for Comparing Two Population Distributions: Independent Random Samples

A statistical test for comparing two populations based on independent random samples was proposed by F. Wilcoxon in 1945 (see the references). Suppose that you were to select independent random samples of n_1 and n_2 observations, each from two populations, call them I and II. Wilcoxon's idea was to combine the $n_1 + n_2 = n$ observations and to rank them, in order of magnitude, from 1 (the smallest) to n (the largest). Then if the observations were selected from identical populations, the rank sums for the samples should be more or less proportional to the sample sizes, n_1 and n_2. For example, if n_1 and n_2 were equal, you would expect the rank sums to be nearly equal. In contrast, if the observations in one population, say population I, tended to be larger than those in population II, then the observations in sample I would tend to receive the highest ranks and would have a larger than expected rank sum. Thus (sample sizes being equal) if one rank sum is very large (and, correspondingly, the other is very small), you may have evidence to indicate a difference between the two populations.

Mann and Whitney (see the references) proposed a statistical test in 1947 that also used the rank sums of the two samples, and their test can be shown to be equivalent to the Wilcoxon test. Because the Mann–Whitney U test and tables of critical values of U occur so often in the literature, we will explain its use in Section 16.5 and will give several examples of its application. In this section we illustrate the logic of the rank sum test and show you how the rejection region for the test and how α are determined.

Example 16.3

The bacteria counts per unit volume are shown below for two types of cultures, A and B. Four observations were made for each culture:

Culture A	Culture B
27	32
31	29
26	35
25	28

Let n_1 and n_2 represent the number of observations in samples A and B, respectively. Then the rank sum procedure ranks each of the $(n_1 + n_2) = n$ measurements in order of magnitude and uses T, the sum of the ranks for A or B, as the test statistic. Thus, for the data given above, the corresponding ranks are:

	Culture A	Culture B
	3	7
	6	5
	2	8
	1	4
Rank sum	12	24

Do these data present sufficient evidence to indicate a difference in the population distributions for A and B?

Solution

Let T equal the rank sum for sample A (for this sample, $T = 12$). Certainly very small or very large values of T will provide evidence to indicate a difference between the two population distributions, and hence T, the rank sum, will be employed as a test statistic.

The rejection region for a given test will be obtained in the same manner as for the sign test. We will commence by selecting the most contradictory values of T as the rejection region and will add to these until α is of sufficient size.

The minimum rank sum would include the ranks 1, 2, 3, 4, or $T = 10$. Similarly, the maximum would include 5, 6, 7, 8, with $T = 26$. We will commence by including these two values of T in the rejection region. What is the corresponding value of α?

Finding the value of α is a probability problem that can be solved by using the methods of Chapter 4. If the populations are identical, every permutation of the eight ranks will represent a sample point and will be equally likely. Then α will represent the sum of the probabilities of the sample points (arrangements) which imply $T = 10$ or $T = 26$ in sample A. The total number of permutations of the eight ranks is 8!. The number of different arrangements of the ranks 1, 2, 3, 4 in sample A with the 5, 6, 7, 8 of sample B is (4!)(4!). Similarly, the number of arrangements that place the maximum value of T in sample A (ranks 5, 6, 7, 8) is (4!)(4!). Then the probability that $T = 10$ or $T = 26$ is

$$P(T = 10 \text{ or } 26) = p(10) + p(26) = \frac{2(4!)(4!)}{8!} = \frac{2}{C_4^8} = \frac{1}{35} = .029.$$

If this value of α is too small, the rejection region can be enlarged to include the next smallest and largest rank sums, $T = 11$ and $T = 25$. The rank sum $T = 11$ will include the ranks 1, 2, 3, 5, and

$$p(11) = \frac{4!4!}{8!} = \frac{1}{70}.$$

Similarly,

$$p(25) = \frac{1}{70}.$$

Then,

$$\alpha = p(10) + p(11) + p(25) + p(26) = \frac{2}{35} = .057.$$

Expansion of the rejection region to include 12 and 24 will substantially increase the value of α. The set of sample points giving a rank of 12 will be all sample points associated with rankings of (1, 2, 3, 6) and (1, 2, 4, 5). Thus

$$p(12) = \frac{2(4!)(4!)}{8!} = \frac{1}{35}$$

and

$$\alpha = p(10) + p(11) + p(12) + p(24) + p(25) + p(26)$$
$$= \frac{1}{70} + \frac{1}{70} + \frac{1}{35} + \frac{1}{35} + \frac{1}{70} + \frac{1}{70} = \frac{4}{35} = .114.$$

This value of α might be considered too large for practical purposes. In this case, we would be better satisfied with the rejection region $T = 10, 11, 25,$ and 26.

The rank sum for the sample, $T = 12$, falls in the nonrejection region, and hence we do not have sufficient evidence to reject the hypothesis that the population distributions of bacteria counts for the two cultures are identical.

16.5
The Mann–Whitney *U* Test: Independent Random Samples

The **Mann–Whitney statistic, *U*,** is obtained by ordering all $(n_1 + n_2)$ observations according to their magnitude and counting the number of observations in sample *A* that precede each observation in sample *B*. The *U* statistic is the sum of these counts.

For example, the eight observations of Example 16.3 are

$$\begin{array}{cccccccc} 25, & 26, & 27, & 28, & 29, & 31, & 32, & 35. \\ A & A & A & B & B & A & B & B \end{array}$$

The smallest *B* observation is 28, and $u_1 = 3$ observations from sample *A* precede it. Similarly, $u_2 = 3$ *A* observations precede the second *B* observation and $u_3 = 4$ and $u_4 = 4$ *A* observations precede the third and fourth *B* observations, respectively (32 and 35). Then,

$$U = u_1 + u_2 + u_3 + u_4 = 14.$$

Or you could count the number of *B* observations that precede each *A* observation and use the sum of the counts, U_B, as the *U* statistic. In either case, very large or small values of *U* will imply a separation of the ordered *A* and *B* observations and will provide evidence to indicate a difference (a shift in location) between the population distributions for *A* and *B*.

As noted in Section 16.4, it can be shown that the Mann–Whitney U statistic is related to Wilcoxon's rank sum. In fact, it can be shown (proof omitted) that:

Formulas for the Mann–Whitney U Statistic

$$U_A = n_1 n_2 + \frac{n_1(n_1 + 1)}{2} - T_A,$$

$$U_B = n_1 n_2 + \frac{n_2(n_2 + 1)}{2} - T_B,$$

where

n_1 = number of observations in sample A

n_2 = number of observations in sample B.

$U_A + U_B = n_1 n_2$ and T_A and T_B are the rank sums for samples A and B, respectively.

As you can see from the formulas for U_A and U_B, U_A will be smaller when T_A is large, a situation that likely will occur when the population distribution of the A measurements is shifted to the right of the population distribution for the B measurements. Consequently, to conduct a one-tailed test to detect a shift in the A distribution to the right of the B distribution, you will reject the null hypothesis of "no difference in the population distributions" if U_A is less than some specified value, U_0. That is, you will reject H_0 for small values of U_A. Similarly, to conduct a one-tailed test to detect a shift of the B distribution to the right of the A distribution, you would reject H_0 if U_B is less than some specified value, say U_0. Consequently, the rejection region for the Mann–Whitney U test would appear as shown in Figure 16.2.

Figure 16.2 Rejection region for a Mann–Whitney U test

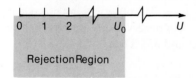

Rejection Region

Table 11, the Appendix, gives the probability that an observed value of U will be less than some specified value, say U_0. This is the value of α for a one-tailed test. To conduct a two-tailed test, that is, to detect a shift in the population distributions for the A and B measurements, in either direction, we will agree to always use U, the smaller of U_A or U_B, as the test statistic and reject H_0 for

$U < U_0$ (see Figure 16.2). The value of α for the two-tailed test will be double the tabulated value given in Table 11, the Appendix.

To see how to locate the rejection region for the Mann–Whitney U test, suppose that $n_1 = 4$ and $n_2 = 5$. Then you would consult the third table in Table 11, the Appendix, the one corresponding to $n_2 = 5$. The first few lines of Table 11, $n_2 = 5$, are shown here in Table 16.1. Note that the table is constructed on the assumption that $n_1 \leq n_2$.

Table 16.1 An abbreviated version of Table 11, the Appendix; $P(U \leq U_0)$ for $n_2 = 5$

| U_0 | \multicolumn{5}{c}{n_1} |
	1	2	3	4	5
0	.1667	.0376	.0179	.0079	.0040
1	.3333	.0952	.0357	.0159	.0079
2	.5000	.1905	.0714	.0317	.0159
3		.2857	.1250	.0556	.0278
4		.4286	.1964	.0952	.0476
5		.5714	.2857	.1429	.0754
⋮		⋮		⋮	⋮

Across the top of Table 16.1, you see values of n_1. Values of U_0 are shown down the left side of the table. The entries give the probability that U will assume a small value, namely, the probability that $U \leq U_0$. Since for our example, $n_1 = 4$, we will move across the top of the table to $n_1 = 4$. Move to the second row of the table corresponding to $U_0 = 2$ for $n_1 = 4$. Then you see the probability that U will be less than or equal to 2 is .0317. Similarly, moving across the row for $U_0 = 3$, you see that the probability that U is less than or equal to 3 is .0556. (This value is shaded in Table 16.1.) So, if you want to conduct a one-tailed Mann–Whitney U test with $n_1 = 4$ and $n_2 = 5$ and would like α to be near .05, you would reject the null hypothesis of equality of population relative frequency distributions when $U \leq 3$. The probability of a type I error for the test would be $\alpha = .0556$. If you use this same rejection region for a two-tailed test, that is, $U \leq 3$, α would be double the tabulated value, or $\alpha = 2(.0556) = .1112$.

Table 11, the Appendix, can also be used to find the observed significance level for a test. For example, if $n_1 = 5$, $n_2 = 5$, and $U = 4$, then from Table 16.1, the p value for a one-tailed test would be,

$$P\{U \leq 4\} = .0476.$$

If the test were two-tailed, the p value would be

$$2(.0476) \quad \text{or} \quad .0952.$$

The Mann–Whitney *U* Test

1. *Null Hypothesis:* H_0: The population relative frequency distributions for *A* and *B* are identical.

2. *Alternative Hypothesis:* H_a: The two population relative frequency distributions are shifted in respect to their relative locations (a two-tailed test). Or H_a: The population relative frequency distribution for *A* is shifted to the right of the relative frequency distribution for population *B* (a one-tailed test).*

3. *Test Statistic:*   For a two-tailed test, use *U*, the smaller of

$$U_A = n_1 n_2 + \frac{n_1(n_1 + 1)}{2} - T_A$$

and

$$U_B = n_1 n_2 + \frac{n_2(n_2 + 1)}{2} - T_B$$

where T_A and T_B are the rank sums for samples *A* and *B*, respectively. For a one-tailed test, use U_A.

4. *Rejection Region:*

   1. For the two-tailed test and a given value of α, reject H_0 if $U \leq U_0$, where $P(U \leq U_0) = \alpha/2$. [*Note:* U_0 is the value such that $P(U \leq U_0)$ is equal to half of α.]

   2. For a one-tailed test and a given value of α, reject H_0 if $U_A \leq U_0$, where $P(U_A \leq U_0) = \alpha$.

* For the sake of convenience, we will describe the one-tailed test as one designed to detect a shift in the distribution of the *A* measurements to the right of the distribution of the *B* measurements. To detect a shift in the *B* distribution to the right of the *A* distribution, just interchange the letters *A* and *B* in the discussion.

When applying the test to a set of data, you may find that some of the observations are of equal value. **Ties in the observations can be handled by averaging the ranks that would have been assigned to the tied observations and assigning this average to each.** Thus, if three observations are tied and are due to receive ranks 3, 4, 5, we would assign the rank of 4 to all three. The next observation in the sequence would receive the rank of 6, and ranks 3 and 5 would not appear. Similarly, if two observations are tied for ranks 3 and 4, each would receive a rank of 3.5, and ranks 3 and 4 would not appear.

Example 16.4

Test the hypothesis that there is no difference in the population distributions for the bacteria-count data of Example 16.3.

Solution

We have already noted that the Mann-Whitney U test and the Wilcoxon rank sum test are equivalent, so we should reach identically the same conclusions as for Example 16.3. Recall that the alternative hypothesis was that there was a difference in the distributions of bacteria counts for cultures A and B and that this implied a two-tailed test. Thus, since Table 11, the Appendix, gives values of $P(U \leq U_0)$ for specified sample sizes and values of U_0, we must double the tabulated value to find α. Suppose, as in Example 16.3, that we desire a value of α near .05. Checking table 11, the Appendix, for $n_1 = 4$ and $n_2 = 4$, we find $P(U \leq 1) = .0286$. Using $U \leq 1$ as the rejection region (see Figure 16.3), α will equal $2(.0286) = .0572$ or, rounding to three decimal places, $\alpha = .057$ (the same value of α obtained for Example 16.3).

Figure 16.3 Rejection region for Example 16.4

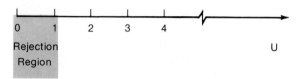

For the bacteria data, the rank sums are $T_A = 12$ and $T_B = 24$. Then

$$U_A = n_1 n_2 + \frac{n_1(n_1 + 1)}{2} - T_A$$

$$= (4)(4) + \frac{4(4 + 1)}{2} - 12$$

$$= 14,$$

and

$$U_B = n_1 n_2 + \frac{n_2(n_2 + 1)}{2} - T_B$$

$$= (4)(4) + \frac{4(4 + 1)}{2} - 24$$

$$= 2.$$

Because the smaller observed value of U is 2 (calculated above), U does not fall in the rejection region. Hence there is not sufficient evidence to show a shift in locations of the population distributions of bacteria counts for cultures A and B.

Example 16.5

An experiment was conducted to compare the strength of two types of kraft papers, one a standard kraft paper of a specified weight and the other the same standard kraft paper treated with a chemical substance. Ten pieces of each type of paper, randomly selected from production, produced the strength measurements shown in the accompanying table. Test the hypothesis of "no difference in the distributions of strengths for the two types of paper" against the alternative hypothesis that the treated paper tends to be of greater strength (i.e., its distribution of strength measurements is shifted to the right of the corresponding distribution for the untreated paper).

	Standard *A*	Treated *B*
	1.21 (2)	1.49 (15)
	1.43 (12)	1.37 (7.5)
	1.35 (6)	1.67 (20)
	1.51 (17)	1.50 (16)
	1.39 (9)	1.31 (5)
	1.17 (1)	1.29 (3.5)
	1.48 (14)	1.52 (18)
	1.42 (11)	1.37 (7.5)
	1.29 (3.5)	1.44 (13)
	1.40 (10)	1.53 (19)
Rank sums	$T_A = 85.8$	$T_B = 124.5$

Solution

The ranks are shown in parentheses alongside the $n_1 + n_2 = 10 + 10 = 20$ strength measurements and the rank sums, T_A and T_B, are shown below the columns. Since we wish to detect a shift in the distribution of the B measurements to the right of the distribution for the A measurements, we would reject the null hypothesis of "no difference in population strength distributions" when T_B is excessively large. Because this situation will occur when U_B is small, we shall conduct a one-tailed statistical test and reject the null hypothesis when $U_B \leq U_0$.

Suppose that we choose a value of α near .05. Then we can find U_0 by consulting the portion of Table 11, the Appendix, corresponding to $n_2 = 10$. The probability, $P(U \leq U_0)$, nearest .05 is .0526 and corresponds to $U_0 = 28$. Hence, we will reject if $U_B \leq 28$.

Calculating U_B, we have

$$U_B = n_1 n_2 + \frac{n_2(n_2 + 1)}{2} - T_B$$

$$= (10)(10) + \frac{(10)(11)}{2} - 124.5$$

$$= 30.5.$$

As you can see, U_B is not less than $U_0 = 28$. Therefore, we cannot reject the null hypothesis. At the $\alpha = .05$ level of significance, there is not sufficient evidence to indicate that the treated kraft paper is stronger than the standard.

A simplified large-sample test ($n_1 \geq 10$ and $n_2 \geq 10$) can be obtained by using the familiar z statistic of Chapter 9. When the population distributions are identical, it can be shown that the U statistic has expected value and variance,

$$E(U) = \frac{n_1 n_2}{2}$$

and

$$V(U) = \frac{n_1 n_2 (n_1 + n_2 + 1)}{12},$$

and the distribution of

$$z = \frac{U - E(U)}{\sigma_U}$$

tends to normality with mean zero and variance equal to 1 as n_1 and n_2 become large. **This approximation will be adequate when n_1 and n_2 are both greater than or equal to 10.** Thus, for a two-tailed test with $\alpha = .05$, we would reject the null hypothesis if $|z| \geq 1.96$.

Observe that the z statistic will reach the same conclusion as the exact U test for Example 16.5. Thus

$$z = \frac{30.5 - [(10)(10)/2]}{\sqrt{[(10)(10)(10 + 10 + 1)]/12}} = \frac{30.5 - 50}{\sqrt{2100/12}} = -\frac{19.5}{\sqrt{175}}$$

$$= -\frac{19.5}{13.23} = -1.47.$$

For a one-tailed test with $\alpha = .05$ located in the lower tail of the z distribution, we shall reject the null hypothesis if $z < -1.645$. You can see that $z = -1.47$ does not fall in the rejection region and that this test reaches the same conclusion as the exact U test of Example 16.5.

The Mann–Whitney U Test for Large Samples, $n_1 > 10$ and $n_2 > 10$

1. *Null Hypothesis:* H_0: The population relative frequency distributions for A and B are identical.

2. *Alternative Hypothesis:* H_a: The two population relative frequency distributions are not identical (a two-tailed test). Or H_a: The population relative frequency distribution for A is shifted to the right (or left) of the relative frequency distribution for population B (a one-tailed test).

3. *Test Statistic:* $z = \dfrac{U - (n_1 n_2/2)}{\sqrt{n_1 n_2 (n_1 + n_2 + 1)/12}}$

   Let $U = U_A$.

4. *Rejection Region:*   Reject H_0 if $z > z_{\alpha/2}$ or $z < -z_{\alpha/2}$ for a two-tailed test. For a one-tailed test, place all of α in one tail of the z distribution. To detect a shift in the distribution of the A observations to the right of the distribution of the B observations, let $U = U_A$ and reject H_0 when $z < -z_\alpha$. To detect a shift in the opposite direction, let $U = U_A$ and reject H_0 when $z > z_\alpha$. Tabulated values of z are given in Table 3, the Appendix.

Exercises

16.6.   Data collected on air pollution at 8 Ohio locations (presented earlier in Exercise 10.17) seem to suggest that air pollution, caused by pollution drifting from urban areas, is greater in rural areas than in urban areas. The data, reproduced from *Environmental Midwest* (December 1976), are shown in the table.

City	Designation	Maximum Concentration (ppm)	Days Exceeding Standard (%)	Number of Violations (Total)
Canton	urban	.14	44	148
Cincinnati	urban	.18	44	54
Cleveland	urban	.14	26	51
Columbus	urban	.15	27	113
Dayton	urban	.13	35	114
McConnelsville	rural	.16	56	239
Wilmington	rural	.18	58	259
Wooster	rural	.17	55	262

a. In Exercise 10.17 we used a Student's t test to test the null hypothesis that there was no difference in the mean number of violations between urban and rural Ohio locations. Use the Mann–Whitney U test to test for a shift in distributions for the two locations. Are the conclusions of the t test and the Mann–Whitney U test in agreement? Explain why you think they should or should not be in agreement.

b. Now use the Mann–Whitney U test to test a similar hypothesis concerning the maximum concentration of pollutants recorded over the time period.

16.7.   We shall reanalyze the data of Exercise 10.19 by use of a nonparametric statistical test. The observations, which represent the dissolved oxygen content in water, measure the ability of a river, lake, or stream to support aquatic life. In this experiment, a pollution-control inspector suspected that a river community was releasing amounts of semitreated sewage into a river. To check his theory, he drew five randomly selected specimens of river water at a location above the town and another five below. The dissolved oxygen readings, in parts per million, are as follows:

Above Town	4.8	5.2	5.0	4.9	5.1
Below Town	5.0	4.7	4.9	4.8	4.9

    a. Use a one-tailed Mann–Whitney U test with $\alpha = .05$ to answer the question.

    b. Use a Student's t test (with $\alpha = .05$) to analyze the data (you may have done this in Exercise 10.19). Compare the conclusions reached in parts (a) and (b).

16.8.   In an investigation of visual scanning behavior of deaf children, measurements of eye-movement rate were taken on nine deaf and nine hearing children. From the data given, does it appear that the distributions of eye-movement rates for deaf children (A) and hearing children (B) differ?

	Deaf Children (A)	Hearing Children (B)
	(15) 2.75	.89 (1)
	(11) 2.14	1.43 (7)
	(18) 3.23	1.06 (4)
	(10) 2.07	1.01 (3)
	(14) 2.49	.94 (2)
	(12) 2.18	1.79 (8)
	(17) 3.16	1.12 (5.5)
	(16) 2.93	2.01 (9)
	(13) 2.20	1.12 (5.5)
Rank sum	126	45

16.9.   The life in months of service before failure of the color television picture tube in eight television sets manufactured by firm A and ten sets manufactured by firm B are as follows:

Firm	Life of Picture Tube (months)									
A	32	25	40	31	35	29	37	39		
B	41	39	36	47	45	34	48	44	43	33

Use the U test to analyze the data and test to see if the life in months of service before failure of the picture tube differs for the picture tubes produced by the two manufacturers. (Use $\alpha = .10$.)

16.10. Is the Mann–Whitney U test appropriate for the data of Exercise 16.4? Explain.

16.11. A comparison of the weights of turtles caught in two different lakes was conducted to compare the effects of the two lake environments on turtle growth. All the turtles were of the same age and were tagged before being released in the lakes. The weight measurements for $n_1 = 10$ tagged turtles caught in lake 1 and $n_2 = 8$ caught in lake 2 are as follows:

Lake	Weight (oz)									
1	14.1	15.2	13.9	14.5	14.7	13.8	14.0	16.1	12.7	15.3
2	12.2	13.0	14.1	13.6	12.4	11.9	12.5	13.8		

Do the data provide sufficient evidence to indicate a difference in the distributions of weights for the tagged turtles exposed to the two lake environments? Use a Mann–Whitney U test with $\alpha = .05$ to answer the question.

16.12.   Cancer treatment by means of chemicals—chemotherapy—utilizes chemicals that kill both cancerous and normal cells. In some instances the toxicity of the cancer drug, that is, its effect on normal cells, can be reduced by the simultaneous injection of a second drug. A study was conducted to determine whether a particular drug injection was beneficial in reducing the harmful effects of a chemotherapy treatment on the survival time for rats. Two randomly selected groups of rats, 12 rats in each group, were used for the experiment. Both groups, call them A and B, received the toxic drug in a dosage large enough to cause death, but in addition, group B received the antitoxin which was to reduce the toxic effect of the chemotherapy on normal cells. The test was terminated at the end of 20 days, or 480 hours. The lengths of survival time for the two groups of rats, to the nearest 4 hours, are shown in the table. Do the data provide sufficient evidence to indicate that rats receiving the antitoxin tend to survive longer after chemotherapy than those not receiving the toxin? Use the Mann–Whitney U test with a value of α near .05.

Chemotherapy Only A	Chemotherapy Plus Drug B
84	140
128	184
168	368
92	96
184	480
92	188
76	480
104	244
72	440
180	380
144	480
120	196

16.6
The Wilcoxon Signed Rank Test for a Paired Experiment

A rank sum test proposed by Wilcoxon can be used to analyze the paired-difference experiment of Section 10.5 by considering the paired differences of the two treatments, A and B. Under the null hypothesis of "no differences in the distributions for A and B," you would expect (on the average) half of the differences in pairs to be negative and half to be positive. That is, the expected number of negative differences between pairs would be $n/2$ (where n is the number of pairs). Further, it would follow that positive and negative differences of equal absolute magnitude should occur with equal probability. If one were to order the differences according to their absolute values and rank them from smallest to largest, the expected rank sums for the negative and positive differences would be equal. Sizable differences in the sums of the ranks assigned to the positive and negative differences would provide evidence to indicate a shift in location between the distributions of responses for the two treatments, A and B.

To carry out the Wilcoxon test, calculate the differences $(y_A - y_B)$ for each of the n pairs. Differences equal to zero are eliminated and the number of pairs, n, is reduced accordingly. Rank the *absolute values* of the differences, assigning a 1 to the smallest, a 2 to the second smallest, and so on. Then calculate the rank sum for the negative differences and also calculate the rank sum for the positive differences. For a two-tailed test, we use the smaller of these two quantities, T, as a test statistic to test the null hypothesis that the two population relative frequency histograms are identical. The smaller the value of T, the greater will be the weight of evidence favoring rejection of the null hypothesis. Hence, we shall reject the null hypothesis if T is less than or equal to some value, say T_0.

To detect the one-sided alternative, that the distribution of the A observations is shifted to the right of the B observations, use the rank sum T^- of the negative differences and reject the null hypothesis for small values of T^-, say $T^- \le T_0$. If we wish to detect a shift of the distribution of B observations to the right of the distribution of A observations, use the rank sum T^+ of the positive differences as a test statistic and reject for small values of T^+, say $T^+ \le T_0$.

The probability that T is less than or equal to some value, T_0, has been calculated for a combination of sample sizes and values of T_0. These probabilities, given in Table 12, the Appendix, can be used to find the rejection region for the T test.

An abbreviated version of Table 12, the Appendix, is shown here in Table 16.2. Across the top of the table you see the number of differences (the number of pairs), n. Values of α (denoted by the symbol P in Wilcoxon's table)

Table 16.2 An abbreviated version of Table 12, the Appendix; critical values of T

One-sided	Two-sided	$n = 5$	$n = 6$	$n = 7$	$n = 8$	$n = 9$	$n = 10$
$P = .05$	$P = .10$	1	2	4	6	8	11
$P = .025$	$P = .05$		1	2	4	6	8
$P = .01$	$P = .02$			0	2	3	5
$P = .005$	$P = .01$				0	2	3

One-sided	Two-sided	$n = 11$	$n = 12$	$n = 13$	$n = 14$	$n = 15$	$n = 16$
$P = .05$	$P = .10$	14	17	21	26	30	36
$P = .025$	$P = .05$	11	14	17	21	25	30
$P = .01$	$P = .02$	7	10	13	16	20	24
$P = .005$	$P = .01$	5	7	10	13	16	19

One-sided	Two-sided	$n = 17$
$P = .05$	$P = .10$	41
$P = .025$	$P = .05$	35
$P = .01$	$P = .02$	28
$P = .005$	$P = .01$	23

for a one-tailed test appear in the first column of the table. The second column gives values of α (P in Wilcoxon's notation) for a two-tailed test. Table entries are the critical values of T. You will recall that the critical value of a test statistic is the value that locates the boundary of the rejection region.

For example, suppose you have $n = 7$ pairs and you are conducting a two-tailed test of the null hypothesis that the two population relative frequency distributions are identical. Checking the $n = 7$ column of Table 16.2 and using the second row (corresponding to $P = \alpha = .05$ for a two-tailed test), you see the entry, 2 (shaded in Table 16.2). This is T_0, the critical value of T. As noted earlier, the smaller the value of T, the greater will be the evidence to reject the null hypothesis. Therefore, you will reject the null hypothesis for all values of T less than or equal to 2. The rejection region for the Wilcoxon signed rank test for a paired experiment is always of the form: reject H_0 if $T \leq T_0$, where T_0 is the critical value of T. The rejection region is shown symbolically in Figure 16.4.

Figure 16.4 Rejection region for the Wilcoxon signed rank test for a paired experiment (reject H_0 if $T \leq T_0$)

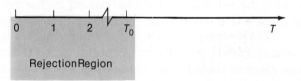

Wilcoxon Signed Rank Test for a Paired Experiment

1. *Null Hypothesis:* H_0: The two population relative frequency distributions are identical.

2. *Alternative Hypothesis:* H_a: The two population relative frequency distributions differ in location (a two-tailed test). Or H_a: The population relative frequency distribution for A is shifted to the right* of the relative frequency distribution for population B (a one-tailed test).

3. *Test Statistic:*
   1. For a two-tailed test, use T, the smaller of the rank sum for positive and the rank sum for negative differences.
   2. For a one-tailed test (to detect the alternative hypothesis described above), use the rank sum T^- of the negative differences.

4. *Rejection Region:*
   1. For a two-tailed test, reject H_0 if $T \leq T_0$, where T_0 is the critical value given in Table 12, the Appendix.
   2. For a one-tailed test (to detect the alternative hypothesis described above), use the rank sum T^- of the negative differences. Reject H_0 if $T^- \leq T_0$.

 * To detect a shift of the distribution of B observations to the right of the distribution of A observations, use the rank sum T^+ of the positive differences as the test statistic and reject H_0 if $T^+ \leq T_0$.

Example 16.6   Test an hypothesis of no difference in population distributions of cake density for the paired-difference experiment, Exercise 10.73.

Solution   The original data and differences in density for the six pairs of cakes are as follows:

	Density (oz/in.^{3})					
y_A	.135	.102	.098	.141	.131	.144
y_B	.129	.120	.112	.152	.135	.163
Difference ($y_A - y_B$)	.006	−.018	−.014	−.011	−.004	−.019
Rank	2	5	4	3	1	6

As with our other nonparametric tests, the null hypothesis to be tested is that the two population frequency distributions of cake densities are identical. The alternative hypothesis, which implies a two-tailed test, is that the distributions are different.

Because the amount of data is small, we will conduct our test using $\alpha = .10$. From Table 12, the critical value of T for a two-tailed test, $\alpha = .10$, is $T_0 = 2$. Hence we will reject H_0 if $T \le 2$ (see Figure 16.5).

Figure 16.5 Rejection region for Example 16.6, $\alpha = .10$

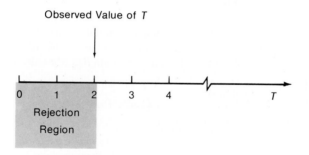

Now let us find the observed value of the test statistic, T. Checking the ranks of the absolute values of the differences, you can see that there is only one positive difference that has a rank of 2. Since for the two-tailed test we have agreed to always use the smaller rank sum as the test statistic, the observed value of the test statistic is $T = 2$. Since $T = 2$ falls in the rejection region, we reject H_0 and conclude that the two population frequency distributions of cake densities differ.

Although Table 12, the Appendix, is applicable for values of n (number of data pairs) as large as $n = 50$, it is worth noting that T^+, like the Mann–Whitney U, **will be approximately normally distributed when the null hypothesis is true and n is large (say 25 or more). This enables us to construct a large-sample**

z **test,** where

$$E(T) = \frac{n(n + 1)}{4},$$

$$V(T) = \frac{n(n + 1)(2n + 1)}{24}.$$

Then the z statistic,

$$z = \frac{T^+ - E(T^+)}{\sigma_{T^+}} = \frac{T^+ - \dfrac{n(n + 1)}{4}}{\sqrt{\dfrac{n(n + 1)(2n + 1)}{24}}},$$

can be used as a test statistic. Thus, for a two-tailed test and $\alpha = .05$, we would reject the hypothesis of "identical population distributions" when $|z| \geq 1.96$.

A Large-Sample Wilcoxon Signed Rank Test for a Paired Experiment: $n \geq 25$

1. *Null Hypothesis:* H_0: The population relative frequency distributions for A and B are identical.

2. *Alternative Hypothesis:* H_a: The two population relative frequency distributions differ in location (a two-tailed test). Or H_a: The population relative frequency distribution for A is shifted to the right (or left) of the relative frequency distribution for population B (a one-tailed test).

3. *Test Statistic:* $z = \dfrac{T^+ - [n(n + 1)/4]}{\sqrt{[n(n + 1)(2n + 1)]/24}}$

4. *Rejection Region:* Reject H_0 if $z \geq z_{\alpha/2}$ or $z < -z_{\alpha/2}$ for a two-tailed test. For a one-tailed test, place all of α in one tail of the z distribution. To detect a shift in the distribution of the A observations to the right of the distribution of the B observations, reject H_0 when $z > z_\alpha$. To detect a shift in the opposite direction, reject H_0 if $z < -z_\alpha$. Tabulated values of z are given in Table 3, the Appendix.

Exercises

16.13. In Exercise 16.2, we used the sign test to determine whether the data provided sufficient evidence to indicate a shift in the distributions of property assessments for assessors A and B.

a. Use the Wilcoxon signed rank test for a paired experiment to test the null hypothesis that there is no difference in the distributions of property assessments between assessors A and B. Test by using a value of α near .05.

b. Compare the conclusion of the test, part (a), with the conclusions derived from the t test, Exercise 10.32, and the sign test, Exercise 16.2. Explain why these test conclusions are (or are not) consistent.

16.14.   The number of machine breakdowns per month was recorded for nine months on two identical machines used to make wire rope. The data are shown in the table.

	Machines	
Month	A	B
1	3	7
2	14	12
3	7	9
4	10	15
5	9	12
6	6	6
7	13	12
8	6	5
9	7	13

a. Do the data provide sufficient evidence to indicate a difference in the monthly breakdown rates for the two machines? Test by using a value of α near .05.

b. Can you think of a reason why the breakdown rates for the two machines might vary from month to month?

16.15.   Refer to the comparison of gourmet mean ratings, Exercise 16.3, and use the Wilcoxon signed rank test to determine whether the data provide sufficient evidence to indicate a difference in the ratings of the two gourmets. Test by using a value of α near .05. Compare the results of this test with the results of the sign test, Exercise 16.3. Are the test conclusions consistent?

16.16.   Two methods for controlling traffic, A and B, were employed at each of $n = 12$ intersections for a period of one week and the numbers of accidents occurring during this time period were recorded. The order of use (which method would be employed for the first week) was selected in a random manner.

Intersection	Method A	Method B	Intersection	Method A	Method B
1	5	4	7	2	3
2	6	4	8	4	1
3	8	9	9	7	9
4	3	2	10	5	2
5	6	3	11	6	5
6	1	0	12	1	1

Do the data provide sufficient evidence to indicate a shift in the distributions of accident rates for traffic control methods A and B?

a. Analyze using a sign test.

b. Analyze using the Wilcoxon signed rank test for a paired experiment.

16.17.   Analyze the data of Exercise 16.5 using the Wilcoxon signed rank test for a paired experiment. Compare with your analysis, Exercise 16.5.

Ψ 16.18.   Eight subjects were asked to perform a simple puzzle assembly task under normal conditions and under conditions of stress. During the stress condition, the subjects were told
that a mild shock would be delivered 3 minutes after the start of the experiment and every
30 seconds thereafter until the task was completed. Blood-pressure readings were taken
under both conditions. The following data represent the highest reading during the
experiment:

Subject	Normal	Stress
1	126	130
2	117	118
3	115	125
4	118	120
5	118	121
6	128	125
7	125	130
8	120	120

Do the data present sufficient evidence to indicate higher blood-pressure readings during
conditions of stress? Analyze the data using the Wilcoxon signed rank test for a paired
experiment.

16.7
Rank Correlation Coefficient

In the preceding sections we have used ranks to indicate the relative magnitude
of observations in nonparametric tests for comparison of treatments. We will
now employ the same technique in testing for a relation between two ranked
variables. Two common rank correlation coefficients are the **Spearman** r_s and
the **Kendall** τ. We will present the Spearman r_s because its computation is
identical to that for the sample correlation coefficient, r, of Chapter 11. Kendall's
rank correlation coefficient is discussed in detail in Kendall and Stuart (1974).

Suppose that eight elementary science teachers have been ranked by a
judge according to their teaching ability, and all have taken a "national teachers'
examination." The data are as follows:

Teacher	Judge's Rank	Examination Score
1	7	44
2	4	72
3	2	69
4	6	70
5	1	93
6	3	82
7	8	67
8	5	80

Do the data suggest an agreement between the judge's ranking and the examination score? Or one might express this question by asking whether a correlation exists between ranks and test scores.

The two variables of interest are rank and test score. The former is already in rank form, and the test scores may be ranked similarly, as shown below. **The ranks for tied observations are obtained by averaging the ranks that the tied observations would occupy as for the Mann–Whitney U statistic.**

Teacher	Judge's Rank (x_i)	Test Rank (y_i)
1	7	1
2	4	5
3	2	3
4	6	4
5	1	8
6	3	7
7	8	2
8	5	6

The Spearman rank correlation coefficient, r_s, is calculated using the ranks as the paired measurements on the two variables, x and y, in the formula for r, Chapter 11. Thus,

---

Spearman's Rank Correlation Coefficient

$$r_s = \frac{SS_{xy}}{\sqrt{SS_x SS_y}},$$

where x_i and y_i represent the ranks of the ith pair of observations and

$$SS_{xy} = \sum_{i=1}^{n} (x_i - \bar{x})(y_i - \bar{y}) = \sum_{i=1}^{n} x_i y_i - \frac{\left(\sum_{i=1}^{n} x_i\right)\left(\sum_{i=1}^{n} y_i\right)}{n},$$

$$SS_x = \sum_{i=1}^{n} (x_i - \bar{x})^2 = \sum_{i=1}^{n} x_i^2 - \frac{\left(\sum_{i=1}^{n} x_i\right)^2}{n},$$

$$SS_y = \sum_{i=1}^{n} (y_i - \bar{y})^2 = \sum_{i=1}^{n} y_i^2 - \frac{\left(\sum_{i=1}^{n} y_i\right)^2}{n}.$$

When there are no ties in either the x observations or the y observations, the above expression for r_s algebraically reduces to the simpler expression

$$r_s = 1 - \frac{6 \sum_{i=1}^{n} d_i^2}{n(n^2 - 1)}, \qquad \text{where } d_i = x_i - y_i.$$

If the number of ties is small in comparison with the number of data pairs, little error will result in using this shortcut formula.

Example 16.7

Calculate r_s for the teacher–judge test-score data.

Solution

The differences and squares of differences between the two rankings are as follows:

Teacher	x_i	y_i	d_i	d_i^2
1	7	1	6	36
2	4	5	-1	1
3	2	3	-1	1
4	6	4	2	4
5	1	8	-7	49
6	3	7	-4	16
7	8	2	6	36
8	5	6	-1	1
Total				144

Substituting into the formula for r_s,

$$r_s = 1 - \frac{6 \sum_{i=1}^{n} d_i^2}{n(n^2 - 1)} = 1 - \frac{6(144)}{8(64 - 1)}$$

$$= -.714.$$

The Spearman rank correlation coefficient may be employed as a test statistic to test an hypothesis of "no association" between two populations. We assume that the n pairs of observations, (x_i, y_i), have been randomly selected and therefore "no association between the populations" would imply a random assignment of the n ranks within each sample. Each random assignment (for the two samples) would represent a sample point associated with the experiment and a value of r_s could be calculated for each. Thus it is possible to calculate the probability that r_s assumes a large absolute value due solely to chance and thereby suggests an association between populations when none exists.

The rejection region for a two-tailed test is shown in Figure 16.6. If the alternative hypothesis is that the correlation between x and y is negative, you would reject H_0 for negative values of r_s that are close to -1 (in the lower tail of Figure 16.6). Similarly, if the alternative hypothesis is that the correlation between x and y is positive, you would reject H_0 for large positive values of r_s (in the upper tail of Figure 16.6).

Figure 16.6 Rejection region for a two-tailed test of the null hypothesis of "no association," using Spearman's rank correlation test

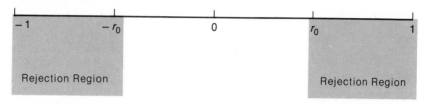

r_s: Spearman's Rank Correlation Coefficient

The critical values of r_s are given in Table 14, the Appendix. An abbreviated version of Table 14 is shown here in Table 16.3.

Table 16.3 An abbreviated version of Table 14, the Appendix, for Spearman's rank correlation test

n	$\alpha = .05$	$\alpha = .025$	$\alpha = .01$	$\alpha = .005$
5	0.900	—	—	—
6	0.829	0.886	0.943	—
7	0.714	0.786	0.893	—
8	0.643	0.738	0.833	0.881
9	0.600	0.683	0.783	0.833
10	0.564	0.648	0.745	0.794
11	0.523	0.623	0.736	0.818
12	0.497	0.591	0.703	0.780
13	0.475	0.566	0.673	0.745
14	0.457	0.545		
15	0.441	0.525		
16	0.425			
17	0.412			
18	0.399	$\vdots$	$\vdots$	$\vdots$
19	0.388			
20	0.377			
$\vdots$	$\vdots$			

Across the top of Table 16.3 (and Table 14, the Appendix) are recorded values of α that you might wish to use for a one-tailed test of the null hypothesis of "no association" between x and y. The number of rank pairs, n, appears at the left side of the table. The table entries give the critical value r_0 for a one-tailed test. Thus, $P(r_s \geq r_0) = \alpha$. For example, suppose you have $n = 8$ rank pairs and the alternative hypothesis is that the correlation between the ranks is positive. Then you would want to reject the null hypothesis of "no association" only for large positive values of r_s and would use a one-tailed test. Referring to Table 16.3

and using the row corresponding to $n = 8$ and the column for $\alpha = .05$, you read $r_0 = 0.643$. Therefore, you would reject H_0 for all values of r_s greater than or equal to 0.643.

The test is conducted in exactly the same manner if you wish to test only the alternative hypothesis that the ranks are negatively correlated. The only difference is that you would reject the null hypothesis if $r_s \leq -0.643$. That is, you just place a minus sign in front of the tabulated value of r_0 to get the lower-tail critical value.

To conduct a two-tailed test, reject the null hypothesis if $r_s \geq r_0$ or $r_s \leq -r_0$. The value of α for the test will be double the value shown at the top of the table. For example, if $n = 8$ and you choose the 0.025 column, you will reject H_0 if $r_s \geq .738$ or $r_s \leq -.738$. The α value for the test will be $2(.025) = .05$.

Spearman's Rank Correlation Test

1. *Null Hypothesis:* H_0: There is no association between the rank pairs.
2. *Alternative Hypothesis:* H_a: There is an association between the rank pairs (a two-tailed test). Or H_a: The correlation between the rank pairs is positive (or negative) (a one-tailed test).
3. *Test Statistic:* $r_s = \dfrac{SS_{xy}}{\sqrt{SS_x SS_y}}$,

   where x_i and y_i represent the ranks of the $i$the pair of observations.
4. *Rejection Region:* For a two-tailed test, reject H_0 if $r_s \geq r_0$ or $r_s \leq -r_0$, where r_0 is given in Table 14, the Appendix. Double the tabulated probability to obtain the value of α for the two-tailed test. For a one-tailed test, reject H_0 if $r_s \geq r_0$ (for an upper-tailed test) or $r_s \leq -r_0$ (for a lower-tailed test). The α value for a one-tailed test is the value shown in Table 14, the Appendix.

Example 16.8

Test an hypothesis of "no association" between the populations for Example 16.7.

Solution

The critical value of r_s for a one-tailed test with $\alpha = .05$ and $n = 8$ is .643. Let us assume that a correlation between judge's rank and the teachers' test scores could not possibly be positive. (Low rank means good teaching and should be associated with a high test score if the judge and test measure teaching ability.) The alternative hypothesis would be that the population rank correlation coefficient, ρ_s, is less than zero, and we would be concerned with a one-tailed statistical test. Thus α for the test would be the tabulated value .05, and we would reject the null hypothesis if $r_s \leq -.643$.

The calculated value of the test statistic, $r_s = -.714$, is less than the critical value for $\alpha = .05$. Hence the null hypothesis would be rejected at the $\alpha = .05$ level of significance. It appears that some agreement does exist between the judge's rankings and the test scores. However, it should be noted that this agreement could exist when *neither* provides an adequate yardstick for measuring teaching ability. For example, the association could exist if both the judge and those who constructed the teacher's examination possessed a completely erroneous, but similar, concept of the characteristics of good teaching.

Exercises

16.19.   In Exercise 11.93 we provided data on the bending stiffness and flex of tennis rackets. In addition to engineering tests, *Tennis Industry Magazine* in Florida, a leading tennis business magazine, obtained a subjective evaluation of racket vibration, stiffness, and torque (twist) by 29 male and female tennis professionals. A summary of the subjective ratings of the professionals for each racket is shown in the table.* Calculate r_s for the following pairs of variables:

Racket	Vibration Center (Sweet Spot) Hits	Vibration Off-Center Hits	Flex	Torque
Wilson T-4000	55	71	46	71
Davis Classic	59	71	43	65
Yamaha YFG-30	55	61	59	65
P.D.P. Fiberstaff	46	55	74	54
Aldila Cannon	50	61	66	53
Dunlop-Fort	56	65	60	60
Garcia "Pro" Royal	59	68	51	66
Head Comp. II	54	66	63	59
Yonex T-7500	51	53	69	58
F. Willys-Devastator	76	75	73	73
Wilson Kramer	58	66	55	65
Garcia 240	60	65	70	63

a. Flex and torque.

b. Vibration from sweet-spot hits and vibration from off-center hits.

c. Torque and vibration from off-center hits.

Test to determine whether the data provide sufficient evidence to indicate a correlation between each of the pairs of variables, (a), (b), and (c). Test using a value of α near .05.

16.20.   A school principal suspected that a teacher's attitude toward a first grader depended on his or her original judgment of the child's ability. The principal also suspected that much of that judgment was based on the first grader's IQ score—which was usually known to the teacher. After three weeks of teaching, a teacher was asked to rank the nine children in his or her class from 1 (highest) to (9 lowest) as to his or her opinion of their ability. Calculate r_s for the following teacher-IQ ranks:

Teacher	1	2	3	4	5	6	7	8	9
IQ	3	1	2	4	5	7	9	6	8

16.21.   Refer to Exercise 16.20. Do the data provide sufficient evidence to indicate a positive correlation between the teacher's ranks and the ranks of the IQs? Use $\alpha = .05$.

* Gillen, R., ed., "Equipment Preview 1976: How to Pick the Right Racquet for Your Game," *Tennis USA*, February 1976, pp. 87–91, 99.

16.22. Two art critics each ranked 12 paintings by contemporary (but anonymous) artists in accordance with their appeal to the respective critics. The ratings are shown in the table. Do the critics seem to agree on their ratings of contemporary art? That is, do the data provide sufficient evidence to indicate a positive correlation between critics A and B? Test by using a value of α near .05.

Paintings	Critic A	Critic B
1	6	5
2	4	6
3	9	10
4	1	2
5	2	3
6	7	8
7	3	1
8	8	7
9	5	4
10	10	9

16.23. Refer to Exercise 16.3, which gives the ratings of two gourmets, A and B, for 20 different meals.

a. Are the ratings of the two gourmets positively correlated? Test by using Spearman's r_s and a value of α near .05.

b. Explain why it would or would not be reasonable to test for a correlation between the two gourmets' ratings using the t test of Chapter 11.

16.24. An experiment was conducted to study the relationship between the ratings of a tobacco leaf grader and the moisture content of the corresponding tobacco leaves. Twelve leaves were rated by the grader on a scale of 1 to 10 and corresponding readings of moisture content were made. The data are as follows:

Leaf	Grader's Rating	Moisture Content
1	9	.22
2	6	.16
3	7	.17
4	7	.14
5	5	.12
6	8	.19
7	2	.10
8	6	.12
9	1	.05
10	10	.20
11	9	.16
12	3	.09

Calculate r_s. Do the data provide sufficient evidence to indicate an association between the grader's ratings and the moisture content of the leaves?

16.8
Does It Pay to Save? A Rank Correlation Analysis

Is the national savings rate y (that is, the percentage of national income saved) negatively correlated with the investment income tax rate? The data of Table 11.1 are reproduced in Table 16.4. Since we suspect that these percentages may not satisfy the assumptions required for the test of correlation given in Chapter 11, the ranks of the y values and x values are shown in parentheses to the right of their respective values.

Table 16.4 A comparison of personal savings rates and taxation rates on investment income

Country	y Personal Savings Rate (%)	Rank of y	x Investment Income Tax Liability (%)	Rank of x
Italy	23.1	(8)	6.4	(1)
Japan	21.5	(7)	14.4	(4)
France	17.2	(6)	7.3	(2)
W. Germany	14.5	(5)	11.8	(3)
U.K.	12.2	(4)	32.5	(6)
Canada	10.3	(3)	30.0	(5)
Sweden	9.1	(2)	52.7	(8)
U.S.	6.3	(1)	33.5	(7)

Source: New York Stock Exchange with assistance of Price Waterhouse

The Spearman rank correlation analysis of the savings rate y and investment tax rate x data of Table 16.4 was performed using the SAS statistical program package. The computer printout is shown in Table 16.5.

The SAS program assigns ranks to the raw scores and computes the quantities shown on the printout. The top two rows of the printout give the

Table 16.5 Computer printout for a Spearman's rank correlation analysis of the savings rate—investment tax rate data

SPEARMAN'S RANK CORRELATION COEFFICIENT
Y = SAVINGS RATE,   X = INVESTMENT INCOME

VARIABLE	N	MEAN	STD DEV	MEDIAN	MINIMUM	MAXIMUM
Y	8	14.27500000	5.96627426	13.34999847	6.30000000	23.10000000
X	8	23.57500000	16.25860212	22.19999695	6.40000000	52.70000000

SPEARMAN CORRELATION COEFFICIENTS / PROB > |R| UNDER H0:RHO=0 /   N = 8

	Y	X
Y	1.00000	−0.88095
	0.0000	0.0039
X	−0.88095	1.00000
	0.0039	0.0000

means, standard deviations, medians, minimum values and maximum values for the two sets of data. The Spearman rank correlation coefficient and the associated observed significance level are shown at the bottom of the printout.

$$r_s = -.88095$$

$$p \text{ value} = .0039$$

Since the program computes the p value for a two-tailed test, the p value for our one-tailed test (corresponding to $H_a : \rho_s < 0$) is one-half of the p value shown on the printout, or .00195.

The small observed significance level, .00195, leaves little doubt that there is a negative correlation between the national savings rate y and the investment income tax rate x of a country.

16.9
Some General Comments on Nonparametric Statistical Tests

The nonparametric statistical tests presented in the preceding pages represent only a few of the many nonparametric statistical methods of inference available. A much larger collection of nonparametric test procedures, along with worked examples, are given by Siegel (1956) and Conover (1971) (see references).

We have indicated that nonparametric statistical procedures are particularly useful when the experimental observations are susceptible to ordering but cannot be measured on a quantitative scale. Parametric statistical procedures usually cannot be applied to this type of data, hence all inferential procedures must be based on nonparametric methods.

A second application of nonparametric statistical methods is in testing hypotheses associated with populations of quantitative data when uncertainty exists concerning the satisfaction of assumptions about the form of the population distributions. Just how useful are nonparametric methods for this situation?

In this chapter, we presented a number of useful nonparametric methods along with illustrations of their applications. The Mann–Whitney U test can be used to compare two populations when the observations can be ranked according to their relative magnitudes and when the samples have been randomly and independently selected from the two populations. The simplest nonparametric test, the sign test, provides a rapid procedure for comparing two populations when the observations have been independently selected in pairs. If the differences between pairs can be ranked according to their relative magnitudes, you can use the Wilcoxon signed rank test for comparing the two populations. This latter test utilizes more sample information than the sign test and consequently is more likely to detect a difference between the two populations if a difference exists. Finally, we presented a nonparametric method, Spearman's rank correlation test, for testing the correlation between two variables when the observations associated with each variable can be ranked according to their relative magnitudes.

References   Conover, W.J., *Practical Nonparametric Statistics*. New York: John Wiley & Sons, Inc., 1971.

Kendall, M.G., and A. Stuart, *The Advanced Theory of Statistics*, Vol. 2, rev. 3rd ed. New York: Hafner Press, 1974.

Mann, H.B., and D.R. Whitney, "On a Test of Whether One of Two Random Variables is Stochastically Larger Than the Other," *Ann. Math. Stat.*, *18* (1947), pp. 50–60.

Noether, G.E., *Elements of Nonparametric Statistics*. New York: John Wiley & Sons, Inc., 1967.

Siegel, S., *Nonparametric Statistics for the Behavioral Sciences*. New York: McGraw-Hill Book Company, 1956.

Wilcoxon, F., "Individual Comparisons by Ranking Methods," *Biometrics*, *1* (1945), pp. 80–83.

Supplementary Exercises

16.25. A psychological experiment was conducted to compare the lengths of response time (in seconds) for two different stimuli. In order to remove natural person-to-person variability in the responses, both stimuli were applied to each of nine subjects, thus permitting an analysis of the difference between stimuli *within* each person.

Subject	Stimulus 1	Stimulus 2
1	9.4	10.3
2	7.8	8.9
3	5.6	4.1
4	12.1	14.7
5	6.9	8.7
6	4.2	7.1
7	8.8	11.3
8	7.7	5.2
9	6.4	7.8

    a. Use the sign test to determine whether sufficient evidence exists to indicate a difference in mean response for the two stimuli. Use a rejection region for which $\alpha \leq .05$.

    b. Test the hypothesis of no difference in mean response using Student's t test.

16.26. Refer to Exercise 16.25. Test the hypothesis that no difference exists in the distributions of responses for the two stimuli, using the Wilcoxon signed rank test. Use a rejection region for which α is as near as possible to the α achieved in Exercise 16.25(a).

16.27. To compare two junior high schools, *A* and *B*, in academic effectiveness, an experiment was designed requiring the use of ten sets of identical twins, each twin having just completed the sixth grade. In each case, the twins in the same set had obtained their schooling in the same classrooms at each grade level. One child was selected at random from each pair of twins and assigned to school *A*. The remaining children were sent to school *B*.

Near the end of the ninth grade, a certain achievement test was given to each child in the experiment. The results are shown in the following table.

School	Twin Pair									
	1	2	3	4	5	6	7	8	9	10
A	67	80	65	70	86	50	63	81	86	60
B	39	75	69	55	74	52	56	72	89	47

a. Test (using the sign test) the hypothesis that the two schools are the same in academic effectiveness, as measured by scores on the achievement test, against the alternative that the schools are not equally effective. Use a level of significance as near as possible to $\alpha = .05$.

b. Suppose it were known that junior high school A had a superior faculty and better learning facilities. Test the hypothesis of equal academic effectiveness against the alternative that school A is superior. Use a level of significance as near as possible to $\alpha = .05$.

16.28.   Refer to Exercise 16.27. What answers are obtained if Wilcoxon's signed rank test is used in analyzing the data? Compare with the answers to Exercise 16.27.

16.29.   The coded values for a measure of brightness in paper (light reflectivity), prepared by two different processes, are given below for samples of size nine drawn randomly from each of the two processes:

Process	Brightness								
A	6.1	9.2	8.7	8.9	7.6	7.1	9.5	8.3	9.0
B	9.1	8.2	8.6	6.9	7.5	7.9	8.3	7.8	8.9

Do the data present sufficient evidence ($\alpha = .10$) to indicate a difference in the populations of brightness measurements for the two processes?

a. Use the Mann–Whitney U test.

b. Use Student's t test.

16.30.   If (as in the case of measurements produced by two well-calibrated measuring instruments) the means of two populations are equal, it is possible to use the Mann–Whitney U statistic for testing hypotheses concerning the population variances as follows:

1. Rank the combined sample.

2. Number the ranked observations "from the outside in"; that is, number the smallest observation 1; the largest, 2; the next-to-smallest, 3; the next-to-largest, 4; etc. This final sequence of numbers induces an ordering on the symbols A (population A items) and B (population B items). If $\sigma_A^2 > \sigma_B^2$, one would expect to find a preponderance of A's near the first of the sequences, and thus a relatively small "sum of ranks" for the A observations.

a. Given the following measurements produced by well-calibrated precision instruments A and B, test at near the $\alpha = .05$ level to determine whether the more expensive instrument, B, is more precise than A. (Note that this would imply a one-tailed test.) Use the Mann–Whitney U test.

Instrument A	Instrument B
1,060.21	1,060.24
1,060.34	1,060.28
1,060.27	1,060.32
1,060.36	1,060.30
1,060.40	

b. Test, using the F statistic of Section 10.8.

16.31. A large corporation selects college graduates for employment, using both interviews and a psychological achievement test. Interviews conducted at the home office of the company were far more expensive than the tests that could be conducted on campus. Consequently, the personnel office was interested in determining whether the test scores were correlated with interview ratings and whether tests could be substituted for interviews. The idea was not to eliminate interviews but to reduce their number. To determine whether correlation was present, ten prospects were ranked during interviews, and tested. The paired scores are as follows:

Subject	Interview Rank	Test Score
1	8	74
2	5	81
3	10	66
4	3	83
5	6	66
6	1	94
7	4	96
8	7	70
9	9	61
10	2	86

Calculate the Spearman rank correlation coefficient, r_s. Rank 1 is assigned to the candidate judged to be the best.

16.32. Refer to Exercise 16.31. Do the data present sufficient evidence to indicate that the correlation between interview rankings and test scores is less than zero? If this evidence does exist, can we say that tests could be used to reduce the number of interviews?

16.33. A political scientist wished to examine the relationship of the voter image of a conservative political candidate and the distance between the residences of the voter and the candidate.

Each of 12 voters rated the candidate on a scale of 1 to 20. The data are as follows:

Voter	Rating	Distance
1	12	75
2	7	165
3	5	300
4	19	15
5	17	180
6	12	240
7	9	120
8	18	60
9	3	230
10	8	200
11	15	130
12	4	130

Calculate the Spearman rank correlation coefficient, r_s.

16.34.   Refer to Exercise 16.33. Do these data provide sufficient evidence to indicate a negative correlation between rating and distance?

16.35.   A comparison of reaction times for two different stimuli in a psychological word-association experiment produced the following results when applied to a random sample of 16 people:

Stimulus	Reaction Time (sec)							
1	1	3	2	1	2	1	3	2
2	4	2	3	3	1	2	3	3

Do the data present sufficient evidence to indicate a difference in mean reaction time for the two stimuli? Use the Mann–Whitney U statistic and test using $\alpha = .05$. (*Note:* This test was conducted using Student's t in the exercises for Chapter 10. Compare your results.)

16.36.   The following table gives the scores of a group of 15 students in mathematics and art.

Student	Math	Art	Student	Math	Art
1	22	53	8	60	71
2	37	68	9	62	55
3	36	42	10	65	74
4	38	49	11	66	68
5	42	51	12	56	64
6	58	65	13	66	67
7	58	51	14	67	73
			15	62	65

Use Wilcoxon's signed rank test to determine if the median scores for these students differ significantly for the two subjects.

16.37.   Refer to Exercise 16.36. Compute Spearman's rank correlation coefficient for these data and test $H_0: \rho = 0$ at the 10 percent level of significance.

16.38.   Calculate the probability that $U \leq 2$ for $n_1 = n_2 = 5$. Assume that no ties will be present.

17. A Summary and Conclusion

The preceding 16 chapters construct a picture of statistics centered about the dominant feature of the subject, statistical inference. Inference, the objective of statistics, runs as a thread through the entire book, from the phrasing of the inference through the discussion of the probabilistic mechanism and the presentation of the reasoning involved in making the inference, to the formal elementary discussion of the theory of statistical inference presented in Chapters 8 and 9. What is statistics, what is its purpose, and how does it accomplish its objective? If we have answered these questions to your satisfaction, if each chapter and section seems to fulfill a purpose and to complete a portion of the picture, we have in some measure accomplished our instructional objective.

Chapter 1 presented statistics as a scientific tool utilized in making inferences, a prediction, or a decision concerning a population of measurements based upon information contained in a sample. Thus statistics is employed in the evolutionary process known as the scientific method—which in essence is the observation of nature—in order that we may form inferences or theories concerning the structure of nature and test the theories against repeated observations. Inherent in this objective is the sampling and experimentation that purchase a quantity of information that, hopefully, will be employed to provide the best inferences concerning the population from which the sample was drawn.

The method of phrasing the inference (that is, describing a set of measurements) was presented in Chapter 3 in terms of a frequency distribution and associated numerical descriptive measures. In particular we noted that the frequency distribution is subject to a probabilistic interpretation and that the numerical descriptive measures are more suitable for inferential purposes because we can more easily associate with them a measure of their goodness. Finally, a secondary but extremely important result of our study of numerical descriptive measures involved the notion of variation, its measurement in terms of a standard deviation, and its interpretation by using Tchebysheff's Theorem and the Empirical Rule. Thus, while concerned with describing a set of measurements—the population—we provided the basis for a description of the sampling distributions of estimators and the z test statistic to be considered as we progressed in our study.

The mechanism involved in making an inference—a decision—concerning the parameters of a population of die throws was introduced in Chapter 4. We hypothesized the population of die throws to be known (that is, that the die was perfectly balanced) and then drew a sample of $n = 10$ tosses from the population. Observing 10 ones, we concluded that the observed sample was highly improbable, assuming our hypothesis to be true, and we therefore rejected the hypothesis and concluded that the die was unbalanced. Thus, we noted that the theory of probability assumes the population known, and reasons from the population to the sample. Statistical inference, using probability, observes the sample and attempts to make inferences concerning the population. Fundamental to this procedure is a probabilistic model for the frequency distribution of the population, the acquisition of which was considered in Chapters 4 and 5.

The methodology of Chapters 4 and 5 was employed in Chapter 6 in the construction of a probability distribution (that is, a model for the frequency distribution) of a discrete random variable generated by the binomial experiment. The binomial experiment was chosen as an example because the acquisition of the probabilities, $p(y)$, is a task easily handled by the beginner and thus gives us an opportunity to utilize our probabilistic tools. In addition, it was chosen because of its utility, which was exemplified by the cold vaccine and lot acceptance sampling problems. Particularly, the inferential aspects of these problems were noted, with emphasis upon the reasoning involved.

Study of a useful continuous random variable in Chapter 7 centered about the Central Limit Theorem, its suggested support of the Empirical Rule, and its use in describing the sampling distributions of sample means and sums. Thus, we used the Central Limit Theorem to justify the use of the Empirical Rule and the normal probability distribution as an approximation to the binomial probability distribution when the number of trials, n, is large. Through examples we attempted to reinforce the probabilistic concept of statistical inference introduced in preceding chapters and induce you, as a matter of intuition, to employ statistical reasoning in making inferences.

Chapters 8 and 9 formally discussed statistical inference, estimation and tests of hypotheses, and the methods of measuring the goodness of the inference. These chapters also presented a number of estimators and test statistics that, because of the Central Limit Theorem, possess sampling distributions that are approximately normal. These notions were carried over to the discussion of the small-sample tests and estimators in Chapter 10.

Chapters 11, 12, and 13 attempted, primarily, to broaden the view of the beginner, presenting two rather interesting and unique inferential problems. We studied the relation between two variables, y and x, in Chapter 11, and expanded these ideas in Chapter 12 to develop an equation to predict y based on information contained in a set of predictor variables, $x_1, x_2, \ldots, x_k$. The analysis of enumerative data, Chapter 13, presents a methodology that is interesting and extremely useful in the social sciences as well as in many other areas.

Chapters 1 through 13 dealt with the philosophy and the methodology for making inferences and for measuring the quantity of information in a sample. Thus they form a necessary foundation for understanding how and why noise and volume affect the quantity of information in an experiment and set the stage for a discussion of experimental design in Chapter 14.

The analysis of variance, Chapter 15, extended the comparison of treatment means in the completely random and blocked designs (Chapter 10) to more than two treatments. Simultaneously, it introduced a completely new way to view the analysis of data through the analysis (or partitioning) of variance. Nonparametric statistical methods, Chapter 16, provided a discussion of some very useful statistical test procedures that are particularly appropriate for ordinal data.

Finally, we note that the methodology presented in this introduction to statistics represents a very small sample of the population of statistical methodology that is available to the researcher. It is a bare introduction and nothing more. The design of experiments—an extremely useful topic—was barely touched. Indeed, the methodology presented, although very useful, is intended primarily to serve as a vehicle for conveying to you the philosophy involved.

Appendix

Tables

TABLE 1   Binomial probability

Tabulated values are $P(y \leq a) = \sum_{y=0}^{a} p(y)$. (Computations are rounded at the third decimal place.)

(a) $n = 5$

						P								
a	0.01	0.05	0.10	0.20	0.30	0.40	0.50	0.60	0.70	0.80	0.90	0.95	0.99	a
0	.951	.774	.590	.328	.168	.078	.031	.010	.002	.000	.000	.000	.000	0
1	.999	.977	.919	.737	.528	.337	.188	.087	.031	.007	.000	.000	.000	1
2	1.000	.999	.991	.942	.837	.683	.500	.317	.163	.058	.009	.001	.000	2
3	1.000	1.000	1.000	.993	.969	.913	.812	.663	.472	.263	.081	.023	.001	3
4	1.000	1.000	1.000	1.000	.998	.990	.969	.922	.832	.672	.410	.226	.049	4

TABLE 1 (*Continued*)

(b) *n* = 10

							P							
a	0.01	0.05	0.10	0.20	0.30	0.40	0.50	0.60	0.70	0.80	0.90	0.95	0.99	*a*
0	.904	.599	.349	.107	.028	.006	.001	.000	.000	.000	.000	.000	.000	0
1	.996	.914	.736	.376	.149	.046	.011	.002	.000	.000	.000	.000	.000	1
2	1.000	.988	.930	.678	.383	.167	.055	.012	.002	.000	.000	.000	.000	2
3	1.000	.999	.987	.879	.650	.382	.172	.055	.011	.001	.000	.000	.000	3
4	1.000	1.000	.998	.967	.850	.633	.377	.166	.047	.006	.000	.000	.000	4
5	1.000	1.000	1.000	.994	.953	.834	.623	.367	.150	.033	.002	.000	.000	5
6	1.000	1.000	1.000	.999	.989	.945	.828	.618	.350	.121	.013	.001	.000	6
7	1.000	1.000	1.000	1.000	.998	.988	.945	.833	.617	.322	.070	.012	.000	7
8	1.000	1.000	1.000	1.000	1.000	.998	.989	.954	.851	.624	.264	.086	.004	8
9	1.000	1.000	1.000	1.000	1.000	1.000	.999	.994	.972	.893	.651	.401	.096	9

(c) *n* = 15

							P							
a	0.01	0.05	0.10	0.20	0.30	0.40	0.50	0.60	0.70	0.80	0.90	0.95	0.99	*a*
0	.860	.463	.206	.035	.005	.000	.000	.000	.000	.000	.000	.000	.000	0
1	.990	.829	.549	.167	.035	.005	.000	.000	.000	.000	.000	.000	.000	1
2	1.000	.964	.816	.398	.127	.027	.004	.000	.000	.000	.000	.000	.000	2
3	1.000	.995	.944	.648	.297	.091	.018	.002	.000	.000	.000	.000	.000	3
4	1.000	.999	.987	.836	.515	.217	.059	.009	.001	.000	.000	.000	.000	4
5	1.000	1.000	.998	.939	.722	.403	.151	.034	.004	.000	.000	.000	.000	5
6	1.000	1.000	1.000	.982	.869	.610	.304	.095	.015	.001	.000	.000	.000	6
7	1.000	1.000	1.000	.996	.950	.787	.500	.213	.050	.004	.000	.000	.000	7
8	1.000	1.000	1.000	.999	.985	.905	.696	.390	.131	.018	.000	.000	.000	8
9	1.000	1.000	1.000	1.000	.996	.966	.849	.597	.278	.061	.002	.000	.000	9
10	1.000	1.000	1.000	1.000	.999	.991	.941	.783	.485	.164	.013	.001	.000	10
11	1.000	1.000	1.000	1.000	1.000	.998	.982	.909	.703	.352	.056	.005	.000	11
12	1.000	1.000	1.000	1.000	1.000	1.000	.996	.973	.873	.602	.184	.036	.000	12
13	1.000	1.000	1.000	1.000	1.000	1.000	1.000	.995	.965	.833	.451	.171	.010	13
14	1.000	1.000	1.000	1.000	1.000	1.000	1.000	1.000	.995	.965	.794	.537	.140	14

TABLE 1   (*Continued*)

(d) $n = 20$

a	0.01	0.05	0.10	0.20	0.30	0.40	0.50	0.60	0.70	0.80	0.90	0.95	0.99	a
0	.818	.358	.122	.012	.001	.000	.000	.000	.000	.000	.000	.000	.000	0
1	.983	.736	.392	.069	.008	.001	.000	.000	.000	.000	.000	.000	.000	1
2	.999	.925	.677	.206	.035	.004	.000	.000	.000	.000	.000	.000	.000	2
3	1.000	.984	.867	.411	.107	.016	.001	.000	.000	.000	.000	.000	.000	3
4	1.000	.997	.957	.630	.238	.051	.006	.000	.000	.000	.000	.000	.000	4
5	1.000	1.000	.989	.804	.416	.126	.021	.002	.000	.000	.000	.000	.000	5
6	1.000	1.000	.998	.913	.608	.250	.058	.006	.000	.000	.000	.000	.000	6
7	1.000	1.000	1.000	.968	.772	.416	.132	.021	.001	.000	.000	.000	.000	7
8	1.000	1.000	1.000	.990	.887	.596	.252	.057	.005	.000	.000	.000	.000	8
9	1.000	1.000	1.000	.997	.952	.755	.412	.128	.017	.001	.000	.000	.000	9
10	1.000	1.000	1.000	.999	.983	.872	.588	.245	.048	.003	.000	.000	.000	10
11	1.000	1.000	1.000	1.000	.995	.943	.748	.404	.113	.010	.000	.000	.000	11
12	1.000	1.000	1.000	1.000	.999	.979	.868	.584	.228	.032	.000	.000	.000	12
13	1.000	1.000	1.000	1.000	1.000	.994	.942	.750	.392	.087	.002	.000	.000	13
14	1.000	1.000	1.000	1.000	1.000	.998	.979	.874	.584	.196	.011	.000	.000	14
15	1.000	1.000	1.000	1.000	1.000	1.000	.994	.949	.762	.370	.043	.003	.000	15
16	1.000	1.000	1.000	1.000	1.000	1.000	.999	.984	.893	.589	.133	.016	.000	16
17	1.000	1.000	1.000	1.000	1.000	1.000	1.000	.996	.965	.794	.323	.075	.001	17
18	1.000	1.000	1.000	1.000	1.000	1.000	1.000	.999	.992	.931	.608	.264	.017	18
19	1.000	1.000	1.000	1.000	1.000	1.000	1.000	1.000	.999	.988	.878	.642	.182	19

TABLE 1   (*Concluded*)

(e) *n* = 25

							P							
a	*0.01*	*0.05*	*0.10*	*0.20*	*0.30*	*0.40*	*0.50*	*0.60*	*0.70*	*0.80*	*0.90*	*0.95*	*0.99*	*a*
0	.778	.277	.072	.004	.000	.000	.000	.000	.000	.000	.000	.000	.000	0
1	.974	.642	.271	.027	.002	.000	.000	.000	.000	.000	.000	.000	.000	1
2	.998	.873	.537	.098	.009	.000	.000	.000	.000	.000	.000	.000	.000	2
3	1.000	.966	.764	.234	.033	.002	.000	.000	.000	.000	.000	.000	.000	3
4	1.000	.993	.902	.421	.090	.009	.000	.000	.000	.000	.000	.000	.000	4
5	1.000	.999	.967	.617	.193	.029	.002	.000	.000	.000	.000	.000	.000	5
6	1.000	1.000	.991	.780	.341	.074	.007	.000	.000	.000	.000	.000	.000	6
7	1.000	1.000	.998	.891	.512	.154	.022	.001	.000	.000	.000	.000	.000	7
8	1.000	1.000	1.000	.953	.677	.274	.054	.004	.000	.000	.000	.000	.000	8
9	1.000	1.000	1.000	.983	.811	.425	.115	.013	.000	.000	.000	.000	.000	9
10	1.000	1.000	1.000	.994	.902	.586	.212	.034	.002	.000	.000	.000	.000	10
11	1.000	1.000	1.000	.998	.956	.732	.345	.078	.006	.000	.000	.000	.000	11
12	1.000	1.000	1.000	1.000	.983	.846	.500	.154	.017	.000	.000	.000	.000	12
13	1.000	1.000	1.000	1.000	.994	.922	.655	.268	.044	.002	.000	.000	.000	13
14	1.000	1.000	1.000	1.000	.998	.966	.788	.414	.098	.006	.000	.000	.000	14
15	1.000	1.000	1.000	1.000	1.000	.987	.885	.575	.189	.017	.000	.000	.000	15
16	1.000	1.000	1.000	1.000	1.000	.996	.946	.726	.323	.047	.000	.000	.000	16
17	1.000	1.000	1.000	1.000	1.000	.999	.978	.846	.488	.109	.002	.000	.000	17
18	1.000	1.000	1.000	1.000	1.000	1.000	.993	.926	.659	.220	.009	.000	.000	18
19	1.000	1.000	1.000	1.000	1.000	1.000	.998	.971	.807	.383	.033	.001	.000	19
20	1.000	1.000	1.000	1.000	1.000	1.000	1.000	.991	.910	.579	.098	.007	.000	20
21	1.000	1.000	1.000	1.000	1.000	1.000	1.000	.998	.967	.766	.236	.034	.000	21
22	1.000	1.000	1.000	1.000	1.000	1.000	1.000	1.000	.991	.902	.463	.127	.002	22
23	1.000	1.000	1.000	1.000	1.000	1.000	1.000	1.000	.998	.973	.729	.358	.026	23
24	1.000	1.000	1.000	1.000	1.000	1.000	1.000	1.000	1.000	.996	.928	.723	.222	24

TABLE 2   Values of $e^{-\mu}$

μ	$e^{-\mu}$	μ	$e^{-\mu}$	μ	$e^{-\mu}$	μ	$e^{-\mu}$
0.00	1.000000	2.60	.074274	5.10	.006097	7.60	.000501
0.10	.904837	2.70	.067206	5.20	.005517	7.70	.000453
0.20	.818731	2.80	.060810	5.30	.004992	7.80	.000410
0.30	.740818	2.90	.055023	5.40	.004517	7.90	.000371
0.40	.670320	3.00	.049787	5.50	.004087	8.00	.000336
0.50	.606531	3.10	.045049	5.60	.003698	8.10	.000304
0.60	.548812	3.20	.040762	5.70	.003346	8.20	.000275
0.70	.496585	3.30	.036883	5.80	.003028	8.30	.000249
0.80	.449329	3.40	.033373	5.90	.002739	8.40	.000225
0.90	.406570	3.50	.030197	6.00	.002479	8.50	.000204
1.00	.367879	3.60	.027324	6.10	.002243	8.60	.000184
1.10	.332871	3.70	.024724	6.20	.002029	8.70	.000167
1.20	.301194	3.80	.022371	6.30	.001836	8.80	.000151
1.30	.272532	3.90	.020242	6.40	.001661	8.90	.000136
1.40	.246597	4.00	.018316	6.50	.001503	9.00	.000123
1.50	.223130	4.10	.016573	6.60	.001360	9.10	.000112
1.60	.201897	4.20	.014996	6.70	.001231	9.20	.000101
1.70	.182684	4.30	.013569	6.80	.001114	9.30	.000091
1.80	.165299	4.40	.012277	6.90	.001008	9.40	.000083
1.90	.149569	4.50	.011109	7.00	.000912	9.50	.000075
2.00	.135335	4.60	.010052	7.10	.000825	9.60	.000068
2.10	.122456	4.70	.009095	7.20	.000747	9.70	.000061
2.20	.110803	4.80	.008230	7.30	.000676	9.80	.000056
2.30	.100259	4.90	.007447	7.40	.000611	9.90	.000050
2.40	.090718	5.00	.006738	7.50	.000553	10.00	.000045
2.50	.082085						

TABLE 3   Normal curve areas

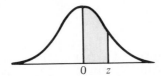

z	.00	.01	.02	.03	.04	.05	.06	.07	.08	.09
0.0	.0000	.0040	.0080	.0120	.0160	.0199	.0239	.0279	.0319	.0359
0.1	.0398	.0438	.0478	.0517	.0557	.0596	.0636	.0675	.0714	.0753
0.2	.0793	.0832	.0871	.0910	.0948	.0987	.1026	.1064	.1103	.1141
0.3	.1179	.1217	.1255	.1293	.1331	.1368	.1406	.1443	.1480	.1517
0.4	.1554	.1591	.1628	.1664	.1700	.1736	.1772	.1808	.1844	.1879
0.5	.1915	.1950	.1985	.2019	.2054	.2088	.2123	.2157	.2190	.2224
0.6	.2257	.2291	.2324	.2357	.2389	.2422	.2454	.2486	.2517	.2549
0.7	.2580	.2611	.2642	.2673	.2704	.2734	.2764	.2794	.2823	.2852
0.8	.2881	.2910	.2939	.2967	.2995	.3023	.3051	.3078	.3106	.3133
0.9	.3159	.3186	.3212	.3238	.3264	.3289	.3315	.3340	.3365	.3389
1.0	.3413	.3438	.3461	.3485	.3508	.3531	.3554	.3577	.3599	.3621
1.1	.3643	.3665	.3686	.3708	.3729	.3749	.3770	.3790	.3810	.3830
1.2	.3849	.3869	.3888	.3907	.3925	.3944	.3962	.3980	.3997	.4015
1.3	.4032	.4049	.4066	.4082	.4099	.4115	.4131	.4147	.4162	.4177
1.4	.4192	.4207	.4222	.4236	.4251	.4265	.4279	.4292	.4306	.4319
1.5	.4332	.4345	.4357	.4370	.4382	.4394	.4406	.4418	.4429	.4441
1.6	.4452	.4463	.4474	.4484	.4495	.4505	.4515	.4525	.4535	.4545
1.7	.4554	.4564	.4573	.4582	.4591	.4599	.4608	.4616	.4625	.4633
1.8	.4641	.4649	.4656	.4664	.4671	.4678	.4686	.4693	.4699	.4706
1.9	.4713	.4719	.4726	.4732	.4738	.4744	.4750	.4756	.4761	.4767
2.0	.4772	.4778	.4783	.4788	.4793	.4798	.4803	.4808	.4812	.4817
2.1	.4821	.4826	.4830	.4834	.4838	.4842	.4846	.4850	.4854	.4857
2.2	.4861	.4864	.4868	.4871	.4875	.4878	.4881	.4884	.4887	.4890
2.3	.4893	.4896	.4898	.4901	.4904	.4906	.4909	.4911	.4913	.4916
2.4	.4918	.4920	.4922	.4925	.4927	.4929	.4931	.4932	.4934	.4936
2.5	.4938	.4940	.4941	.4943	.4945	.4946	.4948	.4949	.4951	.4952
2.6	.4953	.4955	.4956	.4957	.4959	.4960	.4961	.4962	.4963	.4964
2.7	.4965	.4966	.4967	.4968	.4969	.4970	.4971	.4972	.4973	.4974
2.8	.4974	.4975	.4976	.4977	.4977	.4978	.4979	.4979	.4980	.4981
2.9	.4981	.4982	.4982	.4983	.4984	.4984	.4985	.4985	.4986	.4986
3.0	.4987	.4987	.4987	.4988	.4988	.4989	.4989	.4989	.4990	.4990

This table is abridged from Table 1 of *Statistical Tables and Formulas*, by A. Hald (New York: John Wiley & Sons, Inc., 1952). Reproduced by permission of A. Hald and the publishers, John Wiley & Sons, Inc.

TABLE 4   Critical values of t

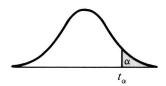

t_α

d.f.	$t_{.100}$	$t_{.050}$	$t_{.025}$	$t_{.010}$	$t_{.005}$	d.f.
1	3.078	6.314	12.706	31.821	63.657	1
2	1.886	2.920	4.303	6.965	9.925	2
3	1.638	2.353	3.182	4.541	5.841	3
4	1.533	2.132	2.776	3.747	4.604	4
5	1.476	2.015	2.571	3.365	4.032	5
6	1.440	1.943	2.447	3.143	3.707	6
7	1.415	1.895	2.365	2.998	3.499	7
8	1.397	1.860	2.306	2.896	3.355	8
9	1.383	1.833	2.262	2.821	3.250	9
10	1.372	1.812	2.228	2.764	3.169	10
11	1.363	1.796	2.201	2.718	3.106	11
12	1.356	1.782	2.179	2.681	3.055	12
13	1.350	1.771	2.160	2.650	3.012	13
14	1.345	1.761	2.145	2.624	2.977	14
15	1.341	1.753	2.131	2.602	2.947	15
16	1.337	1.746	2.120	2.583	2.921	16
17	1.333	1.740	2.110	2.567	2.898	17
18	1.330	1.734	2.101	2.552	2.878	18
19	1.328	1.729	2.093	2.539	2.861	19
20	1.325	1.725	2.086	2.528	2.845	20
21	1.323	1.721	2.080	2.518	2.831	21
22	1.321	1.717	2.074	2.508	2.819	22
23	1.319	1.714	2.069	2.500	2.807	23
24	1.318	1.711	2.064	2.492	2.797	24
25	1.316	1.708	2.060	2.485	2.787	25
26	1.315	1.706	2.056	2.479	2.779	26
27	1.314	1.703	2.052	2.473	2.771	27
28	1.313	1.701	2.048	2.467	2.763	28
29	1.311	1.699	2.045	2.462	2.756	29
inf.	1.282	1.645	1.960	2.326	2.576	inf.

TABLE 5 Critical values of chi-square

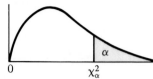

d.f.	$\chi^2_{0.995}$	$\chi^2_{0.990}$	$\chi^2_{0.975}$	$\chi^2_{0.950}$	$\chi^2_{0.900}$
1	0.0000393	0.0001571	0.0009821	0.0039321	0.0157908
2	0.0100251	0.0201007	0.0506356	0.102587	0.210720
3	0.0717212	0.114832	0.215795	0.351846	0.584375
4	0.206990	0.297110	0.484419	0.710721	1.063623
5	0.411740	0.554300	0.831211	1.145476	1.61031
6	0.675727	0.872085	1.237347	1.63539	2.20413
7	0.989265	1.239043	1.68987	2.16735	2.83311
8	1.344419	1.646482	2.17973	2.73264	3.48954
9	1.734926	2.087912	2.70039	3.32511	4.16816
10	2.15585	2.55821	3.24697	3.94030	4.86518
11	2.60321	3.05347	3.81575	4.57481	5.57779
12	3.07382	3.57056	4.40379	5.22603	6.30380
13	3.56503	4.10691	5.00874	5.89186	7.04150
14	4.07468	4.66043	5.62872	6.57063	7.78953
15	4.60094	5.22935	6.26214	7.26094	8.54675
16	5.14224	5.81221	6.90766	7.96164	9.31223
17	5.69724	6.40776	7.56418	8.67176	10.0852
18	6.26481	7.01491	8.23075	9.39046	10.8649
19	6.84398	7.63273	8.90655	10.1170	11.6509
20	7.43386	8.26040	9.59083	10.8508	12.4426
21	8.03366	8.89720	10.28293	11.5913	13.2396
22	8.64272	9.54249	10.9823	12.3380	14.0415
23	9.26042	10.19567	11.6885	13.0905	14.8479
24	9.88623	10.8564	12.4011	13.8484	15.6587
25	10.5197	11.5240	13.1197	14.6114	16.4734
26	11.1603	12.1981	13.8439	15.3791	17.2919
27	11.8076	12.8786	14.5733	16.1513	18.1138
28	12.4613	13.5648	15.3079	16.9279	18.9392
29	13.1211	14.2565	16.0471	17.7083	19.7677
30	13.7867	14.9535	16.7908	18.4926	20.5992
40	20.7065	22.1643	24.4331	26.5093	29.0505
50	27.9907	29.7067	32.3574	34.7642	37.6886
60	35.5346	37.4848	40.4817	43.1879	46.4589
70	43.2752	45.4418	48.7576	51.7393	55.3290
80	51.1720	53.5400	57.1532	60.3915	64.2778
90	59.1963	61.7541	65.6466	69.1260	73.2912
100	67.3276	70.0648	74.2219	77.9295	82.3581

TABLE 5   (*Concluded*)

$\chi^2_{0.100}$	$\chi^2_{0.050}$	$\chi^2_{0.025}$	$\chi^2_{0.010}$	$\chi^2_{0.005}$	d.f.
2.70554	3.84146	5.02389	6.63490	7.87944	1
4.60517	5.99147	7.37776	9.21034	10.5966	2
6.25139	7.81473	9.34840	11.3449	12.8381	3
7.77944	9.48773	11.1433	13.2767	14.8602	4
9.23635	11.0705	12.8325	15.0863	16.7496	5
10.6446	12.5916	14.4494	16.8119	18.5476	6
12.0170	14.0671	16.0128	18.4753	20.2777	7
13.3616	15.5073	17.5346	20.0902	21.9550	8
14.6837	16.9190	19.0228	21.6660	23.5893	9
15.9871	18.3070	20.4831	23.2093	25.1882	10
17.2750	19.6751	21.9200	24.7250	26.7569	11
18.5494	21.0261	23.3367	26.2170	28.2995	12
19.8119	22.3621	24.7356	27.6883	29.8194	13
21.0642	23.6848	26.1190	29.1413	31.3193	14
22.3072	24.9958	27.4884	30.5779	32.8013	15
23.5418	26.2962	28.8454	31.9999	34.2672	16
24.7690	27.5871	30.1910	33.4087	35.7185	17
25.9894	28.8693	31.5264	34.8053	37.1564	18
27.2036	30.1435	32.8523	36.1908	38.5822	19
28.4120	31.4104	34.1696	37.5662	39.9968	20
29.6151	32.6705	35.4789	38.9321	41.4010	21
30.8133	33.9244	36.7807	40.2894	42.7956	22
32.0069	35.1725	38.0757	41.6384	44.1813	23
33.1963	36.4151	39.3641	42.9798	45.5585	24
34.3816	37.6525	40.6465	44.3141	46.9278	25
35.5631	38.8852	41.9232	45.6417	48.2899	26
36.7412	40.1133	43.1944	46.9630	49.6449	27
37.9159	41.3372	44.4607	48.2782	50.9933	28
39.0875	42.5569	45.7222	49.5879	52.3356	29
40.2560	43.7729	46.9792	50.8922	53.6720	30
51.8050	55.7585	59.3417	63.6907	66.7659	40
63.1671	67.5048	71.4202	76.1539	79.4900	50
74.3970	79.0819	83.2976	88.3794	91.9517	60
85.5271	90.5312	95.0231	100.425	104.215	70
96.5782	101.879	106.629	112.329	116.321	80
107.565	113.145	118.136	124.116	128.299	90
118.498	124.342	129.561	135.807	140.169	100

TABLE 6   Percentage points of the F distribution: $\alpha = .10$

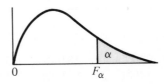

					v_1 (d.f.)				
v_2 (d.f.)	*1*	*2*	*3*	*4*	*5*	*6*	*7*	*8*	*9*
1	39.86	49.50	53.59	55.83	57.24	58.20	58.91	59.44	59.86
2	8.53	9.00	9.16	9.24	9.29	9.33	9.35	9.37	9.38
3	5.54	5.46	5.39	5.34	5.31	5.28	5.27	5.25	5.24
4	4.54	4.32	4.19	4.11	4.05	4.01	3.98	3.95	3.94
5	4.06	3.78	3.62	3.52	3.45	3.40	3.37	3.34	3.32
6	3.78	3.46	3.29	3.18	3.11	3.05	3.01	2.98	2.96
7	3.59	3.26	3.07	2.96	2.88	2.83	2.78	2.75	2.72
8	3.46	3.11	2.92	2.81	2.73	2.67	2.62	2.59	2.56
9	3.36	3.01	2.81	2.69	2.61	2.55	2.51	2.47	2.44
10	3.29	2.92	2.73	2.61	2.52	2.46	2.41	2.38	2.35
11	3.23	2.86	2.66	2.54	2.45	2.39	2.34	2.30	2.27
12	3.18	2.81	2.61	2.48	2.39	2.33	2.28	2.24	2.21
13	3.14	2.76	2.56	2.43	2.35	2.28	2.23	2.20	2.16
14	3.10	2.73	2.52	2.39	2.31	2.24	2.19	2.15	2.12
15	3.07	2.70	2.49	2.36	2.27	2.21	2.16	2.12	2.09
16	3.05	2.67	2.46	2.33	2.24	2.18	2.13	2.09	2.06
17	3.03	2.64	2.44	2.31	2.22	2.15	2.10	2.06	2.03
18	3.01	2.62	2.42	2.29	2.20	2.13	2.08	2.04	2.00
19	2.99	2.61	2.40	2.27	2.18	2.11	2.06	2.02	1.98
20	2.97	2.59	2.38	2.25	2.16	2.09	2.04	2.00	1.96
21	2.96	2.57	2.36	2.23	2.14	2.08	2.02	1.98	1.95
22	2.95	2.56	2.35	2.22	2.13	2.06	2.01	1.97	1.93
23	2.94	2.55	2.34	2.21	2.11	2.05	1.99	1.95	1.92
24	2.93	2.54	2.33	2.19	2.10	2.04	1.98	1.94	1.91
25	2.92	2.53	2.32	2.18	2.09	2.02	1.97	1.93	1.89
26	2.91	2.52	2.31	2.17	2.08	2.01	1.96	1.92	1.88
27	2.90	2.51	2.30	2.17	2.07	2.00	1.95	1.91	1.87
28	2.89	2.50	2.29	2.16	2.06	2.00	1.94	1.90	1.87
29	2.89	2.50	2.28	2.15	2.06	1.99	1.93	1.89	1.86
30	2.88	2.49	2.28	2.14	2.05	1.98	1.93	1.88	1.85
40	2.84	2.44	2.23	2.09	2.00	1.93	1.87	1.83	1.79
60	2.79	2.39	2.18	2.04	1.95	1.87	1.82	1.77	1.74
120	2.75	2.35	2.13	1.99	1.90	1.82	1.77	1.72	1.68
∞	2.71	2.30	2.08	1.94	1.85	1.77	1.72	1.67	1.63

TABLE 6   (*Concluded*)

10	12	15	20	24	30	40	60	120	∞	v_2 (d.f.)
60.19	60.71	61.22	61.74	62.00	62.26	62.53	62.79	63.06	63.33	1
9.39	9.41	9.42	9.44	9.45	9.46	9.47	9.47	9.48	9.49	2
5.23	5.22	5.20	5.18	5.18	5.17	5.16	5.15	5.14	5.13	3
3.92	3.90	3.87	3.84	3.83	3.82	3.80	3.79	3.78	3.76	4
3.30	3.27	3.24	3.21	3.19	3.17	3.16	3.14	3.12	3.10	5
2.94	2.90	2.87	2.84	2.82	2.80	2.78	2.76	2.74	2.72	6
2.70	2.67	2.63	2.59	2.58	2.56	2.54	2.51	2.49	2.47	7
2.54	2.50	2.46	2.42	2.40	2.38	2.36	2.34	2.32	2.29	8
2.42	2.38	2.34	2.30	2.28	2.25	2.23	2.21	2.18	2.16	9
2.32	2.28	2.24	2.20	2.18	2.16	2.13	2.11	2.08	2.06	10
2.25	2.21	2.17	2.12	2.10	2.08	2.05	2.03	2.00	1.97	11
2.19	2.15	2.10	2.06	2.04	2.01	1.99	1.96	1.93	1.90	12
2.14	2.10	2.05	2.01	1.98	1.96	1.93	1.90	1.88	1.85	13
2.10	2.05	2.01	1.96	1.94	1.91	1.89	1.86	1.83	1.80	14
2.06	2.02	1.97	1.92	1.90	1.87	1.85	1.82	1.79	1.76	15
2.03	1.99	1.94	1.89	1.87	1.84	1.81	1.78	1.75	1.72	16
2.00	1.96	1.91	1.86	1.84	1.81	1.78	1.75	1.72	1.69	17
1.98	1.93	1.89	1.84	1.81	1.78	1.75	1.72	1.69	1.66	18
1.96	1.91	1.86	1.81	1.79	1.76	1.73	1.70	1.67	1.63	19
1.94	1.89	1.84	1.79	1.77	1.74	1.71	1.68	1.64	1.61	20
1.92	1.87	1.83	1.78	1.75	1.72	1.69	1.66	1.62	1.59	21
1.90	1.86	1.81	1.76	1.73	1.70	1.67	1.64	1.60	1.57	22
1.89	1.84	1.80	1.74	1.72	1.69	1.66	1.62	1.59	1.55	23
1.88	1.83	1.78	1.73	1.70	1.67	1.64	1.61	1.57	1.53	24
1.87	1.82	1.77	1.72	1.69	1.66	1.63	1.59	1.56	1.52	25
1.86	1.81	1.76	1.71	1.68	1.65	1.61	1.58	1.54	1.50	26
1.85	1.80	1.75	1.70	1.67	1.64	1.60	1.57	1.53	1.49	27
1.84	1.79	1.74	1.69	1.66	1.63	1.59	1.56	1.52	1.48	28
1.83	1.78	1.73	1.68	1.65	1.62	1.58	1.55	1.51	1.47	29
1.82	1.77	1.72	1.67	1.64	1.61	1.57	1.54	1.50	1.46	30
1.76	1.71	1.66	1.61	1.57	1.54	1.51	1.47	1.42	1.38	40
1.71	1.66	1.60	1.54	1.51	1.48	1.44	1.40	1.35	1.29	60
1.65	1.60	1.55	1.48	1.45	1.41	1.37	1.32	1.26	1.19	120
1.60	1.55	1.49	1.42	1.38	1.34	1.30	1.24	1.17	1.00	∞

Header: v_1 (d.f.)

TABLE 7   Percentage points of the F distribution; $\alpha = .05$

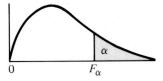

v_2 (d.f.)	v_1 (d.f.)								
	1	2	3	4	5	6	7	8	9
1	161.4	199.5	215.7	224.6	230.2	234.0	236.8	238.9	240.5
2	18.51	19.00	19.16	19.25	19.30	19.33	19.35	19.37	19.38
3	10.13	9.55	9.28	9.12	9.01	8.94	8.89	8.85	8.81
4	7.71	6.94	6.59	6.39	6.26	6.16	6.09	6.04	6.00
5	6.61	5.79	5.41	5.19	5.05	4.95	4.88	4.82	4.77
6	5.99	5.14	4.76	4.53	4.39	4.28	4.21	4.15	4.10
7	5.59	4.74	4.35	4.12	3.97	3.87	3.79	3.73	3.68
8	5.32	4.46	4.07	3.84	3.69	3.58	3.50	3.44	3.39
9	5.12	4.26	3.86	3.63	3.48	3.37	3.29	3.23	3.18
10	4.96	4.10	3.71	3.48	3.33	3.22	3.14	3.07	3.02
11	4.84	3.98	3.59	3.36	3.20	3.09	3.01	2.95	2.90
12	4.75	3.89	3.49	3.26	3.11	3.00	2.91	2.85	2.80
13	4.67	3.81	3.41	3.18	3.03	2.92	2.83	2.77	2.71
14	4.60	3.74	3.34	3.11	2.96	2.85	2.76	2.70	2.65
15	4.54	3.68	3.29	3.06	2.90	2.79	2.71	2.64	2.59
16	4.49	3.63	3.24	3.01	2.85	2.74	2.66	2.59	2.54
17	4.45	3.59	3.20	2.96	2.81	2.70	2.61	2.55	2.49
18	4.41	3.55	3.16	2.93	2.77	2.66	2.58	2.51	2.46
19	4.38	3.52	3.13	2.90	2.74	2.63	2.54	2.48	2.42
20	4.35	3.49	3.10	2.87	2.71	2.60	2.51	2.45	2.39
21	4.32	3.47	3.07	2.84	2.68	2.57	2.49	2.42	2.37
22	4.30	3.44	3.05	2.82	2.66	2.55	2.46	2.40	2.34
23	4.28	3.42	3.03	2.80	2.64	2.53	2.44	2.37	2.32
24	4.26	3.40	3.01	2.78	2.62	2.51	2.42	2.36	2.30
25	4.24	3.39	2.99	2.76	2.60	2.49	2.40	2.34	2.28
26	4.23	3.37	2.98	2.74	2.59	2.47	2.39	2.32	2.27
27	4.21	3.35	2.96	2.73	2.57	2.46	2.37	2.31	2.25
28	4.20	3.34	2.95	2.71	2.56	2.45	2.36	2.29	2.24
29	4.18	3.33	2.93	2.70	2.55	2.43	2.35	2.28	2.22
30	4.17	3.32	2.92	2.69	2.53	2.42	2.33	2.27	2.21
40	4.08	3.23	2.84	2.61	2.45	2.34	2.25	2.18	2.12
60	4.00	3.15	2.76	2.53	2.37	2.25	2.17	2.10	2.04
120	3.92	3.07	2.68	2.45	2.29	2.17	2.09	2.02	1.96
∞	3.84	3.00	2.60	2.37	2.21	2.10	2.01	1.94	1.88

TABLE 7 (*Concluded*)

				v_1 (d.f.)						
10	*12*	*15*	*20*	*24*	*30*	*40*	*60*	*120*	∞	v_2 (d.f.)
241.9	243.9	245.9	248.0	249.1	250.1	251.1	252.2	253.3	254.3	1
19.40	19.41	19.43	19.45	19.45	19.46	19.47	19.48	19.49	19.50	2
8.79	8.74	8.70	8.66	8.64	8.62	8.59	8.57	8.55	8.53	3
5.96	5.91	5.86	5.80	5.77	5.75	5.72	5.69	5.66	5.63	4
4.74	4.68	4.62	4.56	4.53	4.50	4.46	4.43	4.40	4.36	5
4.06	4.00	3.94	3.87	3.84	3.81	3.77	3.74	3.70	3.67	6
3.64	3.57	3.51	3.44	3.41	3.38	3.34	3.30	3.27	3.23	7
3.35	3.28	3.22	3.15	3.12	3.08	3.04	3.01	2.97	2.93	8
3.14	3.07	3.01	2.94	2.90	2.86	2.83	2.79	2.75	2.71	9
2.98	2.91	2.85	2.77	2.74	2.70	2.66	2.62	2.58	2.54	10
2.85	2.79	2.72	2.65	2.61	2.57	2.53	2.49	2.45	2.40	11
2.75	2.69	2.62	2.54	2.51	2.47	2.43	2.38	2.34	2.30	12
2.67	2.60	2.53	2.46	2.42	2.38	2.34	2.30	2.25	2.21	13
2.60	2.53	2.46	2.39	2.35	2.31	2.27	2.22	2.18	2.13	14
2.54	2.48	2.40	2.33	2.29	2.25	2.20	2.16	2.11	2.07	15
2.49	2.42	2.35	2.28	2.24	2.19	2.15	2.11	2.06	2.01	16
2.45	2.38	2.31	2.23	2.19	2.15	2.10	2.06	2.01	1.96	17
2.41	2.34	2.27	2.19	2.15	2.11	2.06	2.02	1.97	1.92	18
2.38	2.31	2.23	2.16	2.11	2.07	2.03	1.98	1.93	1.88	19
2.35	2.28	2.20	2.12	2.08	2.04	1.99	1.95	1.90	1.84	20
2.32	2.25	2.18	2.10	2.05	2.01	1.96	1.92	1.87	1.81	21
2.30	2.23	2.15	2.07	2.03	1.98	1.94	1.89	1.84	1.78	22
2.27	2.20	2.13	2.05	2.01	1.96	1.91	1.86	1.81	1.76	23
2.25	2.18	2.11	2.03	1.98	1.94	1.89	1.84	1.79	1.73	24
2.24	2.16	2.09	2.01	1.96	1.92	1.87	1.82	1.77	1.71	25
2.22	2.15	2.07	1.99	1.95	1.90	1.85	1.80	1.75	1.69	26
2.20	2.13	2.06	1.97	1.93	1.88	1.84	1.79	1.73	1.67	27
2.19	2.12	2.04	1.96	1.91	1.87	1.82	1.77	1.71	1.65	28
2.18	2.10	2.03	1.94	1.90	1.85	1.81	1.75	1.70	1.64	29
2.16	2.09	2.01	1.93	1.89	1.84	1.79	1.74	1.68	1.62	30
2.08	2.00	1.92	1.84	1.79	1.74	1.69	1.64	1.58	1.51	40
1.99	1.92	1.84	1.75	1.70	1.65	1.59	1.53	1.47	1.39	60
1.91	1.83	1.75	1.66	1.61	1.55	1.50	1.43	1.35	1.25	120
1.83	1.75	1.67	1.57	1.52	1.46	1.39	1.32	1.22	1.00	∞

TABLE 8   Percentage points of the *F* distribution: $\alpha = .025$

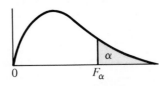

	v_1 (d.f.)								
v_2 (d.f.)	*1*	*2*	*3*	*4*	*5*	*6*	*7*	*8*	*9*
1	647.8	799.5	864.2	899.6	921.8	937.1	948.2	956.7	963.3
2	38.51	39.00	39.17	39.25	39.30	39.33	39.36	39.37	39.39
3	17.44	16.04	15.44	15.10	14.88	14.73	14.62	14.54	14.47
4	12.22	10.65	9.98	9.60	9.36	9.20	9.07	8.98	8.90
5	10.01	8.43	7.76	7.39	7.15	6.98	6.85	6.76	6.68
6	8.81	7.26	6.60	6.23	5.99	5.82	5.70	5.60	5.52
7	8.07	6.54	5.89	5.52	5.29	5.12	4.99	4.90	4.82
8	7.57	6.06	5.42	5.05	4.82	4.65	4.53	4.43	4.36
9	7.21	5.71	5.08	4.72	4.48	4.32	4.20	4.10	4.03
10	6.94	5.46	4.83	4.47	4.24	4.07	3.95	3.85	3.78
11	6.72	5.26	4.63	4.28	4.04	3.88	3.76	3.66	3.59
12	6.55	5.10	4.47	4.12	3.89	3.73	3.61	3.51	3.44
13	6.41	4.97	4.35	4.00	3.77	3.60	3.48	3.39	3.31
14	6.30	4.86	4.24	3.89	3.66	3.50	3.38	3.29	3.21
15	6.20	4.77	4.15	3.80	3.58	3.41	3.29	3.20	3.12
16	6.12	4.69	4.08	3.73	3.50	3.34	3.22	3.12	3.05
17	6.04	4.62	4.01	3.66	3.44	3.28	3.16	3.06	2.98
18	5.98	4.56	3.95	3.61	3.38	3.22	3.10	3.01	2.93
19	5.92	4.51	3.90	3.56	3.33	3.17	3.05	2.96	2.88
20	5.87	4.46	3.86	3.51	3.29	3.13	3.01	2.91	2.84
21	5.83	4.42	3.82	3.48	3.25	3.09	2.97	2.87	2.80
22	5.79	4.38	3.78	3.44	3.22	3.05	2.93	2.84	2.76
23	5.75	4.35	3.75	3.41	3.18	3.02	2.90	2.81	2.73
24	5.72	4.32	3.72	3.38	3.15	2.99	2.87	2.78	2.70
25	5.69	4.29	3.69	3.35	3.13	2.97	2.85	2.75	2.68
26	5.66	4.27	3.67	3.33	3.10	2.94	2.82	2.73	2.65
27	5.63	4.24	3.65	3.31	3.08	2.92	2.80	2.71	2.63
28	5.61	4.22	3.63	3.29	3.06	2.90	2.78	2.69	2.61
29	5.59	4.20	3.61	3.27	3.04	2.88	2.76	2.67	2.59
30	5.57	4.18	3.59	3.25	3.03	2.87	2.75	2.65	2.57
40	5.42	4.05	3.46	3.13	2.90	2.74	2.62	2.53	2.45
60	5.29	3.93	3.34	3.01	2.79	2.63	2.51	2.41	2.33
120	5.15	3.80	3.23	2.89	2.67	2.52	2.39	2.30	2.22
∞	5.02	3.69	3.12	2.79	2.57	2.41	2.29	2.19	2.11

TABLE 8   (*Concluded*)

				v_1 (d.f.)							
10	*12*	*15*	*20*	*24*	*30*	*40*	*60*	*120*	∞	v_2 (d.f.)
968.6	976.7	984.9	993.1	997.2	1001	1006	1010	1014	1018	1
39.40	39.41	39.43	39.45	39.46	39.46	39.47	39.48	39.49	39.50	2
14.42	14.34	14.25	14.17	14.12	14.08	14.04	13.99	13.95	13.90	3
8.84	8.75	8.66	8.56	8.51	8.46	8.41	8.36	8.31	8.26	4
6.62	6.52	6.43	6.33	6.28	6.23	6.18	6.12	6.07	6.02	5
5.46	5.37	5.27	5.17	5.12	5.07	5.01	4.96	4.90	4.85	6
4.76	4.67	4.57	4.47	4.42	4.36	4.31	4.25	4.20	4.14	7
4.30	4.20	4.10	4.00	3.95	3.89	3.84	3.78	3.73	3.67	8
3.96	3.87	3.77	3.67	3.61	3.56	3.51	3.45	3.39	3.33	9
3.72	3.62	3.52	3.42	3.37	3.31	3.26	3.20	3.14	3.08	10
3.53	3.43	3.33	3.23	3.17	3.12	3.06	3.00	2.94	2.88	11
3.37	3.28	3.18	3.07	3.02	2.96	2.91	2.85	2.79	2.72	12
3.25	3.15	3.05	2.95	2.89	2.84	2.78	2.72	2.66	2.60	13
3.15	3.05	2.95	2.84	2.79	2.73	2.67	2.61	2.55	2.49	14
3.06	2.96	2.86	2.76	2.70	2.64	2.59	2.52	2.46	2.40	15
2.99	2.89	2.79	2.68	2.63	2.57	2.51	2.45	2.38	2.32	16
2.92	2.82	2.72	2.62	2.56	2.50	2.44	2.38	2.32	2.25	17
2.87	2.77	2.67	2.56	2.50	2.44	2.38	2.32	2.26	2.19	18
2.82	2.72	2.62	2.51	2.45	2.39	2.33	2.27	2.20	2.13	19
2.77	2.68	2.57	2.46	2.41	2.35	2.29	2.22	2.16	2.09	20
2.73	2.64	2.53	2.42	2.37	2.31	2.25	2.18	2.11	2.04	21
2.70	2.60	2.50	2.39	2.33	2.27	2.21	2.14	2.08	2.00	22
2.67	2.57	2.47	2.36	2.30	2.24	2.18	2.11	2.04	1.97	23
2.64	2.54	2.44	2.33	2.27	2.21	2.15	2.08	2.01	1.94	24
2.61	2.51	2.41	2.30	2.24	2.18	2.12	2.05	1.98	1.91	25
2.59	2.49	2.39	2.28	2.22	2.16	2.09	2.03	1.95	1.88	26
2.57	2.47	2.36	2.25	2.19	2.13	2.07	2.00	1.93	1.85	27
2.55	2.45	2.34	2.23	2.17	2.11	2.05	1.98	1.91	1.83	28
2.53	2.43	2.32	2.21	2.15	2.09	2.03	1.96	1.89	1.81	29
2.51	2.41	2.31	2.20	2.14	2.07	2.01	1.94	1.87	1.79	30
2.39	2.29	2.18	2.07	2.01	1.94	1.88	1.80	1.72	1.64	40
2.27	2.17	2.06	1.94	1.88	1.82	1.74	1.67	1.58	1.48	60
2.16	2.05	1.94	1.82	1.76	1.69	1.61	1.53	1.43	1.31	120
2.05	1.94	1.83	1.71	1.64	1.57	1.48	1.39	1.27	1.00	∞

TABLE 9   Percentage points of the F distribution; $\alpha = .01$

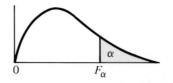

	v_1 (d.f.)								
v_2 (d.f.)	1	2	3	4	5	6	7	8	9
1	4052	4999.5	5403	5625	5764	5859	5928	5982	6022
2	98.50	99.00	99.17	99.25	99.30	99.33	99.36	99.37	99.39
3	34.12	30.82	29.46	28.71	28.24	27.91	27.67	27.49	27.35
4	21.20	18.00	16.69	15.98	15.52	15.21	14.98	14.80	14.66
5	16.26	13.27	12.06	11.39	10.97	10.67	10.46	10.29	10.16
6	13.75	10.92	9.78	9.15	8.75	8.47	8.26	8.10	7.98
7	12.25	9.55	8.45	7.85	7.46	7.19	6.99	6.84	6.72
8	11.26	8.65	7.59	7.01	6.63	6.37	6.18	6.03	5.91
9	10.56	8.02	6.99	6.42	6.06	5.80	5.61	5.47	5.35
10	10.04	7.56	6.55	5.99	5.64	5.39	5.20	5.06	4.94
11	9.65	7.21	6.22	5.67	5.32	5.07	4.89	4.74	4.63
12	9.33	6.93	5.95	5.41	5.06	4.82	4.64	4.50	4.39
13	9.07	6.70	5.74	5.21	4.86	4.62	4.44	4.30	4.19
14	8.86	6.51	5.56	5.04	4.69	4.46	4.28	4.14	4.03
15	8.68	6.36	5.42	4.89	4.56	4.32	4.14	4.00	3.89
16	8.53	6.23	5.29	4.77	4.44	4.20	4.03	3.89	3.78
17	8.40	6.11	5.18	4.67	4.34	4.10	3.93	3.79	3.68
18	8.29	6.01	5.09	4.58	4.25	4.01	3.84	3.71	3.60
19	8.18	5.93	5.01	4.50	4.17	3.94	3.77	3.63	3.52
20	8.10	5.85	4.94	4.43	4.10	3.87	3.70	3.56	3.46
21	8.02	5.78	4.87	4.37	4.04	3.81	3.64	3.51	3.40
22	7.95	5.72	4.82	4.31	3.99	3.76	3.59	3.45	3.35
23	7.88	5.66	4.76	4.26	3.94	3.71	3.54	3.41	3.30
24	7.82	5.61	4.72	4.22	3.90	3.67	3.50	3.36	3.26
25	7.77	5.57	4.68	4.18	3.85	3.63	3.46	3.32	3.22
26	7.72	5.53	4.64	4.14	3.82	3.59	3.42	3.29	3.18
27	7.68	5.49	4.60	4.11	3.78	3.56	3.39	3.26	3.15
28	7.64	5.45	4.57	4.07	3.75	3.53	3.36	3.23	3.12
29	7.60	5.42	4.54	4.04	3.73	3.50	3.33	3.20	3.09
30	7.56	5.39	4.51	4.02	3.70	3.47	3.30	3.17	3.07
40	7.31	5.18	4.31	3.83	3.51	3.29	3.12	2.99	2.89
60	7.08	4.98	4.13	3.65	3.34	3.12	2.95	2.82	2.72
120	6.85	4.79	3.95	3.48	3.17	2.96	2.79	2.66	2.56
∞	6.63	4.61	3.78	3.32	3.02	2.80	2.64	2.51	2.41

TABLE 9   (*Concluded*)

			v_1 (d.f.)							
10	*12*	*15*	*20*	*24*	*30*	*40*	*60*	*120*	*∞*	*v_2 (d.f.)*
6056	6106	6157	6209	6235	6261	6287	6313	6339	6366	1
99.40	99.42	99.43	99.45	99.46	99.47	99.47	99.48	99.49	99.50	2
27.23	27.05	26.87	26.69	26.60	26.50	26.41	26.32	26.22	26.13	3
14.55	14.37	14.20	14.02	13.93	13.84	13.75	13.65	13.56	13.46	4
10.05	9.89	9.72	9.55	9.47	9.38	9.29	9.20	9.11	9.02	5
7.87	7.72	7.56	7.40	7.31	7.23	7.14	7.06	6.97	6.88	6
6.62	6.47	6.31	6.16	6.07	5.99	5.91	5.82	5.74	5.65	7
5.81	5.67	5.52	5.36	5.28	5.20	5.12	5.03	4.95	4.86	8
5.26	5.11	4.96	4.81	4.73	4.65	4.57	4.48	4.40	4.31	9
4.85	4.71	4.56	4.41	4.33	4.25	4.17	4.08	4.00	3.91	10
4.54	4.40	4.25	4.10	4.02	3.94	3.86	3.78	3.69	3.60	11
4.30	4.16	4.01	3.86	3.78	3.70	3.62	3.54	3.45	3.36	12
4.10	3.96	3.82	3.66	3.59	3.51	3.43	3.34	3.25	3.17	13
3.94	3.80	3.66	3.51	3.43	3.35	3.27	3.18	3.09	3.00	14
3.80	3.67	3.52	3.37	3.29	3.21	3.13	3.05	2.96	2.87	15
3.69	3.55	3.41	3.26	3.18	3.10	3.02	2.93	2.84	2.75	16
3.59	3.46	3.31	3.16	3.08	3.00	2.92	2.83	2.75	2.65	17
3.51	3.37	3.23	3.08	3.00	2.92	2.84	2.75	2.66	2.57	18
3.43	3.30	3.15	3.00	2.92	2.84	2.76	2.67	2.58	2.49	19
3.37	3.23	3.09	2.94	2.86	2.78	2.69	2.61	2.52	2.42	20
3.31	3.17	3.03	2.88	2.80	2.72	2.64	2.55	2.46	2.36	21
3.26	3.12	2.98	2.83	2.75	2.67	2.58	2.50	2.40	2.31	22
3.21	3.07	2.93	2.78	2.70	2.62	2.54	2.45	2.35	2.26	23
3.17	3.03	2.89	2.74	2.66	2.58	2.49	2.40	2.31	2.21	24
3.13	2.99	2.85	2.70	2.62	2.54	2.45	2.36	2.27	2.17	25
3.09	2.96	2.81	2.66	2.58	2.50	2.42	2.33	2.23	2.13	26
3.06	2.93	2.78	2.63	2.55	2.47	2.38	2.29	2.20	2.10	27
3.03	2.90	2.75	2.60	2.52	2.44	2.35	2.26	2.17	2.06	28
3.00	2.87	2.73	2.57	2.49	2.41	2.33	2.23	2.14	2.03	29
2.98	2.84	2.70	2.55	2.47	2.39	2.30	2.21	2.11	2.01	30
2.80	2.66	2.52	2.37	2.29	2.20	2.11	2.02	1.92	1.80	40
2.63	2.50	2.35	2.20	2.12	2.03	1.94	1.84	1.73	1.60	60
2.47	2.34	2.19	2.03	1.95	1.86	1.76	1.66	1.53	1.38	120
2.32	2.18	2.04	1.88	1.79	1.70	1.59	1.47	1.32	1.00	∞

TABLE 10   Percentage points of the F distribution: $\alpha = .005$

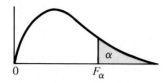

	v_1 (d.f.)								
v_2 (d.f.)	1	2	3	4	5	6	7	8	9
1	16211	20000	21615	22500	23056	23437	23715	23925	24091
2	198.5	199.0	199.2	199.2	199.3	199.3	199.4	199.4	199.4
3	55.55	49.80	47.47	46.19	45.39	44.84	44.43	44.13	43.88
4	31.33	26.28	24.26	23.15	22.46	21.97	21.62	21.35	21.14
5	22.78	18.31	16.53	15.56	14.94	14.51	14.20	13.96	13.77
6	18.63	14.54	12.92	12.03	11.46	11.07	10.79	10.57	10.39
7	16.24	12.40	10.88	10.05	9.52	9.16	8.89	8.68	8.51
8	14.69	11.04	9.60	8.81	8.30	7.95	7.69	7.50	7.34
9	13.61	10.11	8.72	7.96	7.47	7.13	6.88	6.69	6.54
10	12.83	9.43	8.08	7.34	6.87	6.54	6.30	6.12	5.97
11	12.23	8.91	7.60	6.88	6.42	6.10	5.86	5.68	5.54
12	11.75	8.51	7.23	6.52	6.07	5.76	5.52	5.35	5.20
13	11.37	8.19	6.93	6.23	5.79	5.48	5.25	5.08	4.94
14	11.06	7.92	6.68	6.00	5.56	5.26	5.03	4.86	4.72
15	10.80	7.70	6.48	5.80	5.37	5.07	4.85	4.67	4.54
16	10.58	7.51	6.30	5.64	5.21	4.91	4.69	4.52	4.38
17	10.38	7.35	6.16	5.50	5.07	4.78	4.56	4.39	4.25
18	10.22	7.21	6.03	5.37	4.96	4.66	4.44	4.28	4.14
19	10.07	7.09	5.92	5.27	4.85	4.56	4.34	4.18	4.04
20	9.94	6.99	5.82	5.17	4.76	4.47	4.26	4.09	3.96
21	9.83	6.89	5.73	5.09	4.68	4.39	4.18	4.01	3.88
22	9.73	6.81	5.65	5.02	4.61	4.32	4.11	3.94	3.81
23	9.63	6.73	5.58	4.95	4.54	4.26	4.05	3.88	3.75
24	9.55	6.66	5.52	4.89	4.49	4.20	3.99	3.83	3.69
25	9.48	6.60	5.46	4.84	4.43	4.15	3.94	3.78	3.64
26	9.41	6.54	5.41	4.79	4.38	4.10	3.89	3.73	3.60
27	9.34	6.49	5.36	4.74	4.34	4.06	3.85	3.69	3.56
28	9.28	6.44	5.32	4.70	4.30	4.02	3.81	3.65	3.52
29	9.23	6.40	5.28	4.66	4.26	3.98	3.77	3.61	3.48
30	9.18	6.35	5.24	4.62	4.23	3.95	3.74	3.58	3.45
40	8.83	6.07	4.98	4.37	3.99	3.71	3.51	3.35	3.22
60	8.49	5.79	4.73	4.14	3.76	3.49	3.29	3.13	3.01
120	8.18	5.54	4.50	3.92	3.55	3.28	3.09	2.93	2.81
∞	7.88	5.30	4.28	3.72	3.35	3.09	2.90	2.74	2.62

TABLE 10   (*Concluded*)

				v_1 (d.f.)							
10	12	15	20	24	30	40	60	120	∞	v_1 (d.f.)
24224	24426	24630	24836	24940	25044	25148	25253	25359	25465	1
199.4	199.4	199.4	199.4	199.5	199.5	199.5	199.5	199.5	199.5	2
43.69	43.39	43.08	42.78	42.62	42.47	42.31	42.15	41.99	41.83	3
20.97	20.70	20.44	20.17	20.03	19.89	19.75	19.61	19.47	19.32	4
13.62	13.38	13.15	12.90	12.78	12.66	12.53	12.40	12.27	12.14	5
10.25	10.03	9.81	9.59	9.47	9.36	9.24	9.12	9.00	8.88	6
8.38	8.18	7.97	7.75	7.65	7.53	7.42	7.31	7.19	7.08	7
7.21	7.01	6.81	6.61	6.50	6.40	6.29	6.18	6.06	5.95	8
6.42	6.23	6.03	5.83	5.73	5.62	5.52	5.41	5.30	5.19	9
5.85	5.66	5.47	5.27	5.17	5.07	4.97	4.86	4.75	4.64	10
5.42	5.24	5.05	4.86	4.76	4.65	4.55	4.44	4.34	4.23	11
5.09	4.91	4.72	4.53	4.43	4.33	4.23	4.12	4.01	3.90	12
4.82	4.64	4.46	4.27	4.17	4.07	3.97	3.87	3.76	3.65	13
4.60	4.43	4.25	4.06	3.96	3.86	3.76	3.66	3.55	3.44	14
4.42	4.25	4.07	3.88	3.79	3.69	3.58	3.48	3.37	3.26	15
4.27	4.10	3.92	3.73	3.64	3.54	3.44	3.33	3.22	3.11	16
4.14	3.97	3.79	3.61	3.51	3.41	3.31	3.21	3.10	2.98	17
4.03	3.86	3.68	3.50	3.40	3.30	3.20	3.10	2.99	2.87	18
3.93	3.76	3.59	3.40	3.31	3.21	3.11	3.00	2.89	2.78	19
3.85	3.68	3.50	3.32	3.22	3.12	3.02	2.92	2.81	2.69	20
3.77	3.60	3.43	3.24	3.15	3.05	2.95	2.84	2.73	2.61	21
3.70	3.54	3.36	3.18	3.08	2.98	2.88	2.77	2.66	2.55	22
3.64	3.47	3.30	3.12	3.02	2.92	2.82	2.71	2.60	2.48	23
3.59	3.42	3.25	3.06	2.97	2.87	2.77	2.66	2.55	2.43	24
3.54	3.37	3.20	3.01	2.92	2.82	2.72	2.61	2.50	2.38	25
3.49	3.33	3.15	2.97	2.87	2.77	2.67	2.56	2.45	2.33	26
3.45	3.28	3.11	2.93	2.83	2.73	2.63	2.52	2.41	2.29	27
3.41	3.25	3.07	2.89	2.79	2.69	2.59	2.48	2.37	2.25	28
3.38	3.21	3.04	2.86	2.76	2.66	2.56	2.45	2.33	2.21	29
3.34	3.18	3.01	2.82	2.73	2.63	2.52	2.42	2.30	2.18	30
3.12	2.95	2.78	2.60	2.50	2.40	2.30	2.18	2.06	1.93	40
2.90	2.74	2.57	2.39	2.29	2.19	2.08	1.96	1.83	1.69	60
2.71	2.54	2.37	2.19	2.09	1.98	1.87	1.75	1.61	1.43	120
2.52	2.36	2.19	2.00	1.90	1.79	1.67	1.53	1.36	1.00	∞

From "Tables of Percentage Points of the Inverted Beta (*F*)-Distribution," *Biometrika*, Vol. 33 (1943), pp. 73–88, by Maxine Merrington and Catherine M. Thompson. Reproduced by permission of the *Biometrika* Trustees.

TABLE 11   Distribution function of U

$P(U \leq U_0)$; U_0 is the argument; $n_1 \leq n_2$;
$3 \leq n_2 \leq 10$.

$n_2 = 3$

U_0	n_1		
	1	2	3
0	.25	.10	.05
1	.50	.20	.10
2		.40	.20
3		.60	.35
4			.50

$n_2 = 4$

U_0	n_1			
	1	2	3	4
0	.2000	.0667	.0286	.0143
1	.4000	.1333	.0571	.0286
2	.6000	.2667	.1143	.0571
3		.4000	.2000	.1000
4		.6000	.3143	.1714
5			.4286	.2429
6			.5714	.3429
7				.4429
8				.5571

TABLE 11   (*Continued*)

$n_2 = 5$

U_0	n_1				
	1	*2*	*3*	*4*	*5*
0	.1667	.0476	.0179	.0079	.0040
1	.3333	.0952	.0357	.0159	.0079
2	.5000	.1905	.0714	.0317	.0159
3		.2857	.1250	.0556	.0278
4		.4286	.1964	.0952	.0476
5		.5714	.2857	.1429	.0754
6			.3929	.2063	.1111
7			.5000	.2778	.1548
8				.3651	.2103
9				.4524	.2738
10				.5476	.3452
11					.4206
12					.5000

$n_2 = 6$

U_0	n_1					
	1	*2*	*3*	*4*	*5*	*6*
0	.1429	.0357	.0119	.0048	.0022	.0011
1	.2857	.0714	.0238	.0095	.0043	.0022
2	.4286	.1429	.0476	.0190	.0087	.0043
3	.5714	.2143	.0833	.0333	.0152	.0076
4		.3214	.1310	.0571	.0260	.0130
5		.4286	.1905	.0857	.0411	.0206
6		.5714	.2738	.1286	.0628	.0325
7			.3571	.1762	.0887	.0465
8			.4524	.2381	.1234	.0660
9			.5476	.3048	.1645	.0898
10				.3810	.2143	.1201
11				.4571	.2684	.1548
12				.5429	.3312	.1970
13					.3961	.2424
14					.4654	.2944
15					.5346	.3496
16						.4091
17						.4686
18						.5314

TABLE 11   (*Continued*)

$n_2 = 7$

U_0				n_1			
	1	*2*	*3*	*4*	*5*	*6*	*7*
0	.1250	.0278	.0083	.0030	.0013	.0006	.0003
1	.2500	.0556	.0167	.0061	.0025	.0012	.0006
2	.3750	.1111	.0333	.0121	.0051	.0023	.0012
3	.5000	.1667	.0583	.0212	.0088	.0041	.0020
4		.2500	.0917	.0364	.0152	.0070	.0035
5		.3333	.1333	.0545	.0240	.0111	.0055
6		.4444	.1917	.0818	.0366	.0175	.0087
7		.5556	.2583	.1152	.0530	.0256	.0131
8			.3333	.1576	.0745	.0367	.0189
9			.4167	.2061	.1010	.0507	.0265
10			.5000	.2636	.1338	.0688	.0364
11				.3242	.1717	.0903	.0487
12				.3939	.2159	.1171	.0641
13				.4636	.2652	.1474	.0825
14				.5364	.3194	.1830	.1043
15					.3775	.2226	.1297
16					.4381	.2669	.1588
17					.5000	.3141	.1914
18						.3654	.2279
19						.4178	.2675
20						.4726	.3100
21						.5274	.3552
22							.4024
23							.4508
24							.5000

TABLE 11   (*Continued*)

$n_2 = 8$

U_0	n_1							
	1	*2*	*3*	*4*	*5*	*6*	*7*	*8*
0	.1111	.0222	.0061	.0020	.0008	.0003	.0002	.0001
1	.2222	.0444	.0121	.0040	.0016	.0007	.0003	.0002
2	.3333	.0889	.0242	.0081	.0031	.0013	.0006	.0003
3	.4444	.1333	.0424	.0141	.0054	.0023	.0011	.0005
4	.5556	.2000	.0667	.0242	.0093	.0040	.0019	.0009
5		.2667	.0970	.0364	.0148	.0063	.0030	.0015
6		.3556	.1394	.0545	.0225	.0100	.0047	.0023
7		.4444	.1879	.0768	.0326	.0147	.0070	.0035
8		.5556	.2485	.1071	.0466	.0213	.0103	.0052
9			.3152	.1414	.0637	.0296	.0145	.0074
10			.3879	.1838	.0855	.0406	.0200	.0103
11			.4606	.2303	.1111	.0539	.0270	.0141
12			.5394	.2848	.1422	.0709	.0361	.0190
13				.3414	.1772	.0906	.0469	.0249
14				.4040	.2176	.1142	.0603	.0325
15				.4667	.2618	.1412	.0760	.0415
16				.5333	.3108	.1725	.0946	.0524
17					.3621	.2068	.1159	.0652
18					.4165	.2454	.1405	.0803
19					.4716	.2864	.1678	.0974
20					.5284	.3310	.1984	.1172
21						.3773	.2317	.1393
22						.4259	.2679	.1641
23						.4749	.3063	.1911
24						.5251	.3472	.2209
25							.3894	.2527
26							.4333	.2869
27							.4775	.3227
28							.5225	.3605
29								.3992
30								.4392
31								.4796
32								.5204

TABLE 11   (*Continued*)

$n_2 = 9$

U_0	n_1								
	1	*2*	*3*	*4*	*5*	*6*	*7*	*8*	*9*
0	.1000	.0182	.0045	.0014	.0005	.0002	.0001	.0000	.0000
1	.2000	.0364	.0091	.0028	.0010	.0004	.0002	.0001	.0000
2	.3000	.0727	.0182	.0056	.0020	.0008	.0003	.0002	.0001
3	.4000	.1091	.0318	.0098	.0035	.0014	.0006	.0003	.0001
4	.5000	.1636	.0500	.0168	.0060	.0024	.0010	.0005	.0002
5		.2182	.0727	.0252	.0095	.0038	.0017	.0008	.0004
6		.2909	.1045	.0378	.0145	.0060	.0026	.0012	.0006
7		.3636	.1409	.0531	.0210	.0088	.0039	.0019	.0009
8		.4545	.1864	.0741	.0300	.0128	.0058	.0028	.0014
9		.5455	.2409	.0993	.0415	.0180	.0082	.0039	.0020
10			.3000	.1301	.0559	.0248	.0115	.0056	.0028
11			.3636	.1650	.0734	.0332	.0156	.0076	.0039
12			.4318	.2070	.0949	.0440	.0209	.0103	.0053
13			.5000	.2517	.1199	.0567	.0274	.0137	.0071
14				.3021	.1489	.0723	.0356	.0180	.0094
15				.3552	.1818	.0905	.0454	.0232	.0122
16				.4126	.2188	.1119	.0571	.0296	.0157
17				.4699	.2592	.1361	.0708	.0372	.0200
18				.5301	.3032	.1638	.0869	.0464	.0252
19					.3497	.1942	.1052	.0570	.0313
20					.3986	.2280	.1261	.0694	.0385
21					.4491	.2643	.1496	.0836	.0470
22					.5000	.3035	.1755	.0998	.0567
23						.3445	.2039	.1179	.0680
24						.3878	.2349	.1383	.0807
25						.4320	.2680	.1606	.0951
26						.4773	.3032	.1852	.1112
27						.5227	.3403	.2117	.1290
28							.3788	.2404	.1487
29							.4185	.2707	.1701
30							.4591	.3029	.1933
31							.5000	.3365	.2181
32								.3715	.2447
33								.4074	.2729
34								.4442	.3024
35								.4813	.3332
36								.5187	.3652
37									.3981
38									.4317
39									.4657
40									.5000

TABLE 11   (*Continued*)

$n_2 = 10$

U_0	1	2	3	4	5	6	7	8	9	10
0	.0909	.0152	.0035	.0010	.0003	.0001	.0001	.0000	.0000	.0000
1	.1818	.0303	.0070	.0020	.0007	.0002	.0001	.0000	.0000	.0000
2	.2727	.0606	.0140	.0040	.0013	.0005	.0002	.0001	.0000	.0000
3	.3636	.0909	.0245	.0070	.0023	.0009	.0004	.0002	.0001	.0000
4	.4545	.1364	.0385	.0120	.0040	.0015	.0006	.0003	.0001	.0001
5	.5455	.1818	.0559	.0180	.0063	.0024	.0010	.0004	.0002	.0001
6		.2424	.0804	.0270	.0097	.0037	.0015	.0007	.0003	.0002
7		.3030	.1084	.0380	.0140	.0055	.0023	.0010	.0005	.0002
8		.3788	.1434	.0529	.0200	.0080	.0034	.0015	.0007	.0004
9		.4545	.1853	.0709	.0276	.0112	.0048	.0022	.0011	.0005
10		.5455	.2343	.0939	.0376	.0156	.0068	.0031	.0015	.0008
11			.2867	.1199	.0496	.0210	.0093	.0043	.0021	.0010
12			.3462	.1518	.0646	.0280	.0125	.0058	.0028	.0014
13			.4056	.1868	.0823	.0363	.0165	.0078	.0038	.0019
14			.4685	.2268	.1032	.0467	.0215	.0103	.0051	.0026
15			.5315	.2697	.1272	.0589	.0277	.0133	.0066	.0034
16				.3177	.1548	.0736	.0351	.0171	.0086	.0045
17				.3666	.1855	.0903	.0439	.0217	.0110	.0057
18				.4196	.2198	.1099	.0544	.0273	.0140	.0073
19				.4725	.2567	.1317	.0665	.0338	.0175	.0093
20				.5275	.2970	.1566	.0806	.0416	.0217	.0116
21					.3393	.1838	.0966	.0506	.0267	.0144
22					.3839	.2139	.1148	.0610	.0326	.0177
23					.4296	.2461	.1349	.0729	.0394	.0216
24					.4765	.2811	.1574	.0864	.0474	.0262
25					.5235	.3177	.1819	.1015	.0564	.0315
26						.3564	.2087	.1185	.0667	.0376
27						.3962	.2374	.1371	.0782	.0446
28						.4374	.2681	.1577	.0912	.0526
29						.4789	.3004	.1800	.1055	.0615
30						.5211	.3345	.2041	.1214	.0716
31							.3698	.2299	.1388	.0827
32							.4063	.2574	.1577	.0952
33							.4434	.2863	.1781	.1088
34							.4811	.3167	.2001	.1237
35							.5189	.3482	.2235	.1399
36								.3809	.2483	.1575
37								.4143	.2745	.1763
38								.4484	.3019	.1965
39								.4827	.3304	.2179

TABLE 11   (*Concluded*)

$n_2 = 10$ (*Concluded*)

U_0	*1*	*2*	*3*	*4*	*5*	*6*	*7*	*8*	*9*	*10*
						n_1				
40								.5173	.3598	.2406
41									.3901	.2644
42									.4211	.2894
43									.4524	.3153
44									.4841	.3421
45									.5159	.3697
46										.3980
47										.4267
48										.4559
49										.4853
50										.5147

Computed by M. Pagano, Department of Statistics, University of Florida.

TABLE 12   Critical values of T in the Wilcoxon signed-rank test; $n = 5(1)50$

One-sided	*Two-sided*	$n = 5$	$n = 6$	$n = 7$	$n = 8$	$n = 9$	$n = 10$
$P = .05$	$P = .10$	1	2	4	6	8	11
$P = .025$	$P = .05$		1	2	4	6	8
$P = .01$	$P = .02$			0	2	3	5
$P = .005$	$P = .01$				0	2	3

One-sided	*Two-sided*	$n = 11$	$n = 12$	$n = 13$	$n = 14$	$n = 15$	$n = 16$
$P = .05$	$P = .10$	14	17	21	26	30	36
$P = .025$	$P = .05$	11	14	17	21	25	30
$P = .01$	$P = .02$	7	10	13	16	20	24
$P = .005$	$P = .01$	5	7	10	13	16	19

One-sided	*Two-sided*	$n = 17$	$n = 18$	$n = 19$	$n = 20$	$n = 21$	$n = 22$
$P = .05$	$P = .10$	41	47	54	60	68	75
$P = .025$	$P = .05$	35	40	46	52	59	66
$P = .01$	$P = .02$	28	33	38	43	49	56
$P = .005$	$P = .01$	23	28	32	37	43	49

TABLE 12 (*Concluded*)

One-sided	Two-sided	$n = 23$	$n = 24$	$n = 25$	$n = 26$	$n = 27$	$n = 28$
$P = .05$	$P = .10$	83	92	101	110	120	130
$P = .025$	$P = .05$	73	81	90	98	107	117
$P = .01$	$P = .02$	62	69	77	85	93	102
$P = .005$	$P = .01$	55	68	68	76	84	92

One-sided	Two-sided	$n = 29$	$n = 30$	$n = 31$	$n = 32$	$n = 33$	$n = 34$
$P = .05$	$P = .10$	141	152	163	175	188	201
$P = .025$	$P = .05$	127	137	148	159	171	183
$P = .01$	$P = .02$	111	120	130	141	151	162
$P = .005$	$P = .01$	100	109	118	128	138	149

One-sided	Two-sided	$n = 35$	$n = 36$	$n = 37$	$n = 38$	$n = 39$	
$P = .05$	$P = .10$	214	228	242	256	271	
$P = .025$	$P = .05$	195	208	222	235	250	
$P = .01$	$P = .02$	174	186	198	211	224	
$P = .005$	$P = .01$	160	171	183	195	208	

One-sided	Two-sided	$n = 40$	$n = 41$	$n = 42$	$n = 43$	$n = 44$	$n = 45$
$P = .05$	$P = .10$	287	303	319	336	353	371
$P = .025$	$P = .05$	264	279	295	311	327	344
$P = .01$	$P = .02$	238	252	267	281	297	313
$P = .005$	$P = .01$	221	234	248	262	277	292

One-sided	Two-sided	$n = 46$	$n = 47$	$n = 48$	$n = 49$	$n = 50$	
$P = .05$	$P = .10$	389	408	427	446	466	
$P = .025$	$P = .05$	361	379	397	415	434	
$P = .01$	$P = .02$	329	345	362	380	398	
$P = .005$	$P = .01$	307	323	339	356	373	

From "Some Rapid Approximate Statistical Procedures" (1964), 28, F. Wilcoxon and R. A. Wilcox. Reproduced with the kind permission of Lederle Laboratories, a division of American Cyanamid Company.

TABLE 13   Distribution of the total number of runs R in samples of size (n_1, n_2); $P(R \leq a)$

(n_1, n_2)	2	3	4	5	6	7	8	9	10
					a				
(2, 3)	.200	.500	.900	1.000					
(2, 4)	.133	.400	.800	1.000					
(2, 5)	.095	.333	.714	1.000					
(2, 6)	.071	.286	.643	1.000					
(2, 7)	.056	.250	.583	1.000					
(2, 8)	.044	.222	.533	1.000					
(2, 9)	.036	.200	.491	1.000					
(2, 10)	.030	.182	.455	1.000					
(3, 3)	.100	.300	.700	.900	1.000				
(3, 4)	.057	.200	.543	.800	.971	1.000			
(3, 5)	.036	.143	.429	.714	.929	1.000			
(3, 6)	.024	.107	.345	.643	.881	1.000			
(3, 7)	.017	.083	.283	.583	.833	1.000			
(3, 8)	.012	.067	.236	.533	.788	1.000			
(3, 9)	.009	.055	.200	.491	.745	1.000			
(3, 10)	.007	.045	.171	.455	.706	1.000			
(4, 4)	.029	.114	.371	.629	.886	.971	1.000		
(4, 5)	.016	.071	.262	.500	.786	.929	.992	1.000	
(4, 6)	.010	.048	.190	.405	.690	.881	.976	1.000	
(4, 7)	.006	.033	.142	.333	.606	.833	.954	1.000	
(4, 8)	.004	.024	.109	.279	.533	.788	.929	1.000	
(4, 9)	.003	.018	.085	.236	.471	.745	.902	1.000	
(4, 10)	.002	.014	.068	.203	.419	.706	.874	1.000	
(5, 5)	.008	.040	.167	.357	.643	.833	.960	.992	1.000
(5, 6)	.004	.024	.110	.262	.522	.738	.911	.976	.998
(5, 7)	.003	.015	.076	.197	.424	.652	.854	.955	.992
(5, 8)	.002	.010	.054	.152	.347	.576	.793	.929	.984
(5, 9)	.001	.007	.039	.119	.287	.510	.734	.902	.972
(5, 10)	.001	.005	.029	.095	.239	.455	.678	.874	.958
(6, 6)	.002	.013	.067	.175	.392	.608	.825	.933	.987
(6, 7)	.001	.008	.043	.121	.296	.500	.733	.879	.966
(6, 8)	.001	.005	.028	.086	.226	.413	.646	.821	.937
(6, 9)	.000	.003	.019	.063	.175	.343	.566	.762	.902
(6, 10)	.000	.002	.013	.047	.137	.288	.497	.706	.864
(7, 7)		.001	.004	.025	.078	.209	.383	.617	.791	.922
(7, 8)	.000	.002	.015	.051	.149	.296	.514	.704	.867
(7, 9)	.000	.001	.010	.035	.108	.231	.427	.622	.806
(7, 10)	.000	.001	.006	.024	.080	.182	.355	.549	.743
(8, 8)	.000	.001	.009	.032	.100	.214	.405	.595	.786
(8, 9)	.000	.001	.005	.020	.069	.157	.319	.500	.702
(8, 10)	.000	.000	.003	.013	.048	.117	.251	.419	.621
(9, 9)	.000	.000	.003	.012	.044	.109	.238	.399	.601
(9, 10)	.000	.000	.002	.008	.029	.077	.179	.319	.510
(10,10)	.000	.000	.001	.004	.019	.051	.128	.242	.414

TABLE 13   *(Concluded)*

					a					
(n_1, n_2)	*11*	*12*	*13*	*14*	*15*	*16*	*17*	*18*	*19*	*20*
(2, 3)										
(2, 4)										
(2, 5)										
(2, 6)										
(2, 7)				—						
(2, 8)										
(2, 9)										
(2, 10)										
(3, 3)										
(3, 4)										
(3, 5)										
(3, 6)										
(3, 7)										
(3, 8)										
(3, 9)										
(3, 10)										
(4, 4)										
(4, 5)										
(4, 6)										
(4, 7)										
(4, 8)										
(4, 9)										
(4, 10)										
(5, 5)										
(5, 6)	1.000									
(5, 7)	1.000									
(5, 8)	1.000									
(5, 9)	1.000									
(5, 10)	1.000									
(6, 6)	.998	1.000								
(6, 7)	.992	.999	1.000							
(6, 8)	.984	.998	1.000							
(6, 9)	.972	.994	1.000							
(6, 10)	.958	.990	1.000							
(7, 7)	.975	.996	.999	1.000						
(7, 8)	.949	.988	.998	1.000	1.000					
(7, 9)	.916	.975	.994	.999	1.000					
(7, 10)	.879	.957	.990	.998	1.000					
(8, 8)	.900	.968	.991	.999	1.000	1.000				
(8, 9)	.843	.939	.980	.996	.999	1.000	1.000			
(8, 10)	.782	.903	.964	.990	.998	1.000	1.000			
(9, 9)	.762	.891	.956	.988	.997	1.000	1.000	1.000		
(9, 10)	.681	.834	.923	.974	.992	.999	1.000	1.000	1.000	
(10, 10)	.586	.758	.872	.949	.981	.996	.999	1.000	1.000	1.000

From "Tables for Testing Randomness of Grouping in a Sequence of Alternatives," F. Swed and C. Eisenhart, *Annals of Mathematical Statistics*, Volume 14 (1943). Reproduced with the kind permission of the authors and of the Editor, *Annals of Mathematical Statistics*.

TABLE 14   Critical values of Spearman's rank correlation coefficient

n	$\alpha = .05$	$\alpha = .025$	$\alpha = .01$	$\alpha = .005$
5	0.900	—	—	—
6	0.829	0.886	0.943	—
7	0.714	0.786	0.893	—
8	0.643	0.738	0.833	0.881
9	0.600	0.683	0.783	0.833
10	0.564	0.648	0.745	0.794
11	0.523	0.623	0.736	0.818
12	0.497	0.591	0.703	0.780
13	0.475	0.566	0.673	0.745
14	0.457	0.545	0.646	0.716
15	0.441	0.525	0.623	0.689
16	0.425	0.507	0.601	0.666
17	0.412	0.490	0.582	0.645
18	0.399	0.476	0.564	0.625
19	0.388	0.462	0.549	0.608
20	0.377	0.450	0.534	0.591
21	0.368	0.438	0.521	0.576
22	0.359	0.428	0.508	0.562
23	0.351	0.418	0.496	0.549
24	0.343	0.409	0.485	0.537
25	0.336	0.400	0.475	0.526
26	0.329	0.392	0.465	0.515
27	0.323	0.385	0.456	0.505
28	0.317	0.377	0.448	0.496
29	0.311	0.370	0.440	0.487
30	0.305	0.364	0.432	0.478

From "Distribution of Sums of Squares of Rank Differences for Small Samples," E. G. Olds, *Annals of Mathematical Statistics.* Volume 9 (1938). Reproduced with the kind permission of the Editor, *Annals of Mathematical Statistics.*

TABLE 15   Random numbers

							Column							
Line	1	2	3	4	5	6	7	8	9	10	11	12	13	14
1	10480	15011	01536	02011	81647	91646	69179	14194	62590	36207	20969	99570	91291	90700
2	22368	46573	25595	85393	30995	89198	27982	53402	93965	34095	52666	19174	39615	99505
3	24130	48360	22527	97265	76393	64809	15179	24830	49340	32081	30680	19655	63348	58629
4	42167	93093	06243	61680	07856	16376	39440	53537	71341	57004	00849	74917	97758	16379
5	37570	39975	81837	16656	06121	91782	60468	81305	49684	60672	14110	06927	01263	54613
6	77921	06907	11008	42751	27756	53498	18602	70659	90655	15053	21916	81825	44394	42880
7	99562	72905	56420	69994	98872	31016	71194	18738	44013	48840	63213	21069	10634	12952
8	96301	91977	05463	07972	18876	20922	94595	56869	69014	60045	18425	84903	42508	32307
9	89579	14342	63661	10281	17453	18103	57740	84378	25331	12566	58678	44947	05585	56941
10	85475	36857	53342	53988	53060	59533	38867	62300	08158	17983	16439	11458	18593	64952
11	28918	69578	88231	33276	70997	79936	56865	05859	90106	31595	01547	85590	91610	78188
12	63553	40961	48235	03427	49626	69445	18663	72695	52180	20847	12234	90511	33703	90322
13	09429	93969	52636	92737	88974	33488	36320	17617	30015	08272	84115	27156	30613	74952
14	10365	61129	87529	85689	48237	52267	67689	93394	01511	26358	85104	20285	29975	89868
15	07119	97336	71048	08178	77233	13916	47564	81056	97735	85977	29372	74461	28551	90707
16	51085	12765	51821	51259	77452	16308	60756	92144	49442	53900	70960	63990	75601	40719
17	02368	21382	52404	60268	89368	19885	55322	44819	01188	65255	64835	44919	05944	55157
18	01011	54092	33362	94904	31273	04146	18594	29852	71585	85030	51132	01915	92747	64951
19	52162	53916	46369	58586	23216	14513	83149	98736	23495	64350	94738	17752	35156	35749
20	07056	97628	33787	09998	42698	06691	76988	13602	51851	46104	88916	19509	25625	58104
21	48663	91245	85828	14346	09172	30168	90229	04734	59193	22178	30421	61666	99904	32812
22	54164	58492	22421	74103	47070	25306	76468	26384	58151	06646	21524	15227	96909	44592
23	32639	32363	05597	24200	13363	38005	94342	28728	35806	06912	17012	64161	18296	22851
24	29334	27001	87637	87308	58731	00256	45834	15398	46557	41135	10367	07684	36188	18510
25	02488	33062	28834	07351	19731	92420	60952	61280	50001	67658	32586	86679	50720	94953
26	81525	72295	04839	96423	24878	82651	66566	14778	76797	14780	13300	87074	79666	95725
27	29676	20591	68086	26432	46901	20849	89768	81536	86645	12659	92259	57102	80428	25280
28	00742	57392	39064	66432	84673	40027	32832	61362	98947	96067	64760	64584	96096	98253
29	05366	04213	25669	26422	44407	44048	37937	63904	45766	66134	75470	66520	34693	90449
30	91921	26418	64117	94305	26766	25940	39972	22209	71500	64568	91402	42416	07844	69618
31	00582	04711	87917	77341	42206	35126	74087	99547	81817	42607	43808	76655	62028	76630
32	00725	69884	62797	56170	86324	88072	76222	36086	84637	93161	76038	65855	77919	88006
34	69011	65795	95876	55293	18988	27354	26575	08625	40801	59920	29841	80150	12777	48501
34	25976	57948	29888	88604	67917	48708	18912	82271	65424	69774	33611	54262	85963	03547
35	09763	83473	73577	12908	30883	18317	28290	35797	05998	41688	34952	37888	38917	88050
36	91567	42595	27958	30134	04024	86385	29880	99730	55536	84855	29080	09250	79656	73211
37	17955	56349	90999	49127	20044	59931	06115	20542	18059	02008	73708	83517	36103	42791
38	46503	18584	18845	49618	02304	51038	20655	58727	28168	15475	56942	53389	20562	87338
39	92157	89634	94824	78171	84610	82834	09922	25417	44137	48413	25555	21246	35509	20468
40	14577	62765	35605	81263	39667	47358	56873	56307	61607	49518	89656	20103	77490	18062
41	98427	07523	33362	64270	01638	92477	66969	98420	04880	45585	46565	04102	46880	45709
42	34914	63976	88720	82765	34476	17032	87589	40836	32427	70002	70663	88863	77775	69348
43	70060	28277	39475	46473	23219	53416	94970	25832	69975	94884	19661	72828	00102	66794
44	53976	54914	06990	67245	68350	82948	11398	42878	80287	88267	47363	46634	06541	97809
45	76072	29515	40980	07391	58745	25774	22987	80059	39911	96189	41151	14222	60697	59583
46	90725	52210	83974	29992	65831	38857	50490	83765	55657	14361	31720	57375	56228	41546
47	64364	67412	33339	31926	14883	24413	59744	92351	97473	89286	35931	04110	23726	51900
48	08962	00358	31662	25388	61642	34072	81249	35648	56891	69352	48373	45578	78547	81788
49	95012	68379	93526	70765	10592	04542	76463	54328	02349	17247	28865	14777	62730	92277
50	15664	10493	20492	38391	91132	21999	59516	81652	27195	48223	46751	22923	32261	85653
51	16408	81899	04153	53381	79401	21438	83035	92350	36693	31238	59649	91754	72772	02338

TABLE 15 *(Concluded)*

							Column							
Line	1	2	3	4	5	6	7	8	9	10	11	12	13	14
52	18629	81953	05520	91962	04739	13092	97662	24822	94730	06496	35090	04822	86774	98289
53	73115	35101	47498	87637	99016	71060	88824	71013	18735	20286	23153	72924	35165	43040
54	57491	16703	23167	49323	45021	33132	12544	41035	80780	45393	44812	12515	98931	91202
55	30405	83946	23792	14422	15059	45799	22716	19792	09983	74353	68668	30429	70735	25499
56	16631	35006	85900	98275	32388	52390	16815	69298	82732	38480	73817	32523	41961	44437
57	96773	20206	42559	78985	05300	22164	24369	54224	35083	19687	11052	91491	60383	19746
58	38935	64202	14349	82674	66523	44133	00697	35552	35970	19124	63318	29686	03387	59846
59	31624	76384	17403	53363	44167	64486	64758	75366	76554	31601	12614	33072	60332	92325
60	78919	19474	23632	27889	47914	02584	37680	20801	72152	39339	34806	08930	85001	87820
61	03931	33309	57047	74211	63445	17361	62825	39908	05607	91284	68833	25570	38818	46920
62	74426	33278	43972	10119	89917	15665	52872	73823	73144	88662	88970	74492	51805	99378
63	09066	00903	20795	95452	92648	45454	09552	88815	16553	51125	79375	97596	16296	66092
64	42238	12426	87025	14267	20979	04508	64535	31355	86064	29472	47689	05974	52468	16834
65	16153	08002	26504	41744	81959	65642	74240	56302	00033	67107	77510	70625	28725	34191
66	21457	40742	29820	96783	29400	21840	15035	34537	33310	06116	95240	15957	16572	06004
67	21581	57802	02050	89728	17937	37621	47075	42080	97403	48626	68995	43805	33386	21597
68	55612	78095	83197	33732	05810	24813	86902	60397	16489	03264	88525	42786	05269	92532
69	44657	66999	99324	51281	84463	60563	79312	93454	68876	25471	93911	25650	12682	73572
70	91340	84979	46949	81973	37949	61023	43997	15263	80644	43942	89203	71795	99533	50501
71	91227	21199	31935	27022	84067	05462	35216	14486	29891	68607	41867	14951	91696	85065
72	50001	38140	66321	19924	72163	09538	12151	06878	91903	18749	34405	56087	82790	70925
73	65390	05224	72958	28609	81406	39147	25549	48542	42627	45233	57202	94617	23772	07896
74	27504	96131	83944	41575	10573	08619	64482	73923	36152	05184	94142	25299	84387	34925
75	37169	94851	39117	89632	00959	16487	65536	49071	39782	17095	02330	74301	00275	48280
76	11508	70225	51111	38351	19444	66499	71945	05422	13442	78675	84081	66938	93654	59894
77	37449	30362	06694	54690	04052	53115	62757	95348	78662	11163	81651	50245	34971	52924
78	46515	70331	85922	38329	57015	15765	97161	17869	45349	61796	66345	81073	49106	79860
79	30986	81223	42416	58353	21532	30502	32305	86482	05174	07901	54339	58861	74818	46942
80	63798	64995	46583	09785	44160	78128	83991	42865	92520	83531	80377	35909	81250	54238
81	82486	84846	99254	67632	43218	50076	21361	64816	51202	88124	41870	52689	51275	83556
82	21885	32906	92431	09060	64297	51674	64126	62570	26123	05155	59194	52799	28225	85762
83	60336	98782	07408	53458	13564	59089	26445	29789	85205	41001	12535	12133	14645	23541
84	43937	46891	24010	25560	86355	33941	25786	54990	71899	15475	95434	98227	21824	19585
85	97656	63175	89303	16275	07100	92063	21942	18611	47348	20203	18534	03862	78095	50136
86	03299	01221	05418	38982	55758	92237	26759	86367	21216	98442	08303	56613	91511	75928
87	79626	06486	03574	17668	07785	76020	79924	25651	83325	88428	85076	72811	22717	50585
88	85636	68335	47539	03129	65651	11977	02510	26113	99447	68645	34327	15152	55230	93448
89	18039	14367	61337	06177	12143	46609	32989	74014	64708	00533	35398	58408	13261	47908
90	08362	15656	60627	36478	65648	16764	53412	09013	07832	41574	17639	82163	60859	75567
91	79556	29068	04142	16268	15387	12856	66227	38358	22478	73373	88732	09443	82558	05250
92	92608	82674	27072	32534	17075	27698	98204	63863	11951	34648	88022	56148	34925	57031
93	23982	25835	40055	67006	12293	02753	14827	23235	35071	99704	37543	11601	35503	85171
94	09915	96306	05908	97901	28395	14186	00821	80703	70426	75647	76310	88717	37890	40129
95	59037	33300	26695	62247	69927	76123	50842	43834	86654	70959	79725	93872	28117	19233
96	42488	78077	69882	61657	34136	79180	97526	43092	04098	73571	80799	76536	71255	64239
97	46764	86273	63003	93017	31204	36692	40202	35275	57306	55543	53203	18098	47625	88684
98	03237	45430	55417	63282	90816	17349	88298	90183	36600	78406	06216	95787	42579	90730
99	86591	81482	52667	61582	14972	90053	89534	76036	49199	43716	97548	04379	46370	28672
100	38534	01715	94964	87288	65680	43772	39560	12918	86537	62738	19636	51132	25739	56947

Answers

Chapter 2

2.1   1, -1, -1

2.2   1, 1/2, .0625 $= 1/16$

2.3   3, 12

2.4   1, 8, 27

2.5   2, -3, -1

2.6   0, 1, 4

2.7   2, 5, 10

2.8 $y + 1$

2.9   (a) -9
     (b) 65
     (c) $y_1 + y_2 + y_3 + y_4 - 8$
     (d) $3y + 12$

2.10   (a) 0, 3, 8, 15
      (b) $\sum_{y=1}^{4} (y^2 - 1)$
      (c) 26

2.11   (a) 4, 1, 0, 1, 4
      (b) $\sum_{y=1}^{5} (y - 3)^2$
      (c) 10

2.12   (b) $\sum_{i=1}^{5} y_i; \; \sum_{i=1}^{5} y_i^2$
      (c) 3, 3
      (d) 73, 73, 5329

2.13   (b) 3220
      (c) 960; 311, 400; 1,000,000

2.14   (a) 105
      (b) 1300
      (c) 3136
      (d) $105x$

2.15   2

2.16   25

2.17 -6

2.18   16

2.19 $1 + p + p^2 + p^3$

2.20 $4y + 7$

2.21   77

2.22   (a) 1, 5, 11, 19
      (b) $\sum_{y=1}^{4} y^2 + \sum_{y=1}^{4} y - 4$

      (c) 36

2.23   (a) $(y - 1)^2$
      (b) $\sum_{y=1}^{5} (y - 1)^2$
      (c) 30

2.24   (a) 3
      (b) 7
      (c) 11
      (d) -1
      (e) -5
      (f) $4a^2 + 3$
      (g) $-4a + 3$
      (h) $7 - 4y$

2.25   (a) 1
      (b) undefined, division by zero
      (c) 17/5
      (d) $(a^2 + 1)/(1 - a)$

2.26 $p(0) = 1, p(1) = 1 - a$

2.27   (a) 1/4
      (b) -6
      (c) $(3/x^2) - (3/x) + 1$

2.28   (a) 2
      (b) 2
      (c) 2

2.29   14.1, 43.63, 198.81

2.30   yields $5(3.1) = 15.5$ less than original sum; -1.4

2.31   36

2.32   21

2.33   25

2.34 $10 + 10y^2$

2.35 $y_1 + y_2 - 3$

2.36 $\sum_{i=1}^{n} y_i - na$

2.37   36

2.38 -29

2.39   431

2.40   610

2.41   654

2.42   649.077

Chapter 3

3.1 (a) 5.2
(b) .8
(e) .76, .76
(f) all mpg calculations for owner's 1980 Olds Omega
(g) no

3.2 (b) 25/60

3.3 (b) 36/50

3.4 (b) 12/28
(c) they are equal

3.5 (a) 2
(b) 1.414

3.6 (a) 2
(b) 1.581

3.7 (a) approximately 68% of measurements in interval 33–39; approximately 95% of measurements in interval 30–42; almost all of measurements in interval 27–45
(b) at least 3/4 of measurements in interval 30–42; at least 8/9 of measurements in interval 27–45

3.8 (a) .68
(b) .95
(c) ≈ 0

3.9 (a) at least 3/4 of measurements in interval .15–.19; at least 8/9 of measurements in interval .14–.20
(b) approximately 68% of measurements in interval .16–.18; approximately 95% of measurements in interval .15–.19; almost all of measurements in interval .14–.20
(c) no, sample is too small for the data to possess a mound-shaped distribution

3.10 at least 3/4 of measurements in interval 0–140; at least 8/9 of measurements in interval 0–193

3.11 (a) 1/2
(b) at least 3/4

3.12 yes, too many standard deviations away from mean

3.13 $\sigma \approx 100$

3.14 (a) 1.75
(b) $s^2 = .5$, $s = .707$

3.15 (a) 2.8
(b) 3.2, 1.789

3.16 (a) .625
(b) $s^2 = 4.268$; $s = 2.066$

3.17 (a) $s \approx .15$
(b) $\bar{y} = .76$, $s = .165$

3.18 (a) $s \approx 3.5$
(b) $\bar{y} = 6.22$; $s = 3.497$
(c) 96%

3.19 (b) 7.729
(c) 1.985

Interval	Actual	Tchebysheff's Theorem	Empirical Rule
$\bar{y} \pm s$	.714	≥ 0	$\approx .68$
$\bar{y} \pm 2s$	.957	$\geq 3/4$	$\approx .95$
$\bar{y} \pm 3s$	1.000	$\geq 8/9$	≈ 1.00

3.20 (a) 5.2
(b) $s \approx 1.3$
(c) $s = 1.305$

3.21

k	$\bar{y} \pm ks$	Actual	Tchebysheff's Theorem	Empirical Rule
1	29.277, 31.887	.66	≥ 0	$\approx .68$
2	27.972, 33.192	.96	$\geq 3/4$	$\approx .95$
3	26.667, 34.497	1.00	$\geq 8/9$	≈ 1.00

(a) yes
(b) yes

3.22 (a) 32.1
(b) 8.025
(c) 7.671

3.23

Interval	Actual	Tchebysheff's Theorem	Empirical Rule
$\bar{y} \pm s$	.74	≥ 0	$\approx .68$
$\bar{y} \pm 2s$	.94	$\geq 3/4$	$\approx .95$
$\bar{y} \pm 3s$	.98	$\geq 8/9$	≈ 1.00

3.24 (a) 1.4, 19.6
(b) 1.4, 19.6

3.25 (a) 2.04, 2.806
(b) and (c)

k	$\bar{y} \pm ks$	Actual	Tchebysheff's Theorem	Empirical Rule
1	$-$ .77, 4.85	.84	≥ 0	$\approx .68$
2	-3.57, 7.65	.92	$\geq 3/4$	$\approx .95$
3	-6.38, 10.46	1.00	$\geq 8/9$	≈ 1.00

3.26 $\bar{x} = 24$, $s_x = 3$

3.27 $\bar{x} = 40$, $s_x = 6$

3.28 $\bar{x} = 44$, $s_x = 6$

3.29 3,001; 8.5

3.30 .01983, .0000013667

3.31 .01983, .0000013667

3.32 15, 1.714

3.33 7 oz., 1 oz.

3.34 7.729, $s^2 = 3.9398$, $s = 1.985$

3.35 69% scored lower than you

3.36 44th percentile

3.37 67% is the 90th percentile

3.38 80% is the 44th percentile
70% is the 25th percentile

3.41 2.286, 3.905, 1.976

3.46 .68; .96; yes; yes

3.47 .68, .95

3.48 .475 vs. .502

3.49 (a) 7.75
(b) 59.2, 10.369 vs. 7.75

3.51 (2, 26)

3.53 $5/2.5 = 2$ vs. 1.976

3.54 1.333; 1.467

3.55 45

3.56

k	$\bar{y} \pm ks$	Approximate Fraction in Interval
1	415, 425	$\approx$ .68
2	410, 430	$\approx$ .95
3	405, 435	$\approx$ 1.00

3.57 (a) .025
(b) .84

3.58 (b) .66, 1.387
(c) .95; .96; yes; yes

3.59 (a) at least 3/4 have between 145 and 205 teachers
(b) .16

3.60 $.5/4 = .125$ vs. .132

3.61 2.652, .2518

3.62 1.52 and 25.177 yield 2.652 and .2518

3.63 17.105, .0173

3.64 1.05, 1.73 yield 17.105, .0173

3.67 (b) 23.65
(c) 1; 1.368

3.68 (b) approximately 90%

Chapter 4

4.1 A: {4}, B: {2, 4, 6}, C: {1, 2} D: {4}, E: {2, 4, 6}, F contains no sample points; 1/6; 1/2; 0

4.2 (a) .21
(b) .91

4.3 (a) (NDQ), (NDF), (NQF), (DQF)
(b) 3/4
(c) 3/4

4.4 (a) choose 2 of 4 cans without replacement
(b) $\{N_1 N_2\}$, $\{N_1 W_1\}$, $\{N_1 W_2\}$, $\{N_2 W_1\}$, $\{N_2 W_2\}$, $\{W_1 W_2\}$ (N = no water; W = water)
(c) 1/6

4.5 (a) (choice of points 1, 2, 3):
(111), (112), (113), (121), (122), (123), (131), (132), (133), (211), (212), (213), (221), (222), (223), (231), (232), (233), (311), (312), (313), (321), (322), (323), (331), (332), (333)
(b) (123), (132), (213), (231), (312), (321)
(c) $P(E_i) = 1/27$, $P(A) = 2/9$

4.6 (a) rank A, B, C
(b) (ABC), (ACB), (BAC), (BCA), (CAB), (CBA)
(c) 1/3, 1/3

4.7 (a) 2/27
(b) 20/27

4.8 (a) assign four men, one to each of the four jobs
(b) (1234), (1243), (1324), (1342), (1423), (1432), (2134), (2143), (2314), (2341), (2413), (2431), (3124), (3142), (3214), (3241), (3412), (3421), (4123), (4132), (4213), (4231), (4312), (4321); (1 and 2 minority)
(c) 1/6

4.9 (a) choose 2 of 5 cans without replacement

4.9 (*cont.*)

(b) $(N_1\ N_2)$, $(N_1\ N_3)$, $(N_1\ W_1)$,
$(N_1\ W_2)$, $(N_2\ N_3)$, $(N_2,\ W_1)$,
$(N_2\ W_2)$, $(N_3\ W_1)$, $(N_3\ W_2)$,
$(W_1\ W_2)$ (N = no water,
W = water)

(c) 1/10

4.10   1/6, 1/18

4.11   3/10

4.12   (a) (1111), (1112), (1113), (1121),
(1122), (1123), (1131), . . .,
(3333); 81 points in all

(b) (3331), (3332), (3333): 3 from
the latest batch.

(c) (3331), (3332), (3313), (3323),
(3133), (3233), (1333), (2333)

(d) all of (c) plus (3333)

(e) 1/81, 1/27, 8/81, 1/9

4.13   (a) select 2 from 5 without
replacement

(b) (12), (13), (14), (15), (23),
(24), (25), (34), (35), (45)

(c) 1/10

4.14   (a) E_1: A vs. B; C vs. D
E_2: A vs. C; B vs. D
E_3: B vs. C; A vs. D

(b) 1

4.15   (a) 1

(b) 1/2

(c) 1/3

(d) 1/2

(e) 1/3

(f) 0

(g) 0

(h) 1/2

(i) 1

(j) 5/6

4.16   (a) no sample points

(b) 5

4.17   0; 1

4.18   (a) 1/2

(b) 2/3

(c) 1

(d) 0

(e) 0

4.19   (a) dependent

(b) not mutually exclusive

4.20   (a) dependent

(b) mutually exclusive

4.21   (a) dependent

(b) mutually exclusive

4.22   (a) 3/8

(b) 5/8

(c) 3/4

(d) 1/4

(e) 2/5

(f) no; E_4 and E_6 are in A and B

(g) $P(A|B) \neq P(A)$ and
$P(B|A) \neq P(B)$; no

4.23   (a) .4

(b) .37

(c) .10

(d) .67

(e) .6

(f) .33

(g) .90

(h) .27

(i) .25

4.25   (a) .45

(b) .27

(c) .78

4.26   (a) (CW), (CC), (WC), (WW)

(b) .64

(c) .04

(d) .32

(e) .36

4.27   (c) $P(\overline{W}_1\ \overline{W}_2) = .64$,
$P(\overline{W}_1\ W_2) = .16$,
$P(W_1\ \overline{W}_2) = .19$,
$P(W_1\ W_2) = .01$

(d) .64

(e) .83

(f) .99

4.28   (a) .0005

(b) .0000001

(c) $2\ (10^{-11})$

4.29   (a) 5/33

(b) 7/22

(c) 35/132

(d) 35/132

(e) 70/132

4.30   (a) .99

(b) .01

4.31   (a) .000125

(b) .142625

4.32   .016

4.33   .05

4.34   .016

4.35   .335

4.36   .108

4.37   .0214

4.38   .5952

4.39   (a) sample points correspond to a pair of birth dates (d_1, d_2), where d_i, $i = 1, 2$, is the birth date for person i, d_i can assume values 1, 2, 3, . . . , 365

     (b) $1/(365)^2$

     (c) A contains sample points (1, 1), (2, 2), (3, 3), . . . , (365, 365)

     (d) 1/365

     (e) 364/365

4.40   (a) $P(A) = .9918$, $P(B) = .0082$

     (b) $P(A) = .9836$, $P(B) = .0164$

4.41   .012

4.42   .0063

4.43   .25

4.44   .3130

4.45   .9412

4.46   .6667

4.47   .6585

4.48   (a) mn rule

     (b) 9

4.49   (a) combinations

     (b) 15

4.50   (a) permutations

     (b) 30

4.51   720

4.52   12,144

4.53   3,628,800

4.54   5040

4.55   8

4.56   (a) 1/1000

     (b) 1/8000

4.57 $5.720645 \ (10^{12})$

4.58   3; 9; $3^{10} = 59,049$

4.59 $380,204,032 = (52)^5$

4.60   2,598,960

4.61   1/56

4.62   (a) 49,995,000

     (b) .00000012

4.63   (a) 5^{20}

     (b) 5^{-20}

4.64   (a) 25

     (b) 4/25

4.65   (a) .0001049

     (b) learning occurred

4.66   1/120; 7/24

4.67   (a) $1/14 = .0714$

     (b) .2381

     (c) .4396

     (d) .4696

     (e) .4506

4.68 $4/7 = .571$

4.69 $2^{24} = 16,777,216$

4.70   2/3

4.71   (a) (C_1C_2), (C_1A_1), (C_1A_2), (C_2C_1), (C_2A_1), (C_2A_2), (A_1C_1), (A_1C_2), (A_1A_2), (A_2C_1), (A_2C_2), (A_2A_1)

     (b) (C_1C_2), (C_1A_1), (C_1A_2), (C_2C_1), (C_2A_1), (C_2A_2)

     (c) (C_1A_1), (C_1A_2), (C_2A_1), (C_2A_2), (A_1C_1), (A_1C_2), (A_2C_1), (A_2C_2),

     (d) (A_1A_2), (A_2A_1)

4.72   (b) $(HHHT)$, $(HHTH)$, $(HTHH)$, $(THHH)$

     (c) 1/16, 1/4

4.73   identical to coin-tossing problem, exercise 4.7; H represents customer preference for style 1, T corresponds to style 2

     (c) $(HHHH)$, $(TTTT)$

     (d) 1/8

4.74   (a) S contains 21 sample points

     (b) A contains 6 sample points

     (c) 2/7

4.75   1/6

4.76   1/4; 1/12; 0; 1/3

4.77   .3913

4.78   1/2; 2/3; 1/3; 5/6; 1/6; 0; 2/3

4.79   1/27; 8/81; 2/81; 1/9; 1/9; 1/27; 1/9

4.80 $P(AB) = P(A)P(B) > 0$

4.81   0; 1/3

4.82   (a) .73

     (b) .27

4.83   (a) .095

     (b) .855

     (c) .005

     (d) .955

4.84   (a) .000125

     (b) .1354

     (c) .1426

     (d) P(person experiences syndrome) $> .05$

4.85   .5; no

4.86   (a) 5/6
   (b) 25/36
   (c) 11/36
4.87   .999999
4.88   (a) .8
   (b) .64
   (c) .36
4.89   1; .4783; .00781
4.90   .0256; .1296
4.91   .6
4.92   .0625; yes
4.93   .2; .4
4.94   5/6
4.95   .8704
4.96   4
4.97   (a) .128
   (b) .488
4.98   (a) 1/5
   (b) 2/5
   (c) 1/2
4.99   (a) 1/8
   (b) 1/64
   (c) not necessarily; they could
     have studied together, etc.
4.100   36

4.101   12
4.102   20,000
4.104   (a) 6
   (b) 15
4.105   12
4.106   120; 180
4.107   18!/9!
4.108   42
4.109   1/28
4.110   (a) spade royal flush
   (b) royal flush or spade flush
   (c) .0004952
   (d) .000001539
   (e) .0000003848
   (f) .0004964
4.111   .000778; .0000003848; .0004964
4.112   .63
4.113   1/42
4.114   (a) 15/28
   (b) 55/56
4.115   (a) 45
   (b) 10
   (c) 10/45 = 2/9
4.116   8/15; 16/31
4.117   yes; $4!(3!)^4/12!$

Chapter 5

5.1   (a) D
   (b) C
   (c) C
   (d) C
   (e) D
5.2   (a) C
   (b) D
   (c) D
   (d) D
   (e) D
5.3   (a) D
   (b) C
   (c) C
   (d) D
   (e) D
5.4   (a) C
   (b) C
   (c) D
   (d) D
   (e) C
5.5   (a) .2
5.6   (a) (B_1B_2), (B_1W_1), (B_1W_2),
    (B_2W_1), (B_2W_2), (W_1W_2);

   $P(E_i) = 1/6$, $i = 1, 2, \ldots, 6$
   (c) $p(0) = 1/6$; $p(1) = 2/3$;
    $p(2) = 1/6$
5.7   (c) $p(0) = .64$; $p(1) = .32$;
    $p(2) = .04$
5.8   (a) (S), (FS), (FFS), $(FFFS)$
   (b) $p(1) = p(2) = p(3) = p(4) =$
    $1/4$
5.9   (a)

T	$p(T)$	T	$p(T)$
2	1/36	8	5/36
3	2/36	9	4/36
4	3/36	10	3/36
5	4/36	11	2/36
6	5/36	12	1/36
7	6/36		

5.10 $p(0) = 1/5$, $p(1) = 3/5$,
   $p(2) = 1/5$
5.12   (a) $p(0) = .857375$, $p(1) =$
    .136375, $p(2) = .007125$,
    $p(3) = .000125$
   (c) .00725

5.13 (a) .1; .09; .081
      (b) $p(y) = (.9)^{y-1}(.1)$
5.14 $p(3) = .28$; $p(4) = .3744$;
      $p(5) = .3456$
5.15 $p(y) = (.9)^y(.1)$ for $y = 0, 1,$
      $2, \ldots$
5.16 $E(y) = 1.7$; $\sigma^2 = .81$
5.17 $E(y) = 1.5$
5.18 (a) 2.15
      (b) 1.5275
      (d) .95
5.19 $1500
5.20 (a) 7.9
      (b) 2.1749
      (c) .96
5.21 $2050
5.22 2960
5.23 13,800.388
5.24 (a) 4.0656
      (b) 4.125
      (c) 3.3186
5.25 19.559
5.26 (a) 24
      (b) 10
      (c) 1
      (d) 1
5.27 15; (12), (13), (14), (15), (16),
      (23), (24), (25), (26), (34), (35),
      (36), (45) (4, 6), (5, 6)
5.28 1.731×10^{13}
5.29 (a) D
      (b) C
      (c) D
      (d) D

(e) C
5.30 (a) 2.5
      (b) 7.5
      (c) 1.25; 1.118
5.31 $\sigma^2 = 2.9167$; 1.0
5.32 $\mu = 1$; yes
5.33 $\sigma^2 = .4$
5.35 1.0 vs. $\geq 3/4$
5.36 8333.333
5.37 $p(0) = .0256$; $p(1) = .1536$,
      $p(2) = .3456$; $p(3) = .3456$,
      $p(4) = .1296$; .4752
5.38 $p(0) = .008$, $p(1) = .096$,
      $p(2) = .384$, $p(3) = .512$; .992
5.39 2.4; .48
5.40 (a) 7
      (b) 5.8333
      (d) $34/36 = .944$
5.41 187.20
5.42 1023.7; 40.9%
5.43 $p(y) = C_y^2 C_{3-y}^4 / C_3^6$, $y = 0, 1, 2$
5.44 (b) .135
      (c) .947
5.45 no, $y = 18$ lies more than 10
      standard deviations above the
      mean
5.46 (a) .084; .125
      (b) no
5.47 no
5.48 yes
5.51 .5
5.52 1.25
5.53 (a) 2/7
      (b) 5/7

Chapter 6

6.1 not binomial; non-independent
     trials, p varies from trial to trial
6.2 binomial; $n = 2$, $p = 3/5$
6.3 more than two outcomes per trial;
     not binomial

6.4

$p = .1$		$p = .5$		$p = .9$	
y	p(y)	y	p(y)	y	p(y)
0	.5314	0	.0156	0	.000001
1	.3543	1	.0938	1	.0001
2	.0984	2	.2344	2	.0012
3	.0146	3	.3125	3	.0146
4	.0012	4	.2344	4	.0984
5	.0001	5	.0938	5	.3543
6	.000001	6	.0156	6	.5314

6.5

$p = .2$		$p = .5$		$p = .8$	
y	p(y)	y	p(y)	y	p(y)
0	.32768	0	.03125	0	.00032
1	.40960	1	.15625	1	.00640
2	.20480	2	.31250	2	.05120
3	.05120	3	.31250	3	.20480
4	.00640	4	.15625	4	.40960
5	.00032	5	.03125	5	.32768

6.6 binomial; $n = 12$, $p = 1/3$
     (assuming independent injuries)
6.7 .884; .572
6.8 (a) .0081
     (b) 4116

6.9 (*cont.*)
   (c) .2401

6.10 $p(2) = .1240; P[y \geq 2] = .1497;$
   no

6.11 (a) .9606
   (b) .9994

6.12 .0037; .3439; $p < .9$ because
   $P(y \geq 2) = .0523$

6.13 (a) .3277
   (b) .4096
   (c) .0067

6.14 their performance would seem to
   be below the 80% rate of success

6.15 (a) .1681
   (b) .5282

6.16 (a) .0523
   (b) .0486

6.17 (a) 10; 3
   (b) 50; 5
   (c) 90; 3
   (d) 1; .995

6.19 (a) 24
   (b) 4.898
   (c) no, $y = 40$ lies more than 3
     standard deviations above the
     mean

6.20 (a) 3.2
   (b) 1.780
   (c) perhaps $p > .01$

6.21 600; 420; 20.4939;
   $P(y > 700 \leq .04)$

6.22 observed 71% is more than 16
   standard deviations away from
   50%; consequently, probability is
   very small

6.23 no; the probability of observing
   62% is very small, because the
   observed 62% lies more than three
   standard deviations away from
   50%; yes

6.24 $\mu = 1875; \sigma^2 = 468.75;$
   $\sigma = 21.651$

k	$\mu \pm k\sigma$	Tchebysheff's Theorem
2	1832, 1918	$\geq 3/4$
3	1810, 1940	$\geq 8/9$

6.25 (a) 800
   (b) 12.649
   (c) no, $y = 900$ is nearly 8
     standard deviations above the
     mean

6.26 (a) 5000
   (b) 50
   (c) no

6.27 $\mu = 100; \sigma^2 = 90; \mu \pm 2\sigma:$
   $81 \leq y \leq 119$

6.28 (a) .987
   (b) .213
   (c) .788

6.29

a	$\sum\limits_{y=0}^{a} p(y)$
0	.03125
1	.18750
2	.50000
3	.81250
4	.96875

6.30 $p(3) = .312$

6.31 .469

6.32 (a) .590
   (b) .168
   (c) .031
   (d) 1
   (e) 0

6.33 (a) .919
   (b) .528
   (c) .188
   (d) 1
   (e) 0

6.34 (a) .349
   (b) .028
   (c) .001
   (d) 1
   (e) 0

6.35 (a) .736
   (b) .149
   (c) .011
   (d) 1
   (e) 0

6.36 for a given value of n and p,
   P(accept) increases as a increases;
   for a given value of a and p,
   P(accept) decreases as n increases

6.37 (b) 3
   (c) 1

6.38 $a = 1$

6.39 $n = 25, a = 1$

6.40 (a) $H_0: p = 1/2; H_a: p \neq 1/2$

(b) two-tailed

(c) .125

(d) .7518

6.41 (a) $H_0: p = 1/2$; $H_a: p > 1/2$

    (b) one-tailed

    (c) 0625

    (d) .3439

6.42 (a) $H_0: p = .3$; $H_a: p \neq .3$

    (b) two-tailed

    (c) $y = 0$ or $y = 10$ with $\alpha = .028$

    (d) no

6.43 (a) .151

    (b) .302

6.44 (a) .036

    (b) .352

6.45 (a) .118, larger

    (b) .164

6.46 $H_0: p = .5$; $H_a: p > .5$; yes, for rejection region $y \geq 8$ with $\alpha = .055$

6.47 (a) $H_0: p = .48$; $H_a: p > .48$

    (b) reject H_0 using rejection region $y \geq 12$ with $\alpha = .0268$

6.48 (a) $H_0: p = .5$; $H_a: p \neq .5$

    (b) rejection region: $y \leq 4$

    (c) program was ineffective

6.49 (a) $H_0: p = .5$; $H_a: p \neq .5$

    (b) .03125

    (c) .7378

6.50 yes, with $\alpha = .059$

6.51 (a) $y \leq 1$ or $y \geq 6$ with $\alpha = .125$

    (b) .1497

    (c) do not reject H_0

6.52 (a) .077

    (b) $y = 10$ is not in the rejection region; do not reject H_0

6.53 .922; .078

6.54 (a) $H_0: p = .5$

    (b) $H_a: p \neq .5$

    (c) .063

    (d) .410

6.57 (a) $p(0) = .125$; $p(1) = .375$; $p(2) = .375$; $p(3) = .125$

    (c) 1.5; .866

    (d)

k	$\mu \pm k\sigma$	Probablity
1	.63, 2.37	.75
2	$-.23, 3.23$	1.00

6.58 (a) $p(0) = .729$; $p(1) = .243$; $p(2) = .027$; $p(3) = .001$

    (c) .3; .520

    (d) .729; .972

6.59 (a) .9606

    (b) .9994

6.60 (a) .851

    (b) .829

    (c) .011

6.61 (a) .251

    (b) .092

    (c) .205

6.62 (a) .651

    (b) .772

    (c) .909

6.63 (a) 1.000

    (b) .588

    (c) .021

6.64 (a) .234

    (b) .136

6.65 (a) .012

    (b) .630

    (c) trials could be dependent

6.66 $P(y = 8) = .115$; $P(y \geq 8) = .228$

6.67 3000; no

6.68 no; $y = 1373$ lies more than 16 standard deviations away from μ when split is 50–50

6.69 (a) 805

    (b) 22.875

    (c) 759.251 to 850.749

    (d) no; 249 lies more than 24 standard deviations below the mean

6.70 (a) (25, 5)

    (b) (25, 5)

6.71 (a) .983

    (b) .736

    (c) .392

    (d) .069

6.72 (a) 30,000

    (b) 122.474

    (c) highly unlikely if $p = .5$

    (d) yes

6.73 (a) 20

    (b) 4

    (c) .006

    (d) theory is incorrect

6.74 11

6.75   .017; do not support

6.76   (b) $H_0: p = .25$; $H_a: p < .25$

     (c) no; do not reject H_0 with

$\alpha = .032$

6.81   .04

6.82   .987

Chapter 7

7.1   (a) .4192
    (b) .4599

7.2   (a) .2734
    (b) .2734

7.3   (a) .8301
    (b) .1452

7.4   (a) .6826
    (b) .9544
    (c) .9974

7.5   (a) .6753
    (b) .2694
    (c) .2694

7.6   (a) .6260
    (b) .5672
    (c) .2638

7.7   1.96

7.8   .86

7.9 -1.12

7.10 $-.55$

7.11 -1.96

7.12   .44

7.13   .85

7.14   .74

7.15   1.35

7.16 $-.97$

7.17   1.67

7.18 -1.645

7.19   1.645

7.20   2.575

7.21   (a) .0618
    (b) .7794

7.22   (a) 1.27
    (b) .1020

7.23   .1562; .0012

7.24   .0505

7.25   (a) .1335
    (b) .8665
    (c) .1335

7.26   .0668

7.27   (a) .0778
    (b) .0274

7.28   .0401

7.29   .0475

7.30 $\mu = 1940.119$

7.31   63,550

7.33   3.5; .765

7.34   3.5; .541

7.35   (a) 1; .112
    (b) .0367
    (c) ≈ 0
    (d) .0734

7.36   (a) 106; 2.4
    (b) .0475
    (c) .9050

7.37   (b) $\sigma_{\bar{y}} = .4472$
    (c) $\sigma_{\bar{y}} = .3162$
    (d) $\sigma_{\bar{y}} = .2000$

7.38   (b) $\sigma_{\bar{y}} = .1297$
    (c) $\sigma_{\bar{y}} = .0917$
    (d) $\sigma_{\bar{y}} = .058$

7.39   (b) $\sigma_{\bar{y}} = .4472$
    (c) $\sigma_{\bar{y}} = .3162$
    (d) $\sigma_{\bar{y}} = .2000$

7.40   (a) .3758
    (b) no

7.41   approximately normal; 1500;
    35.355

7.42   (a) ≈ 0
    (b) .0764
    (c) .9236

7.43   (a) normal
    (b) .0571
    (c) $\mu = 21.948$

7.44   (a) 1890; 69.282
    (b) .0559

7.46   (a) .546
    (b) .5468

7.47   (a) .245
    (b) .2483

7.48   (a) .9504
    (b) .0708
    (c) no; 30 lies 1.37 standard
       deviations above the mean

7.49   .1251

7.50 ≈ 0; yes

7.51   .0901; no

7.52   (a) $H_0: p = .3$; $H_a: p > .3$
    (b) reject H_0 if $z \geq 1.645$
       ($y \geq 40.9$)
    (c) do not reject H_0

7.53 $z = 1.461$
7.54 .9441
7.55 ≈ 0
7.56 $z = 2.882$; yes
7.57 (a) .3849
　　　(b) .3159
7.58 (a) .4279
　　　(b) .1628
7.59 (a) .3227
　　　(b) .1586
7.60 .7734
7.61 .9115
7.62 $z_0 = 0$
7.63 .8612
7.64 $z_0 = .6745$
7.65 (a) .1056
　　　(b) .8944
　　　(c) .1056
7.66 no
7.67 .16
7.68 5.065
7.69 3.44%
7.70 (a) .5780
　　　(b) .5752
7.71 85.36

7.72 (b) 3.5; .987
7.74 .0217
7.75 $z = -1.26$; no
7.76 (a) no
　　　(b) .0179
7.77 7.301
7.78 $z = -1.96$; $y \leq 28$ or $y \geq 52$
7.79 .117
7.80 every 383.5 hours
7.81 (a) 141
　　　(b) .0401
7.82 .8980
7.83 .9474
7.84 (a) .178
　　　(b) .392
7.85 (a) .1725
　　　(b) .4052
7.86 .1922
7.87 .3557
7.88 limit total number of passengers
　　　to $n = 11$
7.89 (a) 35
　　　(b) 4.770
　　　(c) $z = 2.935$; yes
7.93 .8944

Chapter 8

8.1 1.265; at least 3/4
8.2 (a) range/4 = 2.65
　　　(b) 1.7 ± .608
8.3 39.8° ± 4.768°
8.4 7.2% ± .776%
8.5 1280 ± 27.832
8.6 .1587
8.7 (a) .8098
　　　(b) .9356
8.8 (a) 619.806
　　　(b) approximately .95
8.9 (a) 3.92
　　　(b) 2.772
　　　(c) 1.96
8.10 (a) decreases by $1/\sqrt{2}$
　　　(b) decreases by 1/2
8.11 (a) 3.29
　　　(b) 5.16
　　　(c) increases the width
8.12 5000 ± 828.436
8.13 7.6 ± .806
8.14 64.3 ± 5.239
8.15 .145 ± .0015

8.16 3.7 ± .204; random sampling
8.17 81 ± 3.326
8.18 −1.3 ± .301
8.19 −1.4 ± .342
8.20 26.2 ± 10.737
8.21 −.7 ± .332
8.22 7.9 ± 1.152
8.23 (a) 7.2 ± .751
　　　(b) 2.5 ± .738
8.24 2.8 ± .443
8.25 .74 ± .027
8.26 .95 ± .014; since $\hat{p}$ is closer to 1,
　　　$\hat{p}\hat{q}$ and hence $\hat{\sigma}_{\hat{p}}$ are smaller
8.27 .629 ± .012
8.28 .27 ± .035
8.29 .7 ± .045
8.30 (a) .40 ± .016
　　　(b) .55 ± .017
　　　(c) .2 ± .018
8.31 .63 ± .113
8.32 (a) no; this is not a random
　　　sample
　　　(b) .145 ± .035

8.33   (a) $-.075 \pm .049$
      (b) independent random samples

8.34   (a) $-.06 \pm .120$
      (b) $.07 \pm .091$
      (c) $.13 \pm .084$

8.35 $.15 \pm .064$

8.36 $.11 \pm .093$

8.37 $.16 \pm .099$

8.38 $.08 \pm .0505$

8.39   94

8.40   897

8.41 $n_1 = n_2 = 94$

8.42 $n_1 = n_2 = 1697$

8.43   (a) 2401

8.44   97

8.45   666

8.46   70

8.47   97

8.48   136

8.49 $n_1 = n_2 = 98$

8.50 $n_1 = n_2 = 301$

8.51 $29.1 \pm .9555$

8.52 $29.1 \pm .802$

8.53   234

8.54 $4.2 \pm .333$

8.55 $n_1 = n_2 = 224$

8.56   (a) $.48 \pm .044$
      (b) $.48 \pm .037$

8.57   1083

8.58 $.1375 \pm .1532$

8.59 $n_1 = n_2 = 376$

8.61 9.7 ± 2.274

8.62 9.7 ± 1.908

8.63 $.1 \pm .025$

8.64 $.6170$

8.65   97

8.66   68

8.67 $.8 \pm .047$

8.68   (a) $.0625 \pm .0237$
      (b) 563

8.69 22 ± 1.553

8.70   9604

8.71   65

8.72   193

8.73 $.07 \pm .096$

8.74 $34 \pm .588$

8.75   6147

8.76 $.1 \pm .041$

8.77 $.387; .651$

8.78 $.58 \pm .018$

Chapter 9

9.1   (a) $H_a: \mu > 30$
      (b) $H_0: \mu = 30$
      (c) $\alpha = .05$
      (d) yes; $z = 2.372$

9.2 $H_0: \mu = 30$; $H_a: \mu < 30$; one-tailed

9.3 $H_0: \mu = 30$; $H_a: \mu \neq 30$; two-tailed

9.4   (a) $H_0: \mu = 60$; $H_a: \mu < 60$
      (b) one-tailed
      (c) yes; $z = -1.992$

9.5   (a) $H_0: \mu = 80$
      (b) $H_a: \mu \neq 80$
      (c) $z = -3.75$; reject H_0

9.6   (a) $H_0: \mu = 7.5$
      (b) $H_a: \mu < 7.5$
      (d) $z = -5.477$; reject H_0

9.7   no; $z = -.992$

9.8   (a) $H_0: \mu_1 - \mu_2 = 0$
         $H_a: \mu_1 - \mu_2 > 0$
      (b) one-tailed
      (c) $z > 1.28$
      (e) reject H_0; $z = 2.381$

9.9 $H_0: \mu_1 - \mu_2 = 0$;
     $H_a: \mu_1 - \mu_2 < 0$

9.10   (b) $H_0: \mu_1 - \mu_2 = 0$;
        $H_a: \mu_1 - \mu_2 \neq 0$
      (c) reject H_0: $z = 2.381$

9.11 $\alpha = .20$

9.12   (a) $H_0: \mu_1 - \mu_2 = 0$;
        $H_a: \mu_1 - \mu_2 > 0$
      (b) one-tailed
      (d) reject H_0; $z = 2.074$

9.13   (a) $H_0: \mu_1 - \mu_2 = 0$;
        $H_a: \mu_1 - \mu_2 \neq 0$
      (b) two-tailed
      (c) no; $z = -.954$

9.14   (a) $H_0: \mu_1 - \mu_2 = 0$;
        $H_a: \mu_1 - \mu_2 < 0$
      (b) no; $z = -.426$

9.15   (a) $H_a: \mu_1 - \mu_2 > 0$;
      (b) one-tailed
      (c) reject H_0; $z = 4.899$

9.16   (a) $H_0: p = .4$; $H_a: p < .4$
      (b) one-tailed
      (c) yes; $z = -2.739$

9.17 (a) $H_0: p = .05; H_a: p < .05$
     (b) one-tailed
     (d) do not reject H_0; $z = -1.538$

9.18 (a) $H_0: p = .60; H_a: p > .60$
     (b) yes
     (c) yes; $z = 2.489$

9.19 (a) $H_0: p = 2/3; H_a: p \neq 2/3$
     (b) two-tailed
     (d) do not reject H_0; $z = 1.739$

9.20 yes; $z = -21.033$

9.21 yes; $z = -3.926$; reject H_0

9.22 (a) yes
     (b) no

9.23 (a) $H_0: p_1 - p_2 = 0$;
       $H_a: p_1 - p_2 \neq 0$
     (b) two-tailed
     (c) $z = -.919$; do not reject H_0

9.24 (a) $H_0: p_1 - p_2 = 0$;
       $H_a: p_1 - p_2 < 0$
     (b) one-tailed
     (c) $z = -.919$; do not reject H_0

9.25 (a) $H_0: p_1 - p_2 = 0$;
       $H_a: p_1 - p_2 \neq 0$
     (c) no; $z = 1.621$

9.26 approximately 0; yes; $z = 4.960$;
     reject H_0; yes

9.27 $H_0: p_1 - p_2 = 0$; yes,
     $z = 14.017$; reject H_0

9.28 (a) .0359
     (b) reject H_0

9.29 $p = .0087$; $\alpha = .10$, reject H_0

9.30 $p = .0031$; $\alpha = .10$, reject H_0

9.31 $p = .3576$; $\alpha = .05$, do not reject
     H_0

9.32 (a) $H_0: \mu = 3; H_a: \mu > 3$
     (b) one-tailed
     (c) $p = .0113$
     (d) yes

9.33 (a) no; $p = .1587$
     (b) no
     (c) $p = .0336$

     (d) yes

9.36 (a) $p = .0025$
     (b) reject H_0

9.37 yes, $z = -25.298$; reject H_0

9.38 (a) $p < .001$
     (b) reject H_0; conclude that more
       than 1/3 prefer A

9.39 random sample; either σ known or
     sample size 30 or more

9.40 no, $z = -1.176$; do not reject H_0

9.41 yes, $z = 2.858$

9.42 no, $z = 1.684$

9.43 .1151, .0228

9.44 (a) $p = .0668$
     (b) do not reject H_0

9.45 309

9.46 yes, $z = 4.00$

9.47 yes, $z = 4.00$

9.48 yes, $z = 5.237$

9.49 (a) yes, $z = 4.333$
     (b) 13 ± 5.88

9.50

μ	β
873	.3446
875	.6103
877	.8274

9.51

μ	β
873	.6141
875	.7794
877	.8906

9.52 (a) $p < .002$
     (b) reject H_0

9.53 (a) yes, $z = -10.637$
     (b) $-.093 \pm .017$

9.54 (a) no, $z = -.915$
     (c) $1.37 \pm .111$
     (d) $-.07 \pm .126$

9.55 yes, $z = 2.465$

Chapter 10

10.1 (a) $H_0: \mu = 6; H_a: \mu > 6$
     (b) reject H_0 if $t > 1.44$
     (c) reject H_0; $t = 3.649$

10.2 $6.4 \pm .213$

10.3 (b) do not reject H_0; $t = -.784$

10.4 (a) no, $t = -.942$
     (b) $p > .10$

     (c) no, $t = -1.296$
     (d) $p > .10$

10.5 $85.733 \pm .744$

10.6 no, $t = -.647$

10.7 28.935 ± 3.676

10.8 32.333 ± 30.220

10.9 yes, $t = 5.985$; reject H_0

10.10 $3.782 \pm .1296$

10.11 $s = .1812$; $n = 1273$

10.12 $4.26 \pm .715$

10.13   no, $t = -1.195$; do not reject H_0

10.14 16.7 ± 1.089

10.15   (a) H_0: $\mu_1 - \mu_2 = 0$;
        H_a: $\mu_1 - \mu_2 \neq 0$
        (b) reject H_0 if $|t| > 1.706$
        (c) $t = 2.879$
        (d) $p < .01$
        (e) reject H_0

10.16 2.6 ± 1.541

10.17   (a) yes, $t = -6.127$; reject H_0
        (b) $p < .01$

10.18   (a) yes, $t = -4.535$; reject H_0
        (b) $p < .01$

10.19   (a) no, $t = 1.606$; do not reject H_0
        (b) $.14 \pm .201$

10.20   (a) yes, $t = -3.354$; reject H_0
        (b) $p < .01$

10.21 -6.3 ± 3.946

10.22   (a) approximately 136

10.23   (a) no, sample sizes permit use of z test
        (b) yes, $t = 3.801$; reject H_0

10.24   yes, $t = 3.968$

10.25 3120 ± 1019.183

10.26   (a) no, $t = 1.177$
        (b) $p > .20$
        (c) $.06 \pm .142$

10.27   (b) yes, $t = -1.834$
        (c) $.025 < p < .05$
        (d) $-.063 \pm .062$

10.28   (a) no, $t = -1.772$
        (b) $.10 < p < .20$
        (c) $-.022 \pm .031$

10.29   (a) yes, $t = -4.326$; reject H_0
        (b) -1.58 ± 1.014

10.31   approximately 65

10.32   (a) yes, $t = 2.821$; reject H_0
        (b) 1.4875 ± 1.247 (a 95% confidence interval)
        (d) yes

10.33 42.125 ± 6.280

10.34   (a) no, $\chi^2 = 26.11$
        (b) $p > .10$

10.35 $.00703 < \sigma^2 < .03264$

10.36   (a) no, $\chi^2 = 12.8618$

(b) $p > .10$

10.37 $(.227, 2.194)$

10.38 $\chi^2 = 22.449$, reject H_0: $\sigma^2 = .49$

10.39 $(1.408, 31.264)$

10.40   (a) no, $t = -.232$; do not reject H_0
        (b) yes, $\chi^2 = 19.98$; reject H_0

10.41 $(.00476, .00980)$

10.42   (a) no, 1000 is 5 standard deviations away from 800
        (b) yes, $z = 3.262$; reject H_0
        (c) yes; $\chi^2 = 57.281$; for $\alpha = .05$, reject H_0

10.43   (a) no, $F = 1.333$
        (b) $p > .20$

10.44   (a) yes, $F = 2.486$; reject H_0
        (b) $p < .005$

10.45 $(1.544, 4.003)$ (using $F_{49,49} \approx 1.61$)

10.46   (a) $\sigma_1^2 = \sigma_2^2$
        (b) no, $F = 5.897$

10.47   (a) no, $F = 2.904$
        (b) $(.050, .254)$

10.48   (a) yes, $F = 2.197$

10.50   for $\alpha = .05$, yes; $t = 2.108$; $.025 < p < .05$

10.51 11.3 ± 1.44

10.52 77.763 ± 6.337

10.54   yes, $F = 3.268$

10.55   yes, $t = 2.2$

10.56   no, $t = -1.712$

10.57   yes, $t = 2.497$

10.58 $-.75 \pm .772$

10.59 $-.875 \pm .829$

10.60 $-.875 \pm .967$; yes

10.61 $t = 9.568$; reject H_0

10.62 $.259 \pm .047$ unpaired;
        $.259 \pm .050$ paired

10.63   when within-group variation is less than between-group variation

10.64   no, $t = 2.571$

10.65 $.01 \pm .010$

10.66   (a) yes, $t = -3.408$
        (b) $1.385 \pm .068$

10.68   (a) two-tailed; H_a: $\sigma_1^2 \neq \sigma_2^2$
        (b) lower-tailed; H_a: $\sigma_1^2 < \sigma_2^2$
        (c) upper-tailed; H_a: $\sigma_1^2 > \sigma_2^2$

10.69 $F = 1.922$; $p > .20$

10.70 $(.781, 4.728)$

10.71   yes, $F = 2.407$

10.72   no, $\chi^2 = 7.008$
10.73   (.185, 2.465)
10.74   no, $t = -1.8$
10.75   yes; $t = 2.425$
10.76 15.967 ± 7.141
10.77   (a) $t = -2.657$; $.02 < p < .05$
      (b) -4.542 ± 3.046
10.78 -10.186 ± 1.228
10.79   yes, $t = -2.945$

10.80   no, population distribution not normal
10.81   yes, $t = 2.80$; reject H_0
10.82 $26.4 \pm .877$
10.83   no, population distribution not normal
10.84   yes, $t = 3.038$; reject H_0
10.85   no, $t = 1.862$; do not reject H_0
10.86   (24.582, 73.243)

Chapter 11

11.1 y intercept $= 2$, slope $= 3$
11.2 y intercept $= 2$, slope $= -3$
11.4   same line, but with negative slope; lines are perpendicular
11.5 $y = 1 - 2x$
11.6 $y = -1 + 2x$
11.8   (a) $\hat{y} = 2.6 - .9x$
11.9   (a) $\hat{y} = 1.333 + .314x$
11.10   (a) $\hat{y} = 5.16 - .531x$
11.11   (a) $\hat{y} = 3 + 4.286x$
11.12   (b) $\hat{y} = 34.0092 - .0837x$
      (d) 23.5467
11.13   (a) $\hat{y} = 17.609 - .229x$
      (c) $\hat{y} = 13.029$
11.14   SSE $= 1.1$; $s^2 = .3667$
11.15   SSE $= .419$; $s^2 = .1048$
11.16   SSE $= .0777$; $s^2 = .0194$
11.17   SSE $= 2.5714$; $s^2 = .6429$
11.18   SSE $= 11.098$; $s^2 = 1.585$
11.19   SSE $= 15.400$; $s^2 = 5.133$
11.20   yes, $t = -4.700$
11.21 $-.9 \pm .451$
11.22   yes, $t = 8.124$
11.23 $.314 \pm .107$
11.24   yes, $t = -15.949$
11.25 $-.531 \pm .092$
11.26   yes, $t = 22.361$
11.27 $4.286 \pm .409$
11.28   yes, $t = -4.555$
11.29 $-.084 \pm .035$
11.30   yes, $t = -4.355$
11.31 $-.229 \pm .167$
11.32   (a) $\hat{y} = 3 + .475x$
      (c) 5.025
11.33   yes, $t = 3.791$; no
11.34 $.475 \pm .289$
11.35   (b) $\hat{y} = 5.0032 + .00528x$
      (c) no, $t = .165$
11.36   (a) $\hat{y} = 4.3 + 1.5x$

      (c) 1.531
      (d) yes, $t = 3.834$
      (e) $p < .01$
11.37   (a) $\hat{y} = .9552 + .0462x$
      (c) .00274
      (d) yes, $t = 10.55$
      (e) $.0462 \pm .0079$
      (f) .0462 pounds
11.38   (a) $\hat{y} = .1067 + .0016x$
      (c) .0003239
      (d) no, $t = 1.092$
      (e) $.0016 \pm .00273$
      (f) .0016 pounds
11.39 $1.7 \pm .780$
11.40 $1.019 \pm .294$
11.41 $23.5467 \pm .818$
11.42 12.5 ± 2.002
11.43   (a) $\hat{y} = 452.11935 - 29.40184x$
      (c) 5138.9675
      (d) yes, $t = -5.775$
      (e) $p < .01$
      (f) 305.11 ± 35.773
11.44   (a) 46.343 ± 1.646
11.45   (a) 27.406 ± 3.313
      (b) $76,017.87 \pm 1,341.879$
11.46 1.7 ± 1.625
11.47 $1.019 \pm .7498$
11.48 46.343 ± 4.205
11.49 $1.279 \pm .137$
11.50 $69,988.605 \pm 3,385.344$
11.51   (a) $\hat{y} = -.627 + 1.080x$
      (b) 10.169 ± 2.326
11.52   (a) $\hat{y} = .72 \pm .118x$
      (c) $.118 \pm .008$
      (d) reject H_0; $t = 4.587$
      (e) $7.230 \pm .342$
11.53   (a) $\hat{y} = 4.741 + .096x$
      (b) no, $t = 1.558$
      (d) 7.614 ± 1.498

11.53 (*cont.*)
    (e) 7.614 ± 5.600

11.57 (a) $\hat{y} = 3.0 + 6x$
    (b) positive
    (c) $r = .9487$; $r^2 = .9$

11.58 (a) $\hat{y} = 8.267 - 1.314x$
    (c) $-.982$
    (d) 96.5%

11.59 (a) $\hat{y} = -.933 + 1.314x$
    (c) $.982$
    (d) 96.5%

11.60 (a) $\hat{y} = 307.917 + 34.583x$
    (b) $.874$; $.764$
    (c) 515.415 ± 27.010
    (d) 515.415 ± 97.384
    (e) 76.4

11.61 (a) $\hat{y} = 198.925 + .0385x$
    (c) $.3297$
    (d) 10.9%

11.62 (a) $r = .4101$
    (b) $r^2 = .1682$
    (c) yes

11.63 (a) $r = .6578$
    (b) $r^2 = .4328$
    (c) yes

11.64 (a) $r = -.8801$
    (b) $r^2 = .7746$; only 77% of the variability is accounted for by the linear term, x. Other variables need to be added to the model

11.66 y intercept $= -5/3$, slope $= 2/3$

11.67 (a) $\hat{y} = 2.643 + .554x$

11.68 $.6554$

11.69 yes, $t = 3.618$

11.70 all points fall on a straight line

11.71 σ^2

11.72 $.554 \pm .393$

11.73 $2.089 \pm .880$

11.74 3.75 ± 1.850

11.75 $.851$

11.76 72.36

11.77 (a) $\hat{y} = 46 - .317x$
    (c) 19.033

11.78 $-.317 \pm .102$

11.79 30.167 ± 2.500

11.80 30.167 ± 8.291

11.81 $-.872$

11.82 (a) $\hat{y} = -.333 + 1.4x$
    (c) 1.2410

11.83 yes, $t = 3.974$

11.84 (a) $\hat{y} = -5.497 + .066x$
    (c) $.1800$

11.85 yes, $t = 3.38$

11.86 $.730$

11.87 $2.382 \pm .274$

11.88 $2.382 \pm .816$

11.89 (a) $\hat{y} = .067 + .517x$
    (b) $2.133 \pm .282$

11.90 no, observations are not independent

11.91 (a) $.7738$
    (b) $.5988$
    (c) $\hat{y} = 11.238 + 1.309x$

11.92 (a) $.9803$
    (b) $.9610$
    (c) $\hat{y} = 21.867 + 14.967x$
    (d) yes

11.93 (a) no; $t = 2.066$; $.10 < p < .20$
    (b) $.2992$

11.94 (a) $\hat{y} = 9.590 + .653x$
    (b) 21.346 ± 2.291

11.95 (a) $\hat{y} = .025 + .821x$
    (c) yes, $t = 8.958$
    (d) $.034 \pm .021$

Chapter 12

12.2 (e) changes direction of parabola

12.3 (b) shifts the parabola to the right along the x-axis
    (c) shifts the parabola to the left along the x-axis by the same amount

12.4 (b) parallel lines

12.5 (b) parallel lines

12.6 (b) changes the slopes
    (c) allows the interaction

12.7 (a) negative
    (b) SSE $= 1.0599$; $s^2 = .3533$
    (c) 3
    (d) yes, $F = 332.53$
    (e) $p = .0003$
    (f) yes, $t = -4.49$
    (g) $p = .0206$
    (h) $\hat{y} = -44.192 + 16.334x - .820x^2$
    (i) $R^2 = .9955$

(j) $\hat{y} = 27.3424$

12.10   when $\nu_1 = 1$

12.11   (a) SSE $= .05368$; $s^2 = .00596$

   (b) $\hat{y} = .4381 + .0053x_1 - .0302x_2 + .0007x_1x_2 - .00038x_1^2 + .0004x_2^2$

   (c) $R^2 = .9888$

   (d) yes, $F = 159.23$

   (e) $.0001$

   (f) no

12.12   (a) parabola

   (b) positive

   (c) straight line

   (d) negative

   (e) allows for different curves for different values of x_2

12.13   (a) SSE $= 152.17748$; $s^2 = 8.4543$

   (b) 18

   (c) yes, $F = 31.85$

   (d) $p < .005$; from printout, $p = .0001$

   (e) $R^2 = .898455$; TSS $= SS_y = 1498.625$

   (f) $\hat{y} = 41.262$

12.14   (a) yes

(b) (i) $SSE_1 = 465.1348$; $SSE_2 = 152.1775$

   (ii) $\nu_1 = 3$, $\nu_2 = 18$

   (iii) $F = 12.35$

   (iv) $F_{.05} = 3.16$

   (v) reject H_0

12.15   (a) $E(y) = \beta_0 + \beta_1x_1$

   (b) $E(y) = (\beta_0 + \beta_2) + (\beta_1 + \beta_4)x_1$

   (c) $E(y) = (\beta_0 + \beta_3) + (\beta_1 + \beta_5)x_1$

   (d) β_2

   (e) β_5

   (f) $H_0: \beta_4 = \beta_5 = 0$; $H_a: \beta_i \neq 0$ for some $i = 4, 5$

12.16   (b) 9

   (d) $R^2 = .98397$

   (f) $\hat{y} = 4.10 + 1.04x_1 + 3.53x_2 + 4.76x_3 - .43x_1x_2 - .08x_1x_3$

   (g) yes, $t = -2.613$ (using $s_{\hat{\beta}_4} = .16459$)

   (h) $-.43 \pm .372$

   (i) no, $t = -.486$ (using $s_{\hat{\beta}_5} = .16459$)

12.17   (c) yes, $F = 48.86$

   (d) $.0011$

12.18   reject H_0, $F = 60.506$

Chapter 13

13.2   (a) 4

   (b) $X^2 > 9.4877$

   (c) $H_a: p_i \neq p_j$ for some $i,j = 1, 2, \ldots, 5$.

   (d) $X^2 = 8.00$; do not reject H_0

   (e) $.05 < p < .10$

13.3   yes, $X^2 = 24.48$ ($\chi^2_{.05} = 7.81$)

13.4   no, $X^2 = .658$ ($\chi^2_{.05} = 7.81$)

13.5   yes, $X^2 = 31.77$ ($\chi^2_{.05} = 9.49$)

13.6   no, $X^2 = 13.58$ ($\chi^2_{.10} = 17.275$)

13.7   yes, $X^2 = 173.64$ ($\chi^2_{.05} = 12.59$)

13.8   8

13.9   (a) 2

   (b) $X^2 = 3.059$

   (c) $X^2 > 4.605$

   (d) do not reject H_0

   (e) $p > .10$

13.10   (a) do not reject H_0; $X^2 = 2.344$

   (b) do not reject H_0; $z = 1.531$

13.11   yes, $X^2 = 6.75$ ($\chi^2_{.01} = 6.63$)

13.12   yes, $X^2 = 52.73$ ($\chi^2_{.05} = 5.99$)

13.13   (a) yes, $X^2 = 119.007$ ($\chi^2_{.05} = 5.99$)

   (b) $p < .005$

13.14   (a) yes, $X^2 = 14.496$ ($\chi^2_{.05} = 5.99$)

   (b) $p < .005$

13.15   (a) yes, $X^2 = 22.699$ ($\chi^2_{.05} = 5.99$)

   (b) $p < .005$

13.16   (a) $X^2 = 5.322$

   (b) $X^2 > 5.99$

   (c) do not reject H_0

   (d) $.05 < p < .10$

13.17   (a) yes, $X^2 = 18.53$ ($\chi^2_{.05} = 3.84$)

   (b) yes, $z = 4.30$ ($z_{.05} = 1.645$)

13.18   yes, $X^2 = 8.75$ ($\chi^2_{.05} = 5.99$)

13.19   yes for $\alpha = .10$, $X^2 = 5.49$ ($\chi^2_{.10} = 4.61$)

13.20   yes, $X^2 = 38.43$ ($\chi^2_{.05} = 12.59$)

13.21   (a) no, $X^2 = .685$ ($\chi^2_{.05} = 5.99$)
        (b) no, $X^2 = .670$ ($\chi^2_{.05} = 3.84$)

13.22   no, $X^2 = 4.4$ ($\chi^2_{.10} = 7.78$)

13.23   yes, $X^2 = 21.51$ ($\chi^2_{.01} = 11.34$)

13.24   no, $X^2 = 1.89$ ($\chi^2_{.10} = 4.61$)

13.25   no, $X^2 = 2.87$ ($\chi^2_{.10} = 4.61$)

13.26   yes, $X^2 = 6.18$ ($\chi^2_{.05} = 5.99$)

13.28 $\hat{p} = .5$; 6.25, 25, 37.5, 25, 6.25;
        reject H_0: $X^2 = 8.56$
        ($\chi^2_{.05} = 7.81$)

13.29   yes, $X^2 = 7.48$ ($\chi^2_{.10} = 2.71$)

13.30   (a) yes, $X^2 = 10.27$
            ($\chi^2_{.05} = 3.84$)
        (b) yes, $z = 3.205$ ($z_{.05} = 1.645$)

13.31   yes, $X^2 = 7.62$ ($\chi^2_{.01} = 6.63$)

13.32   yes, $X^2 = 6.19$ ($\chi^2_{.05} = 5.99$)
        .415 ± .057

13.33   yes, $X^2 = 15.85$ ($\chi^2_{.10} = 14.68$)

13.34   yes, $X^2 = 229.78$ ($\chi^2_{.01} = 15.09$)

13.35   yes, $X^2 = 18.82$ ($\chi^2_{.05} = 15.51$)

13.36   yes, $X^2 = 141.03$ ($\chi^2_{.01} = 6.63$)

13.37   (a) yes, $X^2 = 24.776$
            ($\chi^2_{.05} = 9.49$)
        (b) $p < .005$

Chapter 14

14.6   no, people who work during
        these hours will not be included

14.7 $n_1 = n_2 = n/2$, signal amplifica-
        tion

14.8 $n_1 = 34$, $n_2 = 56$

14.9 $n_1 = n_2 = 48$

14.10   for clue to answer, see Figure
        14.3 and its discussion

14.12   yes

14.13   (a) randomized block design

14.14   (a) plots

        (b) six fertilizer combinations
        (c) randomized block design
        (d) noise reduction

14.29   BAC, CBA, ACB

14.30   three observations each at $x = 2$
        and $x = 5$; signal amplification

14.31   (a) 1.46 times as large
        (b) slightly more than twice
            (2.14) as many observations

14.32   test for curvature

14.33   noise reduction (this is a ran-
        domized block design)

Chapter 15

15.1   SSE $= 10.75$; $F = 2.93$; do not
        reject H_0

15.2   (a)

Source	d.f.	SS	MS	F
Method	2	641.88	320.94	5.15
Error	8	498.67	62.33	
Total	10	1140.55		

        (b) yes, $F = 5.15$ ($F_{.05} = 4.46$)

15.3   (a) 76 ± 8.14
        (b) 66.33 ± 10.51
        (c) 9.67 ± 13.30
        (d) no, they are not independent

15.4   (a)

Source	d.f.	SS	MS
Regimens	2	64.31	32.155
Error	25	338.02	13.5208
Total	27	402.33	

        (b) no, $F = 2.38$ ($F_{.05} = 3.39$)

        (c) -5.03 ± 3.48
        (d) 3.69 ± 2.525

15.5   (a) yes, $F = 13.03$ ($F_{.05} = 3.17$)
        (b) -7.5 ± 4.67

15.6   (a) yes, $F = 5.7$ ($F_{.05} = 4.26$);
            $p = .0251$.
        (b) 5.83 ± 5.41
        (c) 60.5 ± 3.54
        (d) yes (note that the observations
            are the means of four
            assembly times)

15.7   (a) each observation is the mean
            length of 10 leaves
        (b) yes, $F = 57.38$ ($F_{.05} = 3.10$)
        (c) reject H_0, $t = 12.105$
        (d) $2.367 \pm .337$

15.8   (a) yes, $F = 63.66$ ($F_{.05} = 3.29$)
        (b) $1.66 \pm .273$

15.9   (a) no, $F = .87$ ($F_{.05} = 2.88$)
        (b) 2.7 ± 3.75

(c) 27.5 ± 2.65

(d) 26

15.10 (a) yes, $F = 6.46$ ($F_{.05} = 5.14$)

(b) yes, $F = 3.75$ ($F_{.05} = 4.76$)

(c) $-1.2 \pm .65$

15.11 (a) yes, $F = 19.44$ ($F_{.05} = 4.76$)

(b) yes, $F = 40.21$ ($F_{.05} = 5.14$)

(c) $-1.4 \pm .697$

15.12 (a)

Source	d.f.	SS	MS	F
Treatments	2	38	19	10.06
Blocks	3	61.667	20.556	10.88
Error	6	11.333	1.889	
Total	11	111.000		

yes, $F = 10.06$ ($F_{.05} = 5.14$)

(b) yes, $F = 10.88$ ($F_{.05} = 4.76$)

(c) 3.5 ± 1.888

15.13 (a)

Source	d.f.	SS	MS	F
Treatments	2	524,177.167	262,088.58	258.19
Blocks	3	173,415	57,805	56.95
Error	6	6,090.5	1,015.083	
Total	11	703,681.67		

(b) 6

(c) yes, $F = 258.19$ ($F_{.05} = 5.14$)

(d) yes, $F = 56.95$ ($F_{.05} = 4.76$)

(e) 22.529

(f) -237.25 ± 55.13

15.14 (a) yes, $F = 5.77$ ($F_{.05} = 3.74$)

(b) 1.25 ± 1.501

15.15 (a) yes, $F = 7.20$ ($F_{.05} = 5.14$)

(b) -2.275 ± 1.194

(c) yes, $F = 16.61$ ($F_{.05} = 4.76$)

15.16

Source	d.f.	SS	MS	F
Rows	2	46.89	43.44	11.11
Columns	2	22.89	11.44	5.42
Treatments	2	91.56	45.78	21.68
Error	2	4.22	2.11	
Total	8	165.56		

7.33 ± 5.105

15.17 (a) yes, $F = 103.62$ ($F_{.05} = 4.76$)

(b) $-2.55 \pm .366$

(c) no, $F = .16$ ($F_{.05} = 4.76$)

(d) no, $F = 2.02$ ($F_{.05} = 4.76$)

15.18 (a)

Source	d.f.	SS	MS	F
Rows	3	80.00	26.67	.41
Columns	3	91.50	30.50	.46
Treatments	3	36.50	12.17	.19
Error	6	394.00	65.67	
Total	15	602.00		

(b) no, $F = .19$ ($F_{.05} = 4.76$)

(c) 1.75 ± 11.133

15.19 (a) no, $F = 3.02$ ($F_{.05} = 4.76$)

(b) no, $F = .62$ ($F_{.05} = 4.76$)

(c) no, $F = .60$ ($F_{.05} = 4.76$)

(d) no

(e) $-.975 \pm 1.4903$

15.20 $n_i = 11$

15.21 $b = 16$

15.22 $b = 21$

15.23 (a) randomized block design

(b) 32

15.24 $b = 18$

15.25 assume $\sigma^2 \approx .03$; $b = 48$

15.27 $SSE = 1196.6$

15.28 (a) no

(b) completely randomized design

15.29 (a) 16

(b) 135

(c) for a 95% confidence interval and $z \approx 2$, half width is 14.14

15.30 (a)

Source	d.f.	SS	MS	F
Treatments	4	1.212	.303	11.67
Error	22	.571	.026	
Total	26	1.783		

reject H_0, $F = 11.67$ ($F_{.05} = 2.82$)

(b) $t = -2.73$, reject H_0: $\mu_D - \mu_A = 0$

15.31

Source	d.f.	SS	MS	F
Treatments	4	.787	.197	27.78
Blocks	3	.140	.047	6.59
Error	12	.085	.007	
Total	19	1.012		

15.31 (*cont.*)

$F = 27.78 \ (F_{.05} = 3.26)$; reject

$H_0: \mu_A = \mu_B = \mu_C = \mu_D = \mu_E$

15.32   yes, $F = 17.34 \ (F_{.05} = 3.26)$;

4.2 ± 2.27 (95% confidence interval)

15.33   no, $F = 1.53 \ (F_{.05} = 6.94)$

15.34   no, $F = .63 \ (F_{.05} = 3.89)$

15.35   (a) yes, $F = 9.84 \ (F_{.05} = 3.03)$

(b) 6.83 ± 6.341

(c) 91.875 ± 4.151

15.36   (a) yes, $F = 5.20 \ (F_{.05} = 3.24)$

(b) no, $t = .88 \ (t_{.05} = 1.746)$

(c) −.348 ± .231

15.37   (a) yes, $F = 18.61 \ (F_{.05} = 2.87)$

(b) yes, $F = 7.11 \ (F_{.05} = 2.71)$

(c) −5.1 ± 1.67

Chapter 16

16.1 $\{0, 1, \ldots, 6, 19, 20, \ldots, 25\}$
for $\alpha = .014$; $\{0, 1, \ldots, 7, 18,$
$19, \ldots, 25\}$ for $\alpha = .044$; $\{0,$
$1, \ldots, 8, 17, 18, \ldots, 25\}$ for
$\alpha = .108$

16.2   (a) $H_0: p = 1/2$; $H_a: p \neq 1/2$;
rejection region: $\{0, 1, 7, 8\}$;
$y = 6$; do not reject H_0 at
$\alpha = .07$; p value = .2891

16.3   rejection region: $\{0, 1, \ldots, 4,$
$13, 14, \ldots, 17\}$ for $\alpha = .049$;
$y = 8$; no

16.4   (a) yes, $y = 2$; rejection region:
$\{0, 1, 2, 8, 9, 10\}$ for
$\alpha = .11$

(b) variance not constant

16.5   no, $y = 2$; rejection region: $\{0,$
$1, 7, 8\}$ for $\alpha = .0703$

16.6   (a) $U_B = 0$; reject H_0; rejection
region: $\{0, 1,\}$ for $\alpha = .0714$

(b) $U_B = 2.5$; do not reject H_0;
rejection region: $\{0, 1\}$ for
$\alpha = .0714$

16.7   (a) $U_A = 6$; do not reject H_0:
rejection region: $\{0, 1, \ldots,$
$4\}$ (one-tailed) for $\alpha = .0476$

(b) do not reject H_0:
$\mu_A - \mu_B = 0$; $t = 1.606$
$(t_{.05} = 1.86)$

16.8   yes, $U_A = 0$; rejection region:
$\{0, 1, \ldots, 18\}$ for $\alpha = .0504$

16.9 $U_B = 12.5$; reject H_0; rejection
region: $\{0, 1, \ldots, 21\}$ for
$\alpha = .1012$

16.10   no, samples are not independent

16.11   yes, $U_1 = 8$; rejection region:
$\{0, 1, \ldots, 18\}$ for $\alpha = .0546$

16.12   yes, $z = 3.49$

16.13   (a) $T = 3$; rejection region:
$\{T \leq 4\}$; reject H_0; consistent with results of Exercise
10.32

16.14   (a) no, $T = 6.5$; rejection
region: $\{T \leq 4\}$

16.15   no, $T = 73.5$; rejection region:
$\{T \leq 35\}$; consistent with results
of Exercise 16.3

16.16   (a) no, $y = 8$, rejection region:
$\{0, 1, 2, 9, 10, 11\}$ for
$\alpha = .0654$

(b) no, $T = 14.5$; rejection region: $\{T \leq 11\}$ for $\alpha = .05$

16.17 $T = 7$; do not reject H_0; rejection region: $\{T \leq 4\}$ for
$\alpha = .05$

16.18   yes, $T = 3.5$; rejection region:
$\{T \leq 4\}$ for $\alpha = .05$

16.19   (a) −.465; do not reject H_0:
$r_s = 0$

(b) .708; reject H_0: $r_s = 0$

(c) .819; reject H_0: $r_s = 0$

16.20   .867

16.21   yes

16.22 $r_s = .903$; yes

16.23   (a) yes, $r_s = .660$

16.24   yes, $r_s = .9118$

16.25   (a) do not reject H_0; $y = 2$; rejection region: $\{0, 1, 8, 9\}$
for $\alpha = .0390$

(b) do not reject H_0; $t = -1.65$;
rejection region: $|t| > 2.306$
for $\alpha = .05$

16.26   do not reject H_0; $T = 10.5$; rejection region: $\{T \leq 6\}$ for
$\alpha = .05$

16.27   (a) do not reject H_0; $y = 7$; re-

jection region: {0, 1, 9, 10}
for $\alpha = .022$

(b) do not reject H_0; $y = 7$; rejection region: {8, 9, 10} for
$\alpha = .055$

16.28 (a) reject H_0; $T = 6$; rejection
region: $T \leq 8$ for $\alpha = .05$

(b) reject H_0; $T = 6$; rejection
region: $T \leq 11$ for $\alpha = .05$

16.29 (a) no, $U = 32$; rejection region: $U \leq 21$ for $\alpha = .094$

(b) no, $t = .30$

16.30 (a) $U_B = 3$; reject H_0; rejection
region: $U \leq 3$ for $\alpha = .0556$

(b) $F = 4.91$; do not reject H_0;
rejection region: $F \geq 9.12$

16.31 $-.845$

16.32 yes; yes

16.33 $-.593$

16.34 yes

16.35 $U = 17.5$; no; rejection region:
$U \leq 13$ for $\alpha = .0498$

16.36 $T = 14$; reject H_0; rejection
region: $T \leq 16$ for $\alpha = .01$ or
$T \leq 25$ for $\alpha = .05$

16.37 .67684, reject H_0; rejection
region: $|r_s| \geq .441$ for $\alpha = .10$

16.38 .0159

Index